PLANT TAXONOMY

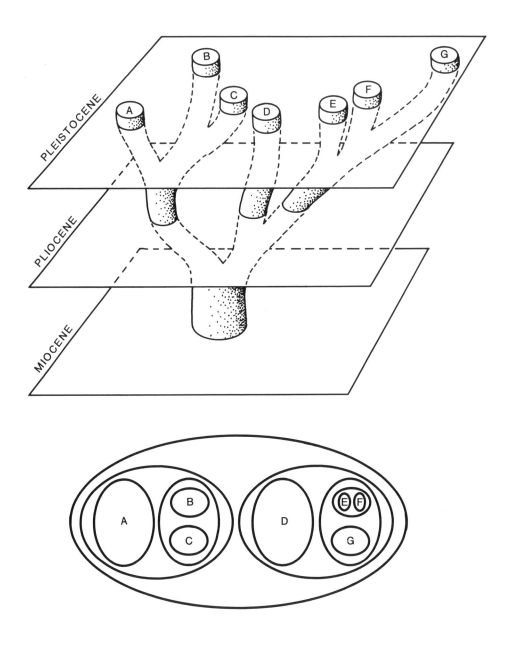

FRONTISPIECE The reconstruction of phylogeny (top) has been, and continues to be, the principal basis for the construction of plant classifications (bottom).

TOD F. STUESSY

Plant Taxonomy
The Systematic Evaluation
of Comparative Data

 Columbia University Press New York

COLUMBIA UNIVERSITY PRESS
New York   Oxford
Copyright © 1990 Columbia University Press
All rights reserved

Library of Congress Cataloging-in-Publication Data

Stuessy, Tod F.
Plant taxonomy : the systematic evaluation of comparative data /
Tod F. Stuessy.
p.   cm.
Includes bibliographical references.
ISBN 0-231-06784-4
1. Botany—Classification.   I. Title.
QK95.S78   1989
581'.012—dc20   89-17401
                                                          CIP

Book design by Jennifer Dossin
Casebound editions of Columbia University Press books are Smyth-sewn
and printed on permanent and durable acid-free paper
∞

Printed in the United States of America
c 10 9 8 7 6 5 4 3 2 1

CONTENTS

Preface   xiii
Acknowledgments   xv

PART ONE. PRINCIPLES OF TAXONOMY

Section One. The Meaning of Classification   3

  Chapter 1. A Few Definitions   5
    Classification, Taxonomy and Systematics   5
    Nomenclature   9
    Identification   9
    Biosystematics   10
    Experimental Taxonomy   11
    The New Systematics   11
    Comparative Biology   13
  Chapter 2. The Relevance of Systematics   14
    The Importance of Systematics in Society   14
    The Contributions of Systematics to Biology   16
  Chapter 3. The Importance and Universality of Classification   20
    The Process of Classification   20
    Hierarchical System of Classes   23
  Chapter 4. Characters   27
    General Terms   27
    Role of Characters and States   30
    Kinds of Characters   31
    Criteria for Selecting Characters and States   38

Section Two. Different Approaches to Biological Classification   43

  Chapter 5. The Anatomy of Classification and the Artificial Approach   45
    Aesthetics and Classification   45
    The Process of Classification   47
    Artificial Classification   49
  Chapter 6. Natural and Phyletic Approaches   51
    Natural Classification   52

Phyletic (Evolutionary) Classification   54
Definitions of "Naturalness"   56
Chapter 7. Phenetic Approach   59
Definitions   59
History of Phenetics   61
Methodology of Phenetics   62
Impact of Phenetics   88
Chapter 8. Cladistic Approach   93
Definitions   93
History of Cladistics   96
Methodology of Cladistics   99
Formal Classification   128
Impact of Cladistics   130
Chapter 9. Evaluation of the Three Major Approaches:
       The New Phyletics   135
Historical Influences   135
Previous Evaluations   136
Evaluation of the Process of Classification   137
Evaluation of the Resultant Hierarchy   139
The New Phyletics   142

Section Three. Concepts of Categories   155

Chapter 10. The Taxonomic Hierarchy   156
History   156
Logical Structure   157
Chapter 11. Species   161
History of Species Concepts   162
Reality of Species   165
Naturalness of Species   168
Species as Individuals   169
Current Species Concepts   171
Recommended Species Concept for General Use   179
Chapter 12. Subspecies, Variety and Form   182
History of Infraspecific Categories   183
Difficulties in Application of Concepts   185
Forms   186
Biosystematic Infraspecific Categories   187
Related Zoological Concepts   188
Recommended Infraspecific Concepts   189
Chapter 13. Genus   194
History of Generic Concepts   195
Types of Data Used to Delimit Genera   198
Numerical Delimitation of Genera   202
Naturalness of Genera   202
Remodeling of Genera   203
Paleontological Genera   205

Monotypic Genera 205
Cladistics and Generic Delimitation 206
Chapter 14. Family and Higher Categories 207
History of Concepts of Higher Categories 208
Naturalness of Higher Categories 208
Higher Categories as Individuals 209
Size of Higher Taxa 209
Vertical vs. Horizontal Classification 210
Practical Difficulties 210
Types of Data Used with Higher Categories 211
Numerical Approaches 211
Evolution of Higher Taxa 212

PART TWO. TAXONOMIC DATA

Section Four. Types of Data 217

Chapter 15. Morphology 218
History of Morphology in Plant Taxonomy 218
General Morphological Texts and References 218
Types of Morphological Data 219
Investments for Gathering Morphological Data 231
Efficacy of Morphological Data in the Taxonomic Hierarchy 232
Chapter 16. Anatomy 233
History of Anatomy in Plant Taxonomy 233
General Anatomical Texts and References 234
Types of Anatomical Data 234
Investments for Gathering Anatomical Data 248
Efficacy of Anatomical Data in the Taxonomic Hierarchy 248
Special Concerns with Anatomical Data 249
Chapter 17. Embryology 251
History of Embryology in Plant Taxonomy 252
General Embryological Texts and References 252
Types of Embryological Data 253
Investments for Gathering Embryological Data 263
Efficacy of Embryological Data in the Taxonomic Hierarchy 265
Special Concerns with Embryological Data 266
Chapter 18. Palynology 267
History of Palynology in Plant Taxonomy 267
General Palynological Texts and References 270
Types of Palynological Data 271
Investments for Gathering Palynological Data 279
Efficacy of Palynological Data in the Taxonomic Hierarchy 279
Special Concerns with Palynological Data 284
Chapter 19. Cytology 288
History of Cytology in Plant Taxonomy 288

General Cytological Texts and References 290
Genomic Organization 290
Types of Cytological Data 291
Investments for Gathering Cytological Data 304
Efficacy of Cytological Data in the Taxonomic Hierarchy 307
Special Concerns with Cytological Data 310
Chapter 20. Genetics and Cytogenetics 313
History of Genetics and Cytogenetics in Plant Taxonomy 313
General Genetic and Cytogenetic Texts and References 314
Types of Genetic and Cytogenetic Data 314
Investments for Gathering Genetic and Cytogenetic Data 321
Efficacy of Genetic and Cytogenetic Data in the
    Taxonomic Hierarchy 324
Special Concerns with Genetic and Cytogenetic Data 325
Chapter 21. Chemistry 329
History of Chemistry in Plant Taxonomy 329
General Chemical and Chemotaxonomic Texts and References 330
Types of Chemical Data 331
Investments for Gathering Chemical Data 347
Efficacy of Chemical Data in the Taxonomic Hierarchy 348
Special Concerns with Chemical Data 349
Chapter 22. Reproductive Biology 351
History of Reproductive Biology in Plant Taxonomy 351
General Reproductive Biological Texts and References 352
Types of Reproductive Data 352
Investments for Gathering Reproductive Data 361
Efficacy of Reproductive Data in the Taxonomic Hierarchy 362
Special Concerns with Reproductive Data 362
Chapter 23. Ecology 364
History of Ecology in Plant Taxonomy 364
General Ecological Texts and References 365
Types of Ecological Data 365
Investments for Gathering Ecological Data 374
Efficacy of Ecological Data in the Taxonomic Hierarchy 375
Special Concerns with Ecological Data 376

Section Five. Handling of Data 381

Chapter 24. Gathering of Data 382
Collection of Data 382
Evaluation of Data 384
Relative Efficacy of Different Kinds of Data 385
Chapter 25. Presentation of Data 390
History of Graphics in Plant Taxonomy 391
General Graphics References 393
Types of Graphics 393
Graphic Design 401

Epilogue	405
Literature Cited	407
Author Index	491
Taxon Index	503
Subject Index	509

PREFACE

THIS BOOK is designed to introduce the upper-level undergraduate or beginning graduate student to the philosophical and theoretical aspects of
plant taxonomy. At the present time there is no text that fills this need. The
closest book in depth and breadth of coverage would be the excellent *Principles
of Angiosperm Taxonomy* by Davis and Heywood (1963), which is now more than
twenty-five years old. In particular, the past decade has seen a proliferation of
articles and books on phenetic and cladistic philosophies and methodologies, to
the extent that there is now a real need for a balanced account of these new
developments for professors and students of plant taxonomy. The literature is
extensive, the debates often acrimonious, and the polarization of the broad
community of systematic biologists acute. Definitions have been changed, historical perspectives and precedents have been ignored or interpreted differently,
and numerous viewpoints have been offered. The challenge is immense to the
teacher and student of plant taxonomy to sort this all out and apply these
concepts and methods to actual situations. The recent books, *Plant Taxonomy
and Biosystematics* (Stace 1980), *Introduction to Principles of Plant Taxonomy*
(Sivarajan 1984), and *Fundamentals of Plant Systematics* (Radford 1986) are steps
in the proper direction, but they lack the detail of coverage of most topics
desirable for advanced students.

The present text is divided into two parts. Part 1 contains the principles of
taxonomy including the importance of taxonomy and systematics, characters,
different approaches to biological classification, and concepts of categories. These
are the basic chapters that tell what taxonomy is and how one goes about doing
it. As will be obvious, evolutionary (= phyletic) taxonomy is favored as the best
approach to biological classification. Considerable attention has also been given
to phenetics and cladistics, however, and a balanced presentation has been
attempted despite my own biases. We are now entering a new phase of biological
classification in which phyletic classifications can be constructed explicitly,
called here "the New Phyletics" (chapter 9), and it is hoped that this book will
stimulate more interest in this direction.

Part 2 outlines different types of data used in plant taxonomic studies with
suggestions on their efficacy and modes of presentation and evaluation. Not all
types of data have been included, but the most commonly used ones are discussed with references given. The equipment and financial resources needed for
gathering each type of data also are listed briefly. The main point has been to

show (by illustrations and references) the incredible diversity of data used for taxonomic purposes in angiosperms and to stimulate their further use by students and workers. Specific case studies in which these data are employed are fewer than the displays and discussions of data themselves.

Many quotes are placed throughout the text to emphasize the historical perspective, which is so important in the development of taxonomic terminology and philosophy. Similarly, the life-span for historically important workers is given to help show the total period in which the individual lived and worked. The literature cited is not exhaustive, but it is extensive so that most topics are covered reasonably thoroughly and can serve as a good springboard for additional readings in a particular area. The cutoff date for new literature additions was July 1, 1988.

The view of taxonomy presented here is primarily a personal one. I have tried to determine what I do operationally as a practicing plant taxonomist and to view these activities within a meaningful conceptual framework. These ideas have been augmented and refined by the concepts of others, which have been cited when they could be recalled. Some ideas that seem original to me now were stimulated no doubt many years ago by miscellaneous readings or comments from colleagues or students, the sources of which have long been forgotten. I have placed particularly heavy emphasis on concepts throughout these chapters, because I believe strongly that the most creative taxonomy is done by those who know (or at least strive to know) conceptually what they actually are doing. I hope this perspective will be stimulating and useful.

ACKNOWLEDGMENTS

ALMOST EVERY author owes debts of gratitude to numerous people for having encouraged and helped bring a book to successful completion. This work is no exception. Drs. Patrick Dugan and Emanuel D. Rudolph, former Dean of the College of Biological Sciences and former Chairman of the Department of Botany, respectively, of Ohio State University, courteously arranged a sabbatical leave for me in Fall quarter, 1982, during which time the first full draft of the book was initiated. At this same time, Drs. William Anderson and Edward Voss of the Herbarium, University of Michigan, made generous arrangements for my stay at their institution which allowed me to work uninterruptedly and put the writing of this book on schedule.

Many individuals have read various drafts of the manuscript and made many helpful suggestions. A very early (and very different) draft was read by W. P. Adams, S. B. Jones, Jr., J. E. Rodman, O. T. Solbrig, B. L. Turner, J. Wahlert, and R. L. Wilbur. The complete final draft was read by V. H. Heywood, S. B. Jones, Jr. and B. L. Turner. Chapters of the final manuscript were read by (chapter numbers in parentheses): W. G. Abrahamson (23), R. E. J. Boerner (23), B. A. Bohm (21), P. D. Cantino (8), D. J. Crawford (1-4), T. J. Crovello (4, 7), R. H. Eyde (15, 16), K. Jones (19, 20), L. W. Macior (22), J. W. Nowicke (18), J. M. Herr, Jr. (17), V. Raghavan (16, 17), F. D. Sack (16), J. J. Skvarla (18), R. R. Sokal (7), D. E. Soltis (19), R. W. Spellenberg (8), W. P. Stoutamire (22), and R. L. Wilbur (10–14). E. D. Rudolph provided valuable bibliographical assistance.

Gratitude is expressed to numerous holders of copyrights of figures and tables reproduced in this book who have given permission to use these materials. These include authors, publishers, and editors of societal journals. Obviously in a book such as this, which depends so heavily on illustrations (especially in part 2), these permissions were essential for successful completion of the project. Credits to the authors are given in the legends to the presented material with full references to place of publication in the Literature Cited. The publishers and journals which have given generously their permissions are: Knopf, New York; Academic Press, London; *American Journal of Botany; American Scientist; American Zoologist; Annals of the Missouri Botanical Garden; Annual Review of Ecology and Systematics;* American Elsevier, New York; *Australian Journal of Botany; Bartonia; Biotropica; Botanical Journal of the Linnean Society; Botaniska Notiser;* Cambridge University Press, Cambridge; *Canadian Journal of Botany; Canadian Journal of Genetics and Cytology; Chromosoma* (Berlin); DLG-Verlags-GmbH,

Frankfurt; *Evolution; Evolutionary Biology; Fieldiana, Botany;* Garrard, Champaign, Ill.; George Allen & Unwin, London; *Grana;* Harper & Row, New York; Hodder and Stoughton, London; *Journal of the Arnold Arboretum; Journal of the Elisha Mitchell Scientific Society;* Wiley, New York; *Kew Bulletin; Madroño;* McGraw-Hill, New York; *Memoirs of the New York Botanical Garden; New Phytologist; Nordic Journal of Botany; Ohio Journal of Science; Oikos; Opera Botanica;* Oxford University Press, Oxford; Pergamon Press, Oxford; *Plant Systematics and Evolution;* Prentice-Hall, Englewood Cliffs, New Jersey; *Proceedings of the Academy of Natural Sciences of Philadelphia; Rhodora; Science; Smithsonian Contributions to Botany;* Springer-Verlag, Berlin; *Systematic Botany; Systematic Botany Monographs; Systematic Zoology; Taxon; University of California Publications in Botany;* University Press of Kansas, Lawrence; University Park Press, Baltimore; Freeman, San Francisco; Junk, The Hague; Wadsworth, Belmont, California; and Collins, London. Permission was also granted by the British Museum (Natural History) to reproduce figure 15.1. All new figures were drawn by David Dennis and Lisa Mary Einfalt.

Parts of this book have been published already in modified form. The history of botanical cladistics in chapter 8 appeared with less detail in Duncan and Stuessy (1985), and some of the material on species concepts in chapter 11 is to be published in Stuessy (in press).

The editors of Columbia University Press were extremely helpful for their combination of understanding, patience, and professional assistance. Ed Lugenbeel was more than an editor—he was a friend and counselor too. With his competent help and that of the Productions and Marketing staffs, a much higher level of quality has been achieved in this book than would have been possible solely through my efforts.

Significant persons in this undertaking have been Jonathan Abels and my wife, Patricia, who helped check inconsistencies between text citations and the Literature Cited. Special thanks also go to John W. Frederick for assistance with various aspects of manuscript preparation.

Finally, and of the greatest importance, have been the many students who initially stimulated me to write this book, and who have worked through the several drafts and offered useful criticisms. Particularly helpful have been Wen Jun, Thomas Lammers, and James Zech. Without this constant prodding, I doubtless would never have finished the task.

PLANT TAXONOMY

In these days when Molecular Biology is beginning to be seen as a restricted science, narrowing our vision by concentrating on the basic uniformity of organisms at the macromolecular level, the need for taxonomists to draw attention to the enormous diversity and variation of this earth's biota becomes more and more pressing.

V. H. Heywood (1973a:145)

In other words, the field of taxonomy in a way epitomizes the work of all other branches of biology centered on the organism itself, and brings the varied factual information from them to bear on the problems of interrelationship, classification and evolution. Thus taxonomy, as has been aptly remarked, is at once the alpha and omega of biology.

R. C. Rollins (1957:188)

Plant taxonomy has not outlived its usefulness: it is just getting under way on an attractively infinite task.

L. Constance (1957:92)

PART ONE

Principles of Taxonomy

The Meaning of Classification

TAXONOMY IS dynamic, beautiful, frustrating, and challenging all at the same time (fig. 1.1). It is demanding philosophically and technically, yet it offers intellectual rewards to the able scholar and scientist. It can be manifested in works of incredible detail as well as in logical and philosophical conceptualizations about the general order of things. It has strong implications for interpreting the reality of the world as we can ever hope to know it.

Because taxonomy has deep historical roots, the past is never escaped. This places an increasing burden upon practitioners to understand old and new material. The past must be dealt with for older results and every new discovery must be digested and incorporated. As Constance has aptly put it, "My ideal taxonomist, therefore, must be very versatile indeed, and should preferably be something of a two-headed [i.e., two-faced] Janus, so that one set of eyes can look back upon and draw from the experience of the past, and the other pair can be focused upon deriving as much of value as possible from developments on the present scene" (1951:230).

Taxonomy is a synthetic science, drawing upon data from such diverse fields as morphology, anatomy, cytology, genetics, cytogenetics, and chemistry. It has no data of its own. Every new technical development in these other areas of science offers promise for improved portrayal of relationships of organisms. This is a demanding aspect of taxonomy for a practicing worker, because it is virtually impossible to understand completely all of these different data-gathering methods, yet highly desirable to be able to master as many as possible. Furthermore, the accumulation of data and their interpretation never cease. Not only

FIGURE 1.1   An example of the challenges facing the plant taxonomist is shown dramatically by this bizarre landscape, which could represent an obscure area of the earth or perhaps even another planet, with completely new and different plant forms. If this scene were on earth, we would have considerable biological information on modes of reproduction, structures, functions, etc., in plants in general and a good background of ideas on how to proceed with classification of these groups based upon historical classificatory records. If on another planet, however, to attempt a predictive classification of these forms would be unbelievably difficult, with nothing known about modes of reproduction, structures and their functions, mechanisms of evolution, or even what is an individual or population. This same type of overwhelming challenge was faced by plant taxonomists on this planet approximately 500 years ago. (From Lionni 1977, frontispiece)

do new techniques of data-gathering provide more information that must be brought to bear on understanding relationships, but also these new interpretations reveal new taxonomic groups which must be understood and utilized. These are some of the reasons why taxonomy (and its parent discipline, systematics) has rightly been called "an unending synthesis" (Constance 1964), "an unachieved synthesis" (Merxmüller 1972), or even more poetically, "the stone of Sisyphus" (Heywood 1974).

CHAPTER 1

A Few Definitions

CLASSIFICATION, TAXONOMY, AND SYSTEMATICS

Taxonomy has had various meanings over the past one hundred and fifty years, and particular confusion has prevailed with systematics. Systematics no doubt very early was used as "a casual self-evident term" (Mason 1950:194), to refer to the ordering of organisms into rudimentary classifications. This activity has occurred ever since man has lived on earth (Raven, Berlin, and Breedlove 1971). The early documented use of the term *systematics* can be traced (as systematic botany) at least as far back as Linnaeus (1737a, 1751, 1754), and it has persisted to the present day although in modified form. Linnaeus (1737a:3) states that "we reject all the names assigned to plants by anyone, unless they have been either invented by the Systematists or confirmed by them." In 1751 he used the term (as "botano-systematici" p. 17) to refer to workers who "carefully distinguish the powers of drugs (in plants) according to natural classes." He makes the definition of a Systematic Botanist even more clear in the preface to the fifth edition of the *Genera Plantarum:*

> The use of some Botanic System I need not recommend even to beginners, since without system there can be no certainty in Botany. Let two enquirers, one a Systematic, and the other an Empiric, enter a garden fill'd with exotic and unknown plants, and at the same time furnish'd with the best Botanic Library; the former will easily reduce the plants by studying the letters [i.e., features of diagnostic value] inscribed on the fructification, to their Class, Order, and Genus; after which there remains but to distinguish a few species. The latter will be necessitated to turn over all the books, to read all the descriptions, to inspect all the figures with infinite labor; nor unless by accident can be certain of his plant. (1754:xiii, as translated anonymously, 1787:lxxvi)

Books using the term *systematic botany* appeared thereafter (e.g., Smith's *An Introduction to Physiological and Systematical Botany,* 1809, and Nuttall's *An Introduction to Systematic and Physiological Botany,* 1827). Mason, although admitting the difficulty of establishing the place of its first use, ventured the opinion that systematics "might possibly have even preceded it" [i.e., the use of taxonomy] (1950:194), and gave Lindley (1830b) as the earliest reference.

A biologist interested in relationships during this early period mostly studied morphological features and accordingly grouped organisms into units. This ordering of organisms into groups based on similarities and/or differences was (and still is) called *classification*. This is a very old term going back to Theophrastus in the third century B.C. (see 1916 translation). The Swiss botanist, Candolle (1813), in the herbarium at Geneva, coined *taxonomy* (as taxonomie)* to refer to the theory of plant classification. It later became more generally used for the methods and principles of classification of any group of organisms, and it is still used basically in this way (e.g., Simpson 1961). From this point to the publication of the theory of evolution by means of natural selection by Darwin (1859), the two words, taxonomy and systematics, were regarded as synonyms, although the latter was used much more frequently. During this time classifications were believed to reflect the plan of natural order created specially by God, and man was simply rediscovering the Divine Plan. Biologists engaged in these activities of classification were called interchangeably either taxonomists or systematists. Since Darwin's time, systematists have not only continued their interest in classification, but also they have attempted to understand evolutionary relationships among the groups so ordered. Furthermore, some systematists have become interested in the process of evolution itself, that is, in the mechanisms that produce the diversity. Consequently, a systematist today may study many different aspects of evolutionary biology that are far removed from the morphological investigations of a century ago.

The basic methodology of modern systematics is outlined in table 1.1. Data are gathered from organisms and their interactions with the environment and used to answer questions about classification, phylogeny, and the process of evolution. A similar and equally legitimate viewpoint has been presented by Blackwelder and Boyden (1952) in which three steps are indicated: (1) recording of data; (2) analysis of the data for making classifications; and (3) synthesis of these generalizations for insights on phylogeny and evolution. Wilson has aptly pointed out:

> most systematists by choice are not problem solvers in their method of working. It might be said that the perfect experimental biologist selects a problem first and then seeks the organism ideally suited to its solution. The systematist does the reverse. He selects the organism first (for reasons that are highly individual) and only secondarily chances upon phenomena of general significance. The chief value of his discoveries is that they are typically of the kind that would not be made otherwise. If the systematist has an ideal program, it runs refreshingly counter to the conventional wisdom: select the organism first, as a kind of totem animal (or plant) if you wish, then actively seek the problem for the solution of which the organism is ideally suited. (1968:1113)

As specific examples of systematic studies, one might analyze the patterns of adaptive radiation within a particular group of species (e.g., Gardner 1976), or

---

*Some workers, e.g., Richter (1938), believe that taxonomy, if properly derived from its Greek roots, should be spelled "taxionomy" (or even "taxinomy"), but these suggestions for change are unfounded and unnecessary (Mayr 1966; Pasteur 1976) and have never been adopted.

TABLE 1.1 Outline of Methodology of Systematics.

I. Accumulation of Comparative Data
　A. From the Organism
　　1. Structures
　　2. Processes (interactions among structures)
　B. From the Organism-Environment Interactions
　　1. Distribution[a]
　　2. Ecology
II. Use of Comparative Data to Answer Specific Questions
　A. Classification (most predictive system of classification at all levels)
　　1. Method and result of grouping of individuals
　　2. Level in the taxonomic hierarchy at which the groups should be ranked
　B. Process of Evolution
　　1. Nature and origin of individual variation
　　2. Organization of genetic variation within populations
　　3. Differentiation of populations
　　4. Nature of reproductive isolation and modes of speciation
　　5. Hybridization
　C. Phylogeny (divergence and/or development of all groups)
　　1. Mode
　　2. Time
　　3. Place

---

[a]Floristics, or the documentation of what plants grow in particular regions, is deliberately not listed in this table as a separate question, nor does it find a specific place in the areas of systematics in figure 1.2. The question of where particular plants grow is a very legitimate and valuable activity within systematics, but it is essentially data-gathering of distributions of plant groups which have already been classified. Some floristic projects, however, especially of poorly known regions (e.g., Rzedowski and McVaugh 1966; McVaugh 1972a, b) involve considerable amounts of classification as well as original historical scholarship. To this extent they become more revisionary, and less floristic, in character (for these and other distinctions, see Stuessy 1975).

compare amino acid sequences of respiratory enzymes (e.g., Boulter et al. 1979), or investigate the genetic basis of intra-populational morphological variation (e.g., Rollins 1958), and so on.

Because of the diverse nature of these studies that were spawned by evolutionary theory, a collective term was needed to designate these different activities and the people so engaged, and another term also was needed for describing the more traditional activities of classification. As a result, the term systematics (or systematic biology) has come to have a meaning different from and broader than taxonomy. The definition used by most people at the present time and preferred here is that of Simpson, who defines *systematics* as "the scientific study of the kinds and diversity of organisms and of any and all relationships among them" (1961:7). Or a slightly simpler way of defining systematics is "the study of the diversity of organisms" (Mayr 1969c:2; cf. Wilson 1985, for his similar definition "the study of biological diversity"). Diversity is such an important concept in systematics that a useful and delightful perspective on this is provided by Constance:

Much as I respect the giant strides that have been made in clarifying basic principles and processes of wide applicability, I have chosen to celebrate

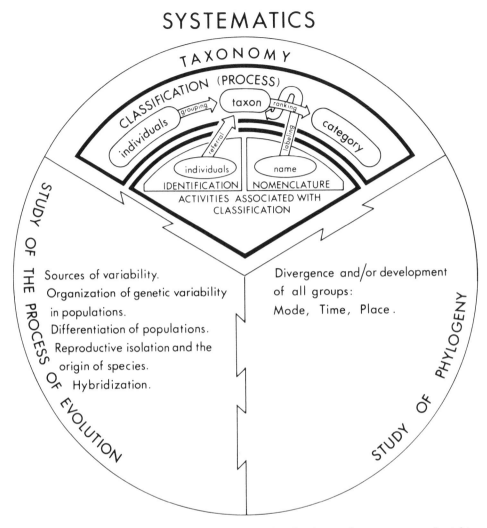

FIGURE 1.2 Diagram of conceptual and procedural relationships among and within areas of systematics. The tighter connection between the bottom two areas emphasizes their closeness as both being aspects of evolutionary processes, short term on one hand, and long term on the other. (From Stuessy 1979b:622)

diversity. It is well enough to know that all music can be reduced to a relatively few notes and a minimum of ways of evoking and receiving them in the human ear. This does not suggest to the music lover that symphonies, sonatas, and operas are redundant because their parts and processes can be analyzed. All literature, after all, is merely spun out of words. Human beings are a lot alike, but it does not necessarily follow that there is no point in knowing more than one of them. Even the most wonderful molecule has its limitations. (1971:22)

Myers has provided a slightly more general definition of systematics: "the study of the nature and origin of the natural populations of living organisms, both present and past" (1952:106). Recently the term *biodiversity* has been used (e.g., Fosberg 1986; Wilson 1988) to refer to the total biotic diversity on earth.

Many biologists still equate taxonomy with systematics (Lawrence 1951; Crowson 1970,* Radford et al. 1974; Jones and Luchsinger 1986; Stace 1980).† Wilson echoes this viewpoint, and he talks about a *pure systematist* who "can be defined as a biologist who works on such a large number of species that he has only enough time to consider classification and phylogeny. If he narrows his focus, his unique knowledge provides him with a good chance to make discoveries in genetics, ecology, behavior, and physiology, as well as in taxonomy. But then we come to know him as a geneticist, or an ecologist, or a behaviorist, or a physiologist" (1968:1113). Swingle (1946) makes a distinction between systematics and taxonomy with the latter regarded as dealing with phylogenetic classification, and the former being broader to include taxonomy and nomenclature. Mason (1950; and followed by Porter 1967) has taken a different approach and regards systematics (specifically systematic botany) as the data-gathering aspect (as Systematic Anatomy, Systematic Cytology, Systematic Genetics, etc.) and taxonomy as the interpretive phase in constructing classifications and in revealing evolutionary relationships. Smith defines systematics as "that branch of science which investigates the philosophy that underlies a classification" (1969:5), which would be closer to the definition of taxonomy used in this book. For agreement with the definitions used here, see Danser (1950), Davis and Heywood (1963), Blackwelder (1967a), Sylvester-Bradley (1968), Mayr (1969c), and Darlington (1971). The relationships among these terms are shown in figure 1.2.

NOMENCLATURE

Two other terms, *nomenclature* and *identification*, are sometimes confused with classification and systematics. After groups of organisms have been classified, names must be given to these groups so that communication about particular units will be facilitated and so that continued progress in classification can be made (Hitchcock 1916). The naming of groups of organisms and the rules governing the application of these names is called *nomenclature*. "Nomenclature cannot, in fact, be wholly separated from the classification that it serves. . . . New taxonomic advances will always pose new nomenclatural problems, whatever form of taxonomy is employed, just as new social forms require new linguistic expression. New nomenclatural procedures will enable new possibilities in taxonomy to be formulated and examined. The process is fruitful, inevitable and in

---

*Although Crowson concludes that "The words classification, systematics, and taxonomy are now commonly treated as synonyms, an example of the confusion and carelessness in the use of words which is prevalent in so much modern writing" (1970:18), he offers the view that "Originally and properly, classification would have denoted the activity of placing things in classificatory groups, whereas systematics would be the body of general theory underlying this activity" (p. 19).

†Griffiths (1974a) has suggested that because systematics and taxonomy have been regarded by "most authors" as synonyms, the term *"metasystematics"* should be used for the more inclusive concept. To my knowledge no one else has yet adopted this terminology.

many respects essential to progress" (Whitehead 1972:216). Many philosophical and methodological topics on nomenclature could profitably be discussed, such as using code numbers (numericlature: Little 1964; Hull 1966) or a descriptive code ("suneg" = "genus" spelled backwards: Amadon 1966) instead of descriptive binominals, possible roles for uninominals (Michener 1963, 1964; Lanham 1965; Steyskal 1965), elimination of Latinization of names (Yochelson 1966), plus the numerous features of the established International Code of Botanical Nomenclature (Greuter et al. 1988). But these topics fall outside the scope of this book. The reader is referred to the discussions in Lawrence (1951) and Jeffrey (1977a) for salient points of the existing Code; case-studies are provided by St. John (1958), but without answers!

IDENTIFICATION

Identification, on the other hand, involves referring an individual specimen to a previously classified and named group (Jardine 1969a). For example, if one walks outside and picks a small branch with leaves from a tree, takes the specimen back to the laboratory, and attempts to find a name for it, what is being sought is an identification of the specimen—not a classification. Many years ago some taxonomist did classify the species represented by the individual tree. It is possible, of course, that the species of tree has never been classified before (unlikely in temperate countries, but entirely plausible in poorly known areas of the tropics), in which case this would need to be done and an appropriate new name provided. Many innovations in identification have been suggested and discussed recently (e.g., Morse 1971; Dallwitz 1974; Pankurst 1974, 1975; Sneath and Chater 1978; Mascherpa and Boquet 1981), especially information content of keys, multiple access keys ("polyclaves," Hansen and Rahn 1969; Simpson and Janos 1974; Duke and Terrell 1974; Westfall, Glen, and Panagos 1986; these also called "tabular keys," Newell 1970, 1972, 1976) and use of computers for automated identification and key construction (Watson and Milne 1972). Details of these topics also fall outside the scope of this book; see Pankhurst (1978) for a good introduction.

These two activities, identification and nomenclature, are directly associated with classification (fig. 1.2). For purposes of communication, groups of organisms without labels are not at all useful, and similarly, without identification, the names given originally to the classified groups cannot be applied to newly collected individuals. These associated activities, therefore, are not only directly related to but are also most important for classification. Despite their importance, however, they must never be regarded as overshadowing the more significant effort of studying the organisms themselves and developing classifications to reflect relationships. Although Myers has overstated the case, it is basically true that "If systematics deals primarily with the nature of populations, such appurtenances as nomenclature are seen in proper perspective, as mere adjuncts to systematics, necessary in speaking of populations but of minor importance" (1952:107). Labels are meaningless and unnecessary unless they refer to individuals or groups with information content such as units within a classification.

BIOSYSTEMATICS

Biosystematics was introduced by Camp and Gilly (as biosystematy, 1943a:327; see also Camp 1951) as an attempt "(1) to delimit the natural biotic units and (2) to apply to these units a system of nomenclature adequate to the task of conveying precise information regarding their defined limits, relationships, variability, and dynamic structure." To determine these natural units involves a program of cytogenetic studies among populations. In this fashion the isolating mechanisms and genetic compatibilities are revealed which lead to discovery of the natural units. As a result of this definition, any studies involving breeding programs between taxa have come to be known as biosystematics, even though the data sought are for determining evolutionary relatedness of taxa rather than for discovering "natural units." Böcher has remarked that "In my opinion the place of biosystematics is closer to cytogenetics and ecology than to taxonomy in a narrower sense. Our main interest is not classification but evolution. This, of course, does not mean that we should never deal with classification. But frequently it will be better to leave problems of taxonomic rank and nomenclature to taxonomists *sensu strictu*, or to cooperate with them. The main goal of a biosystematist is to try to unravel the evolution of a group of taxa, what the evolutionary forces were and how they worked together in each particular case" (1970:3–4; these sentiments also echoed by Merxmüller 1970, and Stebbins 1970b). With the proliferation of other laboratory studies in systematics during the past several decades, however, sometimes biosystematics has been used to refer to any kind of experimental systematic study, i.e., involving any type of data-gathering except traditional morphology and anatomy. Further, because of the very broad definition of systematics now in use by most workers (and used in this book), the need for the term biosystematics has surely lessened and in the minds of some nearly disappeared (Johnson 1970).

EXPERIMENTAL TAXONOMY

Experimental taxonomy (or experimental systematics) is another term most often used for laboratory-based studies other than (or in addition to) breeding data (Hagen 1983, 1984). These are not experiments in the strict scientific sense, but rather the gathering and analyzing of different kinds of comparative data that ordinarily are generated in the laboratory (e.g., cytology, phytochemistry), and which are used to generate and test hypotheses (Gilmartin 1986; Donoghue 1987; La Duke 1987). The term *experimental taxonomy* did derive from actual experimental investigations of the nature of plant species by reciprocal transplants of clones into different environments (Gregor 1930, 1938; Hall, Keck, and Hiesey 1931; Hall 1932; Gregor, Davey, and Lang 1936; Clausen, Keck, and Hiesey 1939, 1941a) which effectively discriminated the genetic vs. environmental components of morphological variation. This was also called genecology by Turesson (1923) and others (e.g., Constance 1957). Although these valuable kinds of studies are still being done (e.g., Müntzing 1969), experimental taxonomy since has come to have a broader (and less precise) usage. Müntzing has emphasized the importance of genetics and cytology in experimental taxonomy and commented that "what experimental taxonomy can do is to establish the nature and occurrence of such intra- and interspecific differentiation that cannot be clarified merely by morphological, ecological and plant-geographical methods" (1969:791). Rollins commented: "The type of experimentation differs, depending upon the objectives. The most frequent are undoubtedly those associated with genetics and cytology in which the reproductive process or the level of polyploidy is under investigation. The most effective experimental approaches in taxonomy have combined work in the herbarium and field with studies in the greenhouse and experimental plot" (1957:192). Some workers (e.g., Stace 1980) regard experimental taxonomy and biosystematics as synonyms.

THE NEW SYSTEMATICS

The term *new systematics* was coined by Hubbs (1934) and popularized by Huxley to stress that "Fundamentally, the problem of systematics, regarded as a branch of general biology, is that of detecting evolution at work" (1940:2). That is, the focus is on understanding the mechanisms by which diversity is produced rather than solely on classification. This was a useful distinction at that time and helped strengthen the existing area of experimental taxonomy and aided spawning of biosystematics. The influence of Huxley's book (1940) was substantial, and this is reflected in Mayr's emphasis on populations and his explanation of the term (in contrast to "the old systematics"):

> *The new systematics* may be characterized as follows: The importance of the species as such is reduced, since most of the actual work is done with subdivisions of the species, such as subspecies and populations. The population or rather an adequate sample of it, the "series" of the museum worker, has become the basic taxonomic unit. The purely morphological species definition has been replaced by a biological one, which takes ecological, geographical, genetic, and other factors into consideration. The choosing of the correct name for the analyzed taxonomic unit no longer occupies the central position of all systematic work and is less often subject to argument between fellow workers. The material available for generic revisions frequently amounts to many hundreds or even thousands of specimens, a number sufficient to permit a detailed study of the extent of individual variation. (1942:7)

Despite its utility at the time, however, new systematics is obviously a dated term and is rarely used today. "Every age has had its own new systematics. As far as I am concerned, there always has been some new systematics and there always will be" (Mayr 1964:13). As Sylvester-Bradley has appropriately put it "the 'new systematics' of today is something very different to the 'new systematics' of thirty years ago. . . . Perhaps the time has come to consider the publication of a 'Newer Systematics' " (1968:176–177).

The label *new taxonomy* was used by Cain (1959a) to refer to anticipated advances in making taxonomic comparisons by more quantitative means. This hope did not materialize in the way Cain envisioned, but the development of what eventually was called numerical taxonomy has yielded many useful results (see chapter 7). Cain also introduced the term *cryptic taxonomy* (1959b) or *cryptotaxonomy* (1962) to refer to taxonomy in which the exact features used for comparisons have not been made explicit (i.e., most of traditional, intuitive approaches). The new taxonomy was to remedy this.

COMPARATIVE BIOLOGY

Comparative biology is a term similar to systematics and regarded as synonymous by some (e.g., Nelson 1970), but it is here viewed as broader, embracing any study that compares particular features of organisms. For comparative biology to be equivalent to systematics involves the asking of questions only about classification and/or evolution (table 1.1). But other very different questions also are sometimes asked which utilize comparative data for answers, such as in genetics, physiology, comparative and developmental anatomy, etc. The focus of these studies is simply descriptive of form and/or function and not interpretive in the context of evolutionary relationships. It is probably true that the most meaningful questions answered with comparative data are, in fact, systematic ones, which may be one of the reasons why some workers (e.g., Mayr 1969c) regard comparative biology as falling completely within systematics. Another and more broad approach is taken by Nelson and Platnick (1981) who regard comparative biology as "the science of diversity" (1981:5) which includes the primary areas of systematics and biogeography, and also the secondary areas of embryology and paleontology.

CHAPTER 2

The Relevance of Systematics

Because of the close relationship of taxonomy to systematics, a few comments on the relevance of the latter are believed appropriate. It would be surprising if systematics, which deals directly with organic diversity, did not relate to every aspect of human endeavor. The impact of systematics upon society is substantial and most easily reflected in satisfying our intellectual curiosity about the world in which we live, formulating principles and methods of classification applicable to many human needs and activities, helping preserve the world's organic diversity for aesthetic and economic reasons, and more directly in developing economic potentials.

THE IMPORTANCE OF SYSTEMATICS IN SOCIETY

There is no general concern more important for an individual during his or her lifetime than to come to grips in some manner with how life came to be. Every person deals with this question in a different way; it is a highly individualized business with different answers approximating the number of questioners. But no matter how one addresses the problem, nor how one offers solutions, four general concerns nearly always come to mind: How did life originate? How has life changed through time? What life now exists on earth? What have been human origins and evolution? Answers to these basic questions of the human condition all come from systematic biology. As Smith has stressed: "It is because the systematist has been able to remove himself from the fragment of earth time that is man-dominated, in order to view this planet and the prior phases of atomic evolution from an effectively external philosophical viewpoint, that he is particularly fitted to appreciate reality" (1969:7–8). Insights from other disciplines are obviously involved too, such as anthropology, biochemistry, geology, and genetics, but systematics is central to these issues. Systematics is involved with studies on the origin of life (e.g., Eigen and Winkler-Oswatitsch 1983; Hartman, Lawless, and Morrison 1987), with investigations of the evolution of life once it was created (e.g., Chaloner and Sheerin 1981), with studies of the biota on earth including the proper storage and documentation of all collected materials (e.g., Cohen and Cressey 1969; Banks 1979), and with all kinds of morphological, anatomical, cytological, and biochemical investigations on the appearance of the human species (e.g., Brown et al. 1982; Blumenberg and Lloyd

1983; Sibley and Ahlquist 1984; Rak 1985; Smouse and Li 1987). Systematics is truly fundamental for satisfying our intellectual curiosity about the nature of the world in which we live.

No one in the past decade can have failed to realize that the environment is of special concern for the continued survival of our species and for the maintenance of a desirable quality of life in our society. Some question our ability to adequately cope with problems of pollution and overpopulation (e.g., Ehrlich 1968) and suggest possible negative scenarios (the "pessimists"; Greuter 1979). Others realistically point to severe problems but hope for future solutions (e.g., Brown 1956; the "optimists," Greuter 1979). Greuter (1979) makes the useful and well considered point that both pessimists and optimists are needed to solve any problem. The pessimist stresses the negative aspects of a situation, which serves to bring attention to it, but no action usually is taken. The optimist, who by nature is a doer, is influenced by the pessimist so that he or she becomes involved and positive action on the problem results. Both types of people are needed. Our present environmental crisis has certainly gone this general route, but so far there have been far more hand-wringing and doomsday forecasting than solutions proposed and accomplished. At the very least, it is obvious that as the human population increases and absorbs more of the energy resources for its own maintenance, fewer total useful resources will be available on earth for future human needs and for most other species as well. In short, severe pressure is being brought to bear on all species in a most serious competition for survival (Prance and Elias 1977). Tropical forests, which hold more undescribed species than anywhere else, are especially under pressure from deforestation and cultivation in developing countries of the world (Croat 1972; Gómez-Pompa, Vázquez-Yanes, and Guevara 1972; Raven 1976; Adams, Dong, and Shelton 1980; Myers 1980). As these habitats disappear, so do the species unique to them. As species disappear, so does their potential to aid our future needs, and we are left with one less weapon in our dwindling arsenal for survival. Further, most human beings are tremendously susceptible to the aesthetics of their environment to the extent that as diversity decreases and the world of our individual experience becomes more monotonous, its mentally therapeutic value declines proportionately with as yet poorly understood impact on the human condition.

The systematist's role in the face of this loss of diversity is obvious: speak out loudly and clearly on conservation issues in which an informed opinion will be helpful (Mosquin 1971), and work for preservation of natural areas (it is far more economical in the long run to preserve a habitat that houses a rare and desirable species than to attempt to maintain it artificially apart from that habitat); collect and inventory vigorously in those areas of immediate danger of destruction (Turner 1971); and help set aside germ plasm resources to be saved for future breeding studies (Hawkes 1978), especially wild relatives of crops already of great economic value (e.g., wheat, rice, corn, soybeans, sunflowers, etc.). As Hedberg has put it well: "In a world with rapidly increasing human population pressures and accelerating exploitation it is imperative to utilize biological resources sagaciously on a sustained yield basis, and to this end we must have an adequate knowledge of its flora" (1978:7). Many species yet to be described will have enormous food and medicinal value, and these are often encountered serendipitously in the course of general floristic work or in field work primarily devoted to other purposes (Iltis 1982). Systematics is essential for helping insure our continued survival on this planet.

Systematics can also help in developing further the economic resources that we already have. Biological control of agricultural pests, especially insects, has been used for decades with frequent success. For such endeavors to work well, systematists must be involved with proper identification of the organisms plus supplying data on their ecology and reproductive habits (Clausen 1942; Sabrosky 1955; Rosen 1986) to avoid unanticipated, unwanted, and economically ruinous results. The proper use of land resources such as the building of new dams, new canals, and strip mines, is another area in which systematists play an indispensable role by advising on the possible ecological impact on organisms living in the region (Hedberg and Hedberg 1972). Further, the knowledge and techniques gained by systematists through study of relationships of wild species can often be used to improve our existing cultivated food crops by similar methods of cytogenetics and artificial selection.

Systematics is also important in an even broader, but less obvious, philosophical way. This perspective has been summarized well by Mayr:

> The study of diversity has perhaps made its most important contribution to the development of new human conceptualizations, to a new approach in philosophy. More than anything else, it is the study of diversity which has undermined essentialism, the most insidious of all philosophies. By emphasizing that each individual is uniquely different from every other one, the students of diversity have focused attention on the importance of the individual, and have developed so-called population thinking, a type of thinking that is of the utmost importance for the interaction of human subgroups, human societies, and human races. By showing that each species is uniquely different from every other species and thus irreplaceable, the student of evolution has taught us a reverence for every single product of evolution, one of the most important components of conservation thinking. By stressing the importance of the individual, by developing and applying population thinking, by giving us a reverence for the diversity of nature, systematic and evolutionary biology have supplied a dimension to human conceptualization which had been largely ignored, if not denied, by the physical sciences. And yet it is a component which is crucial for the well being of human society and for any planning of the future of mankind. (1974a:8–9)

THE CONTRIBUTIONS OF SYSTEMATICS TO BIOLOGY

The contributions of systematics not only extend generally to society, but also more specifically to other areas of biology (Mayr 1968a, 1968b). One of the most pertinent contributions is the role that taxonomy plays as the "data processing system for biology" or in a less eloquent phrasing as the "digestive system of biology" (Heywood 1973a:139, 143). The number of data points that are being collected from organisms every day in biological research laboratories through-out the world is overwhelming. Literally millions of pieces of data are being gathered in the course of studies ranging from the biochemical to the anatomi-cal. Most of these data are tabulated and reduced in some fashion to answer specific questions posed by the biologist. In addition to helping answer these questions, however, these data can be used to assess relationships among organ-isms. In this fashion the isolated pieces of information can be used profitably in a more general way. Although this generality is true, the data so collected are rarely complete by themselves for making systematic inferences; much addi-tional study is almost always needed to develop truly comparative data for all the organisms under study by the systematist. In fact, even the data generated and used in traditional floristic and revisionary work are often deficient in some respects (Watson 1971; Heywood 1973a; Stuessy 1981). This data processing or "digestion" also allows for future specific questions to be asked that are more sophisticated and meaningful than before. The organization of data is enhanced by its attachment to organisms which are arranged hierarchically in a system of classification. "We are engaged in the construction of a framework on which to hang or arrange the total available biological information, the data about life, whether on the molecular or the organismal or the population levels." (Fosberg 1972:632). The informational content of an hierarchical arrangement of data is greatly superior to that in which data are all coordinate to each other.

Systematics also helps us understand the *processes* of evolution, which is information used by many other areas of biology. The microprocesses of evolu-tion, including individual variability, population variation, reproductive isola-tion, modes of speciation, etc., are all revealed through systematic studies. Some workers might prefer to call these kinds of investigations "evolutionary biology" or even more specifically, reproductive biology, population genetics, speciation biology, etc., but they all clearly fall within systematics as broadly defined in this book. The populational data are used by areas such as genetics, developmen-tal biology, and even more distant subdisciplines such as game theory. Broader-scale evolutionary phenomena, sometimes called macroevolution (Stanley 1979), are also revealed through systematic studies, for example, trends in the speciali-zation of seeds and seedlings and many other reproductive features of flowering plants (Stebbins 1967, 1970a, 1971a, 1974). These broader insights are likewise useful for other areas of biology (e.g., anatomy, morphology and developmental genetics in the example of seeds and seedlings given above). Macroevolution has many different meanings and is controversial. Some workers believe that phe-nomena occur over long periods of time that are distinct from populational processes (Stanley 1979), and others believe (including myself) that they are accumulations of effects developed at the population and species levels (Stebbins 1975; Bock 1979; Charlesworth, Lande, and Slatkin 1982) plus extinctions (see Nitecki, 1984). Some workers further claim that the tempo of evolution is punc-tuated (i.e., with periods of stasis followed by periods of rapid divergence; Eldredge and Gould 1972; Gould and Eldredge 1977; Eldredge 1985) as opposed to a more gradualistic view, and some even question the validity of the current modern evolutionary synthesis (Gould 1980). It seems clear that no new synthe-sis is needed (the present one is satisfactory to include all known micro- and macroevolutionary trends; Stebbins and Ayala 1981; Grant 1983a) and that both punctuated and graduated patterns must have occurred during evolution (Rieppel 1983b; Levinton 1988), more one way in some groups or more another way in others. The more interesting question is what kinds of groups show one type of pattern over another. See, for example, studies on divergence in minnows *(Notropis)* and sunfish *(Lepomis)* that suggest change due to gradual as well as punctuated phenomena (Douglas and Avise 1982). See also Malmgren, Berggren, and Lohmann (1983, 1984) for examples of "punctuated gradualism" in foraminifera.

Systematic studies also help reveal *patterns* of evolution that are useful and stimulating to other areas of biology. Patterns resulting from evolutionary processes occur at all levels of organization from the local population to ordinal and class lineages hundreds of millions of years old. The ancestor-descendent and associated patterns of relationship over long periods of evolutionary time, called *phylogeny*, and their reconstruction are of special interest to systematics. Phylogeny is important because it has much to do with the construction of classifications. Different opinions prevail, but most workers (myself included) believe that a classification of a group of organisms should in some measure reflect (or at least not be inconsistent with) its phylogeny. These "phylogenetic" classifications, therefore, contain the information that can be retrieved and used by other areas of biology. These hierarchies also stimulate ideas on the origin of life, the origin and evolution of major groups of organisms (including the human species), the development of ecological zones through geological time, and so forth.

The relationships of some other areas of biology to systematics is so strong that they are virtually dependent upon data generated in systematic studies. Ecology, biogeography, and paleontology depend in this way upon proper classifications of organisms and accurate knowledge of their distributions. Wilson (1971), speaking in the role of an ecologist, has stressed the need for continued support for taxonomic investigations so that the data can be used effectively in ecological work. Throne (1972) has stressed absolute reliance of phytogeography upon sound evolutionary classifications so that meaningful hypotheses on dispersal and vicariant events in particular groups can be formulated. Studies on relationships of organisms, including distributional patterns, have even given rise to new ideas on major earth events. The impact that the documentation of similar fossil biotas in Africa and South America had upon the development of ideas of continental drift was substantial (Schopf 1970; Hughes 1972).

The original conceptualization developed from within systematics (i.e., populational perspectives) "contributes significantly to a broadening of biology and to a better balance within biological science as a whole" (Mayr 1969c:9). This balance is further aided by systematics reminding other biologists that the diversity of life can be (and should be) profitably studied at all levels of hierarchical organization rather than by focusing all attention through myopically reductionistic eyes only on chemical and physical attributes. "Of course, it is quite possible that we could fully account for the properties of each whole if we could know the precise characteristics of all the parts and know in addition all existing relationships between them. Then we could reduce the characteristics of the whole to the sum of the characteristics of the parts in interaction. But this involves intregrating the data not merely for three bodies, but for three thousand, three million, three billion, or more, depending on the whole we are considering" (Laszlo 1972:8). Needless to say, this type of detailed understanding will be a long time in coming (if ever). In the meantime, a broad view of biology with an emphasis on organisms is entirely appropriate for continued advances to be made.

CHAPTER 3

The Importance and Universality of Classification

Of the numerous important contributions that systematics makes to society and biology, none is more significant than that provided by classification (and its theoretical and methodological umbrella, taxonomy). Classification is a pervasive human quality "like the predisposition to sin, it accompanies us into the world at birth and stays with us to the end" (Hopwood 1959:230). Although it cannot be denied that the construction of classifications provides intellectual satisfaction for those who make them (J. A. Moore, in Warburton 1967), and this by itself in my opinion is justification enough, many more positive features of classification also exist. Heywood has suggested that the societal value of taxonomists and their classificatory efforts and products would be negligible: "what effect would a strike of taxonomists have? The immediate effects would be few! A handful more people would die each day as the narcotics bureaux and emergency hospital services were unable to identify plant material; plant introductions might be halted. The papers on some biochemical-taxonomic topics might become even more bizarre without the advice of taxonomists! But it would be a long time before there were any serious consequences" (1973a:143). But Isely (1972) has shown convincingly, with an imaginative example of the hypothetical disappearance of all taxonomists and their works, that the long-term consequences of the cessation of taxonomic efforts would be disastrous for society. This would be especially true in the loss of information content of classifications and identification services derived from them. The term classification has been used in two general senses: (1) the *process* of classification (i.e., the manner in which grouping and ranking of items is accomplished); and (2) the *results* of the process, viz, the resultant hierarchy of classes. The importance and universality of classification derives from both these aspects.

THE PROCESS OF CLASSIFICATION

The significance of the process of classification can be viewed within our innate mental activity and attempts at description and formalization of this activity. The human species has a compulsion for order. This need probably derives from a desire to deal with the environment in a predictable way. As Knight has described it well:

The world into which we are born is a booming buzzing confusion, and we only slowly learn to sort out things of like kind. Instinctively in babyhood, and later more self-consciously, we group things together and attach general terms to them; so that instead of a chaos of endless particular things without apparent order, we come to perceive a world with a finite number of classes of things. We thus begin to feel at home, even though the classes may need to be revised (sometimes painfully) and certainly seem to cut across each other so that everything belongs in more than one. We distinguish parts of ourselves from other things, and we then separate things that are accessible from things that are not, like the Moon, for which there is little point in crying. Some classes have sharp lines, and others have fuzzy ones; the divisions between colours, for example, seem hard to learn and are mastered some time after children have got size relations straight, and differ between cultures. Great scientists are Peter Pans, still anxious to classify and explain at an age when most people are concerned with money, power, and sex. (1981:16–17)

Because of our greater intellect, we alone of all species are smart enough to *worry* about ourselves and our existence on earth (hence the development of science, philosophy, religion, etc.), and the process of classifying everything in our environment (and the utilization of the resultant products) can be interpreted as an attempt to lower this risk of uncertainty of living. "To be confused about what is different and what is not, is to be confused about everything" (Bohm 1980:16). In effect, we are trying to describe and interpret reality (at least as far as we can ever come to know it). We classify everything in our environment including animate and inanimate objects such as furniture, cars, houses, clothes, diseases, and the elements (consider the periodic table). We even rely formally on methods of classification in some non-biological disciplines, e.g., in linguistic analysis: Hoenigswald (1960), Jordan and Swartz (1976), Platnick and Cameron (1977), Halle, Bresnan, and Miller (1978); anthropology: Lomax and Berkowitz (1972); management theory: Goronzy (1969), Higgins and Safayeni (1984); and psychology: Kee and Helfend (1977), Campbell, Muncer, and Bibel (1985), Chiribog and Krystal (1985). Evidence for the pervasiveness and importance of classification for humans comes from studies of primitive peoples of the world who classify objects and ideas in their environment in much the same basic fashion as do professional western taxonomists (Berlin, Breedlove, and Raven 1966, 1974; Raven, Berlin, and Breedlove 1971; Berlin 1973). Obviously many differences exist on the details of resultant classifications produced by aboriginal versus progressive western cultures, but the process of classification is the same. Sometimes even the hierarchy of classes is remarkably similar, except for a tendency in aboriginal societies for many more subdivisions and variations for organisms and things which are economically and/or culturally very important ("over-differentiation"; Berlin 1973).

Because of the importance of the process of classification in our innate mental activity, we have attempted to describe this process more clearly and even to formalize our way of thinking about it. The early attempts go back to Plato and *essentialism* (Hull 1965) and to Aristotle and *logical division*. With Plato's essentialism one could view attempts at classification as simply the search for the *eidos*, or essence of things, which would reveal their true natures and allow for proper communication about them as well as correct alignments together in classification. The search for essences, and in a broader sense the attempt to order the world through the construction of classifications, may be related to the nature of language. Bohm has suggested that this may be due to the subject-verb-object structure of sentences: "This is a pervasive structure, leading in the whole of life to a function of thought tending to divide things into separate entities, such entities being conceived of as essentially fixed and static in their nature. When this view is carried to its limit, one arrives at the prevailing scientific world view, in which everything is regarded as ultimately constituted out of a set of basic particles of fixed nature" (1980:29). Whether language gave rise to early classifications or our innate desire for order led to the structure of language is a familiar chicken-or-egg type problem defying resolution.

Aristotle suggested how objects or classes logically could be divided into subclasses by a process called logical division (e.g., Sinclair 1951). A class composed of rectilineal figures can be divided logically into subclasses such as squares, triangles, hexagons, etc., based on the numbers of angles each contains. This criterion of number of angles is called the *fundamentum divisionis*, or principle of division. The use of this principle several times and the resultant hierarchical arrangement of classes provides the rudiments for the Linnaean hierarchy now in use for organisms (to be discussed fully later in chapter 10). Logical division is still a powerful concept and way of thinking in our attempts at classification, and it is also most conspicuously still used in the construction of taxonomic keys, which are nothing more than a series of contrasting statements of differing features (fundamenta divisionum).

A broader view of the world based on logical division, but in a more hierarchical fashion, has led to the development of *systems theory* (Bertalanffy 1968; Laszlo 1972) or *hierarchy theory* (Pattee 1973; Salthe 1985) in which "we must look at things as systems, with properties and structures of their own. Systems of various kinds can then be compared, their relationships within still larger systems defined, and a general context established. If we are to understand what we are, and what we are faced with in the social and the natural world, evolving a general theory of systems is imperative" (Laszlo 1972:14). To what extent systems theory is significant for management of societal groups, institutions, government, etc., or the interpretation of reality remains to be seen, but it does emphasize the importance of logical division and the process of classification in our society. Systems and hierarchy theory may also be useful in the description and interpretation of ecological structure (Allen and Starr 1982; Auger 1983).

The basics of logical division proposed over 2,000 years ago by Aristotle have been formalized even further into what is known as *set theory*. This is a logical way to determine the precise relationships of classes to each other in coordinate and subordinate fashions, and these can be symbolically shown by Venn diagrams (two-dimensional circle diagrams) and by notation (e.g., Halmos 1960; Carlock and Fensholt 1970; Nanzetta and Strecker 1971). Set theory can explain the logical aspects of the development of hierarchical classification, and it has been used for analyses of the logical structure of the Linnaean hierarchy (to be discussed in more detail in chapter 10; Gregg 1950, 1954, 1967; Buck and Hull 1966; Jardine 1969a). An outgrowth of set theory and related to systems theory discussed above is *categorical system theory* (Louie 1983a, b) in which set notation is used with the mathematical theory of categories to describe natural systems. Related to set theory and with similar notation, *game theory* has evolved as a logically formal way to describe the processes of competition and change among groups in relation to external (usually environmental) factors. It has more application to evolutionary studies of natural selection (e.g., Maynard Smith 1976, 1982; Williams 1987) than classification, but an application to tree construction has been attempted (Marchi and Hansell 1973).

More recent efforts to understand the process of classification can be grouped into the area of *pattern recognition* (e.g., Dunn and Davidson 1968), which is simply a more formal (and in the hopes of some people, potentially automated) way of classifying by ordination or hierarchical division. Long ago Edgar Anderson (1956) stressed that people usually differ in their abilities to recognize patterns in nature, with some having propensity to relate objects in a qualitative and visual way, and others being able to deal more easily with abstract concepts, numbers, and quantitative relationships. The talent for intuitive pattern recognition by taxonomists relates directly to a strong visual memory, and an ability to relate objects to each other in three-dimensional space. More recent developments seek to understand how a person actually does these sorts of analytical activities, the sense perception involved, information processing, feedback mechanisms, etc. Computers are very much used in their simulation (Meisel 1972; Rutovitz, 1973; Sklansky 1973; Kanal and Rosenfeld 1981; Albrecht 1982; Fu 1982; Hunt 1983).The whole area of artificial intelligence is growing rapidly and will doubtless give us many ideas in the future on how people really do classify. One point is already clear: intelligence is based to a large degree on prior accumulated knowledge derived through experience (Siegler 1983; Waldrop 1984) which creates real difficulties with developing machines to handle complex mental problems (such as in playing chess or in constructing classifications!).

HIERARCHICAL SYSTEM OF CLASSES

The significance of the *results* of the process of classification lies in the nature of the hierarchical system developed. Most classifications are hierarchical, but some are not, such as those done by ordination in which variables (such as species or other taxonomic groups) are arranged in two-dimensional space in relation to major axes of variation in the data set used (e.g., DuPraw 1964, 1965). But most hierarchical classifications have higher information content and are clearly more useful for human needs than nonhierarchical ones, and therefore, are of more interest to us here. Definitions for hierarchical classifications are numerous, but four seem most appropriate for discussion (modified after Anderson 1974) with the first two being "general" definitions and the second two being "restricted" definitions: (1) any set of objects grouped into classes (reflecting coordinate relationships); (2) a nested hierarchy of classes (reflecting coordinate and subordinate relationships); (3) a particular scheme of grouping and ranking of a particular set of objects; and (4) the relationships (expressed as some measure of similarity and/or difference) between and among taxa in a particular classification scheme. Hull (1970a:22) has suggested that "some authors use the words *a classification* to refer to the entire taxonomic monograph." But this is not common usage. In the discussion of the importance and universality of hierarchical classification that follows, all of these definitions will be referred to collectively, but more emphasis will be placed on 1, 2, and 4.

An important role of classifications is as devices for information storage and retrieval (Heslop-Harrison 1962; Davis and Heywood 1963). As Bock has put it: "The most fundamental attribute of biological classification is that it must be useful. It must order and summarize biological information, must be heuristic, and must provide the basis for future studies" (1973:379). The hierarchical structure of a system of classification contains a tremendous amount of information that can be tapped and utilized in many ways by many kinds of people. As organisms are newly discovered or as new aspects of already known organisms are revealed, these new data are added to the existing classification for a continual updating of information.

Some workers have attempted to quantify (and maximize) the degree of information content of an hierarchical classification (Gower 1974; Duncan and Estabrook 1976; Farris 1979b; Brooks 1981). For the evaluation of alternative classifications for the same group, such a measure could help pick the one that would be "best" in this context. The difficulty is deciding which information content measure (or "optimality criterion;" Sneath and Sokal 1973) is best, and it is doubtful that any consensus will ever be reached on this issue. If agreement is attained on the best *general* measure (this, in itself, highly doubtful), there will likely be several competing specific measures each with its own positive claims and liabilities. Furthermore, the significance of results from these measures for practical use of the classification, in reflection of evolutionary history, or in use as a predictive device will have to be determined. There is no question that information theory (e.g., Gatlin 1972) is useful in helping explain the hierarchies we recognize in the living world (some of which exist in nature, others we invent for our needs), but much additional work is needed to determine the specific applications to classification. These stored and retrievable data are especially useful for other biologists (especially other comparative biologists) and their own research studies (Darlington 1971; Bock 1973).

Classifications also serve for identification services to be performed (Warburton 1967). "And because identification is a necessary step in both storing and retrieving information, a classification that facilitates identification is essential to all organismic-evolutionary biologists" (Darlington 1971:343). A system of classification serves as a guide for the labeling of unnamed objects by comparison with those items already named. Objects are never "identical" to each other, of course, not even machine-made pieces, but they can be highly similar to each other. The degree of similarity needed for two objects to be regarded as "identical" and thus given the same name depends upon how detailed is the classification—the more detailed or dissected the hierarchy, the more similar the objects must be to carry the same name. In many ways the efforts placed on the identification role of classification are not as highly regarded as those expended on the development of the classification itself, primarily because the former (in the shape of keys) are extracted from the latter. This is in line with the general perspective that service functions in science have a lower priority than those of original contributions to knowledge. Nonetheless, the service role that identification plays in our society is enormous, most significant, and worthy of considerable resource commitment by the systematics community. In some ways, the most visible impact systematics has on society at large is through the production of keys, especially in popular natural history guidebooks (e.g., the Peterson Field Guide series), which base their information on more technical revisions and monographs.

Another significant contribution of classification is in the reflection of evolutionary relationships (Warburton 1967; Nelson 1973c). The degree to which a classification can or should reflect evolutionary relationships is controversial. Suffice it to say at this point that most workers believe that a classification should and can reflect phylogeny in some fashion. This represents an attempt to capture the "reality" of the world as expressed well by Smith (1969) and Darlington (1971). Exactly *how* this is to be done and how *effectively* it can ever be done are areas of extreme debate and differences of opinion (to be discussed more fully later). There are also problems with attempting to reflect three-dimensional data about phylogeny in a two-dimensional hierarchical classification. Classification cannot fully express *all* relationships in a phylogeny, although obviously depending upon the classification constructed and the particular group involved, some hierarchies reflect more of these relationships than others. A minority viewpoint is that since we shall never know any true phylogeny (which must be admitted as a fact of life), then we shouldn't waste time trying to reconstruct them (Sokal and Sneath 1963) and we shouldn't delude ourselves into believing that our classifications really reflect these evolutionary patterns in any precise way (Davis and Heywood 1963; Davis 1978). I do not share this perspective.

Another most significant feature of classifications is their use as summarizing and predictive devices (Rollins 1965; Warburton 1967). This is mentioned by most workers as *the* most important quality of a classification, but it is often phrased in slightly different ways, e.g., "to construct classes about which we can make inductive generalizations" (Gilmour 1951:401; see his similar view in 1940), or "as a basis for predicting a maximum number of unknown characters" (Michener 1978:114). It is worth bearing in mind, however, that "The idea of the predictive value of a classification is ambiguous. It means that one can describe a trait as characteristic of all members of a taxon before it has been verified for all. It also means that if organisms have been classified together as a taxon* because they have all been found to share certain traits, they will later be found to share other traits as well" (Warburton 1967:242). An example of the first meaning of prediction is the following: consider that within class A, defined by features other than leaf arrangement, members of subclasses X and Y are discov-

---

* The term *taxon* refers to a taxonomic group at any rank in the hierarchy, such as species, genus, family, etc. (coined by Meyer-Abich 1926; see also Mayr 1978). The term will be covered later in this book in our discussion of categories, but it is so useful that its early introduction here will facilitate discussion in this and subsequent chapters.

ered to have opposite leaves. We can infer, therefore, that members of a third subclass Z in the same class should also have opposite leaves. To test this prediction in an informal sense, we can look at Z and see what type of leaf arrangement it has. To test the prediction even further, we can look at the leaves of the most distantly related subclass that is still included within class A. The ability to make such predictions is the basis for much of man's search for useful plant materials, especially for chemical compounds that have medicinal value. If a useful compound is discovered in one group of plants, the most closely related groups are looked at next to learn if they, too, might have the same or similar compounds. As an example of the second meaning of prediction, continuing the same example as above, if within class A subclasses X, Y, and Z all have certain morphological and anatomical features the same, then one can predict that their chemical constituents (or lack of them) will be the same too, even though nothing is yet known about this type of data in this group. Warburton stresses the importance of the predictive value of classification: "All other biologists must trust taxonomists to provide them with classifications that maximize this probability [of inductive generalizations], since the validity of all observational and experimental biology depends upon such classifications" (1967:245).

CHAPTER 4
Characters

The construction of classifications with their many positive features depends upon a careful comparison of attributes of the organisms. It is not sufficient to state that "organism X is more similar to organism Y than either is to Z" without also expressing the particular reasons for this conclusion. Such expressions of relationship are based upon a comparison of features of organisms (either clearly stated or intuitively evaluated) called taxonomic characters. As Kendrick has expressed well: "Although a man can visually appreciate a very complex concept almost instantaneously, his much more limited capacity for verbal communication forces him to describe what he sees in a series of words, some of which convey more information than others. The taxonomist often finds himself in an analogous position. He may be able to assimilate the 'facies' of an organism at a glance, but in order to interpret to others what he sees, he must mentally dissect the organism and describe it as a series of characters, some of which may have greater significance than others" (1964:105). The challenge, then, is to understand what characters are, what their role in classification really is, what terms have been applied to them, and how they can be selected for use in making classifications.

GENERAL TERMS

The term character in a taxonomic context has numerous definitions* which need to be discussed and understood before their role in constructing classifications can be appreciated. The definition of a *taxonomic character* used in this book is: a feature of an organism that is divisible into at least two conditions (or states) and that is used for constructing classifications and associated activities (principally identification). Depending upon the type of approach to classification employed, these taxonomic characters may also be used to interpret evolutionary relationships and determine evolutionary pathways as a part of putting together the classification. If the features of the organism are used more to determine the processes of evolution, especially at the populational level, then

---

*The word *character* has many meanings in everyday usage, too. For example, one of my dictionaries (Guralnik and Friend 1953:245) gives 16 definitions for character ranging from "a distinctive mark" to "a person conspicuously different from others; queer or eccentric person," the latter perhaps being more descriptive of some taxonomists than of the attributes of the organisms with which they deal!

we speak of *systematic characters* in a broader sense which would encompass the use of attributes of organisms for any study of diversity. All taxonomic characters, therefore, are also systematic characters, but not the reverse.The conditions or expressions of all types of characters are called *character states*, and it is these that are compared in the construction of a classification. "Characters as such are strictly speaking abstract entities: it is their *expressions* or *states* that taxonomists deal with" (Davis and Heywood 1963:113). For example, corolla color in angiosperms is a character, with red corollas, white corollas, and blue corollas being three different states of that character. Leaf arrangement is a character divisible into three states: alternate, opposite, and whorled.

Many other definitions of character have been proposed by various workers. These are basically the same as that given above in the sense that they stress their importance in delimiting taxa; a few also have more of a practical orientation. Davis and Heywood define a character as "any attribute (or descriptive phrase) referring to form, structure or behavior which the taxonomist separates from the whole organism for a particular purpose such as comparison or interpretation" (1963:113). Jones and Luchsinger give a similar definition: "The *characters* of an organism are all the features or attributes possessed by the organism that may be compared, measured, counted, described, or otherwise assessed" (1986:62) (this basically the same definition as that given in Hedberg 1957). Crowson suggests simply "any feature which may be used to distinguish one taxon from another" (1970:68).

Some workers have reacted against these taxon-oriented definitions on the grounds that they may be circular (Sokal and Sneath 1963; Sneath and Sokal 1973): how can characters be defined in terms of delimiting taxa when the taxa must be defined before we can know what the characters are? To avoid this problem, alternative taxon-free (and more general) definitions of characters have been offered: "any feature of one kind of organism that differentiates it from another kind" (Michener and Sokal 1957:137); "anything that can be considered as a variable independent of any other thing considered at the same time" (Cain and Harrison 1958:89); "any attribute referring to form, structure, or behaviour, which can occur in any one organism as one of two or more mutually exclusive states" (Kendrick 1964:105); "some defined attribute of an organism" (Ross 1974:20); and "the *expression* of the feature in the individual" (Blackwelder 1967b:148).

Jardine (1969a) uses the terms character and character state in a different sense to refer to probability distributions of descriptive terms of individual organisms and taxa. The descriptive terms applied to individual organisms are called attribute states, such as "red" or "2 cm long." Sets of these descriptive terms are called attributes, such as "color" (including the attribute states red, green, etc.). "Probability distributions over the states of an attribute will be called *character states*, and sets of such distributions will be called *characters*" (Jardine 1969a:38). These usages do not seem to aid communication, and in fact, serve to confuse attempts to understand the more commonly accepted definitions. They are not recommended for use, but are mentioned here for interest and completeness of the discussion.

Other workers do not use the designation of character state, and refer only to characters as being "any attribute of a member of a taxon by which it differs or may differ from a member of a different taxon" (Mayr 1969c:121; similar to the definition by Crowson 1970, given above; see Rodrigues 1986, for additional agreement). Mayr (1969c) points out that historically in systematics the term character has been used in this way for centuries, whereas the concept of character state has been only recently introduced explicitly (Michener and Sokal 1957; Cain and Harrison 1958, also used character "value" in the same context). He advocates return to the original usage (as does Blackwelder 1967a).

Other workers agree with this perspective, but not for historical reasons. Those who strive to construct classifications based on explicit ideas of relationship by evolutionary descent and rigid rules, the "evolutionary cladists" (to be discussed fully in chapter 8), view a character as "a feature of an organism which is the product of an ontogenetic or cytogenetic sequence of previously existing features, or a feature of a previously existing parental organism(s). Such features arise in evolution by modification of a previously existing ontogenetic or cytogenetic or molecular sequence" (Wiley 1981a:116). From this perspective, only characters have information of evolutionary value for constructing classifications, and the need for states vanishes. Eldredge and Cracraft also argue that use of character state is unnecessary because character and character state are seen as relative terms and "should be construed to mean relative levels of similarity within a given hierarchy" (1980:30). The "theoretical cladists," also called "transformed cladists" (Hull 1984), or "pattern cladists," (Brady 1985), who view classification systems primarily as informational and organizational systems without direct evolutionary implications, take an even broader view of a character as "a unit of 'sameness' " (Platnick 1979:542). Ghiselin (1984) has even suggested abandonment of character and character state for "feature" for both, but this would be terribly destabilizing and is not at all recommended.

Related to character and character state are several other general terms that need to be discussed, such as attribute, quality, feature, characteristic, descriptive term, property, accident (accidentia), difference (differentia), and essence (essentialia). These descriptors are all somewhat related and will be considered within the context of three general viewpoints: philosophical (or epistemological, which deals with the theory of the origin, nature, and limits of knowledge); logical; and biological. Very generally, a *feature* is defined in modern dictionaries as "a distinct or outstanding part, quality, or characteristic of something" (Guralnik and Friend 1953:531). In a philosophical context, Griffiths treats character and attribute as synonyms and defines them following Kant (from Abbott 1885): "An attribute is that in a thing which constitutes part of our cognition of it;" and "a partial conception so far as it is considered as a ground of cognition of the whole conception" (quoted in Griffiths 1974a:108). An *essence* (essentialia) is a feature that attempts to reflect the philosophical essence of an object, as used in Plato's essentialist philosophy (discussed earlier). The logical viewpoint, deriving from traditional Aristotelian logic and represented clearly in the works of Linnaeus (Cain 1958) and other early classifiers, distinguishes among difference (differentia), property, and accident (accidentia). In a strict logical sense, "Any quality or attribute is regarded as being either a difference or a property or an accident. A quality is said to be a *difference* if it serves to distinguish the class of entities of which it is a quality from other species of the same genus [genus and species used here in a logical sense of set and subset relations], i.e., if it is utilized in the definition of the class. A quality is said to be a *property* if it is a quality necessarily possessed by every member of the class, yet not utilized to distinguish the class from other species of the same genus. A quality is said to be an *accident* if it may indifferently belong or not belong to all or any of the members of the class" (Sinclair 1951:94–95). Within the biological context, a *characteristic* is viewed as: "A particular character state occurring exclusively in certain specimens or species. . . . Thus concerning the character *tail, bushy tail* is a characteristic of squirrels and *scaly tail* is a characteristic of rats. *Bushy* and *scaly* are different states of the character *tail*" (Ross 1974:20). A *descriptive term* is a descriptor referring to a condition (or state) of an organism that has taxonomic import, and it can be equated with characteristic in the sense of Ross (1974). In summary, then: difference, property, accident, and essentialia have precise logical definitions; feature, quality, and attribute all refer generally to some aspect of an organism (whether taxonomically useful or not); and characteristic and descriptive terms refer to aspects that are regarded as taxonomically significant.

A good closing perspective to this discussion of definitions of general characters and related terms is provided by Davis and Heywood: "It follows from what we have just said that no precise general answer can be given to the question 'what is a character?' This can only be considered in individual cases and what we treat as a character will depend on what we want to use it for. Even apparent absence of differential characters or expressions between individuals or groups need not indicate that they are identical: differences may well come to light after detailed study. It has been remarked that one will always find characters for separation if one tries hard enough—and, one might add, find that characters used for separation do not hold when more material is examined!" (1963:114).

ROLE OF CHARACTERS AND STATES

Characters are most important for allowing the numerous significant features of classifications to be realized. From the broadest perspective, characters help us describe and measure the perceived reality of the world so that we can then assess any and all relationships of interest. Our skill in employment of characters and states, therefore, will be directly related to our ability to interpret and appreciate reality in the natural world.

Characters and states constitute the basic data of the living world and of all areas of systematics. These data represent a gigantic *basic data matrix* of features of organisms which can be viewed as a table with all the world's taxa along one side and all the known characters along the other.The descriptors inside the table on the appropriate rows and columns (as descriptive words or numerical values) are the character states, which is the information about each of the particular taxa. As taxonomists we are continually gathering comparative data in the form of characters and states and slowly filling in this basic data matrix for the world's biota.

It is the basic data matrix and its analysis from which all interpretations on classification and evolutionary affinities are derived. All approaches to classification extract information from the basic data matrix: some approaches attempt to take all data; others make a selection. No scientific interpretations about relationships of any kind among organisms can be made without reference to these data. The classifications derived from these data by employment of some method results in the highly desirable predictive value of the classifications, which is their most important feature.

The predictive and information-retrieval values of classifications are given practical expression in the construction and use of keys for identification purposes. This is a valuable extension of the use of characters and their states in making the original classification. Taxonomic keys (usually dichotomous) provide pairs of contrasting statements about the features of the organisms under consideration, and these data are all in the form of characters and states. They are selected to maximize the differences between sets of taxa to allow for rapid and efficient identification of individual organisms. To the extent that the characters and states used in the original classification are properly gathered, described, measured, etc., is also to the degree that the constructed keys will be workable and serve their identification function.

KINDS OF CHARACTERS

The qualifying terms for characters are numerous, complex, and frequent in the literature. These distinctions are necessary for proper appreciation and use. The most comprehensive list is found in Radford et al. (1974). These have been used here, but with comments and additional terms from other sources. The terms are organized into four categories: general types of characters; phyletic (= phylogenetic or evolutionary) characters; cladistic characters; and phenetic characters.

General Characters

Types of general characters are distinguished according to organization of states, variability of characters and states, specific utility, and general validity. The most fundamental distinction often made in the organization of states of a character is into quantitative or qualitative units. *Quantitative characters* are those that can be assessed by counting or measurement and the states are expressed in numbers. Examples include leaf length given in cm, hairs on the undersurface of a leaf given in hairs per mm^2, and number of stamens in a flower. *Qualitative characters* are those that describe shape and form with states given in descriptive words rather than numbers. Examples are leaf shape given in terms such as ovate, obovate, lanceolate, linear, etc., habit with herb, shrub and tree the states, and ovary displacement with the two states being superior and inferior. At first glance the distinction between quantitative and qualitative characters seems striking and meaningful. But "to explain something and to measure it are similar operations. Both are translations. The item being explained is turned into words and when it is measured it is turned into numbers" (Kubler 1962:83). I would go even further and state that *all* characters may be expressed either in quantitative or qualitative terms, although it may be desirable, preferable, or more convenient to treat them one way or another in specific instances. For example, one could describe leaf shape quantitatively by a series of measurements including length, width, area, ratios of length/width, etc., but it would probably be easier simply to treat the character qualitatively and use the states ovate, obovate, lanceolate, etc. Some characters, by their very nature, could be treated either quantitatively or qualitatively. Consider *meristic characters*, generally regarded as a type of quantitative character, which are only describable in discrete states, such as number of stamens in an androecium of 5, 10, 15, and 20. These occur only in a limited number of numerical variations, but the states are clearly numbers rather than words. If words were available for these four conditions, they could just as easily be used as in the leaf shape example above. The contrast in meristic vs. regular quantitative characters is the same as that between *discontinuous* and *continuous* quantitative characters. One lesson has been shown us by experience, however, that quantitative characters tend to be more useful at the lower levels of the taxonomic hierarchy (Davis and Heywood 1963). Both qualitative and quantitative characters can occur in two-state or multistate conditions (Stace 1980). Chapman, Avise, and Asmussen speak of *quantitative multistate characters* as those "expressed by a numerical value which can be arranged in order of magnitude along a one-dimensional axis" (1979:52).

How many states to recognize within a qualitative character depends upon the character, the group being studied (Just 1946), and the type of classification approach used. Kendrick (1964) has suggested no more than six states be used. Although this is an arbitrary rule, it is a useful guideline (especially with qualitative characters). As the number of states increases for a qualitative character, one is tempted to divide this one character into several separate characters with fewer states in each, both to ease comparisons and computations of similarity between taxa but also to describe more clearly the variations believed significant.

Another viewpoint on the organization of states is to view characters as *micro-* and *macrocharacters* (or *cryptic* vs. *phaneritic* characters; Radford et al. 1974). These refer to scaling perspective on the actual size of the states and sometimes also to the type of method of obtaining the data (e.g., use of TEM or SEM, chromatography, electrophoresis). The distinction between these two types of characters is stressed usually by those who believe that one type of data is superior to others in revealing relationships. Davis and Heywood have commented: "Then there is the psychological problem, too, that small-scale characters tend to be considered unimportant, while immediately obvious ones are regarded as valuable. On the other hand, many botanists have considered characters of primary importance just because they are difficult to observe! Neither claim is valid" (1963:117). An excellent recent example is the emphasis on microcharacters of anthers and other floral parts in the tribe Eupatorieae of the sunflower family (Compositae) by King and Robinson (1970) over other more traditionally used macrocharacters. In this case the claim of supremacy of these microcharacters is totally unfounded (Grashoff and Turner 1970; Stuessy 1973; McVaugh 1982). As with all characters, sometimes microcharacters are useful, and sometimes they are not (Stuessy 1979b).

Other terms apply to the variability of characters and their states. Characters which are *invariant* within a taxon under consideration, i.e., occur in only one state, are clearly of no value in determining relationships between subtaxa within that taxon. Only *variant* (or variable) characters with two or more states are useful taxonomically. These same characters may be quite useful, however, in distinguishing these taxa from all others at the next level of the hierarchy provided that groups at that level have contrasting states. A similar distinction prevails between *plastic* and *fixed* characters. Fixed characters are invariant characters and vice versa, and plastic characters are also variant characters, but *not* the reverse. The term plastic refers to environmental modification of characters that has no genetic basis, whereas variant refers to genetically controlled differences. Some characters have states that are *"clear-cut"* and *"major"* as opposed to *"intergrading"* and *"minor"* (Swingle 1946:265, 266), and the former are usually to be preferred: "It is common taxonomic practice, based on experience and reason, to hold more closely to major than to minor characters" (Swingle 1946:266). States of characters that are useful taxonomically tend to be correlated in their patterns of variation with other states of characters also useful; i.e., they are *nonrandom*. This was realized by Sporne (1948, 1954, 1956, 1975, 1976) in studies on the correlation of characters and states in the dicotyledons and particularly in the primitive angiosperms. Meacham (1984b) has shown recently that often states of characters significant taxonomically are correlated in their evolutionary directionality (i.e., they are compatible) which gives further evidence of the nonrandom nature of distribution patterns of useful character states among taxa. A character showing completely randomly distributed states would most probably be under no selection pressure and therefore reveal little about evolutionary tendencies and be of little help in classification.

Terms are also applied to characters to suggest their specific utilities. Characters used to delimit taxa at different levels in the taxonomic hierarchy are sometimes referred to as *specific characters, generic characters, familial characters*, etc. In fact, few characters at any level are absolute for taxa at that rank (to be discussed later), but within a particular group the terms do have relative helpful meaning. In speaking of genera Linnaeus referred to *"essential characters"* which would distinguish them from each other within the same "order" (Cain 1958). *Fundamental characters* that are presumed to have utility even before their patterns of distribution and variation are assessed, are judged invalid. "Characters become important after they are proved valuable by experience. The relative value and utility of characters are empirically determined, and there is no such thing (in a practical sense) as a 'fundamental' character" (Jones and Luchsinger 1986:64). Characters used in identification, characterization and delimitation of a group are called *analytic*, whereas those used in the classification of that same group at a higher level in the hierarchy are called *synthetic* (Just 1946). "There is no inherent difference between analytic and synthetic characters: the difference is more one of the particular use made of them in a particular case" (Davis and Heywood 1963:115). *Diagnostic* or *key characters* are those "used in description, delimitation, or identification" (Jones and Luchsinger 1986:63) and are contrasted with *descriptive characters* that serve to describe some aspect of an organism or taxon but do not reveal how it differs from close relatives. *Biological characters* have some clear function or vital role in the organism as opposed to *fortuitous characters* that have no such function (Wernham 1912, from Davis and Heywood 1963). Diels (1924, in Sprague 1940: 447, 448) went even further and broke biological characters down into three subordinate types: *functional characters*, those "intimately connected with some special function, but uninfluenced by external conditions;" *epharmonic characters*, those "apparently connected with the mode of life of the plant, but nevertheless remaining constant under varying external conditions;" and *adaptive characters*, those "varying according to the external conditions." Just (1946:292; with unreferenced credit to A. de Candolle) recognizes *constitutive* vs. *nonconstitutive characters* with the former being those features not directly affected by environmental factors, and the latter being those "resulting from 'adaptations' to the environment." There is confusion here with regard to adaptations being viewed as heritable traits or as plastic responses (cf. Diels's definitions of the different biological characters given above).

A few general terms for characters refer to their general validity. *Good characters* "(1) are not subject to wide variation within the samples being considered; (2) do not have a high intrinsic genetic variability; (3) are not easily susceptible to environmental modification; (4) show consistency, i.e., agree with the correlations of characters existing in a natural system of classification which was constructed without their use" (Davis and Heywood 1963:119). *Bad characters* obviously lack these qualities. Similar pairs of terms to express these same ideas are *reliable* vs. *unreliable characters* and *meaningful* vs. *meaningless characters* (Sokal and Sneath 1963). Clearly, the "goodness" of a character will depend upon the context in which it is used. A character may be good in one taxon and bad in another. There is no absolute way to know beforehand *(a priori)* whether a character will be good or bad, except through experience one learns general ideas about kinds of characters that *tend* to be useful in particular groups. But "Even if the taxonomic value of a character has been 'proved' in a hundred cases, it still may fail in the hundred and first. The golden rule therefore is: there is no golden rule" (B. Rensch, in Just 1946:296).

Phyletic (Evolutionary) Characters

Phyletic (= phylogenetic* or evolutionary) characters are used primarily in phylogenetic (= phyletic) classification. The most important distinction here is between characters that are *homologous* versus those that are *analogous*. The

---

*The term *phylogenetic* has been used in different ways in recent years, so that its meaning has become confused. Its original usage was as an adjective referring to phylogeny, or the actual evolutionary history of a group (Haeckel 1866). In recent years advocates of a particular approach to classification (cladism) have used phylogenetic to refer to reconstructing only the branching sequence of phylogeny (Hennig 1966; Wiley 1981a) rather than all of its aspects. This more narrow approach is best called cladistics with phylogenetics retained in its original and broader usage (see Mayr 1969c:70, for agreement).

difference between these terms is on the surface simple, but many problems exist philosophically. Richard Owen's original definitions (in Cain 1976: 25) were, for *homologue*, "the same organ in different animals under every variety of form and function;" and, for *analogue* "a part or organ in one animal which has the same function as another part or organ in a different animal." Homologous characters, therefore, were originally viewed prior to Darwin's theory of evolution as simply basic structural differences. After evolutionary theory developed, homologues were viewed as the structural modifications of the same organ, inherited from a common ancestor, in response to different selection processes. Analogues, on the other hand, were those features developed by different organs to the same selection pressures. There is no question that for the proper construction of evolutionary classifications, homologous characters need to be emphasized. The detection of homologous characters in two groups is done by knowing that they have descended from a common ancestor. If this were known, however, there would be no need to use these homologues to reconstruct the phylogeny; it would be known beforehand in order to select the homologous characters. This circularity is a problem which has led some workers to eschew searches for homologous characters (e.g., Davis and Heywood 1963). Philosophically there are solutions to the problem (Ghiselin 1966b; Hull 1967; Cain 1976), such as by considering other features of the organisms for additional signs of similarity (Cain 1976), but they boil down simply to the reliance on structural and ontogenetic similarity as a reflection of the homology. The detection of homologous characters is a difficult problem for any phylogenetic reconstruction and pitfalls can occur (see the problem of interpretation of leaf homologies in *Acacia*, in Kaplan 1984). Because of the complexities of the issue, botanists have tended to deal with the problem obliquely, as indicated in the recent text by Stace: "Homology is usually defined on the basis of common evolutionary origin, a definition which should in theory be uncontentious, but which in fact is usually quite impractical because of our lack of evolutionary data. In practice, therefore, one can only guess at homologies by making as detailed as possible an investigation of the structures concerned. More usually the problem is ignored" (1980:55). Wiley (1981a) has recognized three types of characters: structural, functional, and phylogenetic. The first two are those that *appear* to be similar, but are actually analogous. His phylogenetic characters are the only true homologues. This is a confusing perspective, because the way to determine homologies (or his phylogenetic characters) is in part through detailed structural (or even functional) comparisons.

Other sets of terms also are applied to phyletic characters. *Phylogenetic* and *ontogenetic characters* are simply features which in the first instance are presumed to reflect information about the phylogeny of the group, and in the latter instance deal with developmental features. Ontogenetic sequences can sometimes suggest phylogenetic patterns, but they may not always do so (to be discussed in more detail later). A *regressive character* is one in which loss of appendages or other features has occurred (Mayr 1969c), such as absence of roots in some aquatic angiosperms (e.g., *Ceratophyllum*, Ceratophyllaceae). Use of this type of character requires caution so as not to confuse with original absence of a feature. In the evolutionary context it is common to speak of *adaptive* and *nonadaptive characters.* An adaptive feature is that which contributes to the fitness of an organism (i.e., its ability to leave offspring successfully), whereas a nonadaptive feature does not contribute to fitness. The extent to which characters are adaptive is a contentious issue at present which is not likely to be resolved soon. The extremes range from viewing all characters as adaptive to the persistence of some percentage of neutral traits, that neither aid nor detract from fitness. Part of the difficulty lies in agreeing on an acceptable definition of "adaptive."

Cladistic Characters

Cladistic characters have developed from the cladistic approach to classification, which attempts to determine branching sequences of evolution and base a classification upon them. The branching patterns are revealed through analysis in taxa of distributions of character states which are believed to be significant evolutionarily and contained within homologous characters. A further point is that *only* derived character states are regarded as significant cladistically; primitive conditions are viewed as misleading and noninformative. The pros and cons of this viewpoint will be discussed in detail later in chapters 8 and 9, but often used in this approach, therefore, are *primitive* vs. *derived character states,* or as synonyms, *general* vs. *unique, generalized* vs. *specialized, primitive* vs. *advanced, plesiomorphic* vs. *apomorphic,* and *plesiotypic* vs. *apotypic.* These latter two pairs of terms are those introduced (Hennig 1966) and used by the cladists (see Wiley 1981a, and Wagner 1983, for good definitions of these and other cladistic terms). Shared derived character states between and among taxa are called *synapomorphies* (or *synapotypies*) and shared primitive states are *symplesiomorphies* (or *symplesiotypies*). An *autapomorphy* is a derived character state occurring only in one evolutionary line and thus of no direct use in constructing branching sequences (because only one taxon has the feature). Characters that are useful cladistically are sometimes called *compatible characters* (Estabrook 1978) in which the evolutionary directionality of the states within each character is the same. Estabrook, Johnson and McMorris (1975) make the distinction between *true cladistic characters* in an idealized sense and those that are defined operationally in the course of actual studies. This distinction has philosophical validity and mathematical reality, but is of only passing interest for practicing taxonomists.

Phenetic Characters

Another major approach to biological classification is phenetics, which uses overall similarity to assess relationships (often referred to as numerical taxonomy), and specialized types of phenetic characters have also been proposed. Phenetic classification makes no attempt to reflect evolution; taxa are related based on similarity and difference of character states regardless of the evolutionary content of the characters and states reflected. Much stress in phenetics has been on precision of operations in the process of classification. Therefore, characters are defined in such a way to avoid any circularity as seen earlier with the concept of homologous characters. The character of choice in phenetics is the *unit character*: "a taxonomic character of two or more states, which within the study at hand cannot be subdivided logically, except for subdivision brought about by changes in the method of coding" (Sokal and Sneath 1963:65; the term apparently first used by Gilmour 1940:468). The basis for the search for unit characters is to avoid logical circularity and to obtain data in which each datum represents a new item (or bit) of information (Sneath 1957; Sokal and Sneath 1963; Sneath and Sokal 1973). These unit characters have also been called *"single characters"* by Cain and Harrison (1958). The search for unit characters is laudatory, but as with all other aspects of classification, there are difficulties here in their selection. Attempting to decide what is a logically indivisible character is most difficult, for most all features can be dissected further to a lower level of hierarchical organization down to the atomic level and even beyond. As Griffiths has pointed out: "There now seems no basis for maintaining that the units of perception are minimal units, not capable of further analysis. . . . I do not expect that numerical taxonomists will have any more success at finding their unit characters than did their philosopher predecessors [those advocating logical atomism]. Many seemingly simple character statements can be subdivided logically" (1974a:110). Crowson suggests another definition of unit character, but which is equally difficult to apply in practice: "the amount of phenotypic change which could be brought about by a single continuous bout of natural selection" (1970:70). Another important objective of phenetics is to use *unweighted* characters. Whereas in phyletic and cladistic classification a deliberate selection and weighting of characters believed to be evolutionarily important is made, in phenetics each unit character is given the same (or no) weight. In practice this is difficult, if not impossible, to accomplish, and at the very least the inadvertent exclusion of some feature will lead to "residual weighting" of the others (Stace 1980:54). A final objective in phenetic classification is to use only *noncorrelated characters*. Different kinds of correlation exist, such as evolutionary, developmental, random, etc., but the most important to avoid in phenetics is logical correlation, in which what were regarded as independent unit characters actually are consistently correlated and hence providing redundant information (*"tautological" characters;* Sneath and Sokal 1973:104). Attempts have been made to measure the correlation of characters in a data set (Estabrook 1967; Sneath and Sokal 1973), but the problem only reduces—it never disappears. Sneath and Sokal (1973) refer to *admissible* and *inadmissible characters*. The latter are those not satisfactory for purposes of phenetic classification and include several different subsets of characters (Sneath and Sokal 1973:103–106); *meaningless characters*, "not a reflection of the inherent nature of the organisms themselves;" *logically correlated characters* (already discussed); *partially logically correlated characters* (less than perfect correlation by some measure); *invariant characters* (discussed earlier); and *empirically correlated characters*, i.e., not correlated due to obvious logical connections, but correlated nonetheless developmentally or phyletically (such as in tightly controlled adaptive character complexes) and due to genetic linkage or other factors. Mayr mentions that the problem with selecting individual characters has caused some workers to seek the *"overall character,"* such as from DNA base-pair sequences, which might obviate the need for other data from the organisms (1969c:124). This will be discussed later in the chapter on data-gathering (chapter 24), but even when we will someday have such detailed information for all DNA sequences for entire genomes, the challenge will still be enormous to sort out evolutionary parallelisms and convergences, sequence similarity due to genetic drift, information content of passive vs. active and single-copy vs. repetitive sequences for purposes of classification, the significance of regulatory vs. structural sequences, etc. Although attempts to view characters logically and objectively are useful, in the final analysis:

> Much as one may attempt to maintain a purely objective attitude in order to obtain a random sample of characters, this is extremely difficult if not impossible to do. Even, however, if we could assume that the operator has scaled these virtuous heights, we find him still faced with the difficulty that his materials do not, by any means, reveal all characters equally. Some are readily seen and easily quantified. Others may be recognized but recorded with difficulty. Still others, and this is probably the great majority, are obscure. Selection is inevitable and ease of procurement is undoubtedly a major guide." (Olson 1964:125–126)

CRITERIA FOR SELECTING CHARACTERS AND STATES

The criteria for deciding which characters and states to choose will depend to some extent upon the particular approach to classification used. Nonetheless, a number of viewpoints on selection of characters are sufficiently general that discussion here seems appropriate. These numerous general perspectives can be grouped conveniently into four headings: logical; biological; information theoretical; and practical.

Logical Criteria

The most important logical consideration in the selection of characters is that it be done *a posteriori* rather than *a priori* (Crowson 1970). The former term refers to deciding which characters are important *after* having studied the organisms, whereas the latter term applies to the selection *before* the organisms have been examined carefully (Cain 1959c). For example, if one becomes interested in producing a new classification for a group of angiosperms with which no personal familiarity exists, it could be assumed *a priori* that ovary displacement will be significant and this selection (and weighting) would, in effect, force the classification to develop along these lines, which might be erroneous. Ovary position could be extremely variable within this group and show no correlations with other characters, or it also could be invariant, so that *a posteriori* it would not be a suitable taxonomic character (it could be a useful character in another group, however, if so determined *a posteriori* in that group).

With *a posteriori* selection of characters, logical (or statistical; Adams 1975a) correlations of character states among taxa are sought. This is a very important aspect of selection of taxonomic characters. Those characters which form suites that co-vary, and which give the maximum set of co-varying features, are judged *a posteriori* the most useful taxonomically. The reason for the existence of such constellations of characters is usually descent from a common ancestor, i.e., they share a common evolutionary history. Correlation forms one of Crowson's general principles: "if two characters which are not closely linked functionally, show a marked correlation in their incidence within a group, this is because a particular combination of them characterises a large subgroup" (1970:74). Or in a similar view Cronquist has remarked:

> One of the fundamental taxonomic principles that most of us are comfortable with is that taxonomy proceeds by the recognition of multiple correlations. A corollary of this principle is that individual characters are only as important as they prove to be in marking groups that have been recognized on a larger set of information. It is a natural assumption that once the value of a character in a particular group has been established in this way, it can be applied fairly uniformly across the board in other groups. This assumption is false, and has to be unlearned by each successive generation of taxonomists. There is just enough tendency for consistency in the value of taxonomic characters to mislead the unwary. One of my colleagues in another country has summarized the situation by paraphrasing Orwell: All characters are equal, but some are more equal than others. (1975:519)

The pheneticists stress the concept of logical irreducibility as one of the criteria of selection of characters (the search for "unit characters"). In a general way, characters should be viewed as fundamental units of information (evolutionary or otherwise) that will tell us something about relationships among organisms. It is desirable, therefore, to seek descriptors of variation that logically make sense and are at least reasonably indivisible. But it must be kept in mind, that "However theoretically sound this may be, there are bound to be acute difficulties in making a decision about the logic of dividing up characters into units, a limiting factor being our own state of knowledge, skill and perception. This is one of the most serious weaknesses of any attempts to break down organisms into characters for the purposes of objective comparisons" (Davis and Heywood 1963:113–114).

Biological Criteria

A number of different biological perspectives can be considered in the selection of characters and states. The most important general point is that the biology of the characters (and the organisms in a broader sense) be determined (Mayr 1969c). This point is so simple and self-evident, that it can be deceiving. There has been a tendency in recent decades among zealous pheneticists and cladists to believe that they can understand relationships of any group with their methods—all they need is the basic data matrix. Much valuable information *is* contained in the basic data matrix, but the biology of these characters and states must be understood for a much more accurate and deep portrayal of relationships. Some biological aspects, such as ecological and distributional information, breeding systems, crossing relationships between species, flowering times, sympatric occurrence of taxa, meiotic and mitotic chromosomal configurations and behavior, adaptive significance of characters, their genetic basis, and so on, are most important to know and evaluate. Some of these data can be presented easily in the data matrix and others cannot. Nonetheless, all this biological information is invaluable for realistically and sensibly constructing a classification.

The degree to which the genetic basis of characters and states should be understood before they can be utilized effectively for taxonomic purposes has been a point of past discussion. Ideally, the genetic basis for every character and state would be useful to know. However, we still know very little about the genetics of characters. Some good examples do exist, such as in *Dithyrea* (Cruciferae) in which two species were distinguished by pubescence on the fruits. Rollins (1958) showed that this difference was due to alleles of the same gene, and that this represented merely populational variation rather than specific distinctness. In the past decade numerous electrophoretic studies have been done on isozymes of plant groups (Gottlieb 1977a, 1981a), and the genetics of these metabolically significant enzymes in some cases have been determined. Furthermore, the sequencing of DNA of parts of genomes now has been done (e.g., Palmer and Zamir 1982; Palmer 1986). Technically, therefore, we are approaching the ability to understand much more of the genetics of characters than ever before, although we are still a long way from having such data for the majority of features of taxonomic interest. A good perspective that still applies is offered by Davis and Heywood: "we should rely on those characters that depend on a multiplicity of genes or gene combinations for their ultimate expression, but great caution should be exercised in applying this generalisation since it is not just the genetic basis of a character that has to be assessed but its possibility of change *in practice* and the effects of such changes on the continued integrity of taxa. In other words, *a taxonomic character is only as good as its constancy, no matter what its genetic basis*" (1963:120).

Some workers advocate selecting characters and states for which their adaptive value is obvious. This is a difficult issue because different definitions of adaptive exist. The view taken here is that a character or state is adaptive if it contributes positively to reproductive fitness (i.e., increased numbers of offspring in future generations; Dobzhansky et al. 1977). Cronquist in several papers and books (e.g., 1968, 1969) has emphasized the lack of adaptive value (or at least our inability to detect it) in features of taxonomic value. Davis and Heywood have commented on this point indicating that just because "we cannot comprehend in what way such characters may be functional does not mean they have (or had) no adaptive significance" (1963:126). "In the majority of cases the acceptance of the adaptive nature of particular characters of a plant depends on the ingenuity, perception, skill and indeed imagination of the botanist" (p. 123). Barber (1955) has stressed that the options for the adaptive quality of a character can range from those resulting from genetic drift and showing no adaptive value, through those being developmentally or genetically correlated (or linked) to others which are strongly adaptive, to those that have high adaptive peaks. Much work is needed to understand the adaptive value of characters in each case. Kluge and Kerfoot (1973) suggest that the most variable characters within populations contribute least toward fitness, and therefore are of low adaptive value. If this is so (the basis for this assertion has been questioned; Rohlf, Gilmartin, and Hart 1983), then characters used taxonomically which tend to be constant within populations would be the most adaptive. It seems likely, therefore, that most taxonomic characters are, in fact, adaptive, but more investigations are needed in each specific case to establish correlations from which hypotheses (e.g., Parkhurst 1978; Stuessy and Spooner 1988) and testable predictions can be made (e.g., Stuessy, Spooner, and Evans 1986). A controversy exists at present as to whether some features of organisms, especially electrophoretic protein variants (isozymes and allozymes), may not be selectively neutral, i.e., unaffected by natural selection (e.g., Kimura 1968, 1983; Arnheim and Taylor 1969; King and Jukes 1969; Bullini and Coluzzi 1972). Although some features may be selectively neutral, it is most probable that the majority are adaptive and maintained by selection (G. B. Johnson 1973).

Another biological perspective for selection of characters and states is to pick those with high evolutionary information content. "For the phylogenist, an important character would better be defined as one, a change in respect of which characterised an important step in evolutionary history" (Crowson 1970:75). Caution must be exercised here to avoid characters which have undergone parallelisms (Mayr 1969c). The selection of which characters are significant in evolution and which are not is often most difficult and poses problems (although not insurmountable ones) for the practitioners of phyletic and cladistic classification. More on this topic later.

A final biological viewpoint in character selection is conservatism. Defining what conservatism means is a difficulty, because some workers refer to it as characters that have existed with little change over long periods of evolutionary time (called "stability" by Swingle 1946:265). Others refer to characters that show little variance within (Farris 1966) and between populations. Davis and Heywood stress the "fixity" of characters, i.e., those "little influenced by environmental changes" (1963:119). Mayr (1969c) makes the useful point that characters which change very slowly in evolution of a group are usually more useful at the higher levels of the hierarchy, whereas those which change more rapidly are more useful at the lower levels. The main point is that "good" taxonomic characters do not vary indiscriminately; it is in that sense that they are conservative. This constancy may be due to stability over evolutionary time, strong genetic control, high adaptive value, linkage, or combination of these and other factors.

Information Theoretical Criteria

The information theoretical perspective stresses the selection of characters for their maximum predictive quality. "It is in [this] sense that it is more useful to regard a character as *reliable* since its use will allow us to make a large number of deductions about the other attributes of the groups it separates off" (Davis and Heywood 1963:119). Recognizing the importance of this aspect of characters is one thing, but finding ways to measure this information content for predictive quality is another matter altogether. Some attempts have been made to relate unit characters to basic units of information (Sneath and Sokal 1973). Other attempts have suggested the supremacy of evolutionary characters in leading to maximally predictive classifications (Farris 1979b; Brooks 1981). Much further work is needed before we have a better understanding of how to measure information content of characters. Part of the difficulty will center on finding an acceptable definition (or definitions) for "information."

Practical Considerations

The final perspective on selection of characters and states focuses on practical matters. As all practicing taxonomists well know, it is important theoretically to be aware of and conversant about ways in which characters are important and can be selected, but basically it comes down to choosing characters "which work" (Cronquist 1957, 1964). We rarely know the genetic, adaptive, evolutionary, or information content of characters routinely used in constructing classifications. The practicing taxonomist does make an assessment of the conservative nature of characters, in the sense of intuitively assessing their variations within and between populations, and looks for correlations with other characters to establish the suites needed for taxonomic circumscription. Pheneticists stress no conscious selection of characters, but rather the use of all of them after proper reduction to their "unit" natures (Sneath and Sokal 1973). However desirable this might be in theory, it is not possible to take this approach with the majority of taxonomic efforts in which speed of work is valued. A selection of a handful of constant characters and constructing a classification from them is much more efficient than by carefully reasoning out all the unit characters, which could number more than 100, and developing the classification along these lines. The selection of character states is equally problematical in practice, and this issue has not yet been addressed satisfactorily. The choice of whether to use qualitative or quantitative states is an initial and fundamental problem, although tradition in the taxonomic history of a group offers a guide. But further problems abound: "If a character is to be measured, what set of values should be used to record the state of the character in a given individual? For continuous characters such as leaf length, it might be the nearest millimeter or the nearest centimeter. More difficult decisions involve qualitative attributes, such as leaf shape. Here, an investigator's operational procedure would be simply to recognize those patterns as distinct that best serve the purpose of his analysis. This rule sounds terribly suggestive and unscientific, but, to determine character states for qualitative characters, no formal decision function exists that considers one's material and purpose any better" (Crovello 1974:458).

Different Approaches to Biological Classification

I F WRITTEN scarcely twenty-five years ago, this section would have been brief, if perhaps not even included, in a textbook of plant taxonomy. Consider even the still popular texts such as Lawrence (1951) and Porter (1967) which contain no mention of different approaches to biological classification. During the past decades two new efforts toward classification have been advocated: phenetics and cladistics. A comprehensive discussion of all these and other approaches to biological classification is needed at this time, especially because some of the literature contains pointed (occasionally even personal) commentaries and different "schools" have developed, each with highly definite (and sometimes dogmatic) viewpoints. The student of modern plant taxonomy should be acquainted with and have an informed personal opinion about these issues. In this section of the book are presented brief historical and descriptive accounts of each method, detailed evaluations of the three major approaches (phyletics, phenetics, and cladistics), and a plea for a balanced perspective.

To describe the history and development of all the major approaches to biological classification would require more space than this book permits (for an excellent comprehensive review, see Mayr 1982). The objective here, therefore, is to sketch briefly the historical details that are most important for understanding the approaches themselves. The prinicipal approaches to be considered are: artificial, natural and phyletic, phenetic, and cladistic.

Davis and Heywood (1963) recognize in pre-Darwinian classification those systems based on habit, sexual features (the sexual system), and form relationships. The former two are placed here into the artificial approach, and the latter into the natural approach (mostly *a posteriori* systems such as those of Jussieu and Candolle). Jones and Luchsinger (1986) also recognize two additional historical approaches: form, and utilitarian. The latter stresses the herbalists' contributions (chiefly artificial), and the former refers only to *a priori* approaches based on external form and is therefore also artificial in the context used here. Blackwelder (1964) champions the "omnispective" approach, which would correspond largely to the natural system as used here, in which selected and weighted characters are used to develop a classification but without emphasis on reconstructing the phylogeny. Kavanaugh has mockingly called this the " 'trust me, I know what I'm doing' school" (1978:141). All these perspectives are legitimate alternatives, but I have grouped them here so as not to detract from emphasis on the comparison of the three major (and currently utilized) approaches: phyletics, phenetics, and cladistics. Table 5.1 shows the chronology of these approaches with examples of works primarily from the botanical literature.

TABLE 5.1 Historical Development of the Major Approaches to Biological Classification, with Examples Primarily from the Botanical Literature.

Major approaches to biological classification	B.C.	A.D.	1500	1600	1700	1800	1900	1980
Artificial	Theophrastus[a] (c. 300 B.C.) ⟶		Herbalists (1470–1670)[b] ⟶		Tournefort (1700) ⟶	Linnaeus (1735) ⟶		
Natural and phyletic				Bauhin⟶ (1623)	Ray ⟶ (1686–1704)	Jussieu ⟶ (1789)	Eichler⟶ (1883)	
						Candolle (1824–73)	Engler & Prantl (1887–1915)	
						Bentham & Hooker (1862–83)	Bessey (1915)	
							Hutchinson (1926, 1934, 1969)	
							Cronquist (1968, 1981, 1983)	
							Takhtajan (1969, 1980, 1986, 1987)	
							Thorne (1968, 1976, 1983)	
							Dahlgren (1975, 1980, 1983)	
						NATURAL	*PHYLETIC*	
Phenetic						Adanson⟶ (1763)		⟶ Sokal & Sneath (1963, 1973) ⟶
Cladistic								Hennig (1966) ⟶

NOTE: For a readable and reasonably detailed account of the history of botanical classification, see Core (1955).

[a] Elements of the natural approach to classification are found in Theophrastus' work, in some of the herbals, in Tournefort's publications, and even Linnaeus himself produced a sketch of a natural system published posthumously by his student, Giseke (1792). Nonetheless, these treatments were largely artificial in contrast to the more obviously natural systems beginning with Bauhin and Ray.

[b] Various inclusive dates for the "Age of the Herbals" exist; those used here come from Arber (1988).

CHAPTER 5

The Anatomy of Classification and the Artificial Approach

The process of classification is viewed in different ways by different people ranging from unfathomable, intuitive art to objective, explicit, and testable science. It is believed useful to examine the reasons for the long-standing connection of taxonomy to art, as an aid to clarifying this relationship. The process of classification will also be analyzed in detail, followed by comments on the artificial approach.

AESTHETICS AND CLASSIFICATION

Classification is strongly aesthetic, particularly in phyletic approaches, and therein lies the artistic connection (Friedmann 1966). As Stearn has aptly commented: "Aesthetic appreciation rests, I think, chiefly on a sense of form. The co-ordination of varied elements into a whole which is mentally satisfying and harmonious, giving an impression of completeness, is nevertheless achieved by emphasis and omission. The perception of overall resemblance implies selection by the mind from the multitude of unevaluated details presented to it by the eyes" (1964:83–84). This "sense of form" is very much a part of classification as well as being obviously fundamental to artistic endeavors. Kubler has defined art as "a system of formal relations" (1962:vii), which could apply equally well to the resultant hierarchy of classes constructed in the process of classification. Simpson treats classification as "a useful art" and speaks of "taxonomic art" in which classification is regarded as a combination of art and science (1961:110). Mayr offers this perspective: "It is by reading this chapter [Darwin 1859, ch. 14] that one understands the true meaning of the old saying that classifying is an art. But what is art? To be sure, a superior classification provides a genuine aesthetic pleasure, but the word 'art' in the old saying is used in a somewhat different sense. As in the word 'artisan' it refers to a craft, to a professional competence which can be acquired only through years of practice" (1974a:5–6). Stearn adds appropriately: "The work of a taxonomist is closely linked to that of an artist: both seek patterns within diversity, the one to record those he thinks he finds in nature, the other to record those he finds maybe only in his own head, and one person may be both taxonomist and artist, the danger then being that he makes no clear distinction between these two forms of expression!" (1964:84). There are many examples of excellent taxonomists who are also accomplished artists, e.g., see the personally illustrated works of Hutchinson (1926, 1934, 1969) and Burger (1967). This does not imply in any way, however, that these workers are poorer or more fanciful than nonartistic taxonomists, but rather simply that they have substantial talent in both areas.

Despite the aesthetic ties between classification and art, the former is still unmistakably a science. It is not an experimental science as are chemistry, physics, or molecular biology, and so one cannot repeat in the same sense the construction of classifications and test the predictions derived from them. But this *is* science in the sense of description and arrangement of information in an orderly fashion, the development of hypotheses, and the devising of tests to attempt their disproof (and hence invalidation). The descriptive information in characters of organisms is gathered and evaluated and arranged in an orderly fashion into an hierarchical classification, which is the hypothesis of relationships suggested by the data (characters and states). Predictions from these hypotheses involve finding new data sets which will either correlate with (and not disprove) or be inconsistent with (and hence, disprove, or at least suggest reevaluation of) the hypothesis. Tests are made by gathering these new data and comparing their distributions with already extant data sets of other types. The examination of newly discovered organisms and learning how all their features compare with those characteristics of already classified taxa in the classification is another form of test of the hypothesis. In this fashion, new classifications are often made as new data-gathering techniques become available and as new organisms are discovered through field exploration and museum study.

The close association of the process of classification with art has caused some workers in the past few decades to react negatively against this closeness. One of the reasons for this negativism may be that science is definitely progressive whereas art is less so; in fact the arts have at times been caricatured as a somewhat confused bird "who always flies backward because he doesn't care about where he's going, only about where he's been" (Frye 1981:127). There have been two major efforts to construct classifications on a more objective and repeatable foundation: phenetics and cladistics. In the former, a strong emphasis is placed on the virtues of objectivity and repeatability by taking all characters (unit characters) without subjective selection and by constructing a classification along explicit lines (allowing anyone else to follow clearly what has been done). As Michener and Sokal stressed in one of the earliest phenetic papers: "Taxonomy, more than most other sciences, is affected by subjective opinions of its practitioners. Except for the judgment of his colleagues there is virtually no defense against the poor taxonomist" (1957:159). By the same token, cladistics has developed as a means of revealing branching patterns of evolution more explicitly, and some workers insist that classifications be based directly upon these patterns. "Cladistics has emerged as a powerful analytical tool in comparative biology because it offers most informative (least ambiguous) summations of any set of biological observations represented in a consistent, testable, reproducible framework. Systematics has thus become a truly empirical science, capable of assuming its rightful place as the one indispensible branch of biology —the framework of comparisons for a comparative science" (Funk and Brooks 1981:vi). Despite these laudatory attempts to remove subjectivity, the fact remains that many aspects of classification by whatever approach are still largely based on the sound judgment of the individual worker. The choice of taxa for initial study, the selection of characters, the detection of homologues, the measurement and description of character states, are all aspects that require judgement, creativity, and experience, no matter what approach to classification is used. It is well to keep in mind that these specific areas of uncertainty are backdropped by the general perspective "that science is uncertain in its very nature. With exceptions mostly on a trivial and strictly observational level, its results are rarely absolute but usually establish only levels of probability or, in stricter terminology, of confidence. Scientists must also tolerate frustration because they can never tell beforehand whether their operations, which may consume years or a lifetime, will generate a desired degree of confidence. (If this could be told beforehand, the operations would be unnecessary.) Indeed one thing of which scientists can be quite certain is that they will not achieve a *complete* solution of any worth-while problem" (Simpson 1961:5).

THE PROCESS OF CLASSIFICATION

With all approaches to classification, no matter what the particular bias, the process of classification can be viewed as a series of operations (fig. 5.1). Viewing the classification in this dissected way will facilitate comparison of the different approaches to classification to be discussed next. Earlier in this book, classification was defined as the ordering of individual organisms into groups based on observed similarities and/or differences. When only two or very few groups result from this process, we can treat the resultant units as being coordinate to each other and use the classification system in ways already mentioned. But usually many units are involved, in which case some method is needed for showing the relationships among the groups so that we can communicate more easily about part of the ordered diversity. If many units have been created, we face the same problem as if we were confronting many separate individuals. To solve this difficulty, larger groups composed of smaller units are made and these are given categorical names. In this fashion a taxonomic hierarchy of ranked units results with the largest units being divided into smaller subunits, these being further divided, and so on. The process of classification, therefore, usually involves two separate operations: 1) grouping; and 2) ranking.

Grouping involves three specific operations (fig. 5.1). First, one must select characteristics of the organisms to use in assessing the similarities and differences. It is impossible, in fact, to compare two or more objects without referring to specific features of each (the taxonomic characters). The second operation in grouping involves describing and/or measuring these characters. One cannot use the character "leaf shape" for example, to compare two plants meaningfully, because they both have leaves with shapes. Instead, the kind of particular leaf shapes in the two plants must be compared, such as "obovate" versus "lanceolate." These are the character states which are actually used for purposes of taxonomic comparison and evaluation. The third operation in grouping is to compare the chosen character states to obtain the groups. These comparisons

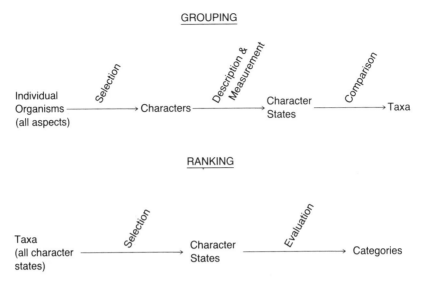

FIGURE 5.1 Representation of the two primary operations of classification. (From Stuessy 1979b:623)

can be made in different ways. A formal method can be used (such as in phenetics or cladistics), or the comparison can be done more intuitively in traditional approaches. In some situations particular character states will be regarded as having more importance than others for the particular grouping, whereas in other situations all the character states will be accorded the same or equal importance. These differences in approach to comparison of character states are important areas of disagreement among some workers regarding taxonomic methodology (to be discussed in detail later).

The next step in classification is ranking of the recognized groups (fig. 5.1). This involves two specific operations. First, all the character states of the groups are examined, and some are selected for use. The character states might be the same and even include all of those used for grouping, but usually not all of them are used for ranking. Other characteristics also might be selected for consideration that were not used for the grouping. Second, these selected character states of the groups are evaluated in terms of the categories available for use in the taxonomic hierarchy. The presence or absence of certain kinds of features usually will suggest an appropriate rank in the hierarchy of classification. A discussion of the kinds of character states of groups that often are used to indicate certain ranks will be taken up later in chapters 10 to 14.

A final point worth mentioning is that most practicing taxonomists, although carrying out the operations just described, are not usually conscious of all the different steps. If one were to ask a taxonomist how he or she classifies, the reply might be: "I simply group things together that look similar to me." Although such an answer implies that characters are not selected nor character states compared, in reality the taxonomist is making many rapid comparisons between features of the organisms selected unconsciously. With prodding, however, even the most pragmatic of workers can give at least some idea of the specific characters used and compared (many of these are presented in the diagnostic keys that accompany the classification).

ARTIFICIAL CLASSIFICATION

The artificial approach to classification is that used by most people today for inanimate objects. With this method only one or at most a few characters are selected for use in making comparisons among objects, and this selection usually is made before the classification is begun (i.e., *a priori*). Because so few characters are involved, the difficulties encountered in describing, measuring, and comparing the character states usually are minimal. Ranking is done subjectively with certain character states being regarded as subordinate to others. The classification systems of libraries serve as excellent examples of the employment of this method. Many libraries have books grouped according to a system of specific subject headings (=character states) that reflect the subject matter (=character) of the organized units. Specific examples of such approaches are the Library of Congress and the Dewey Decimal Systems. These two artificial classifications are based on different sets of subject headings, but they both serve adequately to organize the books in some useful way for proper information retrieval. Other libraries have special artificial classifications ordered by authors' surnames or other subject headings that were designed for specific needs. At least one library exists where the books are classified in part by size (Morton Arboretum Library; seen in 1970)!*

The artificial system was the first to be used for the classification of plants, and these origins begin with the ancient Greeks. Theophrastus (370–285 B.C.), a pupil of Aristotle, made the first elementary groupings of plants, based on distinctions of habit into trees, shrubs, subshrubs, and herbs in his *De Historia Plantarum* (Enquiry into Plants; translated by Hort, 1916). This was followed by the *De Materia Medica* of Dioscorides (ca. A.D. 60), which was the earliest recorded treatise on the medicinal value of plants. These were largely artificially arranged, but some related plants were grouped together (Core 1955).

With the close of the Dark Ages and the development of the Renaissance (about 1450), people began again to look firsthand at the living world instead of relying solely on the observations of the ancient Greek writers. During this period of learning and discovery, it was noticed that certain plants had features in common that for one reason or other seemed to be important. Perhaps due to the great suffering inflicted by contagious diseases during this period, a stronger interest was rekindled in the medicinal value of herbs, and these actual or supposed properties formed the basis for organization of the plants into groups. Workers contributing to this interest were called herbalists and included many in Germany, Italy, and the Netherlands, such as Otto Brunfels, Hieronymus Bock, Leonhart Fuchs, Andrea Cesalpino, and Rembert Dodoens, to name just a

---

*There was good reason for this; elegant custom-made bookcases with fixed shelves were built into the new library and the books could only be arranged in this practical fashion. Most other libraries (such as Ohio State) routinely have oversized collections kept in separate quarters for similar reasons.

few. These early classifications were artificial because only a few characters were used which related to the presumptive medicinal efficacy of the plants.

From these early attempts by the herbalists eventually came a uniform and stable system of botanical classification. One of the most noteworthy of the artificial systems was developed by the French botanist, Joseph Pitton de Tournefort (1656–1708). He published a comprehensive treatise in 1700, *Institutiones Rei Herbariae*, which dealt with nearly 9,000 species of plants in more than 600 genera. His stress on the generic level has earned him the designation as "the founder of the modern concept of genera" (Core 1955:34). Despite its broad coverage, Tournefort's system was still largely artificial with initial emphasis on genera differing in form of the flower and fruit and second-grade genera differing in vegetative features (Davis and Heywood 1963).

The classification (1735, 1753) of the Swedish botanist, Carolus Linnaeus (1707–1778), represents the most complete artificial system ever developed for all plants. Linnaeus was impressed by the discovery by Camerarius (1694) of the existence of sex in plants and the treatise on that same topic by Vaillant (1718). In fact, in his college days at Uppsala, Sweden, Linnaeus wrote a thesis on the sexual habits of plants, the *Praeludia Sponsalia Plantarum* (1729), and it was undoubtedly this interest that led him to develop his sexual system of classification. Linnaeus placed overriding emphasis on the presence, configuration, and numbers of sexual parts of the flower (i.e., the stamens and carpels). All the flowering plants were placed into 23 "classes" based on stamens, such as *Monandria* (one stamen), *Diandria* (two stamens), *Triandria* (three stamens), and so on. These classes were then divided into "orders" based on features of the carpels, such as *Monogynia* (one style or sessile stigma), *Digynia* (two styles or sessile stigmas), *Trigynia* (three styles or sessile stigmas), etc. (description from Stearn 1971). This system had the great advantage of allowing all plants to be easily placed in the classification just by looking at these two parts, and even a botanical novice could make the proper dispositions. As a result of its utility, as well as of the interest generated by its sexual innuendos (e.g., *Diandria* was described as "two husbands in the same marriage," Stearn 1971), the system became very popular and its acceptance virtually worldwide.

CHAPTER 6
Natural and Phyletic Approaches

Despite the widespread dissemination and acceptance of the Linnaean sexual system of classification, some workers never felt entirely satisfied with it. Although the artificial approach could not fail to allow all plants to be grouped with ease, the resulting groups of plants often seemed very different from each other in regard to features other than the sexual parts (fig. 6.1). Bernard de Jussieu (1699–1777), demonstrator of plants in the Royal Botanical Garden in Paris, arranged the plants in the Royal Garden and in Marie Antoinette's garden in La Trianon at Versailles. Earlier the well-known Tournefort had been in charge of the Royal Garden, and in fact, Bernard's brother, Antoine, succeeded Tournefort as director. Linnaeus visited the Jussieu's in 1738 and the meeting was pleasant, but despite this positive personal contact, and the eventual prestige of the Linnaean system, Bernard de Jussieu remained unconvinced. The simple fact was that the sexual system of classification was inadequate for information retrieval and predictive generalization. Bernard attempted to arrange the living plants in the Royal Botanical Garden based upon what he believed to be overall similarity, taking as many characters as possible into consideration. But he was never completely satisfied with the arrangement and

Hello,
big, brown rat.

I am not
a big, brown rat.
I am
a little bunny.

You are not a bunny.
There is a bunny.
He is white.
You are brown.
A rat is brown.
You are a rat.

FIGURE 6.1 Example of how emphasis on single characters can lead to completely erroneous artificial classifications of low predictive value. (From DeLage 1978:6–8; drawings by E. Sloan.)

never himself published his system of classification. It fell to his nephew, Antoine-Laurent de Jussieu (1748–1836), who came to Paris as professor of botany in the Royal Botanical Garden, to rework and elaborate upon his uncle's system until it was finally published in 1789 in the *Genera Plantarum Secundum Ordines Naturales Disposita*. This was a most impressive system containing 100 "orders," which corresponded in large measure to our present concept of families, grouped into 15 classes and these into 3 divisions. The characters used to delimit these orders were numerous but included strong emphasis on ovary displacement (epigyny, hypogyny, and perigyny), the suite of characters distinguishing monocots vs. dicots (parallel vs. net leaf venation, flower parts in threes and sixes vs. fours and fives, absence vs. presence of vascular cambium, etc.), fusion of anthers, and so on. In other words, reproductive features were still stressed, as they had been in Linnaeus' system, but they were selected *a posteriori* based upon their ability to result in multiple correlations with states of other characters.

NATURAL CLASSIFICATION

A *natural system* of classification, therefore, is one based upon states of several to many characters selected *a posteriori* for their value in positively correlating with states of other characters to form a hierarchical structure of groups in ranks containing high information content and predictive value. The characters selected are, in effect, weighted by their selection and employment over those features not selected, and this selection and comparison, and eventual evaluation (for ranking) is done intuitively by the taxonomist. That is, there is nothing explicit about this process; it occurs rapidly in the mind of the maker (herein lies the aesthetics of the process), and diagnostic characters are usually only derived after the classification is constructed and a key for identification purposes is attempted.

A natural classification is therefore *polythetic* (Sneath 1962; from Beckner 1959, as "polytypic"). It "places together organisms that have the greatest number of shared features, and no single feature is either essential to group membership or is sufficient to make an organism a member of the group" (Sokal and Sneath 1963:14). This contrasts with *monothetic* classification, in which groups "are formed by rigid and successive logical divisions so that the possession of a unique set of features is both sufficient and necessary for membership in the group thus defined" (Sokal and Sneath 1963:13). The monothetic quality is characteristic of artificial systems of classification such as exemplified by Linnaeus' sexual system.

It should be noted that even before Linnaeus' time, several workers had begun to struggle with the idea of a more natural system such as Jussieu's, and two are worth special attention (table 5.1). Gaspard Bauhin (1560–1624) of Basel, Switzerland, produced a compendium of all that was known about plants at that time in his *Pinax Theatri Botanici* (1623). This work was divided into 12 books with further subdivisions, and some similar taxa were grouped together such as genera in the easily recognizable families of Cruciferae, Compositae, and Umbelliferae. Despite this useful beginning, much of the *Pinax* was artificially arranged and had its greatest value as a nomenclator (or register) of all names of plants published prior to that time (encompassing the confusing array of monomials, binomials, and polynomials then in use). The English botanist John Ray (1623–1705) also published the rudiments of a natural system in three volumes of his *Historia Plantarum* (1686–1704). He emphasized habit, monocot vs. dicot distinctions, and other features, and several of the "classes" correspond to our easily recognizable modern families such as Labiatae, Leguminosae, Cruciferae, and Gramineae.

Despite the existence of these early natural systems, the success of the artificial sexual system resulted from the forceful personality of Linnaeus, his prolific writings, his numerous students who returned to many different countries as his disciples, the ease of comprehending the system, and its sexual overtones. All these factors combined to overshadow the early natural systems of Bauhin and Ray. It wasn't until the early 1800s that the sexual system passed from common use throughout most of the world. In the United States the sexual system was still in vogue in popular textbooks into the 1830s, when Asa Gray's books provided successful competition using the natural system derived from the works of Candolle in Switzerland (Rudolph 1982). Even though Linnaeus made very impressive contributions to systematic botany and is known as the "Father of Taxonomy," or *"Princeps Botanicorum"* (Core 1955:36), the tremendous success of the artificial sexual system probably retarded the development of the more progressive natural system.

After Jussieu, nearly all subsequent systems of classification were natural until the development of evolutionary perspectives brought about by Darwin's *Origin of Species* in 1859. An important reason for the continued appearance of new natural classification systems was the increasing shipments of new plant specimens from little-explored regions of the world, which occasioned a constant re-evaluation of plant relationships. Two major natural systems should be mentioned: that of Candolle, and that of Bentham and Hooker. The Swiss botanist, Augustin Pyramus de Candolle (1778–1841), was trained at Paris and received part of his instruction from A.-L. de Jussieu. Upon completion of his education, he accepted a position as professor of botany at Montpellier, and eventually returned to Geneva in the Conservatoire de Botanique where he resided for the remainder of his professional career. His new natural system of classification was presented in the monumental *Prodromus Systematis Naturalis Regni Vegetabilis* (1824–1838), a world flora at the specific level (the last one ever completed), the first seven volumes of which were published by him and the remaining ten volumes (1844–1873) by his son, Alphonse (1806–1893). This new system utilized many characters, but it was based upon the foundation laid by Jussieu. George Bentham (1800–1884) and Joseph Dalton Hooker (1817–1911), both working at the Royal Botanic Garden in Kew, England, produced a monumental work of natural classification of all the genera of gymnosperms and angiosperms, *Genera Plantarum*, published in parts between 1862–1883. Their system of classification was based on that of Candolle, who was a close friend of Bentham. Numerous morphological and anatomical characters were used such as numbers of carpels, ovary displacement, nature of perianth, embryo characteristics, fusion of parts, etc. Despite appearing *after* the impact of Darwin's book on evolution (1859), the project had been started about 1857 and was therefore natural rather than avowedly phyletic (i.e., no evolutionary interpretations of any kind were explicitly included). Furthermore, Bentham at the time remained unconvinced of the correctness of Darwin's evolutionary views; but he did accept them later.

PHYLETIC (EVOLUTIONARY) CLASSIFICATION

The question that constantly arose during the period of development of natural systems of plant classification was why some organisms tended to resemble those of one group more than another. The answer given by some was that the order reflected God's plan of creation, whereas others believed that a natural process must be responsible. Lamarck and many other biologists during the early 1800s believed strongly that evolution, or the process of orderly organic change through time, was responsible for the observed patterns of diversity. These ideas, however, were not wholeheartedly accepted by scientific colleagues of the day due to lack of an explanation regarding the mechanisms for such a process. It was Charles Darwin (1809–1882) who provided a plausible solution in his book, *On the Origin of Species by Means of Natural Selection* (1859). From that time on, taxonomists had an explanation other than Special Creation as to why their classified groups were homogeneous—they had descended from a common ancestor. Biological classifications ceased to be just storage-retrieval systems; they now also became illustrations of the patterns of evolution. The classification of plants could now be called "phyletic," "phylogenetic," "evolutionary," "eclectic" (McNeill 1979a), "synthetic," or "syncretistic" (Farris 1979b).

But although an explanation now existed whereby similar individuals were classified together, the theory of evolution by means of natural selection did not alter the process of classification itself (Stevens 1984b). Characters were still selected, described and measured, and character states compared as in the natural system. In reality, the process of classification and the resulting hierarchy of classes had not changed. What was altered was simply the understanding of the origin of similarities and differences among organisms. In other words, the philosophical perspective toward hierarchical classifications changed, but not the process itself.

As a result of the emergence of the theory of evolution, therefore, taxonomists began to look at their finished classifications in a different light. Workers began to emphasize relationships by descent of the groups in their systems, and these relationships were often illustrated diagrammatically by phyletic or phylogenetic "trees" (see Voss 1952, for a history of phyletic trees in biology). The rationale for such evaluations involved subconsciously and/or subjectively assigning ancestral or derived status to various character states which allowed groups to be related in a linear fashion from generally more ancestral to more derived. These phyletic assumptions or "dicta" regarding lineages were explicit or implicit in all major phyletic systems (table 5.1).

The first clearly phyletic system of classification of plants was produced by the German botanist August Wilhelm Eichler (1839–1887). He dealt with the entire plant kingdom and recognized subdivisions that are still part of our botanical language: Cryptogamae, including Thallophyta (algae and fungi), Bryophyta (mosses and liverworts), and Pteridophyta (ferns and fern allies); and Phanerogamae, including Gymnospermae and Angiospermae. From an evolutionary perspective, the Thallophyta were regarded as more primitive than the Bryophyta, these more primitive than the Pteridophyta, and so on.

Based on the Eichler system, a new detailed phyletic system of classification was produced also in Germany by Heinrich Gustav Engler (1844–1930) and his associate, Karl Anton Eugen Prantl (1849–1893). Engler was professor of botany at the University of Berlin and director of the Berlin Botanical Garden from 1889 to 1921. Their new phyletic system was first published by Engler in outline form in 1886 as a guide to the Breslau botanical garden, and more fully in their 23-volume work, *Die Natürlichen Pflanzenfamilien* (1887–1915), which was essentially a world flora at the generic level (a new *Genera Plantarum*).* Many of the groupings in the classification were derived from the natural system of Bentham and Hooker. The conspicuous difference was that very definite ideas were given as to which groups of plants were most primitive and which were more derived. Within the flowering plants, those families with unisexual flowers borne in catkins (or aments), called the Amentiferae, were judged most primitive on the basis of their presumed resemblance to gymnospermous ancestors. Many lines of descent from that basic complex were elaborated. The most important point is that this was a system in which evolutionary interpretations of relationships among groups abounded.† This was, and still is, the dominant feature of the phyletic approach to classification.

Charles Edwin Bessey (1845–1915), a student at Harvard for six months under Asa Gray, worked most of his career at the University of Nebraska, and produced a phyletic system of classification (1915), the concepts of which are still basically followed today. As Cronquist has aptly remarked: "We are all—or nearly all— Besseyans" (1968:52). Bessey departed from the ideas of Engler and instead of viewing the Amentiferae as most primitive, he regarded the Polycarpicae or Ranales as the most primitive group with many separate, helically arranged floral parts with bisexual flowers.

Many additional new phyletic systems of classification for the angiosperms have been published since Bessey's time (table 5.1). These include the systems of Hutchinson (1926, 1934, 1969), Cronquist (1968, 1981, 1983), Thorne (1968, 1976, 1983), Takhtajan (1969, 1980, 1986, 1987), and Dahlgren (1975, 1980, 1983). It is not my purpose here to review these systems in detail, but only to stress that all of them are phyletic in the sense of emphasizing primitive vs. derived character states and drawing lines of descent between and among taxa. (See Lawrence 1951, and Core 1955, for presentations and discussion of many of the older

---

*The impact of this publication was so great that the arrangement of most of the world's herbaria is still based on this scheme, even though more modern phyletic systems of classification exist. None of the more modern systems are so detailed, so well indexed, and so well numerically coded to genus and family, all of which have aided the permanence of the Engler system as useful for storage and retrieval of specimens. Updated editions have also been published regularly (e.g., Melchior 1964).

†Turrill mentions that "Engler did not consider his system as phylogenetic, in the complete sense of the word, but rather as one in which the groups are built up in a step-like manner to form, as far as possible, a generally progressional morphological series. Some of the groups are acknowledged to be probably polyphyletic" (1942:268). It was, nonetheless, phyletic in the context used here in contrast to the pre-Darwinian natural systems.

systems, plus the additional works cited above for the most recent contributions; see also tabular comparisons of some of these systems in Becker 1973, and Swift 1974.)

Because phyletic classification has involved a subjective selection of characters and subjective comparison and evaluation of character states, legitimate differences have arisen among taxonomists even when examining the same set of organisms. Some workers stress certain kinds of characters, and some emphasize others. Even when admitting the same discontinuities in the data, some might evaluate these gaps in terms of a larger difference in ranking than would others. For example, one might believe the observed discontinuity to indicate hierarchical difference at the generic level, whereas another might prefer to recognize the difference only at the specific level. Among practicing phyletic taxonomists, therefore (and these include the majority of workers at the present time), acceptable differences of opinion regarding certain groups occur. When these viewpoints are applied to large numbers of different organisms, the resulting classifications can be very divergent in regard to the number of units recognized at each hierarchical level. Workers who tend to take the broader view of grouping and ranking have been nicknamed *"lumpers"* and those with the opposite viewpoint are called *"splitters"* (e.g., McKusick 1969). Splitters tend to believe that morphological variations of a "minor" nature should be documented formally by the description of new taxa, whereas lumpers may observe the same variations but believe that their formal recognition is neither necessary nor desirable. It is important to emphasize that both these approaches to phyletic classification are legitimate and acceptable, within limits, even though through the years such differences of opinion have been the source for heated (and sometimes personal) debate among the persons involved. Excessive splitting and lumping are to be avoided. Generally speaking, the more different types of data and the more complete the data, the less difference of viewpoint there tends to be.

DEFINITIONS OF "NATURALNESS"

The definition of natural classification used in this book is not shared by all workers, and therefore, a brief discussion is in order. This is important before we consider phenetic and cladistic approaches to classification because some practitioners of each have called their efforts and results "natural." Pre-Linnaean workers sometimes used natural classification in the sense of determining the true "nature" or "essence" of plants, an idea derived from Plato and supported by belief in Special Creation (Davis and Heywood 1963). Post-Linnaean (but pre-Darwinian) systems used natural in the sense of Jussieu's system, i.e., a classification based upon overall similarity (e.g., Lindley 1830a, b). The usage up to this point is clear enough; the problems of interpretation develop after evolutionary thinking and phyletic approaches to classification appear.

Darwin makes very clear his meaning of "natural" in reference to systems of classification:

The Natural System is founded on descent with modification; that the characters which naturalists consider as showing true affinity between any two or more species, are those which have been inherited from a common parent, and, in so far, all true classification is genealogical; that community of descent is the hidden bond which naturalists have been unconsciously seeking, and not some unknown plan of creation, or the enunciation of general propositions, and the mere putting together and separating objects more or less alike. But I must explain my meaning more fully. I believe that the *arrangement* of the groups within each class, in due subordination and relation to the other groups, must be strictly genealogical in order to be natural; but that the *amount* of difference in the several branches or groups, though allied in the same degree in blood to their common progenitor, may differ greatly, being due to the different degrees of modification which they have undergone; and this is expressed by the forms being ranked under different genera, families, sections, or orders. (1859:420)

It is clear, therefore, that Darwin rejects essentialism and also the naturalist's overall similarity (such as Jussieu's) for an emphasis on genealogical relationship and character divergence within lineages as the bases for natural classification.

Gilmour's use of "natural" has been quoted by many workers. He stresses: "A natural classification is that grouping which endeavours to utilize *all* the attributes of the individuals under consideration, and is hence useful for a very wide range of purposes. . . . Phylogeny, therefore, instead of providing the basis for the one, ideal natural classification, is seen to take its place among the other subsidiary classifications constructed for the purpose of special investigations. It may also be regarded as forming a sort of background to a natural classification, since, although natural groups are not primarily phylogenetic, they must, in most cases, be composed of closely related lineages" (Gilmour 1940:472, 473). Davis and Heywood agree with this perspective: "We do not suggest ignoring phylogenetic facts. It is the basing of classification on inferred phylogeny, instead of interpreting classification in phylogenetic terms, to which we are opposed" (1963:68).

The pheneticists (e.g., Sokal and Sneath 1963; Sneath and Sokal 1973), agree with Gilmour's definition of naturalness and use this as their philosophical underpinning for seeking many (upwards of 100) characters to produce their phenetic classifications. Despite Gilmour's use of *"all* the attributes," cited above, he clearly did not mean the extremely large number of characters advocated by the pheneticists. From reading his general paper, one sees clearly that his natural classification "in practice, is the procedure followed in what is sometimes called 'orthodox' taxonomy, and it would seem best to confine the use of the ordinary taxonomic categories of species, genus, family &c., to a natural classification of this type. In so far as it is theoretically possible to envisage a classification on these lines, which does in fact embody all the attributes of the individuals being classified, it can be said that one final and ideal classification of living things is a goal to be aimed at. In practice, however, this aim would never be attained, owing both to the limitations of our knowledge and to the differences of opinion between taxonomists" (1940:472). Gilmour's main stress, therefore, is that use should be made of all correlating characters of whatever type from whichever parts of the organism to produce a natural classification. This is natural in the same sense as with pre-Darwinian authors and provides no new philosophical base for phenetic practitioners.

Cladistic advocates have equated natural with their use of the term "phylogenetic," or in the context of this book, the "cladistic" relationship. Wiley speaks of "phylogenetic naturalness" in which "The members of a phylogenetically natural group share a common ancestor not ancestral to any other group" (1981a:71). This forms a part of Darwin's concept of naturalness, and is essentially the same as that used by Mayr (1969c:78). Wiley continues, however, with a definition of "a natural taxon" as "a taxon that exists in nature independent of man's ability to perceive it" (p. 72). This is a different and much more general usage of "natural."

It is clear from these examples that different definitions of "natural" prevail. The one used in this book is that employed in the development of pre-Darwinian classifications and articulated by Gilmour (1940). This is not to say that phyletic approaches to classification are "unnatural" or "artificial"; only that these classifications are based on selected characters of presumed evolutionary value. As the phyletic taxonomist Simpson has remarked: "One has only to examine a number of widely accepted classifications to see that the point is not so much how many attributes are taken into consideration as what ones are selected and how they are interpreted" (1961:56). Because of all the different perspectives about "natural," therefore, it is important to clarify usage in each particular context.

CHAPTER 7

Phenetic Approach

Not all taxonomists, however, have found the phyletic approach to classification satisfactory. A minority of workers during the past two decades (e.g., Sokal and Sneath 1963; Sneath and Sokal 1973) have regarded this method as too subjective, and as evidence they have pointed to different classifications generated by different taxonomists for the same sets of organisms. There are some striking examples of lumping vs. splitting in phyletic classification, especially in groups that are strongly inbreeding or with asexual modes of reproduction (e.g., in *Crataegus* and *Taraxacum*). In these cases widely divergent views have prevailed even with examination of more or less the same collections and other available evidence. The subjectivity of the phyletic approach is evident in the selection of different characters to be compared, the comparison of character states, and the ranking of the resultant groups. Attempts in recent years have been made to avoid (or at least reduce) this subjectivity, particularly in the process of grouping, by: 1) emphasizing the selection of as many characters as possible (ideally *all* the characters, harking back to the "naturalness" of Gilmour 1940); 2) making the description and measurement of character states as precise as possible; and 3) comparing the character states of the individuals by rigidly defined numerical procedures.

DEFINITIONS

These perspectives have led to the development of the phenetic approach to classification. *Phenetics* is here defined as classification based on numerous precisely delimited characters (with carefully coded states) of equal weight and their comparison by an explicit method of grouping. A general sense is to obtain a measure of overall similarity. The term "phenetic" was introduced by Cain and Harrison to mean a relationship "by overall similarity, based on all available characters without any weighting" (1960:3). Sokal and Sneath used phenetic to refer to a relationship between taxa "evaluated purely on the basis of the resemblances existing *now* in the material at hand" (1963:55) and "the overall similarity as judged by the characters of the organisms without any implication as to their relationship by ancestry" (p. 3). This was redefined to read: "similarity (resemblance) based on a set of phenotypic characteristics of the objects or organisms under study" (Sneath and Sokal 1973:29). Burtt has questioned whether equal weighting should be regarded as a necessary part of phenetics, and he suggests the term *isocratic* for characters with "equal power" (1964:15). Colless (1971) has commented that weighted characters based on "conservative" patterns of variation in populations (Farris 1966) could indeed be used in determining phenetic relationships, and Adams (1975a) has shown how this could be done in classifying species of *Juniperus*. Moss advocates treating phenetics as

> the estimation of relationship due to similarity, but effectively becoming *independent of data base treatment*. Such an interpretation emphasizes that phenetic relationships are similarity relationships obtained when comparing the phenotypes of organisms (or objects) for correspondences of parts; conversely, such relationships are phenetic, regardless of whether the relative weights of characters used to describe these parts are: 1) left unmodified as raw data, with possible unintentional weighting of some characters due to scale factors, 2) equalized due to a process such as standarization by range or variance, or 3) variously modified as the result of logically or biologically valid or invalid assumptions made by the investigator. (1972:237)

Equal weighting is explicit in the definition of phenetics used here, because employment of many differentially weighted characters would be regarded as a complex attempt at natural (if not evolutionary) or phyletic classification (with evolutionary ideas).

Numerical taxonomy was coined by Sokal and Sneath as "the numerical evaluation of the affinity or similarity between taxonomic units and the ordering of these units into taxa on the basis of their affinities" (1963:48). Later they offered a slightly different definition: "the grouping by numerical methods of taxonomic units into taxa on the basis of their character states" (Sneath and Sokal 1973:4). It was the intent that methods of numerical taxonomy would be used to determine phenetic relationships among organisms, and this has been the usual approach. In the minds of some workers, however, numerical taxonomy means simply the use of some quantitative assessment of relationships in classification, usually with help of the computer (Duncan and Baum 1981).* This broader context has also been labeled *"statistical systematics"* (Solbrig 1970b:178), or obvious similar appellations such as *"statistical taxonomy," "mathematical taxonomy,"* or *"quantitative taxonomy."* Hence, some prefer the term *"numerical phenetics"* (Duncan and Baum 1981) for studies employing equal weighting of characters. Other terms in use have been *"taximetrics"* (Rogers 1963), *"taxometrics"* (Mayr 1966),† and *"multivariate morphometrics"* (Blackith and Reyment 1971).

---

*An amusing twist is the recent paper by P. J. H. King (1976) on "taxonomy of computer science" in which organisms (people) are classifying computers and activities associated with them rather than the reverse.

†Mayr (1966) preferred "taxometrics" instead of "taximetrics" because it "is a word in a modern language and formed in analogy to taxonomy" (p. 88) and also, tongue-in-cheek, because "The word taximetrics has the additional disadvantage, as a mischievous friend of mine reminded me, that the name suggests 'the science of taxi meters.' " See also Heywood and McNeill (1964a) and Pasteur (1976) for a discussion of these and related terms.

HISTORY OF PHENETICS

The origin of phenetics can be traced back to the French botanist Michel Adanson (1727–1806). Adanson was acquainted with Bernard de Jussieu (and also with the famous zoologist Cuvier) and was a contemporary of Linnaeus. In 1749 he traveled to Senegal in western Africa and remained there for more than four years studying the natural resources of the country. Perhaps because of the difficulties encountered in attempting to classify the plants of this relatively unknown region (Stearn 1961), he developed a new and highly original system of classification which he published in his *Familles des Plantes* (1763). This contained 65 different classifications, each based upon single characters such as ovary placement, inflorescence type, etc. The groups of genera and families were based upon occurrence of similar positions in each of the different systems. Because of being an unbelievably original system for its time, Adanson's system of classification had little impact on plant classification of his day and has had virtually none since. This early phenetic system was ignored by most workers due to its peculiar nomenclature, the strong influence of the much more easily comprehended Linnaean artificial sexual system, and slightly later the equally strong impact of the well considered natural system in the *Genera Plantarum* of Jussieu (1789), who was a more influential French botanist than Adanson with a position in the Royal Botanical Garden in Paris. Nonetheless, despite the lack of direct impact on the development of subsequent classification systems, the work of Adanson was the beginning of the phenetic approach. Burtt (1965), Jacobs (1966), and Nelson (1979) comment that numerical taxonomists have misinterpreted Adanson's ideas and that he should not be regarded as the originator of the phenetic approach to classification. It is true that Adanson's system is a far cry from those now developed and used by pheneticists, but it does serve as a legitimate philosophical and historical starting point for this approach.

A few additional developments in phenetic approaches to classification occurred since Adanson (see Sneath and Sokal 1973, for a review), but no significant developments can be cited until the independent publication of papers by Michener and Sokal (1957) and Sneath (1957). Charles Michener and Robert Sokal were both at the Department of Entomology of the University of Kansas in Lawrence. Michener, a highly respected phyletic taxonomist working with bees, and Sokal, a biostatistician, interacted with students and other faculty members in an informal Biosystematics Luncheon Group in which concepts and methods of classification were discussed. After repeated claims that surely some improvements toward a more objective approach to classification could be achieved, an attempt was made using data supplied by Michener on groups of bees and Sokal's computational expertise. Even then, though, it was clear that the two authors disagreed on some fundamental aspects of the work, especially in ranking, which was to result in Michener's departure from pure phenetic approaches and continuation of his prior phyletic efforts (see his excellent and balanced review of different approaches to classification, 1970).

At the same time (1957) and completely independently in Leicester, England, Peter Sneath, a microbiologist in the Microbial Systematics Research Unit of the

University of Leicester, attempted a numerical classification of bacteria using the computer. Because many of the characters routinely used in bacterial systematics are chemical ones involving positive or negative responses, and because it is difficult (if not impossible) to evaluate intuitively the morphology of bacteria, except only at the grossest level, it was a natural and meaningful approach to advocate the use of large numbers of characters with comparisons done by computer.

Becoming aware of each other's efforts, Sokal and Sneath published a landmark book in 1963 (coincidentally exactly two hundred years after Adanson's *Familles des Plantes;* in 1957 neither worker was aware of Adanson's contributions) which spawned numerous subsequent investigations. This book was not without its critics, especially because of the provocative tone and the particular section (in chapter two) entitled "The ills of modern taxonomy." A typical reaction from outstanding phyletic taxonomists at that time is indicated by this quote from Rollins:

> The use of digital computers has considerably extended the possibilities for utilizing information in the improvement of systems of classification. There is no doubt that computers should be utilized to the fullest extent that such devices are of practical value. However, we should remind ourselves that the computer is an instrument, just as a microscope is an instrument. The use of either one will not, of itself, insure objectivity or superior results. The fad of "computerism" has taken hold in some quarters where the general popularity of the instrument itself is being used as a promotional lever to gain recognition and to discredit more traditional aspects of taxonomy. I view it as particularly unfortunate that the new cannot be introduced without precipitating conflict through attempts to downgrade and nullify that which has stood the test of time. (1965:4)

METHODOLOGY OF PHENETICS

The operations involved in phenetic classification are basically the same as those shown in the chart in figure 5.1, with a few minor exceptions (in part from Duncan and Baum, 1981): (1) selection of taxa (or individuals) for study (those are usually called *Operational Taxonomic Units* or OTUs, Sokal and Sneath 1963, which are simply the starting point units in phenetic classification; they might be individual organisms, populations, species, genera, etc.); (2) selection of characters (ideally more than 100 unit characters); (3) description and/or measurement of character states; (4) comparison of states to (a) determine a measure of overall similarity (phenetic relationship) between each pair of OTUs, and (b) determine the taxonomic structure, i.e., the detection of possible groups and subgroups among all OTUs; and (5) ranking of all OTUs into the categories of the taxonomic hierarchy.

Introductory Example

Before the details of each of these steps are discussed, it is believed helpful to give a simple example of phenetic classification so that the *general* approach is understood. An example will be used of six OTUs: S, T, W, X, Y, and Z. First, as many characters as possible must be selected. These initially will probably be morphological features, but any other kind of data can be used, such as anatomical features, palynological characters, and chemical aspects. To provide a good indication of phenetic relationships, as many different unit characters as possible should be used, but for this example we will deal with only ten for simplicity. The characters and states are defined and delimited as precisely as practicable, and the characters are usually a mixture of quantitative and qualitative features. In this example, only characters with two states are used (such as alternate vs. opposite leaves, red vs. white petals, or leaves 10 cm vs. 5 cm long, etc.), and these are given arbitrary numerical values of 0 for one state and 1 for the other (table 7.1). Remember that these values have no reference to primitive or derived conditions—they are simply arbitrary numerical designations. Multiple states for characters also can be used and compared.

The character states then are compared from one OTU to the next. A number of statistics (or coefficients) exist for making such comparisons, and one of the simplest is called the *simple matching coefficient* (Sokal and Michener 1958). This is calculated by counting the number of character states for each character in common between two OTUs (i.e., the positive matches) and then dividing this number by the total number of characters used (positive plus negative matches). The higher the coefficient of association (i.e., as the value approaches 1.0), the more closely related are the OTUs. For example, the coefficient between S and T is: 8 character states in common divided by a total of 10 characters equals 0.8. Such calculations are done between all fifteen pairs of OTUs and the resultant figures are displayed in a data matrix of coefficients of association (table 7.2). If many characters and groups are involved, the calculations are usually completed with the aid of computers.

TABLE 7.1 Basic Data Matrix of Characters Selected and Numerical Values Accorded to the Two Character States (either 1 or 0) in OTUs S-Z for Phenetic Analysis. For Simplicity, the States of Only Ten Characters are Shown in the Table, but the Actual Number Might be Well Over 100.

OTU	*Character*									
	1	*2*	*3*	*4*	*5*	*6*	*7*	*8*	*9*	*10*
S	0	0	0	0	0	0	0	0	1	1
T	0	0	0	1	0	0	1	0	1	1
W	0	1	0	1	0	1	1	0	1	0
X	0	1	0	1	0	1	0	0	1	0
Y	1	1	1	0	1	0	0	1	0	0
Z	1	1	1	0	1	1	0	1	0	0

TABLE 7.2   Data Matrix of Simple Matching Coefficients Among OTUs S-Z.

	S	T	W	X	Y	Z
S	1.0					
T	.8	1.0				
W	.5	.7	1.0			
X	.6	.6	.9	1.0		
Y	.3	.1	.2	.2	1.0	
Z	.2	.0	.3	.4	.9	1.0

Because historically in systematics an emphasis has been placed on dendrograms (or trees) for illustrating relationships (Voss 1952), due principally to the influence of phyletic classification over the past century, likewise in the phenetic method a *phenogram* is constructed that graphically expresses the relationships among all the OTUs in a conventional form (fig. 7.1). This reveals the taxonomic structure of groups and subgroups among the OTUs. The vertical lines delineating groups or clusters of OTUs are based on the coefficients of association. S and T are similar at the 0.8 level, and W and X, and Y and Z are similar at the 0.9 level. The former two pairs of OTUs are more similar to each other than either pair is to the latter. The level at which they are connected is based on an average (unweighted arithmetic) of values of the pairs S–W, S–X, and T–X (=0.6). The

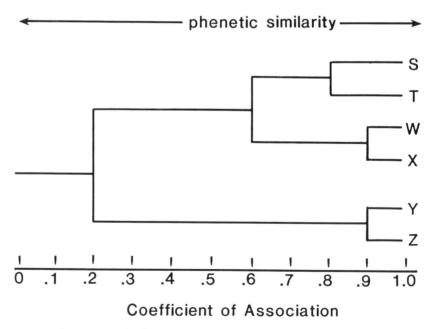

FIGURE 7.1   Phenogram of relationships among OTUs S–Z based on simple matching coefficients and unweighted arithmetic averages.

relationship of Y and Z to the other four OTUs is assessed in the same way and found to be very low (0.2).

Several important points need to be made regarding the resultant pheno-gram. First, the diagram attempts to show only "phenetic similarity" which is based solely on the comparison of character states. Second, evolutionary path-ways are not implied here (strong advocates of the phenetic approach believe that such considerations have no place in classification, per se). Third, the phenogram is a slight distortion of the relationships seen in the matrix of coeffi-cients of association. One reason for this alteration is that this matrix contains multidimensional portrayals of relationships, whereas the phenogram is a con-densation to only two dimensions. The necessary averaging of coefficients of association to produce the phenogram unavoidably creates distortion (although not enough to seriously blemish the method). Finally, formal taxonomic ranking from the phenogram in this example is meaningless, because the initial level of the OTUs is unstated (they could be individuals, populations, species, genera, families, etc.). If they were species, then S–T and W–X might be placed in one genus, as an example, with each pair representing a separate subgenus or sec-tion, and Y–Z might be regarded as belonging to a distinct genus. But all the OTUs could belong to a single series of populations of only one species, in which case varieties or subspecies might be suggested. The proper ranking depends to a considerable degree on the historical context of a particular group, the number and nature of character states, and on the concepts of categories in the mind of the particular pheneticist. In general, therefore, ranking in the phenetic ap-proach is somewhat arbitrary (as it is to some extent also in the other two approaches), with each worker defining at what level of similarity genera, spe-cies, or other taxa should be recognized.

Having now outlined the basic operations in the phenetic approach to classi-fication, we can return to details of the process. Numerous papers and books have been written on various aspects of phenetics, and it is clearly not possible to refer to all of them. The discussion here relies heavily on the ideas of Sneath and Sokal (1973), whose text has been the most influential in the field. Other texts have been consulted also, and some (but not all) will be cited, such as those of Sokal and Sneath (1963); Heywood and McNeill (1964b); Cole (1969); Lock-hardt and Liston (1970); Blackith and Reyment (1971); Jardine and Sibson (1971); Shepard, Romney, and Nerlove (1972); Estabrook (1975); Clifford and Stephenson (1975); Hartigan (1975); Pimental (1979); Chatfield and Collins (1980); Neff and Marcus (1980); Gordon (1981); Dunn and Everitt (1982); Crisci and López (1983); Felsenstein (1983b); Romesburg (1984); and Abbott, Bisby, and Rogers (1985). The recent book of Yablokov (1986), despite its title *(Phenetics: Evolution, Population, Trait)*, deals with microevolution and not classification. For an update of Sneath and Sokal's (1973) book, see the review by Sokal (1986).

Selection of OTUs

The first operation in the phenetic approach to classification involves selection of OTUs for study. This is deceptively simple, but one must choose carefully with a clear idea of the questions of classification being posed. Ideally, to avoid any *a priori* judgments of what any taxa are, all phenetic studies should treat only individuals as OTUs (e.g., Crovello 1968c). This avoids subjectivity on what are populations, species, genera, etc. With plants theoretically this would be possible, but with insects or other organisms with dramatic developmental changes in their life histories, even this perspective would be unworkable. In those cases a particular stage of an individual would better be treated as the OTU (the "character-bearing semaphoront," Hennig 1966). In practice, however, treating individuals as OTUs in all cases is unworkable because of the time and effort needed to approach any problem at any higher level in the hierarchy. A problem of classification at the familial level of angiosperms (e.g., Young and Watson 1970) could only be resolved by first grouping individuals into populations, these then into species and genera, and finally the genera into the families which would then be grouped into orders to determine familial relationships. Clearly, such an approach would prohibit most work of this type from being done—it would be overwhelmingly tedious and time-consuming. Further, in plants which reproduce asexually, it is sometimes difficult to determine the limits even of the individuals. Instead, representative samples must be selected as OTUs that adequately reflect the diversity at a lower level in the hierarchy of the group being treated. In the case of familial relationships, it is sufficient to use several representative genera within each family or to develop familial profiles of characters (average occurrences of characters and values of character states) as OTUs. The better the sample is, the better the assessment of relationships. Some information is lost in this approach, but the trade-off in time and effort at least encourages the work to be done. No doubt statistically it would be possible to calculate the level of variation within characters in each OTU needed to adequately portray the phenetic relationships at some desired level of confidence, but this would be even more time-consuming. The number of OTUs and the particular set of OTUs used in a phenetic study will also affect the assessment of relationships (this is also true in any other approach to classification). Although this may appear to be a most serious problem, in practice minor changes in composition of OTUs in a study group only produce small alterations in the phenograms (Crovello 1968a). Sneath (1976a:437) recommends that "at least 10 and preferably 25 or more" OTUs be used for accurate representation of homogeneous phenetic clusters. This is clearly desirable but obviously not always possible.

Selection of Characters

The second operation in phenetic classification is the selection of characters of the OTUs. This is one of the most important parts of the process. The search here is for *unit characters* which within the particular study group cannot be subdivided logically (Sneath and Sokal 1973; see chapter 4). The main concern is to pick characters that contain or represent single pieces of information both for purposes of avoiding redundancy of data and also for ease of handling in computations. Pratt (1972b) has pointed out that unit characters are difficult to define and there is an infinite number from which to choose. In practice, therefore, some selection must be made and considerable subjectivity prevails in this selection. One method of determining redundancy of information in selected characters (with the view of eliminating or restructuring some of them) is to do statistical correlations of occurrence of their states in the OTUs (e.g., Clifford 1969). In this fashion, the characters can at least be as theoretically close to true unit characters as possible. Some characters correlate due to evolutionary reasons, such as belonging to the same adaptive character complex and/or phyletic line, and this important biological information should be maintained in the data set, while at the same time seeking to eliminate logical or developmental redundancies. Methods of automated character data extraction have been attempted such as the use of perforated computer cards placed over an OTU (Sokal and Rohlf 1966; Rohlf and Sokal 1967), but because of the complexity of most organisms and the utility of internal characters, this automation as a way to avoid problems of unit character selection seems of doubtful value in most cases. Moss and Power (1975) have suggested a method of semiautomatic data recording, but this initially involves the designation of operationally homologous points of data capture (i.e., unit characters) that a technician could then reveal for all OTUs. This does not avoid the problem of selecting unit characters, but it does offer possibilities for more rapid data capture.

Characters must also be selected in a fashion that minimizes the within-OTU variability. That is, the best unit characters are those in which most of the variation can be expressed by the coding values themselves rather than by having to use representative (or exemplary) individuals or mean values. In this way less information is lost. McNeill (1974) has suggested the "character-state frequency procedure" which is basically an expansion of the original single character states into many "secondary" and "tertiary" characters which encompass more completely the observed variation. This has the effect, however, of weighting the more variable characters because there will be more of them contributing to the assessment of the phenetic relationships.

It is also desirable to select characters from any and all parts of the OTUs. This includes external morphology, anatomy, chemistry, ultrastructure, chromosome number, and so on. There is no *a priori* way to prefer one set of data over another; some data sets are more useful in some groups than in others, but this is only determined *a posteriori*. Genetically speaking, it is likely that most phenetic characters are influenced by more than one gene and that most genes influence more than one character (the *"nexus hypothesis;"* Sneath and Sokal 1973), and that this holds for any type of data, morphological or otherwise. Early pheneticists believed that the same classifications would be developed from any data from the same set of OTUs, i.e., they would be *congruent* classifications (Sokal and Sneath 1963). This was called the *"nonspecificity hypothesis."* However appealing it earlier seemed, it is now realized to be true only within limits (Michener and Sokal 1966; Johnson and Holm 1968; Crovello 1969; Sneath and Sokal 1973) and these limits vary from group to group. For example, in studies on *Medicago* (Leguminosae), Small, Lefkovitch, and Classen (1982:2505) found "moderate incongruence" among the sets of vegetative, floral, and fruit characters; flavonoid chemical characters and pollen traits were "highly incongruent" with the other characters. Hu, Crovello, and Sokal (1985), however, found good correspondence of classifications based upon mean values of vegetative characters in *Populus* (Salicaceae) in comparison with previous intuitive work based upon all features (Eckenwalder 1977). Interpreting the reasons for incongruence is difficult because it is clearly due to at least two factors (Farris 1971): 1) mosaic evolution, in which different parts of an organism evolve faster than others; and 2) evolutionary incompatibilities among characters due to convergences, parallelisms, and reversals. Emphasis has shifted away from viewing congruence as a test of the nonspecificity hypothesis and toward its use in testing the "stability" of classifications in comparisons of the different major approaches (e.g., Mickevich 1978). More about this in chapter 9.

Homology

A difficulty with any approach to classification, although somewhat less so with phenetics, is the selection of characters and states that are homologous, i.e., the choosing of features that have descended with modification from a common ancestor. Because early pheneticists eschewed attempts to imbue their efforts with *any* evolutionary import (e.g., Sokal and Sneath 1963; Colless 1967), this may seem a nonproblem for phenetic approaches to classification. Not so. It is significant because even excluding evolutionary considerations, from an information theory point of view, the same kinds of structures must be compared from OTU to OTU to produce any meaningful estimation of phenetic relationship. For example, it serves no purpose to compare features of true leaves in one OTU with those of leaf-like structures in another, which might be shown developmentally to be modified stem tissue. The same kinds of information in each case are not being compared. Sneath regards the problem of homology as one of "the biggest challenges" of the phenetic approach (1976a:447). The usual procedure is to define homology in operational terms, such as those of Sneath and Sokal's "compositional and structural correspondence. By *compositional correspondence* we mean a qualitative resemblance in terms of biological or chemical constituents; by *structural correspondence* we refer to similarity in terms of (spatial or temporal) arrangement of parts, or in structure of biochemical pathways or in sequential arrangement of substances or organized structures" (1973:77). This definition leads to a quantitative view of homology in which characters and their states have different affinities with other such units. Those with the greatest degree of similarity to each other are the ones to be used for comparison (Jardine 1967, 1969b). Despite all cautions and attempts to exclude prior biological or evolutionary considerations in establishing homologies, some circularity seems unavoidable in practice (Inglis 1966), and even to some extent in theory (Sattler 1984). This is even more of a philosophical problem in cladistic and phyletic approaches to classification which depend entirely upon the detection of homologies in an evolutionary context. Sneath and Sokal offer different "phenetic approaches" to determining homology including similarity of complex structures based on unit characters, similarity with undefined characters, and geometric similarity (i.e., attempting "to fit entire images to each other") (1973:89). These all rely on structural and developmental similarity at some level. Rieger and Tyler (1979) have offered three criteria for identifying homologous structures: (1) similarity in a positional hierarchy (i.e., with regard to other structures); (2) arrangement into a transformation sequence of ontogeny (or phylogeny, but this reintroduces circularity); and (3) coincidence with already established homologies (a form of logical correlation; see Crisci and Stuessy 1980). Baum rightly stresses that "The lower the rank, the higher the probability of homology" (1973:330). A phenetic (or any other) approach to classification at the familial level in the angiosperms, for example, poses much more of a problem in determining homologies than at the specific level (see pertinent comments in this regard by Smoot, Jansen, and Taylor 1981; and Young and Richardson 1982). Davis and Heywood also echo this perspective: "In practice, the distinction [between homology and analogy] is easy to make in closely related plants; it becomes increasingly difficult as we consider more distinctly related groups, especially when they hold a very isolated position" (1963:41). Despite these numerous difficulties, however, there is still a clear need to determine homologies correctly in phenetics. Fisher and Rohlf (1969) demonstrate that with only 10% or less of the characters being erroneously interpreted as to their homologies, significant alterations in the phenograms will result.

Number of Characters

How many characters to use has also been of concern in phenetic approaches to classification. Ideally, as many characters as possible should be used so that the phenetic correlation would be strengthened with the addition of more data. However, on theoretical grounds, "as the number of characters sampled increases, the value of the similarity coefficient becomes more stable; eventually a further increase in the number of characters is not warranted by the corresponding mild decrease in the width of the confidence band of the coefficient" (Sokal and Sneath 1963:114). This is called the *matches asymptote hypothesis*. What this maximum ideal number of characters is, of course, would vary significantly depending upon the types of characters used and the natural congruence among the different kinds of data in the OTUs under consideration. Sixty characters have been recommended as a minimum number (Sneath and Sokal 1973), but this is arbitrary. Sneath suggests using 100–200 to give "a large enough sample to keep sampling errors reasonably small" (1976a:440). Steyskal recommends "at least 1,000 characters, at least when working with animals approaching the complexity of insects" (1968:476). Crovello (1969) used 202 characters in a numerical phenetic study of 30 species of *Salix* (Salicaceae) in California, and subsets of this full data set down to 8 characters. He found that with over 60 characters the results from different data sets were very similar, but below that number significant distortions in the relationships of the phenograms occurred. Furthermore, use of vegetative characters alone gave quite different results. Gilmartin (1976) has shown that the within-group phenetic distance remains relatively constant using 50 percent random selection from among 83–180 characters in different OTUs of the Asclepiadaceae, Bromeliaceae, and Leguminosae. Studies below that level resulted in significant differences. Stuessy and Crisci (1984a) showed phenetic relationships among 37 species of *Melampodium* (Compositae) based on 42 reproductive and vegetative characters that correspond well (though not exactly in every detail) with the phyletic classification produced earlier (Stuessy 1972a). Because no general empirical or theoretical means exist to justify fully the matches asymptote hypothesis nor to dictate precisely how many characters should be used, the reasonable approach at this point is the following: "The practical advice that can be given at this time is to take as many characters as is feasible and distribute them as widely as is possible over the various body regions, life history stages, tissues, and levels of organization of the organisms. Since congruence is always less than is expected from random samples of characters, the number of characters used will set a lower limit to the confidence levels of the similarity coefficients. The investigator should therefore employ at the very least as many characters as will give the confidence limits he wishes" (Sneath and Sokal 1973:108). In any event, it is both positive and comforting that "The search for large numbers of unit characters will lead to the discovery of many new facts about plants which, whether taxonomically useful or not, will be important biologically" (Cullen 1968:182).

Weighting of Characters

The question of weighting of characters has been another important issue in phenetic approaches to classification. As defined in this book, the phenetic approach uses unweighted (or equally weighted) characters. Some workers prefer to include weighting within their definition of phenetics, but Sokal and Sneath comment forcefully on this point:

> Even if desirable, there is no rational way of allocating weight to features, and therefore one must in practice give them all equal weight. Even if the entire genetic constitution of a form were known, it would be impossible to find a basis for weighting the genetic units, for these have no fixed adaptive, ontogenetic, or evolutionary significance. . . . Equal weighting can therefore be defended on several independent grounds: it is the only practical solution, it and only it can give the sort of natural taxonomy which we want, and it will appear automatically during the mathematical manipulations. Singly, these arguments are cogent; taken together, we feel that they are overwhelming. (1963:118, 120)

Perhaps the greatest dislike of weighted characters by pheneticists was to avoid using characters that had been judged *a priori* to be important for purposes of classification. R. W. Johnson (1982) has pointed out that *a priori* weighting of characters can be viewed in two different ways: (1) weighting of characters prior to beginning any aspect of the study; or (2) weighting of characters after the study is underway but prior to making the classification. The former use is clearly more objectionable than the latter, but both have been abhorred by many pheneticists. Weighting *a priori* before beginning the study involves determining weights based upon knowledge of characters and their weights in other taxonomic groups (called *"extrinsic weighting"* by Burtt 1965).

A number of workers have stressed that equal weighting of characters in phenetics is also filled with problems (e.g., Mayr 1964; Meeuse 1964). Mayr makes this useful comment: "Indeed, there is doubt that pure non-weighting exists. Any choice of characters is already in itself a weighting process. Any decision whether or not two characters are morphologically equivalent and whether a character is 'zero' or 'inapplicable' [Cain and Harrison 1960] is likewise a form of weighting. Any working out of correlation coefficients and factoring of the characters with these coefficients again is a form of weighting" (1964:27). In a similar vein McNeill (1972) stresses that missing characters, which occur frequently in data sets, particularly above the species level, can cause a weighting effect on the remaining features used in the comparisons.

Some pheneticists have emphasized that equal weighting is not necessary in phenetics and have offered suggestions on how weights might be determined more objectively.* Kendrick (1964, 1965) suggested that the weight of a character be determined by the presumed importance of the organ in which it occurred. Williams (1969) followed up this point and showed that some characters are *serially dependent attributes* in which one primary character influences the use of one or more other secondary characters, and therefore is given more weight in making the classification. For example, the states of the character "fruit" (drupe, capsule, or berry) lead to the second character "type of capsule" (loculicidal, poricidal, or septicidal) only if the state in the first character was "capsule" and not the other types of fruit. A capsule fruit in an OTU, therefore, would be given more weight in the classification than if some other fruit type occurred. Adams (1975a) demonstrated that in using terpenoid data in seven species of *Juniperus* (Cupressaceae), statistical *a posteriori* weighting (three different methods used) gave superior results to equal weighting. Similarly, R. W. Johnson (1982) obtained the "soundest classification" by use of individual attribute weighting based on the variability of characters between and within duplicate pairs of OTUs (plants grown from the same seed source in two consecutive years). The same author made the useful distinction between "individual attribute weighting" and "correlative attribute weighting." The former is the designation of weights to characters based on their importance either *a priori* or *a posteriori* by various means (variance within populations, historical precedence, adaptive value, etc.). The latter is basing weights on the correlations that are determined with other characters, which is done *a posteriori* (this is the sense of *a posteriori* as used by Sneath and Sokal 1973). It is clear, therefore, that objective methods of *a posteriori* weighting of characters have considerable potential.

Description and Measurement of Characters and States

After characters have been selected for use in phenetic classification, the character states must be described and/or measured. This is another difficult phase of the operation, because many different perspectives and options prevail. As Sneath has remarked: "Among the biggest challenges [for phenetics] are those posed by homology and by character coding and scaling. For both of these we will lack comprehensive and practical solutions" (1976a:447). The two major concerns are with the selection and coding of types of character states, and their scaling to reduce distortions in the final assessments of phenetic relationships. The data

---

*Such an approach, with computer handling of many differently and objectively weighted characters, would in this book be called "numerical phyletics" rather than phenetics.

also must be arranged in a basic matrix before correlations among OTUs can be attempted.

The classification of different types of character states within characters can be viewed logically in several ways, but the perspective used here follows Sneath and Sokal (1973): two-state characters; and multistate characters, which are divided into (a) qualitative states and (b) quantitative states, the latter of which can be regarded as either continuous or meristic. Two-state characters (also called binary, presence-absence, or plus-minus characters) are the most commonly used in plant taxonomic studies. Features such as petals white vs. petals pink, etc., are examples. These are usually coded as 0 or 1 for computations of phenetic similarity (or distance).

Qualitative multistate characters are frequently encountered in plant taxonomic studies, but these pose numerous difficulties for satisfactory coding. An example might be petal color which occurs in five states: white, red, orange, yellow, and violet. These states could be coded 0, 1, 2, 3, and 4 in the same order. As Baum (1976) has pointed out, however, the numerical distance between states 0 and 1 and between 3 and 4 is the same as between 1 and 2 and between 2 and 3. This is not intuitively satisfactory because a white petal is reflecting all wavelengths of light and is much more different than those petals reflecting red, orange, and yellow which are wavelengths of similar physical properties and probably represent similar structural and/or chemical qualities of the petals. Likewise, violet is much more different from yellow than yellow is from orange. One solution is to adjust the numerical distance to reflect these known facts, such as using a scale 0, 3, 4, 5, and 8 for the petal colors described above. Another approach is to convert these multistate characters into a series of binary characters. For example, in the case of petal color, the data could be described as three binary characters as presence or absence of (1) white, (2) red, orange, or yellow, and (3) violet. Other techniques for accomplishing the same end, but depending upon the natures of the characters and states are *additive (binary) coding* and *nonadditive coding* (Sneath and Sokal 1973). But whatever method is used, the effect will be the weighting of this feature (here petal color) to the level of three characters instead of only one. Whether this will affect the assessment of phenetic relationships has to be determined in each particular case. With large numbers of characters and only a few treated this way, few changes in the relationships should be noted; with smaller data sets and more additive or nonadditive coding, the greater should be the alteration of of phenetic relationships. Sneath and Sokal emphasize that "In practice it is commonly found that most qualitative multistate characters can be converted into several independent characters if a little thought is given to the problem" (1973:149). True enough, but the question remains whether it is desirable to do this. Each case must be evaluated on its own merits, but this represents yet another difficult area for phenetic approaches and one in which subjectivity of the investigator plays a conspicuous role. The numerous pubescence types commonly encountered on the external surfaces of organs of flowering plants (e.g., hirsute, pilose, sericeous, strigose, etc.) frequently cause problems in their proper coding, particularly because their genetic basis is usually unknown (see Rollins 1958, for a rare case in which the genetics have been worked out) and hence also their degrees of logical dependency and informational redundancy. See Hill and Smith (1976) for suggestions on how multistate characters can be used successfully with some types of phenetic analysis without having to reduce to binary data.

Quantitative multistate characters are more easily treated, but they also can pose problems. Most characters of this type are continuous, i.e., the states equate to the exact data points being measured. In some instances, the data are meristic, in which the data points group conveniently into classes (this begins to approximate a qualitative character with ranges of quantitative values for each state). Depending upon the spread of values, "Quantitative multistate characters are very likely to be caused by more than one genetic factor and several two-state characters may thus be more appropriate" (Sneath and Sokal 1973:148). Sneath (1968b) in a numerical study of *Cytisus* (Leguminosae) showed that a distortion (using a distance measure of a relationship) can be caused by changing a quantitative continuous character into a qualitative binary one.

Another problem that must be addressed with the establishment of character states is how they should be scaled. Consider that the range of state variation within some characters will be many times that in others. This can produce a skewing of the assessment of relationships, with those widely varying characters having more influence than those varying narrowly (e.g., a 0–1 character vs. one with states ranging from 1 to 10,000). Different methods exist for scaling each character so that the effects on the classification will be more or less the same, and they are all some form of mathematical *transformation*. The simplest types are the linear transformations of *translation*, which involves "the addition or subtraction of a constant from all values of a given character," and *expansion*, which is "the multiplication or division of all values by a constant" (Sneath and Sokal 1973:152). Other transformations are more complex such as *geometric transformation* (Proctor 1966) and *logarithmic transformation* (e.g., 1, 10, 100, and 10,000 go to 0, 1, 2, and 4; Sneath and Sokal 1973). Additional methods attempt to equalize the variation and/or size of the transformed characters. One simple way is to subtract the mean from all character states (Sneath and Sokal 1973), which has the effect of equalizing the gross size of each character. *Ranging* (Gower 1971) is another method in which the lowest character state value is subtracted from each value and this divided by the range. This gives values between 0 and 1. Still another approach is to equalize the variability of each character by computing "the mean and standard deviation of each row (the states of each character) and express each state as a deviation from the mean in standard deviation units" (Sneath and Sokal 1973:154). This is called *standardization* and it makes all character means 0 and within character variances equal to 1.

There are, however, some disturbing features about standardization and similar scaling procedures. Characters with small ranges of variability and those with large ranges have equal influence on the resemblance coefficients. We may not be able to distinguish these small variations from variation due to other causes. Clearly one would not wish to employ a character whose variation among the OTU's was principally due to measurement error or environmental effects. Hudson et al. (1966) emphasize this point. Also, we exclude characters that have no variation at all. Therefore one has the anomalous position that as one proceeds to increasingly less variable characters one gives the absolute degree of variation more and more weight until deciding there is no variation, and then giving it zero weight by excluding it. (Sneath and Sokal 1973:155)

After character states have been selected and their coding achieved, the numerical data are placed in a *basic data matrix* (e.g., table 7.1) prior to computation of phenetic resemblance. It is at this point that the data can be seen to be truly comparative. Presenting data in this fashion insures that all data have been accumulated for all OTUs and that the characters and states have been carefully studied and evaluated. One of the greatest strengths of the phenetic approach to classification is that it necessitates a careful consideration of characters and states in their selection, coding, and scaling, and that there are no missing pieces of information (Dale 1968), at least not knowingly. Sometimes specimens are deficient in certain ways, e.g., no fertile or mature structures, which necessitates using them with incomplete data or not using them at all. In traditional phyletic approaches, it is often the case that some data are absent (or at least not mentioned in publication) and the basic data matrix is therefore incomplete. This is not a serious difficulty per se in constructing a classification by intuitive means, but it prohibits further direct phenetic analyses and a fuller understanding of the relationships. Stuessy (1981) has argued that traditional revisionary work should contain complete basic data matrices instead of the usual descriptions, which although valuable, have often not included all characters for all taxa (see Watson 1971, for similar views).

Comparison of Character States

Having selected the characters and described and/or measured the character states, the next step in phenetic classification is the comparison of states to obtain estimates of phenetic resemblance. There are two distinct aspects of this comparison: (1) the calculation of similarity (or dissimilarity) based on some clearly defined coefficient of similarity; and (2) the determination of phenetic structure (usually hierarchical) based on the obtained similarity coefficients by clustering or ordination of the OTUs. These aspects require selection of a specific statistic to measure similarity and selection of an *algorithm* (= procedure; Bossert 1969) to determine the phenetic relationships among all the OTUs being studied. Because there is no general mathematical theory for the assessment of phenetic relationships, for any set of data there are usually several coefficients of similarity and clustering algorithms that will be useful. Evaluation of the results of each method is usually determined empirically; i.e., the relationships obtained are compared with previous intuitive classifications of the same group or with other phenetic assessments. As put well by Sneath and Sokal: "But when all is said and done, the validation of a similarity measure by the scientists working in a given field has so far been primarily empirical, a type of intuitive assessment of similarity based on complex phenomena of human sensory physiology" (1973:146).

Calculation of Affinity. The calculation of affinity between pairs of OTUs is based on some clearly stated statistic. Many statistics exist, and several different outlines of these different measures have been proposed. The one used here is based on that in Sneath and Sokal (1973) in which four basic types are recognized: (1) association coefficients; (2) distance coefficients; (3) correlation coefficients; and (4) probabilistic similarity coefficients. With the case of any of these measures, the values obtained between each pair of OTUs are placed in a data matrix of similarity coefficients (cf. table 7.2) prior to clustering and/or ordination. The discussion below will give examples of the more easily understood statistics only, which should serve as a useful introduction. No attempts to criticize these statistics are provided here. Such viewpoints can be found in Sneath and Sokal (1973), Clifford and Stephenson (1975), and Neff and Smith (1979).

Association coefficients are "pair-functions that measure the agreement between pairs of OTU's over an array of two-state or multi-state characters" (Sneath and Sokal 1973:129). To make the presentation of these methods for binary data more clear, the following chart shows all the different kinds of matches possible in one binary character between two taxa (from Clifford and Stephenson 1975:52).

			Taxon A	
			1	0
T		1	1, 1 (a)	1, 0 (b)
a				
x	B			
o		0	0, 1 (c)	0, 0 (d)
n				

One of the earliest association coefficients used and easiest to comprehend for binary data is the *Jaccard (Sneath) Coefficient* (S_J) from Jaccard (1908) and Sneath (1957):

$$S_J = \frac{a}{a+b+c}$$

The positive matches 1,1 *(a)* are divided by the mismatches 1,0 *(b)* and 0,1 *(c)*. The negative matches 0,0 *(d)* are not considered meaningful. This works well with presence-absence binary data, such as with the presence or absence of chemical compounds, but it is not effective with data in which the 1 and 0 coding refers to different positive data attributes (e.g., leaves alternate vs. opposite). The *Paired Affinity Index* of Ellison, Alston, and Turner (1962) for use with flavonoid spot-pattern data is the Jaccard Coefficient, and here the mutual presence of compounds is much more meaningful than mutual absences, which could derive from different enzymatic steps being disrupted. Another commonly used association coefficient is the *Simple Matching Coefficient* (S_{SM}; Sokal and Michener 1958), also used with binary data. With this coefficient, the sum of the positive (1,1) and negative (0,0) matches is divided by all the possible matches:

$$S_{SM} = \frac{a+d}{a+b+c+d}$$

This coefficient is most effective with all positive attributes and not presence-absence data. A final and much more complex example of association coefficients is *Gower's General Similarity Coefficient* (S_G; Gower 1971) which is useful for binary, multistate, and quantitative data:

$$S_G = \frac{\sum_{i=1}^{n} w_{ijk}\, s_{ijk}}{\sum_{i=1}^{n} w_{ijk}}$$

This gives an assessment of similarity, shown here in character i between OTUs k and j. The weights *(w)* and scores *(s)* for each character in each OTU depend upon the nature of the data, i.e., whether binary, multistate, or quantitative. See Sneath and Sokal (1973) and Dunn and Everitt (1982) for a more detailed discussion with examples. The important point is that it is a flexible and useful coefficient, especially for mixed data sets.

Distance coefficients measure the distance between OTUs in a space that can be defined in a number of ways. They "have inherently the greatest intellectual appeal to taxonomists as they are in many ways the easiest to visualize. It must be pointed out that distance coefficients are the converse of similarity coefficients. They are, in fact, measures of *dissimilarity*" (Sneath and Sokal 1973:119). A simple distance coefficient is the *Euclidean Distance:*

$$D_E = \left[\sum_{i=1}^{n} (X_{ij} - X_{ik})^2 \right]^{1/2}$$

In this statistic X_{ij} is the character state value for character i in taxon J and X_{ik} is that for taxon K. If all data are binary (0,1), the values will be simply 0,1, and -1, which when squared and square root taken, remove all negative numbers and give values of 0 or 1 for the distance between each pair of OTUs for each character. A related statistic is *Manhattan* (or *City-block*) *Distance*, which yields the absolute number of character state differences between two taxa:

$$D_M = \sum_{i=1}^{n} \left| X_{ij} - X_{ik} \right|$$

Here X_{ij} and X_{ik} represent the values of the *i*th character for every pair of OTUs. This has been applied in many plant groups (e.g., Crisci et al., 1979, in *Bulnesia*, Zygophyllaceae) and it also has been used frequently in cladistic studies (Farris 1970; Nelson and Van Horn 1975), to be discussed later. The *Mean Character Difference* (D_{MC}; Cain and Harrison 1958) is also a measure of "the absolute (positive) values of the differences between the OTU's for each character" (Sneath and Sokal 1973:123), but it differs from D_M by being divided by the maximum value of the character in the data set:

$$D_{MC} = \frac{1}{n} \sum_{i=1}^{n} \left| X_{ij} - X_{ik} \right|$$

A final example of a distance measure is the *Coefficient of Divergence* (D$_{CD}$; Clark 1952):

$$D_{CD} = \left[\frac{1}{n} \sum_{i=1}^{n} \left(\frac{X_{ij} - X_{ik}}{X_{ij} + X_{ik}} \right)^2 \right]^{1/2}$$

Here the absolute character value differences between two taxa are divided by their sum, which gives ratios between 0 and 1. This distance measure was used by Rhodes, Carmer, and Courter (1969) in classification of cultivars of horserad-ish (*Amoracia*, Cruciferae).

Correlation coefficients are frequently used in phenetic studies, as well as having been employed in phytosociology (e.g., Greig-Smith 1983). The most commonly used correlation is the *Pearson Product-Moment Correlation Coeffi-cient:*

$$r_{jk} = \frac{\sum\limits_{i=1}^{n} (X_{ij} - \overline{X}_j)(X_{ik} - \overline{X}_k)}{\sqrt{\sum\limits_{i=1}^{n} (X_{ij} - \overline{X}_j)^2 \sum\limits_{i=1}^{n} (X_{ik} - \overline{X}_k)^2}}$$

Here X_{ij} is the character state value of character i in OTU j, X_j is the mean of all state values for OTU j, and n is the number of characters sampled. This correla-tion measure was used for taxonomic purposes first by Michener and Sokal (1957), but subsequently it has been employed frequently in plant groups, e.g., Soria and Heiser (1961) in the *Solanum nigrum* complex (Solanaceae); Morish-ima (1969) in *Oryza perennis* (rice, Gramineae); Crisci et al. (1979) in *Bulnesia* (Zygophyllaceae); and Stuessy and Crisci (1984a) in *Melampodium* (Compositae). It is especially useful where most of the data exist in more than two states.

Probabilistic similarity coefficients attempt to "take into account the distribu-tions of the frequencies of the character states over the set of OTU's. The philos-ophy here is that agreement among rare character states is a less probable event than agreement for frequent character states and should therefore be weighted more heavily" (Sneath and Sokal 1973:140). These are complex coefficients not used very frequently in phenetic studies. One measure designed to handle two-state, multistate, and quantitative characters is *Goodall's Similarity Index* (Good-all 1964, 1966). This measure determines the overall probability that a pair of OTUs will be as similar or more similar than can be observed for each character based on the distribution of states in all the OTUs (Sneath and Sokal 1973). One example of its use is in the Gramineae (Clifford and Goodall 1967).

Which coefficient of similarity to use in a particular instance is not an easy choice, although it has to do often simply with the practical matter of which computer programs are already available in one's home institution. Certain kinds of data, particularly binary vs. multistate and quantitative, are best han-dled with certain coefficients and not others. "When the measurement scale is such that several possible coefficients may be employed, the choice among coef-ficients is often based on the worker's preference in terms of conceptualization of the similarity measure. Thus distances are preferred by some, and association coefficients are preferred by others" (Sneath and Sokal 1973:146). This eclectic approach has bothered some workers, such as Runemark, who raises caution in the taxonomic use of statistics with chemical data from the perspective that: "An unlimited number of coefficients of association can be used. Therefore almost any hypothesis held by the investigator can be supported, provided a suitable coefficient is selected" (1968:29). This is a useful criticism (expressed also by Weimarck 1972), but it can be overcome by using several measures of relationship before drawing conclusions (see R. P. Adams 1972, 1974a, for pertinent comments). Sneath and Sokal favor use of binary data and a correlation coefficient that is suited to their comparison, "not only because of its simplicity and possible relationship to information theory but also because, if the coding is done correctly, there is the hope that similarity between fundamental units of variation is being estimated. There are also the obvious relations of such similarity measures to natural measures of similarity or distance between fundamental genetic units (amino acid or nucleotide sequences)" (Sneath and Sokal 1973:147). These same authors recommend that in the case of potentially competing coefficients for the same data set, the best approach is to choose the simplest one. This is a form of the commonly applied axiom in science that the simplest explanation is most probably the correct one ("Ockam's Razor," or parsimony; Crisci 1982; Kluge 1984).

Determination of Phenetic Structure. The next step in the phenetic approach to classification is the determination of the taxonomic structure among all the OTUs. That is, some algorithm must be chosen to relate the OTUs to each other using the values already determined by the similarity coefficients. This resulting structure will be numerically calculated and graphically displayed so that decisions on classification can be forthcoming. As with similarity coefficients, many different types of algorithms exist. The different classificatory procedures have themselves been classified in various ways (Williams and Dale 1965; Williams 1971; Sneath and Sokal 1973; Clifford and Stephenson 1975) and no consensus prevails. As outline of the different procedures in the form of a dichotomous key is given in figure 7.2. Two basic approaches will be discussed here: (1) clustering; and (2) ordination. These represent the major alternatives at the present time. Because of the numerous calculations necessary for these methods, computers are almost always used. A simple manual technique with edge-punched cards has been developed by Roe (1974) and can be used as a means of initially becoming familiar with phenetic methods.

The most commonly used clustering methods for biological materials are the hierarchical, agglomerative ones. These occur in three basic types (Sneath and Sokal 1973): single linkage; complete linkage; and average linkage. *Single linkage* (also called the "nearest neighbor" technique, Lance and Williams 1967) is a method in which an OTU is linked to an extant cluster through the most similar included OTU. In *complete linkage* (or "farthest neighbor" clustering, Lance and Williams 1967) an OTU is joined to a cluster based on the greatest similarity with the farthest OTU already within the same cluster. With *average linkage* there is an attempt to relate the new OTU to an average value of the extant group rather than to the extreme similarity or difference within it. There are

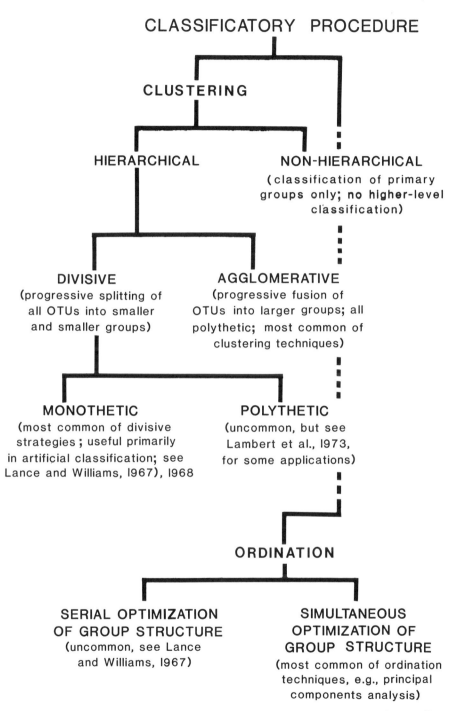

FIGURE 7.2 Outline of major phenetic classificatory algorithms. (Based on Williams 1971:306; Clifford and Stephenson 1975:29; and R. Sokal *in litt.*)

two basic types of average linkage: arithmetic average [both unweighted (UPGMA; this was used in the example shown in fig. 7.1) and weighted (WPGMA)] and centroid (with both weighted and unweighted approaches). In arithmetic averaging the value of a new OTU is related to the average values of all the OTUs already in the cluster. Generally all the OTUs in the extant cluster are accorded equal weight, but sometimes a newcomer (or other OTUs) is given more emphasis (or weight) than others for various reasons. In centroid clustering the new OTU is related to the center OTU which has been calculated as being equidistant in all features from the edges (most extreme value OTUs) of the cluster. Here again, weighting of newly added (or other) OTUs can be employed if so desired.

The most common ordination techniques are those utilizing simultaneous optimization of group structure (fig. 7.2), i.e., the structure in the group is determined all at once rather than by successive comparisons as new OTUs are added. Of these, the principal ones are *principal components analysis* (PCA), *principal coordinate analysis, canonical variate analysis, canonical correlation analysis, factor analysis* (this latter label is sometimes also applied to PCA), and *nonmetric multidimensional scaling* (Clifford and Stephenson 1975). These methods vary in their approaches but they all attempt to calculate multidimensional relationships and to condense them onto a reduced number of planes for more effective visualization and comprehension. For an excellent discussion of several of these methods see Neff and Smith (1979), Neff and Marcus (1980), and Titz (1982). As some workers have stressed, e.g., Sneath and Sokal: "hierarchic classifications often are poor representations of actual phenetic relationships found in nature. Far better representations are often obtained by summarizing the data in an ordination of as few as three dimensions" (1973:201). Moss stresses this point even more forcefully: "Studies on the accuracy of phenograms have shown that these diagrams are extremely poor vehicles by which to represent similarity; in fact, phenograms are generally useful only as rough guides to taxonomic structure. Much more accurate and useful results are obtained from ordination approaches based on multidimensional scaling." (1979:1218). Because the most commonly utilized ordination technique has been principal components analysis, this will be discussed here as an example.

Principal components analysis attempts "to choose axes within the multidimensional space in such a manner that the projections of the entities onto the axes will 'best' display their relationships. The concept of 'best' depends on the outlook of the viewer but it usually means in such a way that the entities are more widely separated one from another in terms of the new than in terms of the original axes" (Clifford and Stephenson 1975:170). The axes of variation (or the factors) presented, therefore, will be selected by their ability to account for the maximum amount of variation among the OTUs of the data set. The percent of variation they actually will account for will vary, but it is not uncommon for 25 to 40 percent of the variation to be contained within the first two factors. Although a great many axes can be calculated by the computer, the major amounts of variation will be accounted for in the first several axes (factors). This is convenient, because for graphic purposes it is most simple to use only two axes (fig. 7.3), and three dimensions are the maximum that can be shown (fig. 7.4; see Rohlf 1968, for other examples of 3-D "stereograms"). The more variation

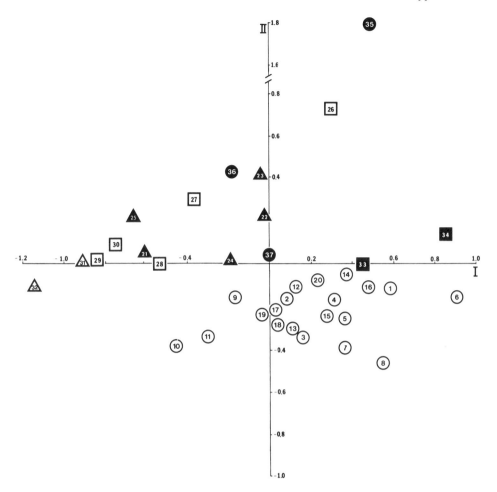

FIGURE 7.3 Principal components analysis of 37 species of *Melampodium* (Compositae) showing relationships of taxa in the first two factor axes. Symbols refer·to taxonomic sections of the genus and numbers refer to individual species. (From Stuessy and Crisci 1984a:12)

contained within the first two or three factors, therefore, the more representative of the real relationships among the OTUs will be the graphic display. PCA has been applied to many plant groups, including *Bulbostylis* (Cyperaceae; Hall, Morton, and Hooper 1976); *Quercus* (Fagaceae; Jensen 1977); *Abies* (Pinaceae; Maze, Parker, and Bradfield 1981); and *Tithonia* (Compositae; La Duke 1982), to name just a few.

Graphic displays must also be selected to express relationships obtained through cluster analysis. Attempts have been made to represent taxonomic structure in multidimensional space (e.g., Jancey 1977), but taxonomists do not relate well to these efforts. In my experience, reality to a practicing plant taxonomist is manifested in 2-D and 3-D images generated during intuitive pattern

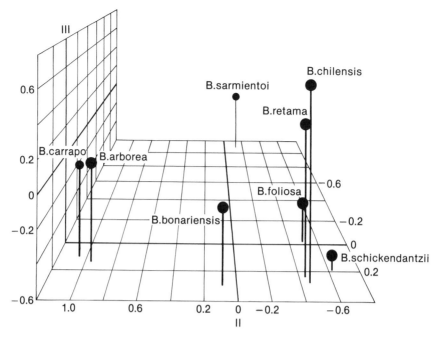

FIGURE 7.4 Principal components analysis of species of *Bulnesia* (Zygophyllaceae) showing relationships among eight taxa in the first three factor axes (total variation accounted for = 80.43 percent). (Redrawn from Crisci et al. 1979:137)

recognition. Graphic displays of phenetic relationships in two or three dimensions, therefore, are usually most effective. Many different kinds of diagrams have been used, and only a representation will be provided here. Phenograms are most commonly used (fig. 7.5). These tend to show close relationships well in contrast to ordination diagrams (figs. 7.3, 7.4, 7.6) which show distant relationships more accurately (Moss 1979). Two-dimensional cluster diagrams of various sorts also have been used, and a wide variety of different types exist (e.g., figs. 7.7–7.9).

Because the representation of taxonomic structure among the OTUs in one, two, or three dimensions necessitates condensation of the relationships expressed in all the dimensions of all characters, a measure of this distortion would be useful. Another way of expressing this is to seek a value that represents the distortion of the relationships in the graphic display from those shown fully between all pairs of OTUs in the matrix of association coefficients. A measure used commonly for this purpose is the *cophenetic correlation coefficient* (Sokal

FIGURE 7.5 Cluster analysis of populations in the *Chenopodium atrovirens* (A), *C. dessicatum* (D), *C. pratericola* (P) complex (Chenopodiaceae). Populations with a question mark were uncertain as to specific identity prior to the phenetic analysis. Dashed line (phenon line) is the suggested level of similarity at which the three series of populations are believed to be specifically distinct. (From Reynolds and Crawford 1980:1358)

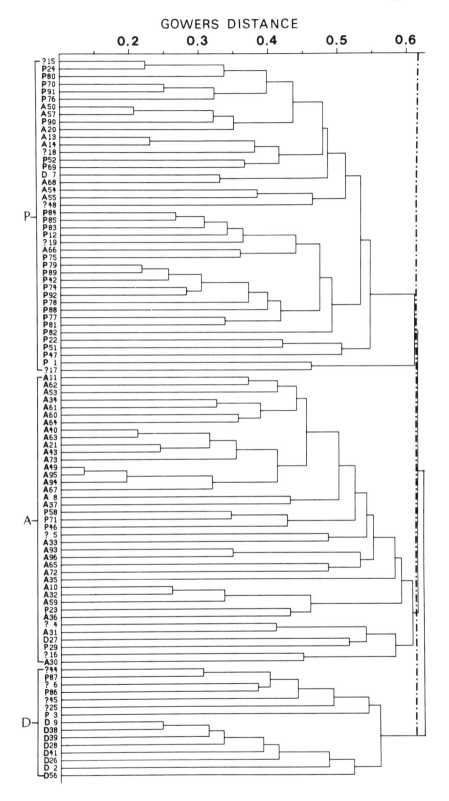

GOWERS DISTANCE

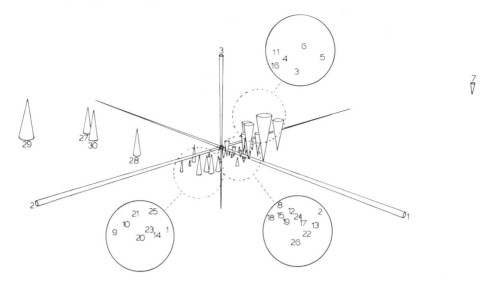

FIGURE 7.6 Three-dimensional plot of factor analysis (IBM Varimax with rotation) of 30 OTUs representing populations of 19 taxa of *Limnanthes* and one of *Floerkea* (Limnanthaceae), based on whole-plant flavonoids. (From Parker and Bohm 1979:196)

and Rohlf 1962). This involves generating a new matrix of values of association derived directly from a graphic display, such as a phenogram. The numerical value of comparison between the two similarity or distance matrices is the product-moment correlation coefficient element by element between them. Values usually range between 1.0 and 0.6. For a phenogram to be a reasonably good representation of a matrix of association, values of 0.85 or higher are desirable. However, a low cophenetic correlation coefficient does not mean that the phenogram (or other graphic display) is invalid, nor that the relationships expressed are lacking in taxonomic value; it only says that distortion has occurred and by how much. It is up to the taxonomist to evaluate the significance of the clustering and/or ordination of the OTUs. Farris (1969) has criticized the use of cophenetic correlation coefficient on several technical grounds, such as sensitivity to cluster size, but it is still regarded by most workers as a useful evaluation for practical work (e.g., Sneath and Sokal 1973).

After the OTUs have been compared and estimates of phenetic relationship have been obtained, the next step in the phenetic approach to classification is to evaluate the expressed relationships to determine the rank of the OTUs. Ranking can be done informally, using such designations as "group," "subgroup," "entity," "cluster," and so on, but usually it involves referring the group recognized phenetically to the categories of the Linnaean hierarchy (e.g., species, genera, etc.). It was hoped in the early days of phenetics that ranking could be derived directly from a phenogram. In the early paper by Michener and Sokal (1957), the latter author preferred "to see uniform and objective standards applied to the recognition of categories" (p. 157), and his approach in this instance was to draw two lines across the phenogram at particular levels of similarity to indicate

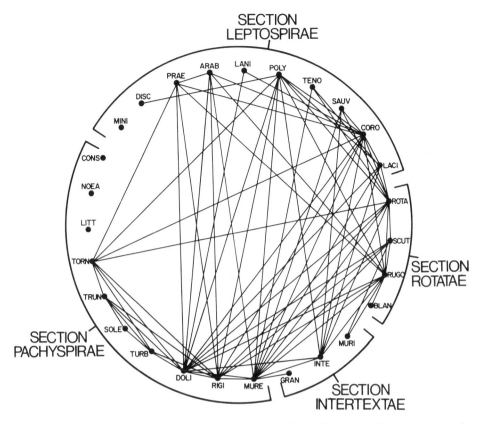

FIGURE 7.7 Maximally connected subgraphs based on cluster analysis showing the relationships of species in subgenus *Spirocarpos* of *Medicago* (Leguminosae) based on phenolic compounds from leaf tissue. A line between two species represents linkage at a dissimilarity level no greater than 0.55. (From Classen, Nozzolillo, and Small 1982:2492)

levels of clustering of genera and subgenera. These are called *phenon lines*. This point of view was reinforced by Sokal and Sneath:

> where a hierarchical tree has been made, the line defining a given rank must be a straight line drawn across it at some one affinity level. The line must not bend up and down according to personal and preconceived whims about the rank of the taxa. We believe that in the foreseeable future each major group will have to be standardized separately. No useful standard can yet be applied to both bees and jellyfish, but within the megachilid bees, or perhaps within all the bees, some worthwhile standardization might obtain. To make this practicable there would have to be agreement on the rank of the whole group under study; we also would have to decide on the rank of the OTU's employed, which will frequently be a category considered to be a species. The other ranks could then be intercalated evenly. (1963:205)

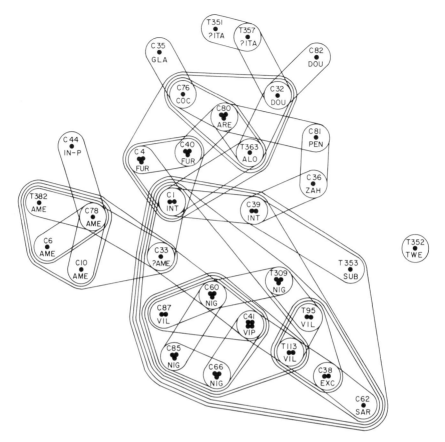

FIGURE 7.8 Cluster analysis of 32 populations (letter and number codes) of 19 species (acronyms) of *Solanum* sect. *Solanum* (Solanaceae) based on a dissimilarity matrix. Dots represent the ploidy level of each taxon (one dot=diploid, two dots=tetraploid, three dots=hexaploid, and four dots=octoploid). OTUs were positioned to facilitate drawing of clusters which are shown by lines enclosing levels of dissimilarity. (From Edmonds 1978:43)

Because of the arbitrary nature of emphasizing lines of similarity to rank OTUs, and because of the obvious variations from group to group and the changes in rank occasioned by shifts in characters, states, coefficients of association, etc., it has now been generally recognized that ranking is not a simple matter solved by drawing lines across a phenogram. The more recent text by Sneath and Sokal offers this more realistic perspective:

> The groups established by numerical taxonomy may, if desired, be equated by the usual rank categories such as genus, tribe, or family. However, these terms have evolutionary, nomenclatural, and other connotations one may wish to avoid. We therefore prefer new expressions (Sneath and Sokal 1962). We call the groups simply *phenons* and preface them with a number indicating the level of resemblance at which they are formed. For example,

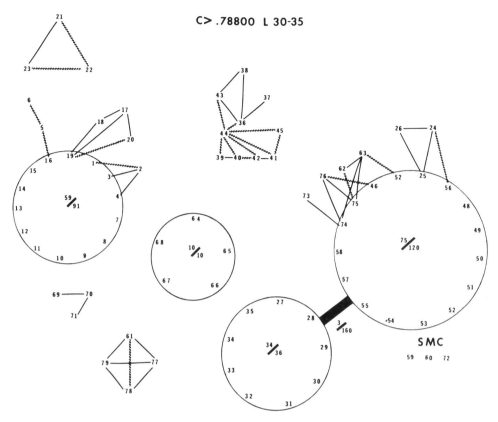

FIGURE 7.9 Two-dimensional plot of results of cluster analysis (called SIMGRA) at similarity value .788 of populations in the *Ranunculus hispidus* complex (Ranunculaceae). Numbers refer to individual populations. Circles represent clusters in which the fraction gives the number of actual similarity connections among the included populations over the total number of possible connections. The wavy lines between populations are connections resulting from the similarity level shown; straight lines are connections made during a previous (higher) similarity level. The dark bar between the two clusters indicates connections (similarity) but not enough for the two to be merged. SMC (=single member clusters) are populations which still have not connected to any others. L=levels of similarity summarized by this figure (total 43 levels were used in this study). (From Duncan 1980:21)

an 80-phenon connotes a group affiliated at no lower than 80 on the similarity scale used in the analysis. Within the context of a given study 80-phenons are a category; *any given* 80-phenon is a group treated as a taxon. ... The term phenon is intended to be general, to cover the groups produced by any form of cluster analysis or from any form of similarity coefficients. Their numerical values will vary with the coefficient, the type of cluster analysis and the sample of characters employed in the study. They are therefore comparable only within the limits of one analysis. Phenons are groups that approach natural taxa more or less closely, and like the term taxon they can be of any hierarchic rank or of indeterminate rank.

> Since they are groups formed by numerical taxonomy, they are not fully
> synonymous with taxa; the term "taxon" is retained for its proper function,
> to indicate any sort of taxonomic group. (1973:294)

The absolute reliance on phenon lines to rank taxa, therefore, has been replaced
by a more tolerant approach in which phenon lines are used as guides to rank
taxa into the traditional Linnaean categories.

IMPACT OF PHENETICS

The development of the phenetic approach to classification has been enormously
stimulating to taxonomy. Old concepts and methods have been scrutinized, new
ideas and techniques have been developed, and new perspectives have been
forged in systematics which are now treated as commonplace. Acrimonious
debates have been replaced by acceptance within narrower and more clearly
defined limits.* Phenetics, which relies heavily on computer assistance for data
analysis, also helped usher in computer applications to taxonomy at a time
when it was just beginning to creep into biology and society as a whole. Computers are now fully established in our everyday lives and destined to become even
more so in the future. In large measure due to phenetic applications, the practicing taxonomist now realizes the positive computational power of the computer
and is beginning to feel more comfortable with it (this positive attitude is being
accelerated now with interactions with word-processing micro-computers for
routine activities such as manuscript and duplicate label preparation). It is now
also recognized that use of the computer in taxonomy is no substitute for sound
judgment; it can only make a good taxonomist better and a bad taxonomist
worse (for a similar balanced view, see Gilmartin 1967).

One of the most important contributions that phenetics has made to taxonomy is the stress on having sound philosophical underpinnings to what we are
doing in classification activities. Most previous work was done intuitively, and
this applies to making classifications as well as to reconstructing phylogenies.
Taxonomy suffered setbacks in prestige and university positions in the 1950s
and 1960s due to attacks from molecular biologists who believed the most
significant work in biology would be done at the cellular and molecular levels
and that taxonomists working with whole organisms in an intuitive mode were
seriously out of step with the modern scientific drummer. We are presently in
another wave of interest in molecular biology. Phenetics offered truly explicit
procedures for classification and was, in that sense, rigorously scientific. As seen
from the inside, however, some practicing workers who were comfortable with

---

*While the intensity of differences of opinion has died back almost completely between early phenetic
practitioners and traditional (phyletic) taxonomists, at the present time debate continues between pheneticists and the cladists who have come on the scene more recently. These are interesting sociological dynamics
which probably reflect the challenge of a new group (cladistics) to the old (phenetics), from which the former
largely developed, and the desire for separation from this background. At the same time cladists are battling
with traditional (phyletic) workers in a fashion similar to what occurred in the early days of phenetics. The
next few years will see the incorporation of useful cladistic ideas (to be discussed in detail in the next
chapter), and the discarding of others, with the intensity of debate again subsiding (pending the next
upheaval)! For a fascinating in-depth analysis of sociological and conceptual developments in both phenetics
and cladistics, see the recent book, by D. L. Hull, *Science As a Process*, Chicago: University of Chicago Press,
1988.

traditional approaches remarked: "No amount of loud proclamations and technical fun and games by numerical pheneticists can turn classification into a science" (Johnson 1970:152). Nonetheless, the fact remains that the efforts of pheneticists did positively project to the outside world a more rigorous posture for classificatory endeavors. Our own philosophical discussions have been so numerous and complex that the criticism of lack of philosophical insight in taxonomy can hardly be hurled by molecular biologists or anyone else. In fact, the plethora of philosophical debate (especially in the pages of *Systematic Zoology*) has caused some to overreact to still more discussion: "I would suggest that instead of having further conferences of this kind to discuss taxonomic philosophies, that we forget about taxonomic philosophy and go back to doing taxonomy. There is a tremendous amount of verbiage which is falling on stony ground" (L. A. S. Johnson 1973:399). This is an especially telling comment because Johnson is one of the most philosophically knowledgeable plant taxonomists in the world, and his critique of numerical taxonomy (1968) is the most detailed and philosophically insightful ever published.

Another very important contribution of phenetics has been the stress on characters. Insights on what is a taxonomic character, what kinds of characters to use, how many, the difference between character and character state, and the importance of examining as many characters as practicable have all been positive results of phenetics. The emphasis on searching for 100 or more meaningful characters, although difficult to achieve in many plant groups (especially in the angiosperms), does cause the practicing taxonomist to seek more character information from his or her organisms whether treated phenetically or not. This can only lead in the long run to classifications with greater stability, predictiveness, and information content. Phenetics also has aided our understanding of correlations of characters. In traditional taxonomy it is difficult (if not impossible) to keep in mind all character state distributions in all taxa in a group under study. Some of the character states may be developmentally correlated, such as length of leaves and length of bracts, or length and width of fruits. Correlation of characters is routinely done in phenetic studies, because it is simple to do by computer from the data already arranged in the basic data matrix (this character by character comparison over all OTUs is called the *R-technique* in contrast to the usual OTU by OTU comparison over all characters which is called the *Q-technique*; Sokal and Sneath, 1963). These analyses allow for a much greater understanding of the characters being used in the taxonomic study. Correlated characters can be identified which may be correlated for a number of reasons: (1) they really are developmentally correlated; (2) they are genetically linked; or (3) they represent adaptive character complexes. General size quantitative characters in the same type of plant structure (such as leaves and bracts) are good possibilities for developmental correlations. Correlations between or among seemingly unrelated features may be due to some kind of genetic linkage either fortuitous or adaptive. If the latter, sometimes clues are evident in the characters themselves, such as reduction in petal length and a shift to self-pollination, or reduction in number of flowers per plant and increase in size and/or number of seeds. Crovello has advocated Key Communality Cluster Analysis "as a context for the generation of ideas and hypotheses on the origin and selective role of the observed character-clusters" (1968b:241), but the same point applies to principal components and other types of phenetic analyses too. In any event, these correlations can be examined and considered for their biological import. This interpretation is most helpful in understanding as broadly as possible the characters and states that are used to construct the classification by whatever means (explicit or otherwise).

Phenetics has also come to be recognized as the prefered way to handle large amounts of data, especially those in which complex patterns of variation prevail. Johnson has remarked: "They, and more traditional statistical approaches, may well illustrate processes or nature of variation and their value is in this field, not in stabilising or regularising formal taxonomic procedures" (1976:158). This has been accepted to the extent that some workers, such as Blair and Turner, comment in reviewing the "integrative approach" to biological classification: "In fact, we have not treated numerical or computer methods as a *new* approach to systematics generally since in all of the approaches listed above [chromosomal, chemical, ultrastructural, etc.], where data assemblage is considerable and complex, numerical treatment is accepted as part of the parcel" (1972:208).

Phenetics is especially useful in the analysis of variation over broad geographical areas. When data are sufficiently large and the distributional relationships so complex that one simply cannot see any meaningful patterns, then the computer analyses are extremely valuable in pointing to potential useful structure. Various statistical methods for this purpose have been developed such as the *simultaneous test procedures* of Gabriel and Sokal (1969) and *contour mapping, shading by overprinting, contoured factor analysis, contoured surface trend analysis* (Adams 1970, 1974b), *canonical trend surface analysis* (Wartenberg 1985), and *spatial autocorrelation analysis* (Sokal and Oden 1978a, b; Sokal, Crovello, and Unnasch 1986). Frequently the questions being asked focus on the infraspecific level with populations scattered over wide areas and varying morphologically (e.g., Jardine and Edmonds 1974) or chemically (e.g., Adams and Turner 1970; Adams 1975b, 1983; Comer, Adams, and van Haverbeke 1982) and sometimes reflecting suspected clinal variation (e.g., Flake, von Rudloff, and Turner 1969) or patterns of hybridization (e.g., Jensen and Eshbaugh 1976; Neff and Smith 1979).

The general acknowledgment of the utility of phenetics with complex data has aided its greater acceptance at the lower levels of the hierarchy (specific and infraspecific) rather than at higher levels (genus and above). Some workers have believed that phenetics is equally effective at all levels: "the numerical approaches to the classification of higher taxonomic categories differ in no ways from those appropriate to lower categories" (Clifford 1977:93–94), but this is simply not true. As higher taxa are compared phenetically, the probability that non-homologous characters and states are being used increases dramatically (Sneath 1976a). At the familial level and above the difficulties become profound (as they also do with any other approach to classification), and it is not surprising that the majority of phenetic studies have dealt with the specific level and below (for a rare and sweeping study of the families of dicots, see Young and Watson 1970). Such a study as that of Young and Watson is so broad that any results might be viewed with suspicion, but it is worth noting on the positive side that the analyses do point to strong clustering of families into dicot subclasses similar to those of the Cronquist-Takhtajan system except for weak differentiation between the Rosiidae and Dilleniidae, a fact which is evident to anyone who has attempted to teach the system and which has even been mentioned by Cronquist (1968, 1981) himself. For a similar approach and results in the monocots, see Clifford (1977).

Although early pheneticists deliberately sidestepped phylogeny as being inappropriate for classification (e.g., Colless 1967), over the years phenetics nonetheless has contributed significantly to efforts to understand evolution both at the populational level and for constructing branching sequences of taxa. In fact, the newer thrust of cladistics has been derived in large measure from much of the earlier phenetic work. It is worth noting that in 1963, Sokal and Sneath stated that: "In developing the principles of numerical taxonomy, we have stressed repeatedly that phylogenetic considerations can have no part in taxonomy and in the classificatory process" (p. 216). This statement was altered in the new edition of their book to read: "As soon as phylogeny becomes a consideration to be dealt with during the classification of a group of organisms, we must be concerned not only with the phenetic relationship among the end points of the branching sequence, but also with phenetic relationships among any points that have at one time or another been occupied by organisms belonging to the phyletic branch under consideration. We have furthermore to concern ourselves with the sequence of branching as well as the time dimension" (Sneath and Sokal 1973:309). This is a substantial change of viewpoint, and it foreshadowed explicit approaches to phyletic (=evolutionary) classification.

The applications of phenetics to solving taxonomic problems have been numerous—too numerous to list comprehensively here. The best recent review of botanical examples is Duncan and Baum (1981), and this should be consulted for particular studies of interest. The extensive general bibliography in Sneath and Sokal (1973) should also be perused. It is important here, however, to give some impression of the breadth of taxonomic problems covered and in what general groups these have been done by citing some recent papers treating the plant kingdom in the traditional broad sense. Numerous studies have been done on bacteria, as might be expected from the obvious utility of presence and absence data in groups for which intuitive *Gestalten* are not easily forthcoming. Some recent examples are: Kaneko and Hashimoto (1982); Shaw and Keddie (1983); Pahlson, Bergqvis, and Forsum (1985); and Quasada et al. (1987). Phenetic studies of fungi have been conducted by Dabinett and Wellman (1973, 1978) and D. S. King (1976, 1977). Algae have been analyzed for phenetic relationships only rarely, by Ducker, Williams, and Lance (1965), Cullimore (1969), McGuire (1984), Rice and Chapman (1985), and the same is true for the bryophytes, studied by Seki (1968), and the pteridophytes, studied by Duek, Sinha, and Muxica (1979). Many investigations have been done with angiosperm taxa. For monocots one can cite Clayton (1971); Melkó (1976); Baum (1977, 1978a, b, c); Badr and Elkington (1978); Davey and Clayton (1978); Dahlgren and Clifford (1981, 1982); Hilu and Wright (1982); and Doebley (1983). As limited examples from the numerous studies on the dicots we can mention: Bisby and Polhill (1973); Crawford and Reynolds (1974); Crisci (1974); Rahn (1974); Pettigrew and Watson (1975); Denton and Moral (1976); Duncan (1980c); La Duke (1982); and Stuessy and Crisci (1984a).

These phenetic investigations have also covered questions at various levels in the hierarchy from the infraspecific level (in *Eucalyptus*, Myrtaceae, Kirkpatrick 1974; and in *Daucus carota*, Umbelliferae, Small 1978b), to the specific level (in *Cannabis*, Cannabaceae, Small, Jui, and Lefkovitch 1976; in *Allium* subg. *Rhiziridium*, Liliaceae, El-Gadi and Elkington 1977; and in *Solanum* sect. *Solanum*, Solanaceae, Edmonds 1978), the infrageneric level (in *Medicago*, Leguminosae, Small 1981), the generic level (in the Chrysobalanaceae, Prance, Rogers, and White 1969; in *Coelorhachis* and *Rhytachne*, Gramineae, Clayton 1970; in yeasts, Campbell 1971, 1972; in Agaricales, Machol and Singer 1971; in the Portulacaceae, McNeill 1975; and in the Triticeae, Gramineae, Baum 1978c), and to higher levels (e.g., orders of Basidiomycetes, Kendrick and Weresub 1966; families of Ericales, Watson, Wiliams, and Lance 1967; among dicot families, Young and Watson 1970; subgroupings within the subtribe Nassauviinae, Compositae, Crisci 1974; and among monocot families, Clifford 1970, 1977).

Numerous other miscellaneous taxonomic applications have also been completed using phenetics. Cultivated plants have been analyzed in this fashion with success, and in fact, these methods are sometimes one of the best ways to make sense of confusing patterns of morphological variation which have been made so by centuries of artificial selection (e.g., Goodman and Bird 1977; Small 1978a, b; Brunken 1979a, b). Along these same lines, Sneath (1976b) has stressed the importance of phenetics in plant breeding. Phenetic analyses have also been used with fossil groups (e.g., Niklas and Gensel 1978), and in some ways this makes good sense for purposes of classification because fossil forms are largely invariant morphologically simply due to the usual small sample size (for an example of a case in which infraspecific variation was assessed, see Stuessy and Irving 1968), which conveniently avoids one of the persistent problems (i.e., variation and plasticity in character states) so common in most extant plant groups. Gingerich (1979a, b) has used phenetics in extinct animal groups with dense fossil records to reveal phylogeny, an approach he calls "*stratophenetics.*" Chemical data have been used with phenetic analyses with good success from the comparisons of presence and absence of compounds, especially with flavonoids (e.g., Challice and Westwood 1973; Hsiao 1973; Mascherpa, 1975; Parker 1976; Parker and Bohm 1979; Classen, Nozzolillo, and Small 1982; and Pryer, Britton, and McNeill 1983) and with monoterpenoids (e.g., Flake, von Rudloff, and Turner 1969; Adams 1983; and Palma-Otal et al., 1983). Macromolecular data are also analyzed phenetically in studies on populational differentiation (e.g., Jensen et al. 1979) and in attempts to reconstruct phylogeny (really more appropriately called caldistics; e.g., Fitch 1977). Phenetic techniques have also been used in studies on plant morphology and anatomy (Li and Phipps 1973; Lubke and Phipps 1973; Hill 1980).

Phenetics has also had a strong impact on related disciplines, and Crovello and Moss (1971) and Sneath and Sokal (1973) offer bibliographies of many of these applications. Suffice it to say that numerous fields have been affected, such as biogeography (Elsal 1985), archaeology (Sneath 1968a), geology (Sneath 1979), psychology (Cattell and Coulter 1966; Prior et al., 1975) and textual criticism (Griffith 1968, 1969).

CHAPTER 8
Cladistic Approach

Some workers have been dissatisfied with both the phyletic approach because of its intuitive (and presumably "less scientific") nature and with phenetics because it has not been based on evolutionary thinking. For example, Bremer and Wanntorp (1978:322) comment on the traditional phyletic approach to classification that "such a system is not falsifiable, not truly part of science according to Popper [a philosopher of science], and in fact more a work of art, and as such highly personal and not repeatable." Farris (1977a:848) has commented that "there does not appear to be any justification for phenetic taxonomy as it is currently practiced." To overcome these problems, workers have sought an explicit approach that directly reflects evolutionary relationships. In response to these concerns, cladistics has developed. Reaction to this new approach has varied, with strong proponents regarding cladistics as heralding a "revolution" in systematics (e.g., Wiley 1981a:1; Kluge 1982:51; Humphries and Funk 1984:323). Others have opined that "cladistics, insofar as it is something reasonable, is nothing new" (Guédès 1982:95).

DEFINITIONS

Cladistics can be defined as the concepts and methods for the determination of branching patterns of evolution (Stuessy 1980). Its use derives from Rensch (1954, 1959) who contrasted two principal modes of evolution: *cladogenesis* (as "kladogenesis"), or the branching events of phylogeny; and *anagenesis,* the progressive change within the same evolutionary line over time (sometimes referred to as *phyletic evolution;* e.g., G. Simpson 1953, 1961; B. Simpson 1973). Another related term is *stasigenesis* (Huxley 1957), in which lineages persist in time without splitting or changing. Cain and Harrison (1960) used cladistics to refer to a relationship expressing recency of common ancestry, or as Sokal and Sneath (1963:220) described it: "Cladistic relationship refers to the paths of the ancestral lineages and therefore describes the sequence of branching of the ancestral lines. . . ." As applied to a method of classification, Mayr (1969c:70) called it "cladism."* Most workers now refer to this approach as cladistics (e.g., Eldredge

---

*Funk and Brooks (1981:v) claim that this was a "term of derision," but it was apparently adopted by Mayr to be compatible with the endings of well established terms for other "theories of classification" discussed in his book (essentialism, nominalism, and empiricism). He now (1974b) favors the term "cladistic."

and Cracraft 1980; Funk and Brooks 1981; Nelson and Platnick 1981; Rieppel 1983a; Duncan and Stuessy 1984). Some workers, however, have preferred the term *"phylogenetic systematics"* or *"phylogenetics"* instead of cladistics (e.g., Hennig 1950, 1965, 1966; Bremer and Wanntorp 1978, 1982; Farris 1979b; Wiley 1981a) to emphasize the reliance on phylogeny for classification. This has created a confusion with the phyletic (or evolutionary) approach, which for the past century has also relied on phylogeny as the basis for classification. It is recommended that this new usage be avoided so as to reduce confusion (for agreement, see Sneath and Sokal 1973; Mayr 1974b, 1985; Michener 1978).

Several terms are unique to cladistics and need to be discussed briefly. Much emphasis is placed in cladistics on distinguishing primitive vs. derived character states, and specialized terms for these conditions have been coined by Hennig (1966) and largely adopted by practicing cladists. *Plesiomorphic* (or plesiomorphous) designates primitive character states* and *apomorphic* (or apomorphous) refers to derived states. Plesiomorphies shared by two or more taxa are called *symplesiomorphies* (or as an adjective, symplesiomorphic), and shared apomorphies are termed *synapomorphies* (as an adjective, synapomorphic). Derived character states (apomorphies) that are found in only one evolutionary line are called *autapomorphies* (as adjective, autapomorphic). A branching diagram (dendrogram) that is constructed by cladistic principles and methods is called a *cladogram* (Camin and Sokal 1965; Mayr 1969c). The taxa compared and evaluated in a cladistic study are usually called OTUs (following phenetic terminology, e.g., Funk and Stuessy 1978) or *EUs* (= evolutionary units; Estabrook 1972; Estabrook and Anderson 1978). The most closely related group cladistically to a taxon is called the *sister group* (Hennig 1966). Two other terms from Hennig, used infrequently but deserving mention, are *semaphoront* and *holomorphy*. The former is "the organism or the individual at a particular point of time, or even better, during a certain, theoretically infinitely small, period of its life. We will call this element of all biological systematics, for the sake of brevity, the *character-bearing semaphoront*" (Hennig 1966:6). This would be useful in groups such as the insects (Hennig was an entomologist), in which dramatic changes in characters occur during ontogeny, but its use with plants is more limited. Holomorphy is the totality of all characters of the semaphoront, including morphology, physiology, ethology (in animals), chemistry, etc.

Other terms are used frequently in cladistics as well as in phyletic approaches to classification, and are profitably mentioned here. The reconstruction of evolutionary history, by whatever means, results in the need to discriminate certain kinds of evolutionary trends. The most important are homology, homoplasy, parallelism, and convergence (the following definitions from Simpson 1961:78, 79). *Homology* is "resemblance due to inheritance from a common ancestry." *Homoplasy* "is resemblance not due to inheritance from a common ancestry," and includes *parallelism* and *convergence*. The former "is the development of similar characters [or states] separately in two or more lineages of common ancestry and on the basis of, or channeled by, characteristics of that ancestry."

---

*Wagner (1984) prefers the term *basimorphic* because it more clearly refers to a basal, or primitive, condition.

Convergence is the development of similar characters or states in different lineages but without a common direct ancestry. Two branches diverging on dendrograms of any type, including cladograms, is called a *dichotomy*, three branches from one point a *trichotomy*, four branches a *tetrachotomy*, and five or more branches a *polychotomy* (= polytomy; Wiley 1981).

A few definitions have been modified by cladists in recent years, and to avoid confusion it is important to understand the alterations in perspectives that these reflect. *Monophyletic* during the past century has referred to groups of organisms that have a common evolutionary ancestor. Recently, this term has been modified to refer only to a group that includes *all* of the descendents of a common ancestor (Hennig 1966). A group that has a common ancestor, but which does not include *all* the descendents of the ancestor, according to this definition is no longer monophyletic but *paraphyletic* (Hennig 1966), and paraphyletic groups to some cladists are not useful for classification (e.g., Platnick 1977b; Farris, Kluge, and Mickevich 1979). There are people who disagree with this viewpoint, however (e.g., Cronquist 1987; Meacham and Duncan 1987), including myself. As Ashlock (1971, 1984) has pointed out, such a restricted concept of complete inclusion is better given a separate term, and he proposed *holophyletic*. As one of the goals of phyletic classification has always been to construct monophyletic groups, such a change in definition obviously can have profound (and irritating) effects. I am in sympathy with Mayr that: "The transfer of such a well-established term as monophyletic to an entirely different concept is as unscientific and unacceptable as if someone were to 'redefine' mass, energy, or gravity, by attaching these terms to entirely new concepts" (1981:516). It is important to maintain the conventional definition of monophyletic, with holophyletic being employed if a more restrictive concept is needed (for agreement, see Bock 1977, and Gauld and Mound 1982).

The terms "character" and "character state" have had clear definitions now for several decades (e.g, Davis and Heywood 1963), due in part to the stress on careful analysis of characters and their divisible properties by the pheneticists in the early 1960s (e.g., Sokal and Sneath 1963). As defined earlier in this book, a character is a feature of an organism that can be used for taxonomic purposes, and a state is a divisible property of that character. In cladistic analysis of any type, the characters and the states of the characters are judged homologous, the evolutionary directionality of the states is determined (i.e., which are primitive and which are derived), and they are then used for the basis of constructing branching sequences. Some cladists, however, reject the use of character state and have redefined character, e.g., "a character is identical with an apomorphy [= derived character state] as defined by Hennig (1966)" (Watrous and Wheeler 1981:5). Such a definition would restrict use of characters in systematics only to Hennigian methods of reconstructing phylogeny. Other recent definitions of character by some cladists are: "an original form plus all of its subsequent modifications" (Watrous and Wheeler 1981:4), or "a unit of 'sameness' " (Platnick 1979:543).

Homologous is a term that has always been difficult to define (e.g., see discussion in Davis and Heywood 1963) primarily due to the danger of circularity (Hull 1967). Structures are homologous that are structurally and developmentally the same and that have derived these similarities from a common ancestor (Simpson 1961). However, this term has been modified recently by Wiley to read: "Two characters are homologues if one is derived directly from the other. . . . Such a pair of homologues is termed an evolutionary transformation series. The original, preexisting character is the plesiomorphic member of the pair." (1981a:9). This is a completely different definition of homologous and refers to different states of the same character rather than to the same states of the same character but in different taxa. To avoid further confusion, the original meaning should be retained.

HISTORY OF CLADISTICS

Because the origins of cladistics are relatively recent, it seems appropriate to sketch these events in some detail with special mention of how they developed in botany (following Duncan and Stuessy 1985). The earliest cladistic methods developed from needs to determine the shortest routes between points, e.g., so that the smallest amount of cable would be used between telephone stations (Kruskal 1956; Prim 1957). These techniques were picked up and used by phenetic workers, and the networks of relationships that were derived were phenetic in the sense of being based on many unweighted characters. The ideas of patterns of branching relationships among taxa were developed, however, that led to cladistics in later applications (by being rooted to make trees).

In the mid-1960s, developments in cladistics were rapid. Edwards and Cavalli-Sforza (1964) offered the first "method of minimum evolution" using continuous human blood-group data. In the following year, Camin and Sokal (1965) conceived the first numerical cladistic technique for discrete data. Wilson (1965) emphasized the selection of unique and unreversed character states in his "consistency test" for constructing cladograms. Throckmorton (1965) provided the first clear contrast between phenetic and cladistic approaches (in *Drosophila).*

In 1965 Willi Hennig (1913–1976) of Germany published his first summary paper in English of his own manual cladistic method. Hennig earlier (1950) had published fully his ideas on cladistics in German, but the impact on the English-speaking systematics community was negligible. The appearance in 1966 of the English translation (and revision) of his book, *Phylogenetic Systematics,* marked the first fully documented statement of the philosophy and methods of cladistics done manually by shared derived character states (synapomorphies). Although Hennig's book is poorly written (obscure, repetitive, poorly organized, and with numerous digressions), it had a strong impact because it was a fully reasoned exposition of his philosophy and methods (see Kavanaugh 1972, 1978, for readable summaries). Gareth Nelson of the American Museum of Natural History must be credited for bringing much of the attention to Hennig's ideas (e.g., Nelson 1971, 1972a, b, 1973c). For a full assessment of Hennig's contributions, see Dupuis (1984).

While these events were occurring, W. H. Wagner at the University of Michigan had developed in the 1950s another manual cladistic technique especially for use in teaching phylogeny in the classroom. This method was first published by his student, James Hardin, in 1957, and then by Wagner himself in 1961 and more fully elaborated in 1962 (and 1980). His *Groundplan/divergence* method was used by other students and workers (e.g., Mickel 1962; Scora 1967; and Fryxell 1971) and represented the principal thrust of cladistics among botanists at that time. Coincidentally, this method has proven virtually identical to Hennig's except for differences in some of the assumptions regarding primitive character states and in the final graphic display. Both methods rely entirely on shared derived character states for delimiting groups and showing lines of affinity (for agreement, see Churchill and Wiley 1980; Churchill, Wiley, and Hauser 1984).

During the late 1960s, developments in cladistics proceeded even more rapidly. In 1967 Walter Fitch of the University of Wisconsin at Madison and Emanuel Margoliash, then of Abbott Laboratories, published a method of tree construction using molecular sequence data (amino acid sequences of cytochrome *c*) which was essentially the building of a network that was then rooted by reference to other data, such as the fossil record. Estabrook (1968) refined the Camin-Sokal approach by offering a mathematical solution to the problem of selecting the most efficient (or parsimonious) trees, and further improvement was made by Nastansky, Selkow, and Stewart (1974). The first generalized method (based on Manhattan distance, originally from Kruskal 1956, and Prim 1957) for numerical cladistics was presented by Kluge and Farris (1969) under the rubric of *"quantitative phyletics."* These so-called "Wagner methods"* were elaborated by James Steve Farris in a paper the following year (1970), and a good example of their application is found in Farris, Kluge, and Eckardt (1970a). Whiffin and Bierner (1972) and Nelson and Van Horn (1975) both offered manual adaptations of the Farris tree algorithms. More recently Jensen (1981) has summarized many of these methods.

Solbrig (1970a) used the Prim-Kruskal and Wagner groundplan/divergence methods to reconstruct the phylogeny of *Gutierrezia* (Compositae). This laid the foundation for the "eclectic" approach to cladistics (especially common among botanists) which encouraged use of different methods to gain the maximum insights on the phylogeny of a group (e.g., Funk and Stuessy 1978; Duncan 1980a).

LeQuesne (1969) stressed selection of the "uniquely derived character" for reconstruction of phylogeny, and Estabrook developed the idea of using a suite of compatible characters that have uniquely derived states (Estabrook 1972; Estabrook, Johnson, and McMorris 1975, 1976a, b). This led to the character compatibility approach to phylogeny reconstruction (e.g., Estabrook, Strauch, and Fiala 1977; Estabrook 1978; Estabrook and Anderson 1978; Gardner and La Duke 1978; La Duke and Crawford 1979; Duncan 1980b; Meacham 1980). Meacham (1981) has presented a summary of how character compatibility can be done manually.

Felsenstein (1973) offered statistical techniques of maximum likelihood for tree construction, following up on suggestions made earlier by Edwards and Cavalli-Sforza (1964), but this time using discrete characters. This has been

---

*Funk and Stuessy (1978) point out that Wagner's Groundplan/divergence method (1961,1962) was based on shared derived character states (as with Hennig 1966) and not on a distance measure. To avoid confusion it seems best to call trees generated by these distance measures *Farris trees* rather than Wagner trees.

followed more recently by papers comparing the various methods (especially parsimony and character compatibility) and by stressing the value of the maximum likelihood algorithm (Felsenstein 1978, 1979, 1981, 1984).

During the late 1970s, many events took place, and the period was essentially one of polarization of workers with differing viewpoints. In 1977 the first symposium (actually a mini-symposium consisting of contributed papers with a common theme) on cladistics for botanists was held at the AIBS meetings in East Lansing, Michigan, with proceedings published in 1978 (Stuessy and Estabrook 1978). The second symposium aimed at botanists was held in 1979 at the AIBS meetings in Stillwater, Oklahoma (published the following year; Stuessy 1980). In 1979 appeared the second book on cladistics, *Phylogenetic Analysis and Paleontology*, edited by Joel Cracraft and Niles Eldredge, but this time with a focus on fossil forms. In November of that year, the annual Numerical Taxonomy meeting was held at Harvard and very strong points of view were voiced. The disagreement between pheneticists and certain cladists was especially acrimonious. As a result, in 1980 many of the more zealous cladistic workers established the new Willi Hennig Society, and its first meeting was held in Lawrence, Kansas. (Interestingly enough, this was also the birthplace of phenetics!)

In the early 1980s an even greater polarization of hard-line cladists and other systematists occurred. This period also saw the publication of many books on the subject. In 1980 there appeared *Phylogenetic Patterns and the Evolutionary Process* by Eldredge and Cracraft. Four more books on cladistics followed: *Advances in Cladistics* (which contained the results of the First Hennig Society Meeting in 1980; Funk and Brooks 1981); *Phylogenetics: The Theory and Practice of Phylogenetic Systematics* (Wiley 1981a); *Systematics and Biogeography: Cladistics and Vicariance* (Nelson and Platnick 1981); and *Insect Phylogeny* (Hennig 1981), an English translation of the original book of 1969. A symposium on phylogenetic studies was held in Cambridge, England, in April 1980 and the proceedings were published in 1982 in two parts as a book, *Problems of Phylogenetic Reconstruction* (Joysey and Friday 1982) and as a journal run (Patterson 1982b) under the title *Methods of Phylogenetic Reconstruction*. Another symposium was held at Berkeley in April 1981 which covered broad aspects of cladistics, published as *Cladistics: Perspectives on the Reconstruction of Evolutionary History* (Duncan and Stuessy 1984). Volume two of *Advances in Cladistics* (Platnick and Funk 1983) has appeared, as has the text by Rieppel (1983a). A new journal, *Cladistics*, sponsored by the Willi Hennig Society, has begun publication as of 1985. There has also appeared a Benchmark collection of classical papers on cladistics entitled *Cladistic Theory and Methodology* (Duncan and Stuessy 1985). Ridley (1986) has a useful text with a glossary, which compares schools of classification and evaluates the status of pattern (or transformed) cladistics. The book by Schoch (1986), despite its title, *Phylogeny Reconstruction in Paleontology*, is the most balanced cladistics treatment yet published, even though the author is himself of the more rigid school. Another interesting series of symposium papers on *Biological Metaphor and Cladistic Classification* (Hoenigswald and Wiener, 1987) has appeared.

At the present time, therefore, cladistics has attracted much attention and has taken its place as one of the major approaches to biological classification. There are many different viewpoints among cladists on nearly every aspect of the procedures and even on the concepts and philosophy behind the methods. Some workers are rigid to the point of being dogmatic and even evangelical (especially the adherents of "New York cladism;" Van Valen, 1978a); others are eclectic (e.g., Duncan, 1980a), and some have even come to view cladistics as a system of organizing information that does not even necessarily have to deal with evolution. These are the "transformed cladists" (e.g., Platnick 1979; Patterson 1980), also called the "natural order systematists" (Charig 1982:369), or "pattern cladists" (Beatty 1982; Brady 1985; Kemp 1985; Platnick 1985). Saether (1986) has used the term *"neocladistics"* to include all non-Hennigian methods but this term was used earlier by Cartmill (1981:73). There is also some obvious strong disagreement among some practitioners (e.g., Farris 1985), and in some ways the approaches are more diverse now than ever before. Attempting to digest these different viewpoints and present a unified perspective is difficult, if not impossible, at this time. What will of necessity be done, therefore, is to offer a balanced perspective on cladistics that gives a good introduction to the more generally held ideas as well as enough diversity of opinion to show some of the areas of contention and controversy.

METHODOLOGY OF CLADISTICS

Depending upon the type of data and algorithms for tree construction that are used, the procedures for cladistic analysis and classification will vary. Nonetheless, there is a general set of procedures that are used by most workers with conventional types of data, especially morphology (Stuessy 1980):

1. Make evolutionary assumptions (select EU's, determine monophyletic groupings, etc.);
2. Select characters of evolutionary interest;
3. Describe and/or measure character states;
4. Ascertain homologies of characters and character states;
5. Construct character state networks;
6. Determine polarity of character state networks (primitive vs. derived conditions); i.e., "root" the character state networks to form character state trees;
7. Construct basic data matrix;
8. Select algorithm and generate trees (cladograms); and
9. Construct classification based upon cladograms.

Introductory Example

To help clarify these steps, a simple example will be used. Six taxa (EUs) will be considered, arbitrarily labeled S, T, W, X, Y, and Z. Assumptions must be made before the analysis can begin, such as the initial rank of each of the EUs (i.e., whether populations, species, genera, etc.), and that the six EUs do form a monophyletic unit. The characters that appear to provide information of evolutionary significance are selected, and in the example used here, ten characters have been used (nos. 1–10). In practice it is usually more, but it must be minimally one character less than the number of EUs for all branches of the tree to be resolved. The characters then must be divided into states, and usually these are qualitative, although this does not have to be the case (e.g., Stuessy 1979a). A check on the homologies of the characters and states at this point is essential to be certain that the same kinds of features are being compared. The next step is to place the character states in a logical sequence based primarily on parsimony of state changes, i.e., the simplest way possible. This is obvious with only two states, but with three or more, it becomes more complicated and more discerning judgment is needed. For example, if three states of corolla color (e.g., white, red, and pink) occur in the study group, the simplest way to connect the states logically is red—pink—white rather than white—red—pink. The degree of color change between each state is the smallest in the first approach and more drastic in the second. The *character state network* must be "rooted" to form a *character state tree* by deciding which state is most primitive. Numerous criteria exist to help guide in this determination, and these will be reviewed in detail later. Suffice it to say that by applying some concepts of primitiveness, one of the states (e.g., white, in this example) is selected as most primitive. This roots the network and turns it into a tree. The states are then placed in a basic data matrix, coded to reflect their primitive or derived status (usually P vs. D, 0 vs. 1, - vs. +, etc.; table 8.1). The most commonly used algorithm (= procedure) for constructing the rooted tree of taxa (cladogram) is by shared derived character states. Deriving this from the hypothetical data, and beginning with the EU with the fewest derived states (Y), gives the cladogram in figure 8.1. Changes in characters 1, 2, 4, and 5 to the derived state are synapomorphies (shared derived character states) and changes in all the other characters are autapomorphies (derived character states in the line to one taxon only). The terminal EUs can be rotated at their branching points (e.g., EU X could be drawn on the left with EU W on the right, instead of as shown). Construction of a classification involves recognizing groups and establishing relative ranking followed by absolute ranking with reference to the Linnaean hierarchy (Hennig 1966). In the example provided, Y–Z form one group coordinate with the collective group S–T and W–X. These latter two pairs of EUs each represent a subgroup within the more inclusive taxon S–T–W–X. Exactly how these EUs would be absolutely ranked depends upon many factors such as their initial rank (i.e., individuals, populations, species, genera, families, etc.), the nature of the characters and states used, historical precedence in ranking within the group (and in related groups), and so on. If the EUs in this instance represent species, it *might* be the case that two genera could be recognized, with Y and Z in one genus and S–X in another, with two subgenera being recognized in the latter. On the other hand, all could be treated as a single genus with two subgenera being recognized initially and two sections within S–X. The main point is that the characters and states are clearly indicated, the reasons for polarity determination are stated, the manner of cladogram construction is explicit, and the classification is derived directly from the cladogram. This is truly an explicit approach to classification that is at the same time based on evolutionary reasoning. This particular example has followed more the less the methods of Hennig (1966).

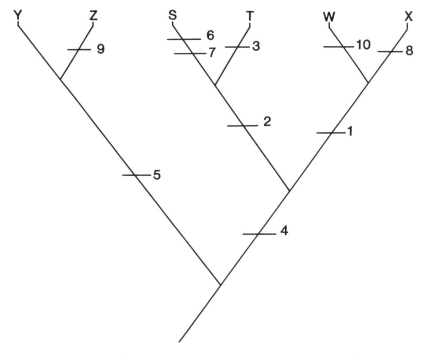

FIGURE 8.1 Cladogram illustrating relationships among EUs S–Z based on binary characters 1–10. Place of change from primitive to derived condition for each character shown by horizontal bar and number.

Another simple example with the same data and EUs uses the Groundplan/divergence method of Wagner (1961, 1962, 1980), which is virtually identical to that of Hennig.* As this has been the manual method used most frequently by botanists to date (Funk and Stuessy 1978; see also references in Funk and Wagner 1982), its use as a further example seems justified. The characters and states are again put in a basic data matrix for EUs, and again the primitive states are assigned a value of "0" and the derived states a value of "1" (table 8.1). In addition, a total of the derived states for each taxon is calculated: S, 4; T, 3; W, 3; X, 3; Y, 1; and Z, 2. The format of graphic display (fig. 8.2) is to use a series of concentric hemicircles, with each representing a level of total derived character states going from the center (most primitive) to the outside (most derived). The EUs are placed on the hemicircles based on their total number of derived character states, and EUs at different levels are joined by lines based upon the maximum number of shared derived states. The exact position of the taxa laterally on each of the hemicircles is arbitrary in Wagner's (1980) approach, but recently Emig (1985) has recommended positioning based on the number of apomorphies scaled from the horizontal. The relationships shown correspond, therefore, to those in figure 8.1 with the exception that the relative advancement

---

*There is no difference between the two methods in the basic aspects. There is a conspicuous difference in the final graphic display, but the cladistic relationships shown are identical (see Churchill and Wiley 1980, for concurrence).

TABLE 8.1 Basic Data Matrix of Binary (Two-State) Characters 1–10 in EUs S–Z for Cladistic Analysis. 1 = Derived State; 0 = Primitive State.

EU	\multicolumn{10}{c}{Character}

EU	1	2	3	4	5	6	7	8	9	10
S	0	1	0	1	0	1	1	0	0	0
T	0	1	1	1	0	0	0	0	0	0
W	1	0	0	1	0	0	0	0	0	1
X	1	0	0	1	0	0	0	1	0	0
Y	0	0	0	0	1	0	0	0	0	0
Z	0	0	0	0	1	0	0	0	1	0

of each EU is more conspicuously displayed. The same measure of evolutionary advancement for each EU (the patristic distance) can also be obtained from figure 8.1 simply by counting the number of shared derived character state changes on the cladogram leading up to each EU.

Evolutionary Assumptions

Having presented a simple example of a cladistic analysis, it is now appropriate to turn attention to details of the process. The first issue is to make evolutionary assumptions without which any cladistic analysis cannot justifiably be at-

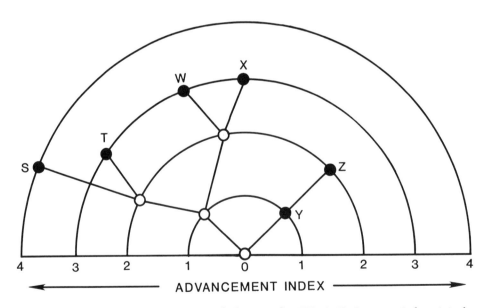

FIGURE 8.2 Groundplan/divergence cladogram for EUs S–Z. Concentric hemicircles indicate relative degrees of ancestral (0) to derived (4) conditions (Advancement Index). Open circles represent presumed extinct or unknown ancestors.

tempted. EUs must be selected that represent appropriate units for the questions being asked, and often these are species. Another important concern is that the entire group of EUs being considered represents a monophyletic unit. These problems may appear simple on the surface, but they are very difficult to resolve with confidence. Many groups are so poorly understood that a basic evolutionary understanding of the more inclusive as well as the individual groups simply is not yet available (this is common in many families of angiosperms). This point is worth stressing, because cladistic analysis is not easy to do (nor should it probably even be attempted) in taxa that are very poorly known and for which no clear generic-level understanding is available. Because cladistics cannot be used to determine the individual monophyletic units, due to obvious problems with circularity, a phenetic (or intuitive phyletic) approach is first employed. These defined groups then become the EUs for the explicit cladistic analysis. After the cladistic analysis, some refinement may occur of initial concepts of which EUs in the study group are really monophyletic. This is useful feedback information which allows for an improved second round of cladistic analysis.

Selection of Characters

The selection of characters of evolutionary import for cladistic analysis is a very important step that to date has received very little attention in the literature. The principal viewpoints so far have been offered by Sober (1986) and in the symposium organized by Funk and Wheeler (1986) with papers by Neff (1986), Shaffer (1986), and Wheeler (1986). Because few characters usually are employed in cladistic analysis (at least few in contrast to phenetics), this is clearly a most important step in the whole approach. Characters are usually selected for which primitive and derived states can be recognized with confidence. This is based on the distribution of states within the study group and in related taxa and other considerations (to be discussed in more detail below). One way to begin is to first consider all the characters which have been used historically within a group and which have been regarded as taxonomically significant. To these can be added any and all other features of the organisms (morphological, anatomical, chemical, etc.) that are revealed through careful study of the group. One way to force a careful and exhaustive examination of many characters is to complete a phenetic study of the group first before attempting the cladistic analysis.* This has the value not only of insuring study of many characters, but also of generating ideas about groups that *may* be monophyletic in an evolutionary context. In fact, in groups that are very poorly known, it makes good sense to complete a detailed phenetic analysis before the cladistic work is attempted. Histograms of characters and their state distributions can also be generated easily by the computer from the data used in the phenetic analysis via any number of standard statistical packages (e.g., BMDP; Dixon 1981) to reveal more clearly the characters that

---

*The use of phenetics and cladistics together (e.g., Varadarajan and Gilmartin 1983; Geesink 1984; Burgman 1985) has the added benefit of allowing a comparison and contrast of the results of the two approaches and for a synthesis of these measures of relationship in more explicit reconstructions of phylogeny (Stuessy 1983; more on this in chapter 9). It is fair to point out, however, that the more zealous cladistic workers would abhor such a suggestion.

may be important to use in the cladistic analysis. Characters that have states clearly distributed into discrete units and which also occur in related groups outside the study group are usually the best to choose. Quantitative characters can also be used (e.g., Stuessy 1979a), but usually these are transformed into discrete units by "generalized gap-coding" (Archie 1985) or other methods (Goldman 1988). Characters that show high levels of conservatism (Sober 1986; or low levels of variance within populations; Farris 1966) are usually best (this it true with any approach to classification). Characters which show high genetic heritabilities have also been advocated (Shaffer 1986). The importance of this step in cladistic analysis cannot be overemphasized, because the characters employed usually are few relative to the number of EUs and thus the inclusion or exclusion of one character can have a dramatic effect on the construction of the final cladogram upon which the classification might be based. Doyle and Donoghue (1986b) used 62 characters in their analysis of angiosperm origins, but this is an exceptionally high number (but desirable *if* many evolutionarily significant characters can be found).

Homology

Also important in the selection of characters is determination at some level of satisfaction that the same characters in each of the EUs are homologous. The concept of homology is most difficult philosophically, and it has been a problem to provide an acceptible definition (Sattler 1964). As Davis and Heywood aptly stress: "Systematics is widely pervaded and influenced by the concept of homology. Both the affinity of groups and considerations of phylogeny are based upon supposed homologies. As this is the case, it is a serious flaw that the terms homology and analogy are so difficult to define, or at least their results are so difficult to distinguish" (1963:40). Characters in two or more EUs are homologous in an evolutionary sense if they represent the same structure which has descended from the immediate common ancestor. The difficulty lies with the implied circularity that to be certain of homologies presupposes knowledge of the phylogeny which vitiates the need for its reconstruction. Bock (1969, 1973, 1977) emphasizes that the circularity is broken by defining homology in terms of phylogeny and defining the latter in terms of evolution. This helps, but evolution itself is in part defined by phylogenetic considerations such as the idea of change through time. Homology is clearly recognized or "tested" by shared similarity (Bock 1977). In a practical sense the detection of homologies lessens due to the feedback from choosing new characters, examining the new reconstructed phylogenies, and so on until a satisfactory view of homology of the utilized characters emerges (for additional comments see p. 35 under the discussion of homologous characters). There is really no alternative but to look very carefully and thoroughly for structural and developmental similarities among the characters prior to their use in cladistic analysis. As Sattler puts it: "two organs are homologous only when they are essentially similar (or identical)" (1966:424). Even while acknowledging their importance for cladistic studies, some authors (e.g., Kaplan 1984) prefer to regard homology as only structural and developmental similarity without attributing common ancestry. This agnostic concept with regard to ancestry has been called *parology* by Hunter (1964).

Character State Networks

After the characters have been selected and homologies ascertained, they must be divided into character states. Here again, the homologies of the states must be determined in the same fashion as with characters. It is conceivable that chosen characters could well have descended from the immediate common ancestor, but parallel developments in the evolution of state diversity in the different populations of EUs could cause resemblances that were analogous rather than homologous. It is, in effect, the same problem but at another level. Only a careful search for perceived structural and developmental similarities can be used to detect the homologies of the states. Congruence of states among several presumptively homologous characters can also be helpful (Patterson 1982a). With the homologies of the states in different EUs affirmed, the next step is to order the states among all EUs of the study group into a *character state network* (fig. 8.3; = *morphocline;* Maslin 1952; or *phenocline;* Ross 1974). As discussed earlier, this is the establishment of a logical connection among the states of the characters that presumably represents the evolutionary sequence within the study group. Parsimony, or the simplest logical arrangement of states, is a guiding criterion here. This does not necessarily mean that the evolution of states in a particular collection of EUs has gone that way, but if there is no evidence to the contrary it is a useful viewpoint and way of proceeding with the analysis. As Bock has correctly noted: "the arrangement of features into a transformation series [= morphocline] depends upon our judgment of how the features could change during evolution. Transformation series are not arranged by chance or by the caprice of the investigator" (1977:885). Cladistically significant characters are often judged to be binary, but sometimes they are best treated as multistate. In that case, construction of the character state networks becomes more complex, and numerical codings must be assigned according to some stated procedure. Several options exist; for a review see O'Grady and Deets (1987).

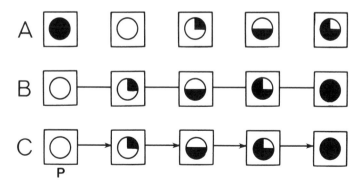

FIGURE 8.3 Hypothetical states of one character (A; e.g., types of leaf margin: entire, sinuate, crenate, dentate, serrate, etc.) arranged into a character state network (B), and a character state tree (C) by rooting the network at the state on the left (P=primitive). Quadrangles refer to taxa, circles to one character, and shaded variations to character states of that character.

Character State Trees: Determination of Polarity

The determination of primitive character states to root the character state networks and turn them into character state trees is another difficult yet important feature of cladistic analysis. Many earlier works grappled with this problem in phyletic approaches to classification (e.g., Frost 1930a, b, 1931; Sporne 1948, 1954, 1956; Danser 1950; Maslin 1952; Eyde 1971). Recently several papers have reviewed these critieria for determining primitiveness with special reference to cladistics (Crisci and Stuessy 1980; Jong 1980; Stevens 1980a; Arnold 1981; Watrous and Wheeler 1981; Wheeler 1981; Stuessy and Crisci 1984b). Different opinions prevail as to what criteria are valid for the recognition of primitive conditions of character states. The perspective offered here stressing an eclectic approach is believed the most reasonable. Nine criteria can be used to determine primitiveness (fig. 8.4). Of these the first six are called "first level" (Crisci and Stuessy 1980) because they require no prior knowledge of evolutionary directionality. The first three criteria deal with distributions of character states within and outside the study group, whereas the second three stress developmental

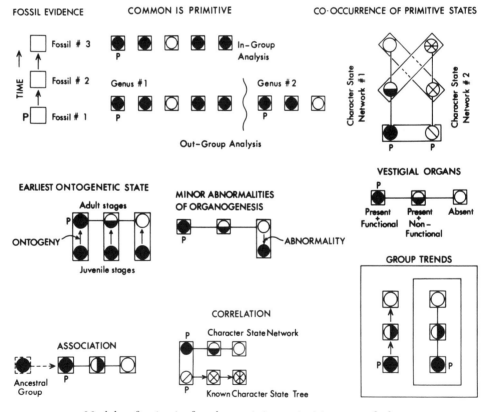

FIGURE 8.4 Models of criteria for determining primitiveness of character states. Meaning of symbols as in figure 8.3. (From Crisci and Stuessy 1980:118)

considerations. The last three criteria all rely on having some determination of primitiveness already established. Other criteria no doubt can be validly formulated. Every group has unique features of its own, not only as to structures, but also as to modes of speciation and the shape of the actual phylogeny. Different critieria, therefore, will have different resolving powers depending upon the group being studied. The more known about the biology of the group, the more carefully the criteria can be selected and applied, and the closer to the real phylogeny will be the reconstruction.* Details of these criteria are presented in Crisci and Stuessy (1980; see also Stuessy and Crisci 1984b) and will not be repeated here, but a brief survey is given below. If the discussion seems more detailed than that for other aspects of cladistic analysis, this is deliberate. Without sound reasoning at this early point, the construction of branching sequences cannot be done effectively.

The criteria for determining primitiveness all have one thing in common; none is infallable. The only concept which must always be true is that the oldest character state is the most primitive. Although this is a useful philosophical perspective, it is not terribly helpful in practice because we never know with certainty which state of a character is the oldest. Even if we have an extensive fossil record, we are not certain that it is the oldest fossil that exists (or oldest form in the lineage that ever existed). At least it does give us the minimum age of the group and its character states. All other criteria will work well in some instances and fail in others. Each one "will not be absolute or universal but simply a statement of high explanatory value in which exceptions should be expected and tolerated. It is impossible to do better than this" (Crisci and Stuessy 1980:116).

The first criterion for determining primitiveness is that in a series of fossils the stratigraphically oldest fossil will be regarded as the most primitive of the lineage, and all of its states will be treated as primitive ones. Obviously the more fossils that exist for a group, the stronger will be the probability that this criterion holds true. Even this criterion, however, is not free from problems. Difficulties lie in the quality of preservation of features in forms at different time zones and especially in the sampling error that may be represented (see Raup 1979, for a good discussion of biases inherent in the fossil record). Schaeffer, Hecht, and Eldredge (1972) and Wheeler (1981) stress that fossils *can* be misleading, especially if only few are at hand, because their position on the actual

---

*Some workers have reacted very negatively to this approach. There seems to be a great desire to have one criterion of polarity (usually out-group analysis) that will explain all polarities in every character in all groups. The idea of exceptions to any criterion seems bothersome to some, e.g., Stevens: "If all criteria have exceptions then everything has to be qualified, caveats added, arms waved vigorously, and yet, when all is said and done, there is no way of knowing either how to proceed in any one instance or if the case under study is really one of the exceptions" (1981:188). The point that must be remembered is that each group has had a different evolutionary history and different mechanisms of speciation (see White 1978, or Grant 1981, for numerous examples of the diversity of speciation processes). The same criterion of investigating the evolution of characters and states cannot possibly reveal phylogeny equally well in all groups. Depending upon the nature of speciation mechanisms (including phyletic and reticulate evolution), the degree of extinctions, the parallelisms and reversals of character states, the background of phenetic and cladistic understanding already at hand for the group and its near relatives, and the level of biological insights on characters, some criteria will more closely reveal the true character state evolution than others. As the true phylogeny is not known, and cannot ever be known with more than the lowest level of probability, we can only bring as many different insights as possible to bear on a particular group and state clearly those ideas that we decide to apply. Only in this fashion can we maximize approximation of the true phylogeny.

phylogeny is unknown. A very young fossil may have been the terminus of a very slowly evolving lineage and hence still closely resembles the ancestor of the complex (i.e., it may still have many primitive features). An older fossil, on the other hand, may be off the main line of evolution of the group and, although stratigraphically older, may have derived features due to rapid divergence. Some workers, therefore, suggest that fossils cannot serve as reliable indicators of phylogeny: "time *by itself* is no proof of primitiveness of a character state or ancestral position of a taxon" (Hecht and Edwards 1977:11; see also Nelson 1978). Patterson comments that "I conclude that instances of fossils overturning theories of relationship based on Recent organisms are very rare, and may be nonexistent. It follows that the widespread belief that fossils are the only, or best, means of determining evolutionary relationships is a myth" (1981:218). I agree that caution is needed with use of fossils just as with any other criterion for determining polarity, but I believe strongly that it is a very useful criterion, nonetheless (see Delevoryas 1969, for many positive comments on the utility of fossils for determining origins and relationships within different plant groups). Bock points out correctly: "As a general rule the resolving power of this test is dependent upon the age span between the stratigraphic position of the organism showing the plesiomorphic condition and that of the organism showing the apomorphic condition with respect to the total age of the group" (1977:887). The more fossils one has, and the closer the gaps in time between them, the higher is the probability that the true phylogeny will be represented.

The second criterion deals with commonality of character states within study groups and within the closest relatives. These concepts, more than any other, have generated considerable discussion and dogmatic positions. There are two main concepts. *In-group analysis* involves looking at the distribution of character states among EUs of the study group, and those features most prevalent are judged to be most primitive. *Out-group* (or "ex-group;" Ross 1974) *analysis* involves looking at the distribution of character states in the most closely related group (or groups) at the same rank (including but not necessarily restricted to the sister group) and the common (ideally, the uniform; Saether 1986) conditions there are regarded as the most primitive for the study group. Numerous opinions have been offered on the validity of one or the other of these concepts (e.g., more recently Crisci and Stuessy 1980; Stevens 1980a; Wagner 1980; Watrous and Wheeler 1981; Wheeler 1981; Stuessy and Crisci 1984b). The perspective offered by Crisci and Stuessy (1980) and reiterated here is that both are important for the determination of polarity. Consideration of in-group character state distributions without investigating the out-group is as short-sighted as the reverse negligence. The reliance and importance of one to the other is the reason why they are treated together here. The perspective of Watrous and Wheeler is appropriate to stress here: "The topic of character polarity cannot be effectively dealt with in the absence of some reference to the commonality principle, which has and continues to receive support in systematics" (1981:10). Or, as noted by Stevens, "in-group and out-group analysis, may give very different results, although the 'common' state of the character is 'primitive' in both" (1980a:335). Despite these viewpoints, the above authors reject in-group analysis and embrace out-group analysis.

In-group analysis has been advocated and used profitably by numerous workers (e.g., Kluge 1967, 1976; Marx and Rabb 1972; Luteyn 1976; Stuessy 1979a; Kirkbride 1982; Bremer 1987, there called "the most parsimonious interpretation," p. 224). The perspective of Crisci and Stuessy (1980) is that during evolution of a taxon, there exists a set of features which was present in the ancestor and from which divergence in specific traits has occurred as speciation has progressed. These are the primitive shared character states, or symplesiomorphies, and are useful for delimiting the taxon from close relatives (these same character states can also be used as synapomorphies, if so desired, depending upon whether one looks at the phylogeny from the bottom-up or top-down). During evolution of the group, the core of primitive states is still recognizable because these conditions are still possessed by a majority of the taxa. As more and more speciation occurs within the group, the less probable it is that these features will be held in common. Further, certain character states which show high adaptation in a particular environment may become pervasive rapidly and give the impression that they are primitive rather than derived conditions (Stebbins 1974). With pinnate phylogenies (i.e., those in which species come off sequentially) and with few characters, this criterion can be misleading.

Because of these difficulties, it is also useful to consider the out-group and its character state distribution. The idea is that primitive character states were present in the common ancestor of the two groups and will still be found prevalently in the out-group. Out-group analysis is a good concept to employ, but some authors have gone overboard and regarded it as the only reliable criterion to the virtual exclusion of all other ideas (e.g., Stevens 1981; Wheeler 1981). This perspective should be avoided because it hampers search for additional valid ideas that will advance our understanding of determining evolutionary polarity (such as concepts of ecological specialization and others). As with in-group analysis (and all other criteria of polarity), there are problems with employing out-group concepts (after Stuessy and Crisci 1984b). The first problem is selection of the proper out-groups (Colless 1985). In angiosperms this is often a serious difficulty. Selecting a more inclusive out-group (such as a family, order, etc.) is not usually helpful due to the parallelisms that are common in almost all plant groups and the reduced likelihood of finding pervasive states (one is forced even more into considerations of majority occurrence). Watrous and Wheeler (1981) recommend that if no out-group is clearly identifiable, the in-group can be divided into major evolutionary lines in a preliminary cladogram. This presupposes knowledge of at least a few character state polarities based on some type of more inclusive out-group before attempting the analysis. This cladogram then is used to provide polarities for the states of all the other characters by using the primitive line within the in-group as a "functional out-group." The other lines become the "functional in-group." This procedure can be helpful, but all the polarities are based entirely on the few characters used for constructing the initial cladogram (caution is needed here). Donoghue and Cantino (1984) have recommended substituting different out-groups or out-group combinations and looking for areas of agreement in the resulting in-group cladograms. Frohlich (1987) suggests determining which character states are most common within all possible out-groups and obtaining a probability of primitiveness for each one. A related problem (to which Frohlich's approach also applies) is the frequent occurrence of two states of the same character in the out-group (fig. 8.4, instead of the more desirable pervasive condition). There is no reason to assume that this kind of stability in character evolution is more likely in the evolution of the out-group than within the in-group. Although some workers stress the importance of having complete occurrence of one state in the out-group (e.g., Watrous and Wheeler 1981), in practice this rarely happens (see Stevens 1980a, for agreement). If character state evolution has proceeded within the in-group (and hence its systematic interest and potential value), it is likely that parallel adaptations have also occurred within the out-group with two states occurring there also. Searching for a uniform set of states at the higher levels of more inclusive taxa (such as tribes, families, orders, etc.) only reveals more state diversity. A further problem is the need to assume no reversals in the character states in evolution of the out-group (Stuessy and Crisci 1984b). Rather than giving absolute answers, out-group comparison gives a statistical estimate which we hope will be correct (Ridley 1983). Some workers (e.g., Farris 1982) have emphasized that out-group comparison works well because it is based on parsimony. This "simple parsimony" has been enlarged upon into the "global parsimony" idea of Maddison, Donoghue, and Maddison (1984) by which one seeks simultaneously the simplest explanation for both the in-group and out-group (or groups) to determine the pleisiomorphic condition of character states. The term "global parsimony" is inappropriate, as pointed out by Meacham (1986) and as expressed well by Clark and Curran that "life lacks a sister group" (1986:425) and hence true global parsimony is impossible. Maddison, Donoghue and Maddison (1984) offer a series of rules to help determine the primitive state among several out-groups when it is not uniform. Wiley (1987a) adds even more rules. Despite the utility of such guildelines, they are arbitrarily created and of dubious value in interpreting the real evolutionary situation. The important point is that neither in-group nor out-group analysis provides absolute answers —both should be examined closely and carefully (as well as with other criteria) before making final judgments on polarity (see Vilgalys 1986, for largely corroborating results using in-group and out-group polarizations independently).

Co-occurrence of primitive states is the third criterion of primitiveness based on distributions of character states in the study group (and relatives). This criterion is less efficacious than the others (Crisci and Stuessy 1980), but it can be helpful in some cases. The evolutionary reasoning behind the criterion is that primitive states will tend to co-occur in particular EUs within a study group, and this can be a guide to selecting the most primitive EUs. This idea forms part of Sporne's (1948, 1956, 1975, 1976) method of correlation of primitive features in the angiosperms as a guide to selecting the most primitive angiosperm families. The criterion has merit, but it can be misleading if strong selection for adaptive complexes occurs (Stebbins 1974). Knowledge of the age of the group and its reproductive biology will be helpful in deciding whether this criterion should be employed in a particular instance.

The next three first-level criteria deal with developmental processes in determining polarity. These can be powerful indicators of polarity *if* such data are available. Some workers regard these as simply modifications of out-group comparison (e.g., Wheeler 1981), but most (including myself) view them as useful independent criteria (e.g., Eldredge and Cracraft 1980; Fioroni 1980).* The criterion used most frequently is *earliest ontogenetic state,* which is a modification of the more familiar "Haeckelian recapitulation" (i.e., "ontogeny recapitulates phylogeny;" Lovtrup 1978). This states that during the ontogeny of related EUs, the state of a character revealed early is likely to reflect the primitive condition. For example, some species of *Acacia* (Leguminosae) bear only phyllodes (narrowed leaf blades with unified growth and vertical planation) at maturity, but seedlings of these species have bipinnate leaves typical of those found in the rest of the genus. As the bipinnate condition occurs earlier in all the ontogenies, it is assumed to be the primitive character state (see Kaplan, 1984, for additional examples and explanation). However useful this might be in some cases, its use also necessitates caution due to *paedomorphosis,* in which a standard point of an ancestral ontogeny develops later in the descendents rather than earlier by slowing down of somatic development *(neoteny)* or speeding up of development of reproductive organs *(progenesis;* see Takhtajan 1976, and Gould 1977, for these and other viewpoints). Many discussions have taken place on the pros and cons of this criterion (e.g., Fink 1982; Alberch 1985; Kluge 1985; Kluge and Strauss 1985; Nelson 1985; Queiroz 1985), and suffice it to say here that it is still viewed as a valid approach to determining polarity. For a recent positive evaluation see Kraus (1988).

The second criterion that uses developmental data is *minor abnormalities of organogenesis.* If a morphological (or other) condition arises in a taxon that ordinarily has a contrasting condition, it can be inferred in some cases that the abnormal condition might be a primitive state. Heslop-Harrison (1952) recognizes three types of developmental anomalies: minor abnormalities of growth; abnormalities of development; and minor abnormalities of organogenesis. The first is viewed as having no phyletic import but may help in understanding normal patterns of development. The second involves the production of structures that result from hormonal changes such as features intermediate to leaves and flowers, and although these may give clues to homologies of characters between taxa, it likewise will not reveal ancestral conditions. The third type, however, may be a reflection of an ancestral condition, but it could also be a portrayal of a potential future condition. As Carlquist (1969b) and Eyde (1971) have stressed, this may only indicate another facet of the total genetic potential ("totipotency") of the genome and not reflect an ancestral condition at all. An important concern seems to be the frequency of occurrence of the abnormality within populations of a taxon. If occurring within more than one population or even in many scattered populations, the feature may well be a relictual one in

---

*All these developmental criteria do have an element of out-group concept in them, especially in the abnormality of organogenesis and vestigial structures. The fact that we can recognize a structure as abnormal or vestigial implies that related (sometimes only very distantly related) groups have different (and normal) conditions. This is use of the out-group in the very broadest sense, much beyond the sister groups discussed earlier. If we assume these developmental criteria to be nothing more than out-group analysis, which can be applied more conveniently and readily with other characters, then the effort needed to obtain the developmental data would not be worthwhile in most cases, and important information might be lost (or never gathered). This is an additional reason why out-group analysis should not be viewed as the only criterion of primitiveness. All information should be brought to bear on the problem for maximum insights.

the process of being eliminated from the populations (see Stuessy 1978, for an example of use of this criterion in *Lagascea*, Compositae). Conversely, of course, it might be arising in parallel within all these populations, but if they are widely distributed in different habitats, this seems unlikely.

The third developmental criterion is *vestigial organs*. Generally plant organs are present and functional (e.g., stamens), but sometimes they are present but not functional (e.g., staminodes), and sometimes they are absent altogether (stamens lacking, flower carpellate). Whether this absence of a structure is primitive or derived depends upon the group concerned and its near relatives [e.g., carpellate ray florets (i.e., without stamens) in the Compositae are derived in relation to discoid hermaphrodite florets (Koch 1930a, b) but primitive with reference to neuter rays (complete absence of sexual parts)]. Some workers regard vestigial organs (especially vascular traces) as unhelpful indicators of primitiveness (e.g., Carlquist 1969b), but others emphasize utility with caution (Kaplan 1971; Stebbins 1973; Naylor 1982; Wilson 1982). Still other workers view this criterion as nothing more than out-group comparison (e.g., Scadding 1981, 1982; Wheeler 1981), e.g., " 'vestigial organs' criterion—aside from the fact that reduction characters are subject to convergence, which is difficult to detect—seems valid *if* handled as out-group comparison" (Wheeler 1981:303). Viewing this criterion as nothing more than out-group comparison is to use this concept in a different way than discussed previously (and as used by Wheeler 1981, and others). It is true that within the biotic world vestigial structures are not the normal situation. In this sense, the rest of life with normal functioning structures or even the rest of a family in which normal conditions occur is an out-group, but this is a much broader context than selecting the sister group or the two most closely related groups for comparison of character states. That is, it distorts the concept of out-group beyond useful limits and encourages a narrower perspective that can lead to ignoring potentially useful additional information. In practice the difficulty of using this (and other) developmental criteria is that the data are simply not available or do not pertain to the particular groups (e.g., no vestigial structures are known).

The remaining three criteria are second level, and are used after application of one or more of the first-level criteria. That is, the second-level arguments cannot be applied by themselves—they rely entirely upon some prior decision about primitive conditions. They are essentially logical arguments which are very useful for determining primitiveness in some characters once decisions have been made regarding others. Wheeler has taken the extreme position that these criteria "are so dependently crippled that they are valueless in character analysis" (1981:303), but this is an overreaction that places blinders on obtaining useful insights. The criterion of *association* assumes knowledge of primitive states by hypothesizing ancestors in groups for which the first-level criteria cannot be applied (i.e., no fossils, no identifiable out-group, ambiguous in-group distributions, and no developmental data). Rooting networks of relationships within isolated taxa would be the most likely situation for which association might be used. Also, this helps focus on the need for more data to test the inferred directionalities by first-level criteria.

Correlation is useful if one character state tree has been determined based on application of the first-level criteria. The terminal state of another character state network that is found also in the same taxon that has the primitive state of the character state tree is judged primitive also. This is a very helpful criterion and has been used repeatedly (e.g., Frost 1930a, b, 1931; Sporne 1948, 1954, 1956, 1975, 1976; Strauch 1978; Crisci 1980; Jansen 1985). Correlation has been done most commonly with morphological features, but it can be efficacious with polyploidy and this provides strong evidence of polarity (from diploid to polyploid; Stuessy and Crisci 1984b). Correlation can also be done with whole taxa using habitats (Hennig 1966; Saville 1954, 1968, 1971, 1975). Caution again is needed here (as with all the criteria), however, because much divergence of habitat tolerances and distributions can occur as well as non-correspondence of rates of co-evolution between parasites and hosts.

Group trends is another useful second-level criterion in which parallel evolutionary trends within groups are used to suggest evolutionary directionality. For example, numerous parallelisms in many features have occurred in the Compositae and most other angiosperm families (Leppik 1968b, 1977). One trend, the development of secondary aggregations of heads (i.e., heads of heads) has occurred in more than 40 genera in 11 different tribes (Good 1956; Crisci 1974). These have occurred in parallel and the secondary level heads in each case are not structurally or developmentally homologous to each other. However, once a decision has been made in one case that primary heads are primitive and secondary heads are derived, it suggests that the polarity has been the same in other parts of the group, even though these other trends have occurred in parallel. Caution must be exercised here, too, because reversal in these trends can occur, especially in large groups in the midst of active evolutionary development (such as the Compositae or Gramineae). As Ross (1974:158) appropriately states: "[Group] Trends are therefore useful but require additional evidence from other sources before they can be considered applicable to a particular group."

Once the polarities of the character state networks have been determined, an appropriate coding (letters or numbers) must be assigned to the states so that they can be placed conveniently in a basic data matrix. Usually 0 is used for the primitive state and 1 for the derived condition. Once this has been done, the character states can be compared and evolutionary trees constructed. Exactly how trees should be developed has been a source of extreme differences of opinion. The viewpoint presented here will be that each of the methods for tree construction provides a different opportunity to learn something about the probable phylogeny of the EUs under study. Some methods are easier to employ than others, some are manual and others require use of the computer, and some are better suited for use with a particular type of data (e.g., molecular sequence data require particular algorithms). The important point is that no single method is the only correct one to use; as with all facets of phylogeny reconstruction, there is no simple solution.

Comparison of States: Tree Construction

The different algorithms for tree construction can themselves be classified in different ways (Funk and Stuessy 1978; Felsenstein 1982, 1983b, 1984). One viewpoint is to group the methods based on whether they use similarities or differences of character states. For example, the methods of Prim-Kruskal, Farris, Nelson, and Van Horn, and Whiffin and Bierner all rely on Manhattan distance, and the resulting network/tree is formed in some manner by using the minimum amount of distance between EUs. On the other hand, methods such as those of Hennig, Wagner, Camin, and Sokal, and Estabrook are based on shared derived character states, and the maximum number of these shared states between EUs is used to construct the tree. Another viewpoint, and the one used here, is to regard the methods as grouping themselves into parsimony, character compatibility, and maximum likelihood. Parsimony methods seek trees of minimum evolution among all EUs in the study group, whereas character compatibility seeks trees based on the maximum number of characters that have evolved (changed) in the same direction. Maximum likelihood is a statistical method that seeks trees offering the highest probability of yielding the observed data. Further breakdown into subdivisions of these types will follow Felsenstein (1984). Despite this apparent simplicity of categorization of methods, some of the algorithms are really more similar to each other than seems apparent at first glance (Felsenstein 1984). As Wagner has remarked: "It provides a salutary experience in the teaching of systematic botany to compare these methods with one another and assess the differences in their underlying philosophies. More often than not, what appear to be differences are actually similarities" (1980:190).

Parsimony Algorithms. The types of cladistic algorithms for tree construction used most frequently have been those of parsimony, which attempt to minimize the number of character state changes among EUs. Five basic subtypes can be recognized (modified from Felsenstein 1984): (1) Hennig and Wagner; (2) Camin and Sokal; (3) Farris; (4) Dollo; and (5) polymorphism. The Hennig and Wagner methods are essentially identical (see earlier discussion) in which character state changes from 0 (primitive)→1 (derived) occur as well as 1→0 (reversals), but with a minimization of the latter. As these two approaches have been discussed in detail already, it suffices to stress that they both are manual methods which have had considerable usage. For example, the Wagner Groundplan/divergence method was first used by Hardin (1957; one of Wagner's students) and subsequently by several dozen workers in different taxonomic groups (fig. 8.5; see reviews by Funk and Stuessy 1978; Wagner 1980; and Funk and Wagner 1982; see also the works of Bacon 1978; Olsen 1979; and Judd 1982). Hennig's methods (fig. 8.6) have been used by several botanists also (see Dupuis 1978, and Funk and Wagner 1982, for reviews, and the papers of Bremer 1976, 1978; Bremer and Wanntorp 1978, 1981, 1982; Ehrendorfer et al. 1977; Humphries 1979, 1981).

The *Camin and Sokal* (1965) *parsimony algorithm*, the first numerical cladistic procedure to be developed, allows only 0→1 with the fewest number of these changes and without permitting reversals. This was implemented for computer use by Bartcher (1966). Because reversals of character states are common occurrences in phylogeny, this procedure is less appealing than those that permit reversals. One application is by Kethley (1977) in treating higher taxa of some parasitic mites.

The *Farris parsimony algorithm* (Eck and Dayhoff 1966; Kluge and Farris 1969;

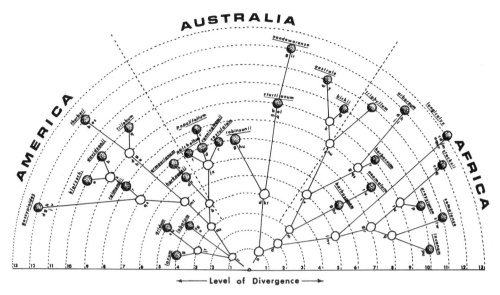

FIGURE 8.5 Example of the Wagner Groundplan/divergence graphic display of cladistic relationships among species of *Gossypium* (Malvaceae). From Fryxell (1971:558). See earlier discussion and figure 8.2 for comparison.

Farris 1970, 1972) allows 0→1 and 1→0 with a minimization of both these types of changes (fig 8.7). The computer program developed by Farris, PHYSYS, is available for purchase. Another available program, and the most popular one, is PAUP (Phylogenetic Analysis Using Parsimony), developed by David Swofford of the Illinois Natural History Survey, Urbana (fig. 8.8). This is available for use with microcomputers. A Farris parsimony algorithm is also available for use on microcomputers (along with other cladistic algorithms) in the package called PHYLIP by Joe Felsenstein of the University of Washington in Seattle. Manual adaptations are those of Whiffin and Bierner (1972), Nelson and Van Horn (1975; fig. 8.9); Bierner et al. (1977), and Stuessy (1979a; fig. 8.10). Jensen (1981) has a clear discussion of this approach which is based on Manhattan distance, rather than shared derived character states. Luteyn (1976) used both Farris networks and trees in reconstructing the phylogeny of *Cavendishia* (Ericaceae). For a good comparison of some of the different microcomputer algorithms, see Fink (1986), Carpenter (1987) and Platnick (1987).

The *Dollo parsimony algorithm* (LeQuesne 1974; Farris 1977b) assumes that 0→1 once only for a particular character (i.e., it allows no parallelisms) and it also allows 1→0 (reversals) but minimizes them. This technique has had limited applications with morphological data, perhaps because of the acknowledged frequency of parallelism in most (especially angiosperm) groups. It is being used more now with analysis of DNA data (DeBry and Slade 1985; see chapter 21). Similarly, the *polymorphism* parsimony algorithm of Farris (1978) and Felsenstein (1979) is useful if genetic data are involved in which both states of a character (01) can occur in one EU. Further, 0→01 can occur once only, reversals

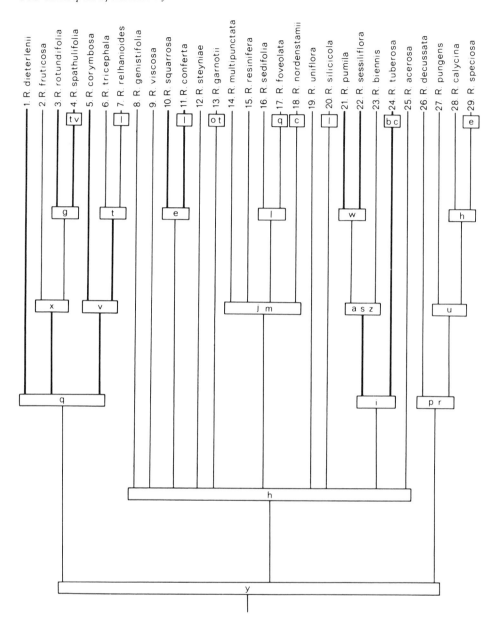

FIGURE 8.6 Example of the original Hennig-type graphic display of cladistic relationships with quadrangles showing synapomorphies among EUs of species of *Relhania* (Compositae). Most Hennigian cladists now prefer the graphic format shown in figure 8.8. (From Bremer 1976:25)

are allowed (01→0 or 01→1), and the retention of the polymorphic condition is minimized.

Character Compatibility Algorithm. The other most commonly used cladistic algorithm, and one that relies on very different assumptions, is character compati-

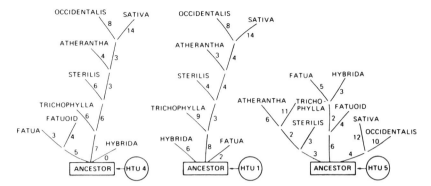

FIGURE 8.7 Examples of the Farris parsimony algorithm showing three trees (cladograms) for the hexaploid species of *Avena* (Gramineae). (From Baum 1975:2125)

bility. As described well by Meacham: "Character compatibility analysis is a technique that reveals patterns of agreement and disagreement among characters in a data set. It is based on facts first noticed by Wilson (1965) and Le Quesne (1969, 1972) which were subsequently provided with a mathematical

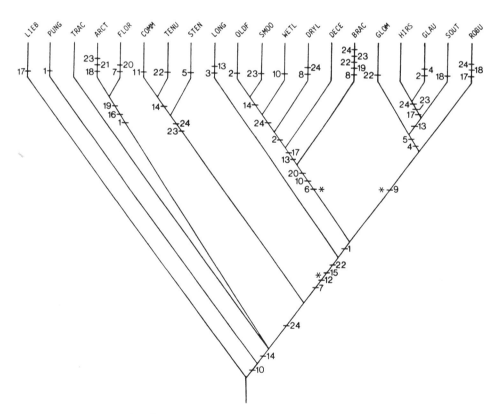

FIGURE 8.8 Example of a Farris tree (cladogram; parsimony algorithm) generated by PAUP showing cladistic relationships among taxa of the *Andropogon virginicus* complex (Gramineae). Asterisks designate internodes with unique synapomorphies. (From Campbell 1986:285)

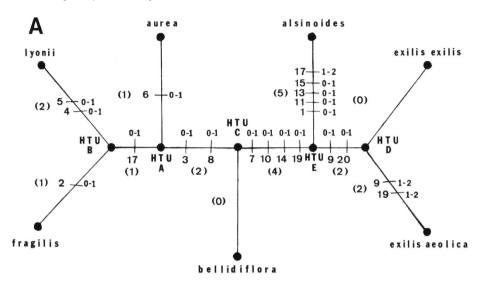

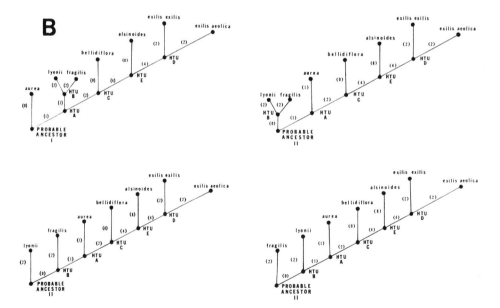

FIGURE 8.9 Examples of the Farris parsimony algorithm done manually to produce a network (A) of relationships among species of *Pentachaeta* (Compositae), and four possible rooted trees (B) from the network with different hypothetical taxonomic units (HTUs) selected as ancestral. Numbers on branches in (A) refer to characters and their state changes (e.g., 0–1). (From Nelson and Van Horn 1975:368, 371)

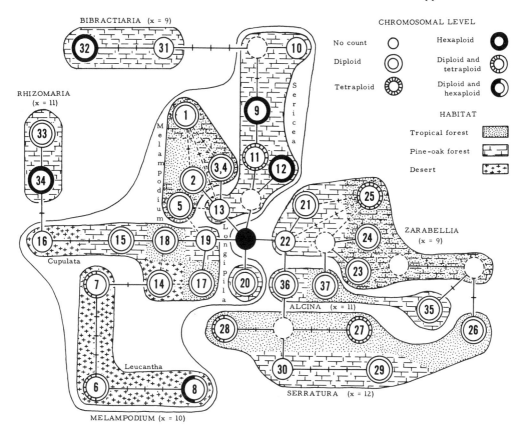

FIGURE 8.10 Cladogram of all species (and sections and series) of *Melampodium* (Compositae) based upon selected qualitative and quantitative characters. Numbered circles refer to taxa. Dashed circles refer to hypothetical taxonomic units; dashed lines refer to equally probable cladistic pathways. The dark circle represents the ancestral taxon of the genus with all primitive conditions of character states. Length of lines (with bar markers) connecting taxa reveals the number of evolutionary character state changes. Chromosomal level and habitat occurrence of all taxa are also shown (see key). (From Stuessy 1979a:183)

foundation by the work of Estabrook and others (Estabrook 1972; Estabrook, Johnson, and McMorris 1975, 1976a, b; Estabrook and Landrum 1975; McMorris 1975, 1977; Estabrook and Meacham 1979)'' (Meacham 1981:591). Because Wagner groundplan/divergence and Hennig parsimony methods have been presented in this book in some detail as examples of the parsimonious approach to cladistic analysis, it seems useful also to present details of character compatibility analysis. The discussion here will follow Meacham (1981). See also Meacham and Estabrook (1985) for a good review.

The steps involved with character compatibility analysis are basically the same as with parsimony techniques in the selection of EUs, selection of characters and states of evolutionary import, creation of a data matrix, and the drawing of trees. Two intervening steps, however, are different from parsimony and

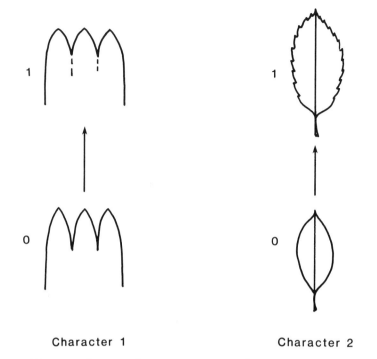

Character 1   Character 2

FIGURE 8.11 Diagram showing character state trees from primitive (0) to derived (1) states in fusion of corolla lobes (character 1) and degree of dissection of the leaf margin (character 2).

these reflect the basic difference between the approaches: (1) determining compatibilities of characters; and (2) finding the cliques (= sets of characters that are mutually compatible). Characters are compatible if they are correlated in their change in evolutionary directionality within a particular study group (i.e., they are congruent). Characters are never correlated (or congruent) in any absolute sense, but only with reference to the distribution of states within a particular group of taxa.

The concept of compatibility of characters can be elusive and further comment is needed. Consider the example in figure 8.11. Within the ancestral complex of a study group, now possessing two states of two characters, the character state combination 0,0 would be found in all of them. (It would be assumed that they differ in other characters not shown here.) As speciation proceeds, character 1 will evolve from 0→1.* Character 2 may not change (at least not initially) and hence the new character state combinations would be 1,0. The more primitive taxa would still have 0,0. As speciation continued and character 2 eventually evolved (0→1), the resulting new taxa would have the new character state combination of 0,1 if the taxon came from the ancestral complex (0,0), or 1,1 if it

---

* Most of the literature dealing with character compatibility has used "A" and "B" for the two states of a character, but here they have been converted to 0 and 1 for correspondence with the example of parsimony algorithms presented earlier.

came from one of the more derived taxa that already had one derived condition (1,0). But *both* 1,1 *and* 0,1 could not be present at the same time. For this to happen, state 1 in character 2 would have to have evolved twice independently (i.e., a non-unique event, from an ancestor with 1 in character 1 in one case and from an ancestor with 0 in character 1 in another case, or alternatively there could have been a reversal in character 1 from 1,1 to 0,1). If one or both of these two characters within a particular study group are uniquely derived and unreversed, then not all possible combinations of states within taxa (0,0; 0,1; 1,0; and 1,1) will be encountered; at least one combination will be absent. If this is the case, the two characters are said to be compatible. As with other approaches, a note of caution must be introduced. Le Quesne stresses that "if three or less of the possible combinations are found, it does not necessarily prove that characters 1 and 2 are both uniquely derived characters, but only that they may possibly be. If, for example, character 2 changed from 2_A to 2_B on two or more occasions on each of which character 1 was in the same state, only three combinations will be found. Moreover, if the four combinations have been evolved during the history of the group, one of these may have died out again or alternatively not be represented in the material studied" (1969:201).

For purposes of illustrating further the compatibility approach, five taxa (A-D, plus a hypothetical ancestral taxon "X" with all primitive states) with three characters will be used, all with binary data with polarities already determined by application of some criteria, as shown below in this small basic data matrix:

	Characters		
	1	2	3
Taxa			
A	1	0	1
B	1	1	0
C	0	0	1
D	1	1	0
X	0	0	0

An easy way to visualize the compatibilities among these three characters is to establish a three-way matrix comparing one to each other, as shown below, with Xs showing the existence of a combination of character states in the four taxa being considered:

	Character 1		Character 2	
Character	0,0	1,0		
2		1,1		
Character	0,0	1,0	0,0	1,0
3	0,1	1,1	0,1	

Clearly characters 1 and 2, and 2 and 3 are compatible whereas 1 and 3 are not.

The next step is to detect cliques among the characters. A compatibility graph derived from the above matrix showing compatibility ties between characters is as follows:

Two (and only two) cliques of characters exist here: 1 and 2, and 2 and 3. The next step is to construct trees that correspond to each clique. To do this requires examining the character state trees for each of the three characters, as shown below [refer to the basic data matrix; the squares refer to the taxa, given on the left, which possess the particular state (0 or 1) of each of the characters]:

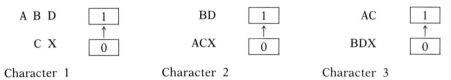

Next, one (any one) of the character state trees is selected, regarded as a taxon tree, and compared with a second one from the same clique in the same fashion (here 1 with 2 or 2 with 3), and they are joined ("popped," Meacham 1981) as follows:

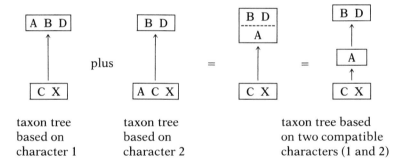

The comparison of the taxon trees of the clique with compatible characters 2 and 3 is as follows:

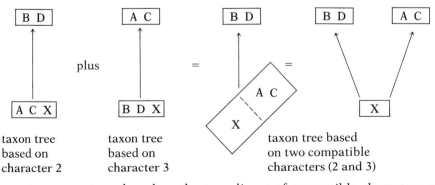

Hence, two taxon trees based on the two cliques of compatible characters are generated from these data. Ordinarily, with more characters and EUs there are

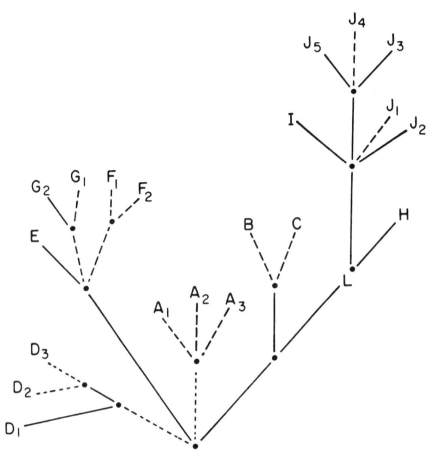

FIGURE 8.12 Cladogram of the *Ranunculus hispidus* complex (Ranunculaceae), constructed by character compatibility analysis. Letters and subscripts refer to individual taxa. Dashed lines are used to move EUs to the tips of the line segments instead of placing them at the nodes. (From Duncan 1980b:451)

only a few very large cliques of compatible characters and therefore only a limited number of taxon trees for evaluation. The graphic form of trees derived from character compatibility analyses has varied but two examples are shown in figures 8.12 and 8.13. Applications of character compatibility for cladistic analysis have been fewer than with parsimony algorithms, but some examples are Estabrook, Strauch and Fiala (1977), Baum and Estabrook (1978), Estabrook and Anderson (1978), Gardner and La Duke (1978), La Duke and Crawford (1979), Duncan (1980b), Meacham (1980), and Landrum (1981).

Despite the apparent great differences between compatibility analysis and parsimony algorithms, the former can be viewed in a way as a kind of parsimony "in which, instead of counting the changes of state, we count the number of characters which require one or more extra changes of state. By minimizing the resulting quantity over all possible phylogenies, we are in effect maximizing the

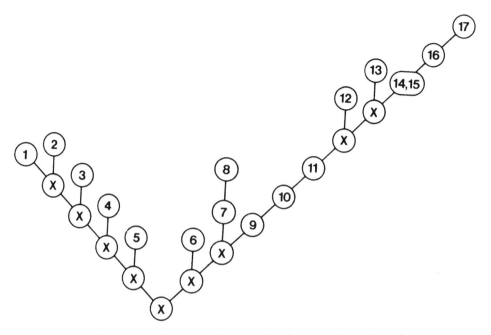

FIGURE 8.13 Cladogram of species of *Crusea* (Rubiaceae), based on character compatibility analysis. Numbers refer to individual taxa; Xs, to hypothetical nodal ancestors. (From Estabrook and Anderson 1978:188)

size of the largest 'clique' of mutually compatible characters. Thus, there are two ways of describing the compatibility method: as a method of analyzing conflict between characters or as a method, similar to parsimony methods, for inferring phylogenies" (Felsenstein 1984:172). Furthermore, compatibility methods are similar to Hennigian and Wagner Groundplan/divergence parsimony methods in that they all deal with uniquely derived character states instead of a distance measure, such as in the Farris parsimony methods.

Maximum Likelihood Algorithm. A third approach to cladistic analysis includes *maximum likelihood methods*. These choose the particular tree that gives the highest probability of yielding the observed data (and not the reverse). These are not simple algorithms; "computation of the likelihood can be difficult. We are attempting to compute the probability of the data, given the tree. This probability is a sum over all the ways that the data could have originated on the given tree. Thus, we must sum the probabilities of all different 'scenarios,' to use Gareth Nelson's useful term (Hull 1979), which lead to the observed data at the tips of the tree" (Felsenstein 1984:178). Maximum likelihood methods, although offering many possibilities for the development or improvement of future tree construction methods, have been little used so far. As Boulter has remarked: "maximum likelihood procedures, whilst giving s[t]atistically consistent results, suffer from the need to have an agreed probabilistic model of character evolution and this is usually not forthcoming" (1980:236).

Taxon Networks

The cladistic methods discussed so far have relied upon the determination of primitive and derived states of characters before the analyses are done. These lead directly to evolutionary trees. An alternative to this approach, however, is to construct first a *network* of relationships among the EUs and then root the network by application of some criteria to select the ancestral EU and turn the network into a tree (e.g., Meacham 1984b). Most of the methods can be dealt with in this fashion. For example, the Farris parsimony algorithm based on Manhattan distance can be used to produce unrooted networks (Farris 1970; Lundberg 1972; Nelson and Van Horn 1975). Likewise, compatibility methods can be done to yield undirected (or "unordered;" Felsenstein 1984) networks (Estabrook and Meacham 1979; Estabrook 1980). One important reason for opting for network construction, at least initially, may be difficulties with out-group selection. If an out-group is not clearly identifiable, a network of relationships among the EUs can be rooted at different points to produce different trees, depending upon the root chosen. Different out-groups can be selected and the different resulting trees compared and evaluated, either intuitively or quantitatively (Estabrook, McMorris, and Meacham 1985; Day 1986). With initial decisions on polarity of character states, one change in one character will usually cause an alteration in the branching patterns of the resulting tree. Choice of different out-groups to help in polarity decisions, therefore, will cause completely new branching diagrams. With network construction, the branching relationships remain the same no matter which EU is selected as ancestral; only the evolutionary directionality of the tree changes. Examples of this kind of application are Cantino (1982a) and Sanders et al. (1987). For general perspectives on networks, see also MacDonald (1983).

The concept of network rather than tree generation had its origin with phenetic rather than cladistic approaches to classification. The early methods of Kruskal (1956) and Prim (1957) were network-generating techniques which were viewed as largely phenetic at that time, because they dealt with many equally weighted data points with no evolutionary interpretations. This same approach later became part of the phenetic repertoire as *minimum spanning networks* (usually called "minimum-spanning trees," or "shortest-spanning trees;" Crovello 1974).* They are now sometimes used to show added information on 3-D ordination plots (e.g., Jackson and Crovello 1971). Prim–Kruskal networks have also been used to generate ideas on phylogeny (e.g., Solbrig 1970a). To add to the confusion, some workers have commented that phenograms contain much cladistic information and in a way can be regarded as cladograms (Colless 1970; Presch 1979)! The viewpoint taken here is that a network that is never rooted remains a phenetic estimate of relationship because no evolutionary interpretations are implied. Once it is rooted it becomes a reflection of evolutionary directionality and can be regarded as showing cladistic relationships.

*There is a confusion regarding use of the terms tree and network. In a mathematical or graph theoretical sense, any series of interconnected points is called a tree (Crovello 1974). In logic and grammar, trees simply show relationships among parts of a sentence (Byerly 1973). For biological purposes, however, it seems worthwhile and clear to use network to refer to an unrooted branching diagram and tree to one that is rooted.

Another kind of network application in cladistics has been with molecular sequence data, in which numerous unweighted data points are available for the EUs. As Crovello has remarked: "The computer and numerical cladistic indices enter into molecular phylogeny because of the large amount of calculation required to obtain the 'best' cladogram from a table of similarity values between taxa" (1976:182). Some people have viewed these methods as phenetic (cf. Fitch 1984), but they are cladistic in the sense discussed above because the networks are almost always rooted to yield a tree. Rooting of networks based on molecular sequence data is based on correlations with other data, such as fossil evidence, presumed relatives, and ideas on rates of molecular evolution (Penny 1976). Numerous studies have been done using molecular sequence data involving cytochrome *c* (Fitch and Margoliash 1967; Fitch 1971; Boulter 1972; Beyer et al. 1974; Estabrook and Landrum 1975; Moore and Goodman 1977), plastocyanins (Boulter 1974, 1980; Boulter et al. 1979), other proteins (Fitch 1977; Goodman and Pechere 1977; Schulz 1977), and DNA-RNA (Praeger and Wilson 1978; Fitch 1980; Gingeras and Roberts 1980; Templeton 1983a, b). Crowson (1972) and Cronquist (1976) have provided cautious evaluations of the efficacy of macromolecular sequence data in plant systematics (to be discussed in more detail in chapter 21).

Efficacy of Algorithms

The existence of numerous algorithms for tree construction has led to an interest in determining which ones might be the most accurate reflections of phylogeny. Some cladists believe that because evolution is largely parsimonious (see comments by Niklas 1980, and Sober 1983), it suffices to adopt one of the parsimony methods.* Evolution is clearly not always parsimonious in flowering plants, however (and perhaps not in other groups either; Dunbar 1980; Crisci 1982), as evidenced by the high levels of hybridization (Grant 1981), polyploidy (Grant 1981, 1982a), and known parallelisms (Cronquist 1963; Stebbins 1974). For example, Gastony (1986) has shown convincingly a more complex and non-parsimonious mode of origin of the autopolyploid fern *Asplenium plenum* via unreduced spores. As pointed out clearly by Funk and Stuessy (1978) and Crisci and Stuessy (1980), no single approach can be relied upon to provide absolute answers. We do not know the true phylogeny for any group of organisms, nor will we ever know it. The best that can be done is to produce trees by several methods and compare the results (such as was done by Baum 1984). A statement of probability on approximating the true phylogeny can also be made, which throws us into methods of statistical inference. As Felsenstein comments: "A great advantage of a statistical approach is that it makes explicit the connections

*R. Johnson, from a purist's perspective, has pointed out that parsimony has been used in different contexts (a "chameleon concept") and that "a thorough re-appraisal of the value of parsimony in the reconstruction of phylogenies should preface its continued use" (1982:79). Since this statement, other studies have probed the role of parsimony in classification and tree reconstruction (Panchen 1982; Kluge 1984; Sober 1985; and Thompson 1986). Some workers are content to accept parsimony as the overriding conceptual and methodological approach (e.g., Kluge 1984), whereas others seek newer solutions, especially from maximum likelihood methods (e.g., Thompson 1986). Because evolution is clearly not always parsimonious, parsimony can only be viewed as a temporary methodological approach until further detailed studies on the group in question reveal more clearly the actual modes of evolution.

between the biological assumptions and the choice of a method of analysis" (1984:187). For example, hybridization obviously can have a profound effect on the development of a phylogenetic tree (see statistical confirmation by Astolfi, Piazza, and Kid 1978) and attempts at its reconstruction, and this should be understood through careful investigations before attempting the phylogenetic reconstruction (Wagner 1983). A compromise approach that is gaining in interest is the development of *consensus trees* (E. N. Adams 1972; Rohlf and Sokal 1981; Rohlf 1982; Day 1983; Neumann 1983; McMorris, Meronk, and Neumann 1983; Smith and Phipps 1984; Stinebrickner 1984a, b; Day and McMorris 1985; Mc-Morris 1985). These attempt to take different trees for the same set of EUs and produce statistically one tree with maximum retention of information. Because in many cases it is impossible to pick the one and only tree to represent the branching pattern of phylogeny, a more satisfying alternative will be to produce a consensus tree from those several (or many) trees that seem most likely. Carpenter (1988) suggests that this is negative, because the consensus tree will of necessity have less information than *any one* of the individual trees. Other workers (e.g., Miyamoto 1985) have criticized consensus trees on the grounds that they make all data of equal value in the analysis. They recommend that the data be combined first and a single analysis then be done on these combined data. Both approaches have merit, depending upon the type and quantity of data involved. Another statistical approach is the "bootstrap" (Felsenstein 1985), which is useful to help give a quantitative estimate of a confidence value for a particular phylogeny. Basically the technique involves simultaneously eliminating and duplicating characters in a data set and developing trees from each one. Branching points which remain in 95 percent of the trees of the altered samples (usually 50 or more) are judged to be statistically significant.

A concept in parsimony analysis worth mentioning is the inference of phylogenetic relationships on the basis of "nonuniversal derived states" (Cantino 1985), earlier referred to as "apomorphic tendencies" (Cantino 1982b). This was offered to assess relationships in a study group in which the same derived state occurs in several EUs but not in every member of each. Cantino argued that these taxa might be regarded as more closely related to one another than to other ingroup taxa not possessing the derived condition. Three explanations for the character state pattern were offered and Cantino maintained that all imply the same phylogenetic conclusion. One of the explanations invoked an argument that related taxa share a genetically based tendency for parallel development of the same apomorphy, an idea espoused by Cronquist (1963, 1968) in a noncladistic context. Rasmussen (1983) has criticized the idea, and it remains unpopular among cladists, although a few others have advocated a similar approach (Tuomikoski 1967; Brundin 1976; Saether 1979, 1983).

A difficulty with cladistic analysis of any type, which relates to the interpretation of evolutionary processes within particular groups, is the assumption that evolutionary mechanisms have been only by cladogenesis in which an ancestral species gives rise to two daughter species and itself becomes extinct. How common this strictly dichotomous mode of speciation is, however, is uncertain. In plants, it may be the minority of modes, for certainly many different mechanisms of speciation occur (e.g., allopolyploidy, budding off of peripheral isolates, polychotomies, etc.; White 1978; Atchley and Woodruff 1981; Grant 1981). It is believed that widespread taxa often fragment into several populational systems at more or less the same time, e.g., in section *Nocca* of *Lagascea* (Stuessy 1978, 1983). It has also been suggested that at least 70 percent of plants are polyploid (Goldblatt 1980) and perhaps higher. This means that the polyploid (or reticulate) origin of plant species is commonplace (e.g., in *Picradeniopsis*, Compositae; Stuessy, Irving, and Ellison 1973; or the classic case of *Tragopogon*, Compositae, Ownbey 1950), and this makes cladistic analysis much more difficult, at least at the intrageneric level. Funk (1981, 1985a) and Bremer (1983a) have offered suggestions for dealing with these difficulties, but they remain problems nonetheless. The important point, therefore, is that many different modes of speciation have occurred in the evolution of species or organisms, and especially in plants. If we have no evidence to the contrary, it does no harm to assume a dichotomous mode. It is better, however, to accumulate as much biological information as possible about the taxa so that the real modes can be determined. Such evolutionary information as types of speciation, rates of evolution, and nature of homoplasious events will all have an effect on cladogram estimation (Fiala and Sokal 1985). Cladistic analysis, therefore, must never be viewed as a panacea or final answer for phylogeny.

FORMAL CLASSIFICATION

Once trees have been generated, the next step in cladistic analysis is to use them in some way to construct a classification. Although seemingly a simple enough step, it has become a source of heated debate among cladists and between cladists and other workers. There are two basic viewpoints: (1) the cladogram should be used directly in constructing the classification and this should be done explicitly so that the cladogram can be derived from the classification and vice versa; and (2) the cladogram should be used as a guide to construct the classification, but other aspects of phylogeny (such as autapomorphies or phenetic divergence) should also be considered. These different viewpoints impact both grouping and ranking of the EUs. The more rigid cladists (e.g., Wiley 1981a; Janvier 1984) believe that only holophyletic groups should be used in classification (fig. 8.14; viewpoint 1 classifications), i.e., only those that include *all* the descendents of a common ancestor. Paraphyletic groups are rejected. The eclectic cladists (e.g., Duncan 1980a; Meacham and Duncan 1987) prefer to consider also the degree of change along the lineages, or the patristic distance, in constructing the classification. For example, in figure 8.14 (viewpoint 2 classification), EU E is placed in its own subgroup because of the many autapomorphies leading to it. These two viewpoints lead to no large differences in classification, unless the cladogram is shaped as in the bottom of figure 8.14, in which all EUs come off sequentially from the ancestor. In this case, a rigid approach (viewpoint 1) leads to a proliferation of groups needed for classification. With regard to ranking in the Linnaean hierarchy (called "absolute ranking" by Hennig 1966), it calls for a "cornucopia of categories" (Colless 1977:349) beyond what are available. As Throckmorton expressed earlier, "Any attempt systematically and

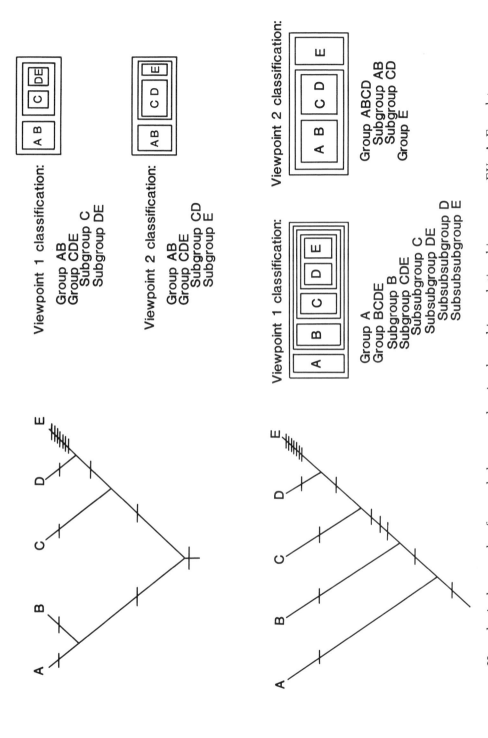

FIGURE 8.14 Hypothetical example of two cladograms showing branching relationships among EUs A–E and two different classifications derived from each of them. Bars on the cladograms indicate positions of apomorphies.

consistently to impose a hierarchical arrangement on such a [cladistic branching] pattern will result in a wildly asymmetrical product that rapidly exhausts the category and subcategory names available" (1965:233). In short, the eclectic approach is advocated as the most reasonable (for concurrence, see Mayr 1974b). Hennig (1966) advocated using absolute geological ages as the criterion for ranking, such that groups that diverged as early as the Cambrian would be treated as divisions, from the Devonian as classes, and so on. This has never been taken seriously, at least by plant workers (see Funk and Stuessy 1978, for comments).

The incorporation of data from fossils in cladistics has also generated considerable discussion. Again, widely divergent viewpoints prevail. Areas in which fossils may be considered for use include determination of evolutionary polarity of character states, inclusion of fossil taxa as EUs in the construction of the cladogram, and use of fossils to test the validity of a generated cladogram (see Stein 1987, for a good review). Use of fossil evidence in polarity determinations has been considered already in this book (p. 107) and will not be discussed further. Whether or not fossils should be included as EUs along with recent (and mostly extant) taxa depends upon the nature of the fossil evidence. In most angiosperm groups, the fossil evidence is usually limited to unconnected organs of the plant, such as leaves or flowers, etc., and therefore is inadequate for meaningful comparisons with EUs with full data available. Hence, for practical reasons alone, cladistic work with angiosperms usually suggests the testing with fossils of cladograms based on recent EUs rather than their inclusion in the full analysis. Doyle and Donoghue (1986b, 1987), however, include fossils directly in their cladistic analyses. Patterson (1981) stresses that almost all estimates of relationship derive from studies of recent forms that can only be secondarily refined by referral to fossils. Farris advocates use of fossils "that are well enough known to be classified at all" (1976:281) as terminal EUs along with recent taxa. This seems reasonable if the fossils are complete enough for full analysis (not always possible in angiosperms with detached organs; Hill and Crane 1982). Another perspective called *"stratophenetics"* is offered by Gingerich (1979a, b) in which the striatigraphic positioning of fossil EUs is used as a beginning structure in which phenetic relationships among the forms are determined. Although avowedly a phenetic method, it yields a branching diagram with evolutionary time and directionality shown, and in that sense is more of a cladistic rather than phenetic approach.

IMPACT OF CLADISTICS

The impact of cladistics on systematics has been marked, and the effects have gone a route similar to those seen for phenetics a decade earlier. Because the cladistic approach to classification is newer, the full and lasting contributions will not be seen clearly for another decade. The claim that cladistics is ushering in a "revolution" (e.g., Funk and Brooks 1981) is clearly exaggerated. Nonetheless, some positive contributions can be seen already and one can guess as to what the lasting values might be. In my opinion, the two most important contributions seem likely to be: (1) if phylogeny is to be attempted as part of a systematic study, it should be done in an explicit manner; and (2) it should be remembered that classifications are hypotheses, just as exist in other areas of science, and they are testable. If phylogeny is to be attempted, it should be done so that others can follow its construction and replicate its methods in all aspects from the selection of characters and states, to the determination of polarity, to the construction of branching sequences. This can only lead to improved communication with other workers. As pointed out by Funk and Stuessy: "Because the development of phylogenies by traditional methods often involves personal or 'intuitive' judgments, differences between phyletic schemes for the same plant taxa are difficult to resolve because of a lack of understanding of the evolutionary assumptions used and the ways in which the trees were generated. In contrast, cladistic methodology facilitates discussion by a clear presentation of evolutionary assumptions and operational procedures" (1978:174–175). Cladistics also helps point more clearly to the existence of parallelisms and convergences (Mickevich 1978).

The cladistic approach has also helped make the point that classifications can be viewed as hypotheses of relationships about organisms that can be "tested" in a reasonably rigorous fashion. It has been a frequent criticism by experimentally oriented biologists that taxonomy has not been rigorously testable, and therefore, not fully scientific. In response to this criticism, many cladists have adopted the " 'deductivist' or 'falsification' " (Cartmill 1981:74) philosophy of science of Karl Popper (1959, 1962). This approach regards scientific reasoning as wholly deductive and maintains that "a scientific hypothesis . . . must imply, in a rigorously deductive fashion, some other statement which is subject to empirical refutation or (to use Popper's word) falsification" (Cartmill 1981:74; see also Kitts 1977). The application by cladists of this philosophy to reconstruction of phylogeny and construction of classifications is well outlined by Gaffney (1979). The first step is to establish hypotheses of holophyly (his monophyly) among taxa (by the cladograms) and then to test these by looking for synapomorphies with parsimony as an underlying guiding principle. If the synapomorphies do not yield a particular branching sequence, then that hypothesis is falsified. If they do yield one of the cladograms, then it is corroborated and remains the diagram of choice until perhaps falsified at some later date by the addition of new synapomorphies (such as by the acquisition of new data). The importance of deriving a classification *directly* from the non-falsified cladogram is obvious; the classification can then also be viewed as a hypothesis capable of future testing. This perspective leads to the rejection of the phenetic and phyletic approaches to classification because they are viewed as nontestable (e.g., Wiley 1975; Platnick 1977c; Settle 1979). There is every reason to welcome the idea of viewing classifications as scientific hypotheses, but it is completely unacceptable that the hard-line cladists have used a rigid way of viewing testability (by synapomorphies) that excludes all other approaches to classification. As the philosopher Frank Fair (1977:90–91) has put it, this type of reasoning

appears to me to be an example of a way of using philosophy of science which is rather widespread and which has, I believe, a constricting effect on inquiry that can be very detrimental. I am uneasy at the way in which philosophers of science such as Kuhn, Popper, or others can be treated as if they had created a definition or *prescription* for what is to be authentically scientific. . . . Rather than provide prescriptions, I believe that philosophers of science can propose different 'ideas of method' that reveal different angles of view that a scientist may adopt toward his work. He may find one of these angles more congenial or helpful, and if so fine. Comments on methodological propriety seem to me to be best offered in something like the spirit of comments on how to view a painting—"Perhaps if you look at it from this angle, it will appear more interesting, and then notice how these colors are in tension with each other, etc., etc." (1977:90–91)

The regarding of classifications as testable hypotheses, therefore, is an important contribution of cladistics, but this general concept applies equally well to those classifications derived from phenetics and phyletics which are tested by different criteria (new degrees of overall similarity, and this plus new derived character states, respectively, all brought about by the inclusion of new data). More on this topic in the next chapter.

Another useful benefit from developments in cladistics has been the emergence of *vicariance biogeography* (Croizat 1962, 1978; Nelson 1973b, 1984; Croizat Nelson, and Rosen, 1974; Rosen 1975, 1978; McDowall 1978; Platnick and Nelson 1978; Nelson and Platnick 1978, 1981; Nelson and Rosen 1981; Platnick 1981; Bremer 1983b; Wiley 1987b). Humphries and Parenti (1986) have gone a step further by formally combining cladistics with vicariance biogeography to yield what they call *"cladistic biogeography."* This approach to biogeography assumes that "species distributions are a result of the following process: (1) a species spreads over a large geographic range; (2) the range fragments several times into allopatric populations (each splitting is called a vicariance event; the fragmented populations are ecological vicars—Udvardy 1969); (3) allopatric populations speciate; and (4) the new species may spread into the range of other species; sympatry indicates dispersal (after Rosen 1978)" (Endler 1982:443). The main emphasis here is in the explanation of patterns of distribution of organisms based on major earth events that cause populational divergence (e.g., continental drift, mountain building, the changing of river courses, etc.), and dispersal, although acknowledged to have occurred, is held to be non-informative and hence largely ignored. Major earth events can be suggested by (1) constructing cladograms of many different plant and animal groups that are found in the same general geographic areas; (2) substituting areas for taxa in these cladograms yielding area (rather than taxon) cladograms; and (3) determining degrees of congruence among these area cladograms, which then point to major areas of earth disturbance. The generalized common pathways reflected by common areas of distribution are called *tracks* (Croizat 1962, 1978).* This is a useful perspective in which to consider biogeographical problems, and it has been used for decades traditionally along with dispersal considerations (e.g., Cain 1944;

---

*Although Croizat's works served as a stimulus in the development of vicariance biogeography, his emphasis was slightly different, as he himself strongly pointed out (1982). His contribution was "panbiogeography," which was a method of analysis by looking for "tracks" between areas of the world, especially between continents, which suggested geological as well as biogeographical affinities (Craw 1984). For an attempt to quantify this approach, see Page (1987).

Raven and Axelrod 1974). Cladistics, however, has given impetus to its development in offering more rigorous approaches which appear more "testable" (Cracraft 1975, 1982, 1983; Mickevich 1981), at least more explicitly so. Some workers have all but ignored the important and obvious effects of dispersal (see Carlquist 1981a, for a good commentary on the importance of chance dispersal), so very evident in the origin of oceanic island floras. The proper perspective surely is to view both vicariance and dispersal as significant aspects of biogeography, and the challenge is to determine which factors have been operative in particular groups (Davis 1982).

Many different applications of cladistics to plant groups have been published already (for the most complete review, see Funk and Wagner 1982). Only a few references will be cited here to stress the clear impact that cladistics has already had in plant systematics. The types of plant groups studied include: fungi (Basidiomycetes; Petersen 1971; Vilgalys 1986); bryophytes (Mniaceae, Koponen 1968, 1973, 1980; *Orthotrichium*, Vitt 1971; Endodontaceae, Buck 1980; and Grimmiaceae, Churchill 1981; see also Mishler and Churchill 1984, and Mishler 1985a); pteridophytes *(Cystopteris*, Blasdell 1962; *Anemia*, Mickel 1962; *Equisetum*, Hauke 1963; *Woodsia*, Brown 1964; *Polypodium*, Evans 1968; *Platycerium*, Hoshizaki 1972; *Athyrium*, Seong 1972); angiosperms, primarily dicotyledons (Hippocastanaceae, Hardin 1957; *Monarda*, Labiatae, Scora 1966, 1967; *Stylisma*, Convolvulaceae, Myint 1966; *Clematis*, Ranunculaceae, Keener 1967; Proteaceae, Johnson and Briggs 1975; *Anacyclus*, Compositae, Ehrendorfer et al. 1977, and Humphries 1979, 1980; Berberidaceae, Meacham 1980; *Spilanthes*, Compositae, Jansen 1981; *Myrceugenia*, Myrtaceae, Landrum 1981; *Agastache*, Labiatae, Sanders 1981; *Physostegia*, Labiatae, Cantino 1982a; *Montanoa*, Compositae, Funk 1982; *Pieris*, Ericaceae, Judd 1982; *Tithonia*, Compositae, La Duke 1982; *Viburnum*, Caprifoliaceae, Donoghue 1983a, b). These various applications occur at different levels in the hierarchy, such as infraspecific taxa (Baum and Estabrook 1978), the specific level (as in most of the papers cited above), the generic level (e.g., Koponen 1968; Meacham 1980; Bolick, 1983; Burns-Balogh and Funk 1986) and the familial and higher levels (e.g., Petersen 1971; Parenti 1980; Rodman et al. 1984; Dahlgren, Clifford, and Yeo 1985). Most applications have been with morphological and anatomical data, but palynological (e.g., Donoghue 1985a) and chemical data have also been used (e.g., Estabrook 1980; Humphries and Richardson 1980; Richardson 1982, 1983a, b; Richardson and Young 1982; Seaman and Funk 1983; Stuessy and Crawford 1983; Scogin 1984). The difficult question of the origin of angiosperms has also been approached cladistically (Hill and Crane 1982; Crane 1985; Doyle and Donoghue 1986a, b; Doyle 1987).

Cladistics has been and will continue to be useful in plant systematics. Just as with the development of phenetics, the early claims to supremacy and absolutism (the "rigid dogma;" Burger 1979) are dying back and leaving several useful contributions. Due to serious problems of homology and parallelism, the most useful application in the hierarchy for cladistic analysis is at the specific and generic levels. The numerous difficulties reflected in the works of Parenti (1980), Mishler and Churchill (1984, 1985a, b, 1987), Sluiman (1985), and Bremer et al. (1987), in which cladograms of all the land plants have been attempted (see criticisms by Smoot, Jansen, and Taylor 1981; Young and Richardson 1982; Robinson 1985; and Whittemore 1987), underscore problems at the higher taxonomic levels. However, these efforts are laudatory and useful especially to point out problems of homologies and lack of truly comparative data, but they are a long way from giving useful results. Formal cladistically based classifications are particularly inappropriate (e.g., Sluiman 1985; Bremer et al. 1987). The only exception to this rule is the cladistic evaluation of nucleotide data which is best suited for determining higher level relationships (see chapter 21). The complexity of structural and developmental patterns in plants and the difficulties of determining homologies is well stressed by Favre-Duchartre (1984). Despite these and other difficulties pointed out by Funk and Stuessy: "For the practicing plant taxonomist, cladistics offers a method of constructing phylogenies by objective and repeatable means. This presents certain advantages over inferring phylogeny by the conventional method, the most important of which is to facilitate discussion by a clear presentation of procedures and evolutionary assumptions" (1978:159).

CHAPTER 9

Evaluation of the Three Major Approaches: The New Phyletics

Before beginning a detailed evaluation of the three major approaches to biological classification (i.e., phyletics, phenetics, and cladistics), it is useful to set the stage by outlining the factors which have influenced their development.* Some of the ideas for the origin of phenetics and cladistics have been presented already in the previous chapters, but a summary for all approaches is needed here.

HISTORICAL INFLUENCES

Figure 9.1 shows the influences that each of the major approaches to biological classification has had upon the others. Four principal influences can be documented. First is the need to cope with increasing levels of knowledge about organic diversity. This has continued to be of concern ever since humans began classifying plants and animals in their environment. It has obviously been a stimulus for the development of all major approaches, but it can be highlighted best in the development of the artificial system. The great artificial classification of Linnaeus (1735) was created principally to provide order and catalogue all the newly collected materials coming back from many parts of the world (especially from tropical regions). An artificial system of classification is perfectly satisfactory for ordering this diversity, but such a system lacks the information content that leads to higher levels of predictivity. This search for increased information content is the second need in biological classification, and it resulted in the natural approach ("polythetic classification") in which many characters were considered, rather than only one or just a few ("monothetic classification"). The development of natural approaches to classifications was pre-Darwinian, and, therefore, evolutionary considerations were not included. There was a third need, therefore, for a system which not only had high information content but which also offered evolutionary interpretations. With the appearance of the theory of evolution by means of natural selection (Darwin 1859), a rationale existed for the construction of classifications that reflected these interpretations. Thus was born the phyletic (= evolutionary) approach to biological classifica-

---

*People who have contributed to these developments have been numerous, as attested in this and previous chapters. For a delightful tongue-in-cheek self-quiz on people and their contributions, see Neticks ("Filo G.," 1978) to read about "Bygeorge Mylord's Winsome, Sir Corns R. Popped, Flames S. Harrass, Garish Telson," and others.

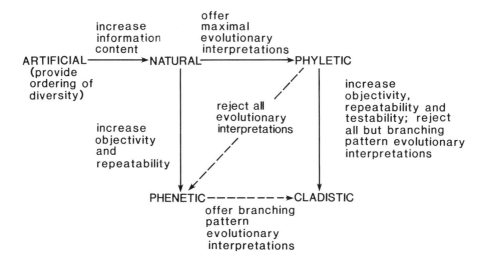

FIGURE 9.1 Diagram showing major approaches to biological classification and the perceived needs that influenced their development. Dashed lines represent less significant influences.

tion. However satisfying the phyletic approach was for the past 100 or so years, the fourth need, for increased objectivity and repeatability in classification, has arisen in recent decades. This need represents a desire to break away from the "art" of natural and phyletic classification and place it on a more "scientific" basis. Both phenetics and cladistics resulted from this need as the primary factor in their developments. Phenetics derived directly from natural classification by increasing objectivity and repeatability. A secondary influence prevailed, however, as a reaction against phyletic classification from the perspective that all evolutionary interpretations were rejected as too difficult to determine and hence inappropriate as a basis for classification. Likewise, cladistics developed primarily from the need to have an objective and repeatable (and testable) approach that stressed at least some aspect of evolution (i.e., branching patterns). Because numerical phenetic approaches developed prior to cladistics, the former served as a strong influence toward stimulating numerical methods in the latter. Nelson and Platnick (1981) suggest that cladistics had its roots in pre-Darwinian natural approaches to classification, but evidence is lacking for this contention. At the present time some workers are wondering if a phyletic approach to classification, which offers more evolutionary information than cladistics, might be made more objective, repeatable, and testable. In effect, there would be strong influences going back from both phenetics and cladistics toward the development of a new phyletic approach. More on this in the second part of the chapter.

PREVIOUS EVALUATIONS

Because there now exist three legitimate approaches to biological classification, it is important to offer an evaluation of the efficacy of all three to serve as a guide to the reader. Numerous commentaries have already been published, sometimes with strong wording and even personal attacks. As Moss has aptly expressed: "the relative merits of these approaches have been debated extensively on theoretical grounds. The continued existence of such diversity in outlook is clearly an indication of health in systematic biology, but the bitterness of the current debate seems to me to be of questionable benefit" (1979:1217). Questionable or not, the debate continues. A few of the papers will be cited here briefly to give an idea of the degree of interest generated. With regard to phenetics, some of the most noteworthy evaluations have been: Meeuse (1964); Mayr (1965); Sokal et al. (1965); Williams and Dale (1965); Blackwelder (1967b); Gilmartin (1967); Williams (1967); Johnson (1968); Colless (1971); Sneath (1971); Moss and Henrickson (1973); Ruse (1973); Sneath and Sokal (1973); Van der Steen and Boontje (1973); and Pratt (1972b, 1974). Cladistics has also generated its share of enthusiasts and critics and a selection of their commentaries includes: Bock (1968); Watt (1968); Colless (1969); Darlington (1970, 1972); Michener (1970); Nelson (1972b, 1973c); Mayr (1974b); Hennig (1975); Simpson (1975); Banarescu (1978); Gasc (1978); Van Valen (1978b); Ashlock (1979); Hull (1979); Simpson (1980); Meeuse (1982); Cronquist (1987); Donoghue and Cantino (1988); and Humphries and Chappill (1988). It is certainly true that virtually every conceivable perspective on phenetics and cladistics has already been offered. The papers of Mayr (1965, 1974b) cover most of the important points. Therefore, the points stressed in this section of the book have been largely adapted from these numerous commentaries and woven into a unified presentation for the systematic botanist.

To evaluate the three approaches to classification, we must consider both the efficacy of the process and the product (the resultant nested set of classes). Several perspectives will be offered for both. Regarding the process of classification, objectivity, repeatability, and efficiency will be considered. With reference to the resulting hierarchy, importance will be placed on stability, the information content (and predictivity), reflections of evolutionary relationships, precision and clarity of expressed relationships, and heuristic value.

EVALUATION OF THE PROCESS OF CLASSIFICATION

Objectivity

The objectivity of classification refers to clear concepts and methods so that the element of personal bias is reduced. The search for objectivity in classification has been so valued that both phenetics and cladistics were born in part from this interest. To some workers, the extent to which the process of classification can be made more objective is the degree to which it becomes more scientific. There is no question that both phenetics and cladistics are more objective approaches to classification than phyletics. In these former two approaches, the steps for making the classifications are entirely clear. With phyletics, although both phenetic and cladistic relationships are a part of the synthesis, the final assessment of relationships is still done largely subjectively (Michener 1970; Stevens 1986). But some workers have questioned whether objectivity is the most important concern. As Burtt has remarked with regard to phenetics (but in this context applying equally well to cladistics): "Is it objectivity, or lack of courage to evaluate?" (1964:14). Although both phenetics and cladistics are more objective than phyletics, none is objective in any absolute sense. Numerous subjective decisions must be made in phenetic and cladistic classification, such as the number and kinds of characters to use, the selection and coding of states, the determination of evolutionary polarity of states, homologies, algorithms to use for clustering or branching sequence construction, and so on. The subjectivity is reduced, but it is certainly not eliminated. Furthermore, cladistics (and to a lesser extent, phyletics) has the additional problem of circularity caused by the search for and use of homologies. In cladistics, almost always "certain cladistic relationships are provisionally assumed before beginning construction of the cladogram. Such a procedure can avoid a vicious circularity only if it ultimately yields results that are independent of the initial assumptions" (Cartmill 1981:85).

Repeatability

Repeatability is another important criterion by which to judge the process of classification. There is no question whatever that phenetics and cladistics are completely repeatable whereas phyletics is not. Several studies have been done with phenetics to show that even an untrained individual (the "intelligent ignoramus;" Sokal and Rohlf 1970) can make reasonably good classifications (see also Moss 1971). This has encouraged pheneticists to hope eventually for automated classification in some groups of economic importance (such as malaria-carrying mosquitoes; e.g., Rohlf and Sokal 1967), but despite these earlier hopes, no real practical gains have been made in recent years. Whether fully automated procedures of classification will ever be developed remains to be seen, but as optical-scanning devices become more sophisticated, the probability will increase of using easily scanned external morphological data of organisms in classification and even more probably for identification purposes.

Efficiency

With regard to the third criterion of efficiency of producing classifications, phyletics is by far the approach of choice. A taxonomist can produce a classification by intuitive phyletic means many times (probably up to 10 times) faster than with phenetic approaches. As Mayr has aptly stated: "Even though the computer calculation itself may be only a matter of minutes, gathering and tabulating the information for proper programming is a very time-consuming operation and is uneconomical when the proper answer is evident to the experienced specialist from a thoughtful inspection of the raw data" (1965:95). Likewise, phyletic approaches are probably at least 3 to 5 times faster than cladistic classification. It is not surprising, therefore, that the majority of practicing taxonomists, despite the furor of past decades over phenetics and cladistics, still make classifications phyletically. The human mind has an incredible capacity to assess pattern data and process these for relationships. People who are especially good at storing visual images and relating them to one another mentally to produce balanced classifications are most valuable members of the taxonomic community. One of the reasons for the lack of appeal of phenetics and cladistics to at least some of these workers is the inordinate amount of time needed to develop classifications in comparison to the intuitive mode. Phenetics and cladistics can (and should) be used selectively in routine taxonomic work (e.g., Gilmartin and Harvey 1976; Funk and Stuessy 1978), but these more time-consuming approaches will never wholly replace the more efficient intuitive efforts until the day that all life is ordered at a base-line (alpha) level of taxonomic understanding.

EVALUATION OF THE RESULTANT HIERARCHY

Stability

As for evaluation of the resulting hierarchy, one measure of a classification is its stability. This attribute can be measured by the resistance to change of a classification with the addition or subtraction of data, taxa, or both.* With regard to the former, Mickevich has expressed the opinion: "A classification with high internal stability is one in which different selections of reasonable size from the set of attributes initially selected are expected to give rise to fairly similar classifications" (1978:143). To determine the resistance to change of a classification requires having some measure of alteration of the classification as data and OTUs are altered. Potvin, Bergeron, and Simon (1983) showed phenetic classifications derived from separate morphological and biochemical data in *Citrus* (Rutaceae) to be congruent, and hence reasonably stable. Mickevich (1978, 1980) compared classifications obtained for several groups of animals based on different types of data (morphological, macromolecular, life histories, etc.) and found presumed stability of classifications based on cladograms over those derived from phenograms. Rohlf and Sokal (1980), however, point to statistical difficulties in these analyses and suggest that no certain conclusion can be made at this time. Colless (1981) also points to difficulties, particularly in comparing phenograms, because of the acknowledged distortion of data from the matrix of correlation coefficients. These concerns have been rebutted by Schuh and Farris (1981). Rohlf and Sokal discuss "the fit of a summary representation to a similarity matrix, stability, general utility, fit to a known cladistic relationship, and optimality criteria of numerical phylogenetic [= cladistic] methods" (1981:459) and again conclude that neither phenetic nor cladistic classification can be shown to be superior. Schuh and Polhemus (1980) address the issue of stability in terms of adding or subtracting taxa, and purport to show superiority of cladistic approaches. This study has been criticized on numerous computational

---

*Rohlf and Sokal also recognize two other types of stability: the "robustness of a data set to different methods of clustering;" and the "robustness of a given classificatory method to changes in character coding of the data" (1980:97). The robustness of a data set, i.e., its discontinuities and correlations that are used to construct the classifications, will affect the stability of classifications as OTUs and data are added or subtracted, and therefore, this is not listed as a separate factor. Likewise, the manner of coding of characters and states will obviously affect the resulting classification but this will have the same effect as adding or subtracting data.

grounds by Sokal and Rohlf (1981b). "But whenever we ask whether cladistic classifications are more congruent than phenetic ones, or more predictive, or better fitting to the original data, we run into difficulties because the definitions of these terms are different in different taxonomic schools" (Rohlf and Sokal 1981:481). Sokal, Fiala and Hart (1984) have shown that cladistic classifications are more stable with few OTUs, whereas phenetic classifications are more stable with larger numbers of OTUs. From my own perspective, it seems that from the standpoint solely of stability (i.e., resistance to change, and ignoring information content or predictivity), phenetics would have to be more stable because more characters are used. Adding or subtracting a few characters will have little effect on the resulting classification. On the other hand, because cladistics usually relies on only a few carefully selected characters, a change in them (in their state codings and/or polarities) will have a great effect on the classification. As Peters has aptly stressed: "Indeed, according to the logic of cladistic classification we shall know the definite number of ranks only after the part of the cladogram with the greatest number of branches is known, and the final stable classification will be identical with the knowledge of the entire genealogy of organisms. A rather difficult task, if one realizes that for objective reasons great parts of the genealogy probably cannot be reconstructed at all!" (1978:226). Phyletic classifications would seem less stable than phenetic ones but more stable than those derived from cladistics, because phyletics has a strong phenetic component and usually utilizes more characters than in the cladistic approach.

Information Content

The information content (or predictivity) of a classification is another way in which it can be evaluated. Although some workers, e.g., Jardine and Sibson, believe that: "The internal stability of a classification is related to two features which biologists have generally felt to characterize 'natural' classifications: 'information content' and 'predictive power' " (1971:155), I regard these latter attributes as different from stability and best considered separately. It is no doubt true that "A classification with low internal stability will in no case serve as a basis for reasonable predictions" (Jardine and Sibson 1971:155), but they are nevertheless different issues. The principal difficulty with evaluating classifications on the basis of information content is how to define this attribute in measurable terms. Estabrook (1971) and Duncan and Estabrook (1976) suggest information theory optimality criteria for evaluating classifications and show applications in the *Ranunculus hispidus* (Ranunculaceae) complex. Manischewitz considers "character predictiveness" by calculating the "prediction error" by finding the "absolute value of the difference between the predicted value and the actual value" (1973:179, 180). Gower (1974) offers a mathematical approach to predicting characters correctly. Michener (1978) suggests that classifications which include phenetic gaps (i.e., phenetics and phyletics) have a higher predictive quality than those that are based on strict holophyly. Platnick (1978) rejects this argument and stresses that only cladistic classifications are most predictive, as does Farris (1979b), and that the highest predictive quality comes from the most parsimoniously derived cladogram (i.e., shortest number of steps). Archie

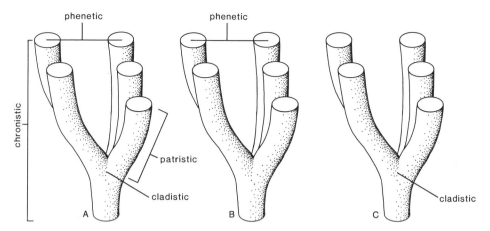

FIGURE 9.2 Kinds of information about phylogeny obtained and utilized in phyletic (A), phenetic (B), and cladistic (C) approaches to classification.

discusses different measures of predictive value of classifications and rejects phenetic correlations, taxonomic congruence, parsimony, and partitions of taxa in favor of "the degree to which states of characters are constant within and restricted to taxa in the classification" (1984:30). Suffice it to say that the returns are far from complete on this issue, and it is likely that no single definition or way of measuring the predictive quality of a classification will be agreed upon in the near future (if ever). It does seem clear, however, that a classification based on evolutionary considerations will be the most predictive, simply because suites of characters of organisms have evolved together. Therefore, both cladistic and phyletic classification of organisms should probably be judged to have greater information content, with the latter even more so because if reflects *all* known aspects of phylogeny (i.e., the actual evolutionary pathways) instead of only branching point information (as in cladistics).

Reflection of Evolution

Another yardstick by which to measure the efficacy of a resultant classification is its ability to reveal or reflect evolutionary relationships. The kinds of relationships revealed in a true phylogeny can best be shown in a three-dimensional diagram (fig. 9.2A): chronistic (time); cladistic (branching patterns); patristic (character divergence within lineages); and phenetic (over-all similarity among extant taxa).* That phylogeny contains these relationships has been mentioned over many years by numerous workers, including Lam (1936), Cain and Harrison (1960), Gisin (1966), Sneath and Sokal (1973), and Stuessy (1983). The kinds of evolutionary information that can be expressed by phyletic, phenetic and cladistic classifications are shown in fig. 9.2A-C, and it is clear that the only approach

---

*Wagner (1980) refers to "forms of cladistic trees" as "patrocladistics, chronocladistics, and topocladistics," which are really partially complete phylograms showing, respectively, degrees of divergence for each lineage, absolute time scale of divergence, and distributional relationships.

that maximizes the information about phylogeny is phyletics, which is a combination of phenetic, patristic, and cladistic information. It is very clear, therefore, that in terms of evolutionary information, phyletics is clearly more desirable than either cladistics or phenetics alone. Of the latter two, obviously cladistics has more information of evolutionary content than phenetics because the branching relationships are most important in phylogeny. Szalay has stressed that: "Phylogenetic trees are preferred over cladograms because the former always contain more information since they can express both ancestor-descendent relationships (anagenesis) as well as sister group relationships (cladogenesis)" (1976:12). Heywood (1983) echoes the importance of anagenesis in classification.

Heuristic Value

Another criterion by which to evaluate classifications is in their heuristic value, i.e., their ability to stimulate additional research, and also in their educational potential. The ability to stimulate future work is so dependent upon the impact on the individual who considers a classification that no generality seems possible. All types of classifications would serve (either positively or negatively) as a stimulus to other workers. The ability to stimulate other workers is probably more dependent upon the skill in presenting and discussing the classification, and in the highlighting of additional problems that might be investigated further, than in the means of constructing the classification.

Testability

A final criterion by which to evaluate classifications is "testability." Cladists regard their classifications as more "testable" and hence more "scientific" because attempts to falsify the relationships of the cladogram can be made (e.g., with new data sets) and if changes seem warranted, then the classification is modified accordingly. Phenetic classifications, however, can also be very easily "tested" by addition of new data, and are of no lesser value in this sense. Because phyletic classifications are usually not explicitly constructed, it is more difficult to test them explicitly, but it can still be done with the addition of new data sets and comparing the new classification with the old.

THE NEW PHYLETICS

The perspective presented here is that the phyletic approach to classification is the most useful one for biological (and more specifically botanical) systematics. Not only is it more useful in practical terms, because it is often done intuitively and therefore is more rapid than either cladistics or phenetics, but also because it is the most meaningful reflection of evolutionary relationships, which should be the basis for biological classification. Virtually the only real problem that faces phyletics is its lack of explicit methods (regarded as "ill-defined;" McNeill 1979a:465). This lack does not hinder the positive results of the intuitively generated classifications, but communication would be greatly fostered if the methods were made explicit. Bock (1977) has given a thorough and detailed review of the methods of phyletic classification, but some explicit means of synthesis of the phenetic and cladistic information is lacking. The need here has been expressed by several workers; e.g., Moss, who has mentioned that "attempts to combine these aspects [phenetics and cladistics] of relationship operationally into a single field (phyletics) are developing rapidly" (1978:88), and Michener, who has remarked: "For the future, we should look toward more formal integration of cladistics and phenetic information to provide a more operational method of producing evolutionary or phylogenetic classifications" (1978:117–118).

Methodology

What is needed now, therefore, is an explicit synthesis of all relationships, especially cladistics and patristics, to arrive at a phyletic classification. Several have already been provided, but these have never been brought together for comparison in one place.

The earliest explicit phyletic approach (one which was brought to my attention through reading the book by Schoch 1986), appears to be that of Hanson (1977). His method is to determine the "phyletic distance" between taxa based on primitive and derived character states. He uses the term "-seme" as a substitute for "-morphous," and hence he uses plesiosemic and aposemic instead of plesiomorphic and apomorphic. He also uses the term neosemic to refer to a relationship in which one taxon has a character and the other is completely lacking it. Relationships are assessed explicitly between each pair of taxa by the following formula:

$$R = \left(\frac{-p + (2a)^2 + (3n)^2}{t} \right) + 1$$

where R is the phyletic distance between two taxa, p the number of plesiosemic affinities, a the number of aposemic affinities, n the number of neosemic relationships, and t the total number of characters used. The formula has the effect of weighting the derived states much more than the primitive ones. Phyletic distance values are obtained between all pairs of taxa and these can be plotted as a 2-D graph or adapted to produce dendrograms. I personally do not favor this particular method, primarily because it does not include data on autapomorphies except in the rare instance when there are new characters (not just new states) involved. It is important, however, as an early attempt to assess symplesiomorphies and apomorphies quantitatively for explicit phyletic classification.

Estabrook (1986) suggested an explicit method for phyletic (= evolutionary) classification using what he called "convex phenetics." In this approach, a basic data matrix is developed for all characters and states of the study group. From these data, an unrooted network of relationships is produced among all taxa. Next, a clustering algorithm is used but before the clusters are finalized, the position of the taxa on the network is compared. For the cluster to be completed, the group must pass the convexity test, i.e., the taxa must all be interconnected

TABLE 9.1 Relationships, Methods, and Resultant Graphic Displays for Explicit Phyletic Classification. (From Stuessy 1987)

| Relationships | Methods for combining relationships | | | |
	Graphic elaborations of cladograms	Resemblance matrix additions	Consensus trees	Two-dimensional vector additions
Cladistics and patristics	Wagner groundplan/ divergence and others[a]			
Cladistics and phenetics	3-D phylogram		Dendrogram-style phylograms	2-D plots
Cladistics, phenetics and patristics	3-D phylogram showing autapomorphies			

[a] Such as in Johnson and Briggs (1984) and Emig (1985).

on the network without intervening members (e.g., Meacham and Duncan 1987). If the newly clustered group is not convex, the algorithm passes to the next most similar cluster with a check on convexity, and so on, until a cluster is found that is also convex. From the clusters, a phylogram is then produced which is basically phenetic but with cladistic modification.

Stuessy (1987) recommended five explicit approaches for phyletic classification (table 9.1). The first is the already familiar construction of phylograms by graphic elaborations of cladograms in which at least the patristic distances are also shown. An example of this approach would be the Wagner Groundplan/ divergence graphic display (Wagner 1980), which superimposes a cladogram on the "advancement index" of hemicircles based on patristic distances (cf. figs. 8.2 and 8.5). Other examples would include the similar graphic techniques of Johnson and Briggs (1984) and Emig (1985).

A second approach involves the reconstruction of three-dimensional phylograms, using phenetic, cladistic, patristic, and chronistic information (fig. 9.2), and done in an explicit way. The cladistic and patristic relationships are derived from a cladistic analysis. The phenetic affinities result from a clustering or ordination of taxa in a two-dimensional plane which represents the relationships in Recent time. Chronistic data are derived from distributional, geological and paleontological evidence. The overall result is a three-dimensional phylogram that has been derived explicity (fig. 9.3). This has been called a "cladophenogramme" by Genermont (1980:39).

A third approach to explicit phyletic classification involves resemblance matrix addition (table 9.1). This involves defining cladistic, patristic and phenetic relationships quantitatively. Values of phenetic relationships are typically quantitative and hence pose no difficulty. The cladistic distance can be defined as the absolute number of nodes separating two taxa on a cladogram, and the patristic

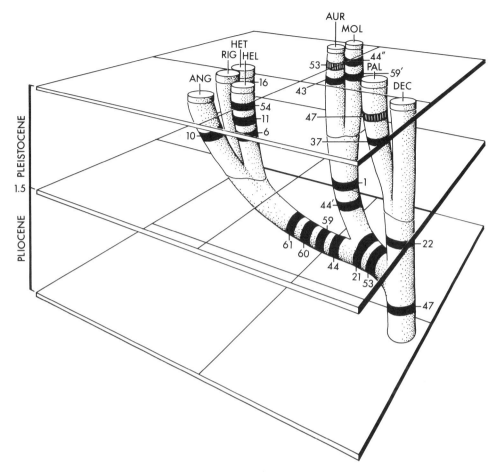

FIGURE 9.3 Three-dimensional phylogram of the genus *Lagascea* (Compositae). The upper plane shows phenetic relationships derived from PCA. Rings represent apomorphies; dashed rings indicate reversals. (From Stuessy 1983:7)

distance as the number of apomorphies (synapomorphies and autapomorphies) between two taxa on the same cladogram (fig. 9.4). These values can then be combined in different explicit (typically phenetic) ways, to yield dendrogram-like phylograms (fig. 9.5) or plots in two-dimensional space. These summary diagrams, therefore, can be made explicitly and can be used as a basis for constructing a phyletic classification.

A fourth approach also results in dendrogram-style phylograms, and this is the use of consensus trees (table 9.1). Here information contained in different dendrograms, such as phenograms, cladograms and "patrograms," can be blended into one consensus phylogram. Consensus techniques have already been discussed with reference to cladistics (cf. chapter 8) and will not be detailed again here.

The fifth approach to explicit phyletic classification is called "two-dimen-

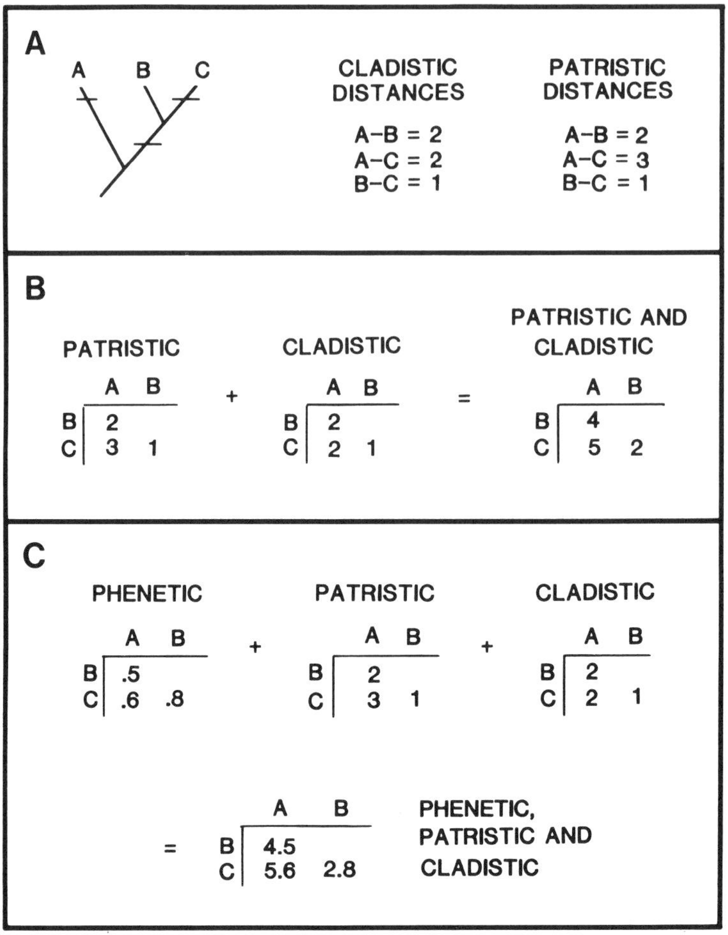

FIGURE 9.4 Resemblance matrix addition among hypothetical taxa A–C. A, hypothetical cladistic and patristic distances; B, addition of patristic and cladistic distances; C, addition of arbitrary phenetic distances to patristic and cladistic distances. (From Stuessy 1987:253)

sional vector addition" (table 9.1). Here one starts with a two-dimensional phenetic plot of the taxa. Their positions are then modified on the basis of cladistic and/or patristic distances. These values are treated as vectors and used to move the taxa away from each other geometrically (fig. 9.6). This results in a modification of the *relative* spatial relationships and gives a different view for purposes of phyletic classification (fig. 9.7). The overall result is somewhat similar to the "convex phenetics" of Estabrook (1986), in which phenetic relationships are also modified by reference to cladistic data.

Hall (1988) offers an explicit approach for phyletic classification also involving modification of phenetic clustering by means of cladistic data. He uses a homogeneity measure to define the initial groups but employs a "cladistic mod-

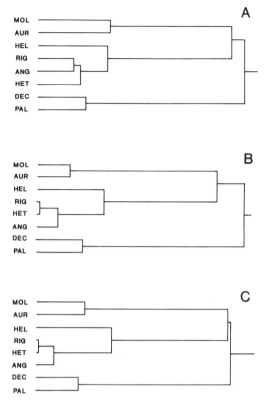

FIGURE 9.5 Resemblance matrix addition to yield different types of phylograms in *Lagascea* (Compositae). A, cladistic and patristic relationships combined; B, cladistics and phenetics; C, cladistics, patristics, and phenetics. (From Stuessy 1987:260)

ulation" to "de-weight" some of the characters involved (the plesiomorphies). This has the effect of emphasizing the more cladistically significant apomorphies and the phenetic relationships are altered accordingly.

Two other less explicit phyletic approaches should also be mentioned. From the perspective of organizational science, McKelvey (1982) advocated the combination of phyletic and phenetic approaches to classification. In procedures approximating consensus techniques, he suggested combining phenograms and intuitively generated phylograms for a final dendrogram of higher informational content. Inglis (1986, 1988) advocates developing a "stratigram" which is a branching diagram based on phenetic relationships but with the shared character states shown as horizontal bars similar to the early Hennigian cladistic graphic display (cf. fig. 8.6). He indicates that the stratigram can be interpreted cladistically to yield a cladogram or with even further biological information it can be modified into a phylogram.

The methods discussed above have all stressed the phyletic *grouping* of taxa, but some mention must also be made of *ranking*. Absolute measures or conventions for ranking in phenetics (phenon lines) and cladistics (phyletic sequencing)

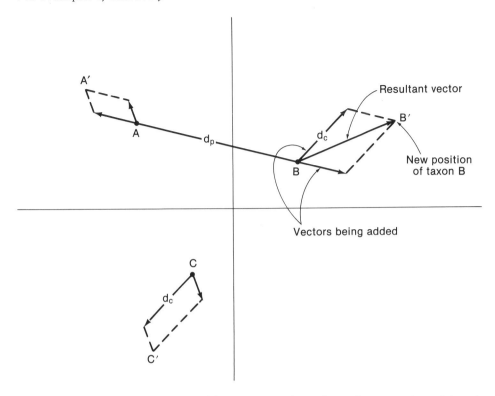

FIGURE 9.6 Example of vector addition among hypothetical taxa A–C positioned arbitrarily in two-dimensional phenetic space. d_c = cladistic distance; d_p = phenetic distance. (Redrawn from Stuessy 1987:254)

have both failed due to over-precision, which trespasses on the biological and evolutionary reality. Hence, no absolute measures for ranking in phyletic classification are offered here, either. The phylograms serve as a basis for making such decisions and paraphyletic groups are completely acceptable (cf. frontispiece). A guiding principle might be: the greater the distance separating taxa on the diagrams (of whatever nature), the higher the level in the hierarchy at which they should be recognized.

All these approaches have one thing in common: they all attempt to include more information about phylogeny together in an explicit way for the purpose of phyletic classification. Which particular phyletic method (or methods) will be regarded as yielding the most predictive classifications will have to await future comparative statistical studies. The main point is that several methods now exist for explicit phyletic classification. Their use should be encouraged.

Efficacy of New Phyletics

An example of the efficacy of the New Phyletics can be seen in studies on relationships of the Lactoridaceae, an endemic angiosperm family of the Juan

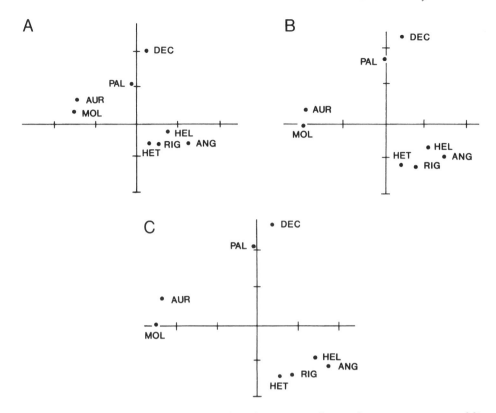

FIGURE 9.7 Two-dimensional plots of evolutionary relationships using vector addition in *Lagascea* (Compositae). A, 2–D plot of phenetic distances; B, phenetics modified by addition of cladistic distances; C, phenetics modified by addition of cladistic and patristic distances. (From Stuessy 1987:262)

Fernandez Islands, Chile (Lammers, Stuessy, and Silva 1986). Phenetic, cladistic and chronistic relationships were combined to produce a 3-D phylogram (their fig. 5) in which the Lactoridaceae were shown to branch from their relatives about 69 million years ago in the Late Cretaceous. More recently the first fossil

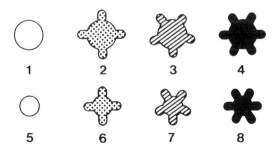

FIGURE 9.8 Set of geometrical objects showing variations in shape, size, and shading that can be classified in several different ways. (From Bell 1967:8)

Lactoris pollen was found in borehole cores off the west coast of Africa (Zavada and Benson 1987) in deposits also of Cretaceous age (or even slightly older). These new data help confirm the relationships contained in the original phyletic reconstruction of Lactoridaceae and relatives, and they give support to the New Phyletics as a valid and useful approach.

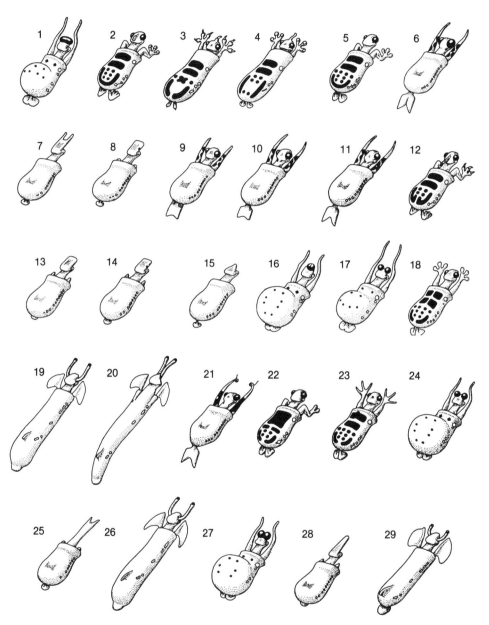

FIGURE 9.9 Set of 29 hypothetical "Caminalcules" that are useful for exercises on the different approaches to classification. (From Sokal 1966:106, 107)

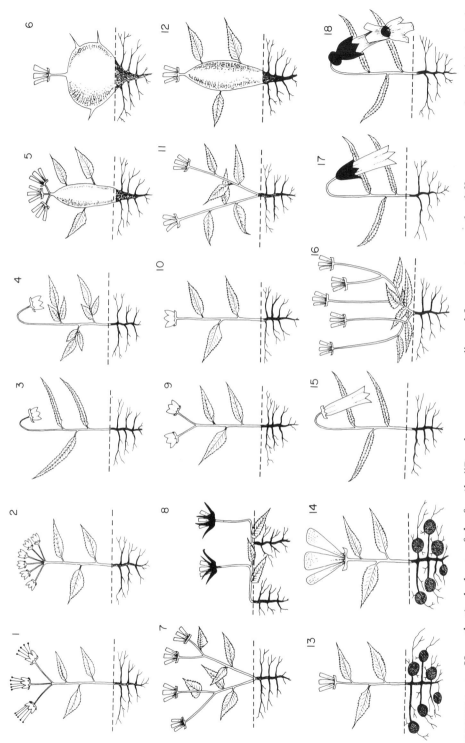

FIGURE 9.10 Hypothetical plants of the family "Dendrogrammaceae" used for exercises on the different approaches to biological classification. (From W. H. Wagner, in Duncan, Phillips, and Wagner 1980:266)

151

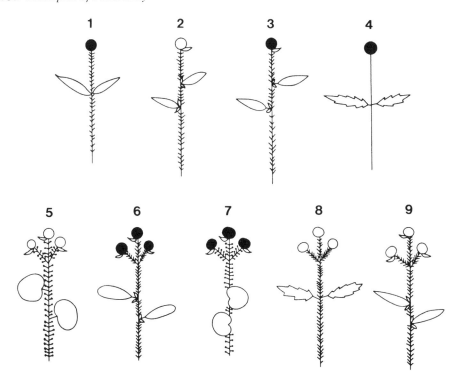

FIGURE 9.11 Hypothetical plants of the "Cookophytes" also used for understanding different approaches to biological classification. (From C. D. K. Cook, in McNeill 1979b:485; also 1979a:470)

Despite the general desirability of phyletic classification in most situations, certain types of taxonomic problems are well suited for phenetic or cladistic analysis alone. The most compelling situation for use of phenetics is in groups displaying confusing character variation that cannot be handled intuitively. An example might be flavonoid chemical variation (20 or so compounds) throughout a broad geographical range of a taxon in several hundred populations. There may well be patterns of variation that simply cannot be seen except by clustering or ordination. For a very appropriate use of phenetics at the infraspecific level, see Furlow (1987). In groups that show obvious discontinuities, phenetics serves only to document the obvious and would be ill-advised in such cases. Likewise, in groups that are well understood phenetically and for which phylogeny is of interest, cladistics is most important for unravelling the branching aspect of the total phylogenetic pattern. Cladistics is rarely useful in groups for which a good phenetic or phyletic understanding is not yet available.

As an aid to understanding the pros and cons of different approaches to biological classification, sets of hypothetical taxa can be used, and these are especially appropriate as classroom exercises. One of these is the "nuts and bolts exercise" recommended by Burns (1968) and Sundberg (1985). Students are asked to classify these objects by the various methods, including the artificial approach (and offering allied information, such as out-groups, etc., as needed). An important follow-up to this exercise, however, is a discussion of how organisms differ from the nuts and bolts, especially in their populational variation, and what problems this creates for biological classification. This is very important lest the students leave with the impression that organisms can be grouped as easily as inanimate objects. A set of geometrical shapes (fig. 9.8) has been used in the same way to illustrate basic classificatory principles and methods (Bell 1967). Perhaps the most interesting set of hypothetical organisms are the "Caminalcules" (fig. 9.9) devised by Joseph Camin. These were constructed by selecting a hypothetical ancestor and making structural modifications into different lineages corresponding to the origin of new species in real animals. These have been used in several studies on classification (e.g., Sokal 1966; Rohlf and Sokal 1967; Moss 1971; and Sokal 1983a, b, c, d, e). Another set of hypothetical animals can be found in Brooks et al. (1984). Also available for use are the collections of hypothetical plants of the family "Dendrogrammaceae" (fig. 9.10; W. H. Wagner, in Duncan, Phillips, and Wagner 1980; see also Duncan 1984) and the "Cookophytes" (fig. 9.11; C. D. K. Cook, in McNeill 1979a, b).

SECTION THREE
Concepts of Categories

I N THE previous section of this book, the discussions of classification were centered primarily on grouping. Ranking is extremely important, because without this second part of the process of classification, none of the benefits of the system can be realized. Ranking has at least two important functions: (1) it allows for a more efficient and consistent method of communication about the taxonomic units; and (2) it allows for the taxonomic units to be placed in categories that reflect degrees of differences in characters and character states. Differences at the higher levels of ranking are great, and those at lower levels are small, and in this fashion ranking allows for expression of the predictive value of the classification. Sneath and Sokal have commented further:

> It seems to be a general human tendency to stress the sharpness of distinctions between classes and to overemphasize gaps in the spectrum of phenetic variation. Mutually exclusive classes are frequently used conceptually by humans, although we are repeatedly warned against stereotyping events and individuals. Nevertheless we succumb to a natural tendency to avoid intersecting sets, which would result in some individuals being members simultaneously of more than one set. And we are so obedient to the Linnean system, which requires mutually exclusive and hierarchically ordered classes, that the process of classification has become synonymous in the minds of many biologists with a mapping of the diversity of nature into the Linnean system. This is reinforced by a third trend, the evolutionary explanation of the hierarchic arrangement of the Linnean system. If natural taxa are to be monophyletic (sensu Hennig [= holophyletic, as used in this book]), a dendritic pattern of organismic diversity has heuristic and intrinsic value and the aim of taxonomy would be to arrange organisms into those nested, mutually exclusive taxa that correspond most closely to the actual clades. A fourth reason for classification of organisms in nested hierarchies has been the achievement of economy of memory, which, though not necessarily restricted to such a taxonomic system, is conveniently associated with it. (1973:200)

CHAPTER 10

The Taxonomic Hierarchy

In order for ranking to be achieved, a hierarchy of categories must be provided into which taxonomic units can be placed.* For organisms, such a structure is called the Linnaean hierarchy, after the Swedish botanist Carolus Linnaeus (e.g., 1753, 1754), who first consistently used many of the categories we now employ. With inanimate objects, many different types of hierarchies, or sets of classes, are available for use. For the biotic world, we use principally one, and this rigidity prevails for at least three reasons: (1) for purposes of efficient and exact communication on a worldwide basis, one standard hierarchy is essential; (2) one particular category, the species, is fundamental to our understanding of the organization of organic diversity, and therefore, all other categories in the hierarchy relate directly or indirectly to this level (this relationship imposes limitations on the numbers and kinds of categories available); and (3) it is assumed that all life originated in the same general way through evolution by natural selection, and therefore, the resultant units of diversity recognized should apply equally well to all or any part of the living world. Griffiths (1976) suggests that biologists abandon the Linnaean hierarchy and substitute an "unclassified hierarchy" (following Hennig 1969). Such a system would use only synapomorphies to construct unnamed classes, which, in my opinion, is not acceptable as a general approach to biological classification (see chapter 9).

HISTORY

The Linnaean hierarchy was developed long before evolution was discussed as a possibility. During the fifteenth and sixteenth centuries, there prevailed a strong belief in the divine creation of organisms. In a sense, this view of the origin and diversity of life recognized a pervasive similarity of organization throughout the living world. The early taxonomists believed their classifications were simply reflections of the Grand Plan of the Creator. Although early workers eagerly sought this Divine Blueprint, it took several centuries for a taxonomic hierarchy

---

*DuPraw (1964, 1965) has advocated nonhierarchical classifications, i.e., ordinations, but despite considerable utility, these do not provide the needed aspects of communication and predictive characteristics of classes that result from hierarchical classification. Jancey (1977) stresses the utility of hyperspatial models, but these seem most useful not in formal classification but rather to understand why some taxa, especially those at the specific and subspecific levels, are more difficult to circumscribe than others.

for organisms to be developed and used consistently, because different workers had their own ideas as to what God's design was really all about. Linnaeus helped stabilize these various attempts to achieve a taxonomic hierarchy of organisms. Because of his influence in botany at that time (1700s), and the simultaneous and almost universal recognition of the need for stabilization of systems of classification, his set of classes was readily adopted. Linnaeus' hierarchy was modified slightly in ensuing years, and the resulting scheme of basic categories of units, still in use today, is as follows:

> Division
>> Class
>>> Order
>>>> Family
>>>>> Genus
>>>>>> Species
>>>>>>> Subspecies
>>>>>>>> Variety
>>>>>>>>> Form

Many additional intermediate categories in the hierarchy exist, such as subfamily, subgenus, section, tribe, etc., but those listed above are the basic units. See the latest edition of the International Code of Botanical Nomenclature (Greuter et al. 1988) for the more generally accepted categories.

LOGICAL STRUCTURE

The Linnaean hierarchy can be viewed as a system of classes within classes (or boxes within boxes, called a system of "nested classes"; Buck and Hull 1966). Because this system was devised by Linnaeus (and his predecessors) by deliberate reasoning, a clear and full understanding of its construction can best be obtained by logical analysis. Some workers might question the utility of attempting to analyze the taxonomic hierarchy in such detail, but I believe that a deeper insight into the nature of the taxonomic hierarchy should clarify and make more meaningful the relationship of the process of classification to the development of the resulting nested sets.

Before the logical structure of the Linnaean hierarchy can be examined, two terms need to be defined and clarified: *category* and *taxon*. The former term refers to a particular level (or rank) in the taxonomic hierarchy, such as genus or class, and taken collectively all these available categories represent all the different levels in the classification system. All the categories may not always be used, of course, but they are available if needed. The second term, taxon, refers to a cluster of individuals grouped together based on the sharing of features in common. These initial groups may then be clustered again to form taxa of higher orders. Based on the characters shared by all members of these groups, the taxa are referred to particular categories (such as species or varieties) available in the hierarchy. To illustrate further the use of these terms, the sequence of activities in the process of classification involves first grouping individuals into taxa and subordinate taxa into taxa of higher levels, followed by evaluation of the characters of the taxa and subsequent referral of these taxa to appropriate categories in the taxonomic hierarchy (Scott 1973; fig. 1.2).

The utility of the taxonomic hierarchy developed by Linnaeus did not result fortuitously (Cain 1958; Larson 1971; Jonsell 1978). One reason for its successful construction and employment was that it built upon less elaborate hierarchies in use at that time, rather than being a radical departure from contemporary schemes. These earlier hierarchies, in turn, were based on concepts of relationship developed by the ancient Greeks, and in particular by Aristotle (Cain 1958). His principle of "logical division" maintained that any group of objects could be divided into subgroups based upon a single criterion called a *fundamentum divisionis*. The larger unit was called the "genus" and the smaller units the "species." (For more discussion of this topic, see chapter 3.) This method of classification and the limited hierarchy of objects that resulted was a powerful precedent for a way to look at the world in an organized fashion. A second reason for the successful development of the Linnaean hierarchy, however, must be credited to the encyclopedic genius of Linnaeus himself (Heywood 1985), who realized the utility of a reasonable and fixed number of categories and who ably demonstrated this utility by successfully classifying all the plants and animals known to him (approximately 7700 species of plants; Stearn 1957).

But a third and more fundamental reason exists for the efficacy of the Linnaean hierarchy. Whether culturally "primitive" or "advanced," people classify objects in their environment (Berlin, Breedlove, and Raven 1966; Raven, Berlin, and Breedlove 1971; Berlin 1973). The rationale for this circumstance is the need to provide a framework and a mechanism for communication of concepts. The human brain cannot assess relationships (or connections) without assigning coordinate and subordinate roles to objects and ideas. As Riedl has stressed: "all the products of man, his knowledge, his tools, his institutions, even his scientific theories and all his associations are structured hierarchically" (1984:87). This intellectual ability most probably mirrors the actual relationships of animate and inanimate objects, since man's cranial powers have also evolved on earth. It seems likely, therefore, that physical and biotic elements and interactions are of coordinate and subordinate natures. Certainly it is clear that hierarchical structure is more diverse with biological data than with random data (Rohlf and Fisher 1968).

As a result of these several reasons, the Linnaean hierarchy has a definite structure and the relationships among its parts, the individual object or organism, the taxon, and the category, can be analyzed logically (Gregg 1954; Buck and Hull 1966; Jardine, Jardine, and Sibson 1967). Two basic types of relationships exist: (1) subordinate, and (2) coordinate. In considering the subordinate relationships, two types are recognized: (a) those between an individual and its taxon, and between a taxon and its respective category; and (b) those between a higher taxon and a lower taxon. The first type of subordinate relationship is one of membership. For example, an individual human being, John Doe, is a member of various taxa ranging from the high taxon Mammalia to the low taxon *Homo sapiens* (table 10.1). In turn, these are each members of single categories with Mammalia a member of the category "class" and *Homo sapiens* a member of the

TABLE 10.1 Relationships of Membership (ϵ) and Inclusion (c) Between Components of the Linnaean Hierarchy.

Individuals		Taxa		Categories
John Doe	ϵ	Mammalia	ϵ	Class
		$\subset$		
John Doe	ϵ	Primata	ϵ	Order
		$\cup$		
John Doe	ϵ	Hominidae	ϵ	Family
		$\cup$		
John Doe	ϵ	*Homo*	ϵ	Genus
		$\cup$		
John Doe	ϵ	*Homo sapiens*	ϵ	Species

category "species." The second type of subordinate relationship, that of inclusion, applies only to the connection between a higher and lower taxon. For example, *Homo sapiens* is included within *Homo* which is likewise included within the Hominidae (table 10.1). Coordinate relationships exist between individuals and between categories. Obviously, for the purposes of the hierarchy, all individuals are equivalent, and similarly, the categories are equivalent to each other because they are each simply points of reference or "steps" in the hierarchy. To have subordinate relationships of membership or inclusion for the categories would essentially destroy the ladder-like structure of the hierarchy and its utility.

One of the difficulties that arises with the above logical analysis of the Linnaean hierarchy is called "Gregg's Paradox" (Buck and Hull 1966), which is the logical acceptance of monotypic taxa (i.e., taxa with only one subordinate unit, such as a genus with only one species). Logically the problem reduces to the following set of relationships: *Ginkgo* = genus, *Ginkgo biloba* = species, *Ginkgo* = *Ginkgo biloba*, and therefore, genus = species. If these two categories are identical, then the ladder-like quality of the hierarchy collapses with negative results. This problem has been discussed by several workers (e.g., Sklar 1964; Buck and Hull 1966; Farris 1967, 1968; Gregg 1967; Ruse 1971), and the most reasonable solution seems to be to bestow upon such monotypic taxa intentional definitions that admit the existence of at least one more unknown or extinct taxon. In this fashion the "paradox" is avoided.

With these generalities as a background, it is worth stressing that the categories of the Linnaean hierarchy have been given definitions over the past two centuries, and it is therefore important to examine these definitions in detail. Some workers, however, believe that categories are not defined (e.g., Mason 1950; Davis and Heywood 1963), but that they are instead used "by convention" (Heywood, pers. comm.) or "international agreement" (Mason 1950:202). This is not an especially helpful perspective, because to refer a taxon to a category "by convention" or "agreement" requires a definition of the category to be able to make the referral. Bock correctly points out: "Categories, such as species, genus, family and order, are words and hence are defined. . . . Good, clear definitions exist for all categories, although only the species definition can be affixed to definite biological phenomena. Taxa are groups of organisms within the scope of classificatory hypotheses and hence are real objects in nature which are recognized, delimited and described, but never defined" (1977:878). Muir (1968) contends that taxa are also defined, but I agree with Bock: categories are defined, and taxa are circumscribed. Within a certain broad viewpoint, one could assert that taxa are *specifically* defined, but this would still differ from the *general* definition applied to the category. Kitts (1983) has attempted to show that names of species are defined by essential properties (a Platonic viewpoint). I do not agree at all with this contention. It seems clear that categories are defined, taxa are circumscribed, and names of taxa are simply assigned arbitrarily (by "christening;" Hull 1976, or "baptism;" Kitts 1983) with proper suffixes added to indicate at which rank the taxa are to be referred (not done in a strict way below the family level, although conventions do exist for the overall formation of Latinized names at the generic level and below).

It is important to realize, however, that no *absolute* means currently exist for the definition of categories in the Linnaean hierarchy. Furthermore, not all botanical taxonomists would define the categories in the same way. In fact, if asked to give criteria for each of the commonly employed units, ten different systematists would probably give ten slightly different answers. It is also true that categories in different divisions (or even classes or orders) are not always comparable in terms of numbers of species, numbers of individuals, phenotypic and genotypic diversity, etc. (Van Valen 1973). Despite these variations, much similarity of definition would prevail, and it is this similarity that is emphasized in this book. In the following sections, in addition to discussions of the current definitions of categories, the history of the botanical usage of each category is discussed briefly.

CHAPTER 11

Species

The species is the fundamental category of the taxonomic hierarchy. Species are the "building bricks" in biological classification (Davis 1978:325) from which concepts of higher and lower groups are developed. "The species is a biological phenomenon that cannot be ignored. Whatever else the species might be, there is no question that it is one of the primary levels of integration in many branches of biology, as in systematics (including that of microorganisms), genetics, and ecology, but also in physiology and in the study of behavior" (Mayr 1957a:iii). The species category is doubtless one of the oldest concepts used historically by people in any consistent way, with the only challenge coming from the easily recognized traditional kingdoms of organisms (i.e., plants and animals). This relates to the ease of recognition of species in the natural environment by both culturally primitive and modern societies (Berlin 1973). It is, therefore, also the lowest category in the hierarchy that is *consistently* used and recognized by all peoples of the world. Because of this consistent usage, the species category has been defined more explicitly and successfully than any other unit. Species are also viewed by most workers as the basic units of the evolutionary process (e.g., Stebbins 1977), although there have been challenges to the contrary (e.g., Levin 1979a). Within biology, species inventories provide a foundation for and even shape other research efforts (Harper 1923). As Mayr has put it: "Whether he realizes it or not, every biologist—even he who works on the molecular level—works with species or parts of species and his findings may be influenced decisively by the choice of a particular species" (1957b:1). And finally, the strong emphasis on species recognition relates to our own existence. Most people accept without question that all the present individuals and races of humans in the world constitute a single biological species, and hence, we interpret our observations of the rest of the diversity from within this perspective. It is natural, then, that we would seek species elsewhere in biotic diversity. We even seek species among viruses (Milne 1985), which may or may not be regarded as living.

Because of the importance of species in biological classification, much attention has been given in the literature to this issue. Darwin (1859) devoted considerable space to the nature of species in his classic work. Several symposia have considered this question in detail (e.g., see papers by or edited by Bessey 1908; Shull 1923; Sylvester-Bradley 1956b; Mayr 1957c; Lewis 1959; Heiser 1963a), and hundreds of additional papers have been written on the topic. Some of these have even been reprinted as a separate volume in the Benchmark series (Slobodchikoff 1976). There have been so many papers, in fact, that Edgar Anderson once remarked (quoted in Heiser 1963a:123): "We need more studies of species and less discussions of them." One of the recurring problems has been the widely different species concepts that have been used in some groups, such as in *Rubus* (Rosaceae), in which apomixis and interspecific hybridization are common. For example, a range of 24 to 381 species has been recognized in this genus within the northeastern United States and adjacent areas by different workers (Camp 1951).

HISTORY OF SPECIES CONCEPTS

To understand well the current usage of concepts of species, it is necessary to review in some detail the historical development of the concept. Many authors have reviewed these ideas (e.g., Britton 1908; Mayr 1957b; Beaudry 1960; Grant 1960; Heywood 1967; Ruse 1969; and Mayr 1982), and from these discussions have come several perspectives of value. Stress will be placed on the changes in species concepts from Plato to Mayr with emphasis on understanding historically the typological, evolutionary, nondimensional, biological, and nominalistic species concepts. Following this will come a discussion of some of the logical and philosophical difficulties with all species concepts, a description and commentary on the principal species concepts in use at the present time, and finally a personal recommendation for a useful species concept for practicing plant taxonomists dealing with angiosperms.

To gain a full understanding of the progression of species concepts (as well as to understand almost any significant aspect of Western culture), we need to turn first to the philosophers of the Greek civilization, and in particular to Plato. Plato's philosophy dealt in part with the organization of all things and contained the concept of the *eidos* or "species" to refer to any different kind of thing. All objects were considered as being only shadows of the *eidos*, and consequently, variation was overlooked in favor of the *typological species* approach that would develop greater biological proportions under Aristotle and Cuvier, and which would also influence both Ray and Linnaeus (Mayr 1957b).

The principle of logical division used by Aristotle, based in part upon the ideas of Plato, was to be the basis of taxonomy for many years to come, and it served as a *schema* upon which his species concept was framed. According to the principles of logical division, a species was any unit possessing a common essence (i.e., an abstract idea or concept that makes the unit what it is), and this logical species was a relative term, being applicable to various levels in a classification scheme. A logical relationship existed between the genus and species, this connection being determined by the use of the species differentia and a *fundamentum divisionis* that was both mutually exclusive and exhaustive. Consequently, a species was defined on an *a priori* basis and was regarded as fixed and unchanging. Lennox argues that Aristotle had two concepts of species (and genera), the one just described and another one based on relative differences, i.e., "to see biological differentiae as ontologically on par with qualities varying along a continuum (e.g., temperature, color, tone, or texture)" (1980:322). While logical division worked relatively well for material objects (books, tables, houses, etc.), it was not adequate to cope with the variability of living organisms. Organisms not fitting exactly into the scheme were forced into it anyway due to the principle of exhaustion. Other problems involved the difficulty of telling if a character was a differentium, a property, or even an accident; and the possibility existed that some species could belong to different genera according to the different *fundamenta divisionum* applied. Logical division surely was a factor in the further development of the typological species concept (equivalent to the "classification type-concept" of Farber 1976), as this latter idea maintained that each class of organisms possessed an essence, its members of which were delimited by differentia. It also doubtless led to formulation of the "type concept" as used in nomenclature whereby names of taxa are rigidly associated with selected specimens designated as types (called the "collection type-concept" by Farber 1976); this in no way vitiates attribution of variation to the taxa—it is simply a useful system to allow consistent labeling. Any deviants from this essence were discarded as being accidental occurrences. Obviously, this scheme neither coped with the variability of life nor necessarily showed true affinities.

John Ray was greatly influenced by the writings of Aristotle as was every scholar of the seventeenth century. But Ray's species concept (as illustrated in his *Historia Plantarum*, 1686–1704) was based upon the belief that species bred true, and any variations that occurred were to be treated as accidents resulting from either environmental factors or factors inherent within the plant itself (Davis and Heywood 1963). These variants were to be disregarded in the consideration of species. As a result, even though the species was treated as possessing morphological distinctions, this distinction was still rigid. Although Ray did not alter the rigidity of the species concept, he added new dimensions by trying to objectively define the species in terms of morphology and reproductive relationships instead of continuing to utilize the "essence" of Aristotle. Ray also believed that in the delimitation of the various groups within a classification scheme, all parts of the plant should be considered. Also, the basic characteristic distinguishing the species was that of reproductive isolation. Consequently, his species concept was completely nonarbitrary, but also completely nonvariable; the emphasis was on "limits" (Davis and Heywood 1963:95).

Linnaeus' species concept was based upon the idea that originally all species had been created by God and each one possessed an "essence" in the Aristotelian and religious senses (Thompson 1952; Svenson 1953; Larson 1971; Stafleu 1971). "The technique Linné uses to describe or define the species is adequate only to his early belief that the elements of order consist of the fixed, discrete, 'natural' kinds created by God" (Larson 1971:121). Due to his gardening experience and contact with the tulip trade, however, he became cognizant of the differences that could result in a species due to the efforts of man; he also became aware of variations and aberrant forms in the specimens he collected (Mayr 1957b). As he aged, he began to recognize that variations could and did occur (Ramsbottom 1938; Engel 1953). He believed, however, that these examples of nature in her "sportive mood" were also God-produced. It was clear to Linnaeus that these deviant forms were not deserving of the rank of species but rather of variety (so first expressed in the *"Methodus,"* dated 1736, a broadside included with most copies of the *Systema Naturae*, 1735; Schmidt 1952).

Filling in the gap between Linnaeus and Darwin, Cuvier (1835) approached the species concept in a different light (although still influenced by Plato), considering it as a specific form "conditioned by function" and governed by laws of metaphysics and mathematics thus acquiring "the coherence and necessity of a geometrical definition from which we can deduce all the attributes of the object of the definition" (Thompson 1952:16). Conversely, each species could logically be deduced from the parts of the organism. An important new idea, however, was the consideration of all parts of the organism for indicating the whole and eventually for indicating the species itself (a corollary of Cuvier's principle of organic integration or correlation; Simpson 1961). These ideas gave more momentum to the typological species concept maintaining the immutability of species, although at the higher levels in the hierarchy Cuvier used more of a "morphological type-concept" based on general morphological plans (Farber 1976) or archetypes (Pratt 1972b).

As with Linnaeus, Darwin's species concept changes during his lifetime, also becoming more plastic probably due to the variation he saw in his travels as well as his observations on domesticated animals (especially pigeons). Combined with his vision of evolution, he began to feel strongly that species had no distinct boundaries, and in fact, could not adequately be delimited at all: "it will be seen that I look at the term species, as one arbitrarily given for the sake of convenience to a set of individuals closely resembling each other, and that it does not essentially differ from the term variety, which is given to less distinct and more fluctuating forms" (Darwin 1859:52). However, Darwin accompanied this statement with the idea that only the competent "naturalist" could adequately judge between a species and variety (cf. Darwin 1859:47). It seems evident that while Darwin believed species were extremely plastic and mutable, he also seemed to believe that they were real entities and not simply mental constructs. He also obviously stressed the evolutionary integrity of species through descent from a common ancestor, and this is the beginning of the *evolutionary species* concept still in use today (Simpson 1961; Wiley 1978; Grant 1981). While Darwin's influence on taxonomic procedures was negligible, his effect upon the species concept was considerable and may be considered the second most influential idea since Plato inadvertently laid the foundations for the typological concept.

Jordan's (1905) concept of the species was based upon the idea of distinct breeding groups existing together within a community (Mayr 1955, 1957b). As a result of this concept, at times he became carried away with his descriptions and began describing species that had resulted from hybridization in cultivation and consequently ended up with many different species based upon his single criterion. Jordan's was the most explicit expression of the *nondimensional species* concept as recognized by Mayr (1957b).

Although many workers have contributed to the development of the species concept, Ernst Mayr has had the greatest impact in recent times. He stressed that "the noninterbreeding of natural populations rather than the sterility of individuals be taken as the decisive species criterion" (1963:15). The definition of a species that results from this viewpoint is "groups of actually or potentially interbreeding natural populations, which are reproductively isolated from other such groups" (Mayr 1942:120). This is called the *biological species* concept and is the one still most widely held today. (For the very early historical roots of this concept, see Mayr 1968b.) This formulated concept was first stressed by Dobzhansky (1935, 1937) and was preceded and accompanied by many studies on the genetical, cytological, and reproductive nature of species such as Shull (1923), Babcock (1931), Goddijn (1934), Gates (1938), Bremekamp (1939), and Clausen, Keck, and Hiesey (1939). What had developed during the 1920s and 1930s was an intense interest in the nature and causes of biological diversity, especially at the populational level. Genetic, cytological, and cytogenetic studies were making rapid advances in understanding the structure of species; it was natural that a reproductively oriented species concept should emerge from these efforts. Problems arise with the biological species concept, in that often it is difficult to determine whether or not a species is actually interbreeding in nature. Another problem is trying to induce organisms of two populations to interbreed in the laboratory to determine potential gene exchange. Furthermore, it is well known that hybridization between otherwise quite morphologically distinct species is a common phenomenon in plants (Grant 1957). As this concept is still widely used, much more will be said about it later in this chapter.

Because of the difficulties encountered in application of the biological species concepts in some groups, a few workers believe that species are "arbitrarily erected, man-made constructs" (Burma 1954:209). Burma, and more recently Ehrlich and Raven (1969), Sokal and Crovello (1970), Levin (1979a), and Raven (1986) follow the reasoning that the only valid evolutionary unit is the local geographic, isolated, breeding population. Although admitting that the label of species is needed to facilitate communication and discussion, they believe that any concept that treats the species as possessing objective reality should be rejected. This is referred to as the *nominalistic species* concept (Mayr 1957b, 1969a). The question of reality of species is a logical and philosophical concern of some importance and therefore is discussed further below.

REALITY OF SPECIES

The reality of species has been a difficult and complicated issue for many decades. Because we as humans constitute a species, the reality of the concept obviously is of some concern to most all educated people and to taxonomists in particular. That there exists a need taxonomically for such a concept to facilitate communication and to allow organization of information about biotic diversity has not been seriously disputed; even those supporting the nominalistic school accept this perspective (e.g., Ehrlich and Raven 1969; Sokal and Crovello 1970; Levin 1979a). The source of difference of opinion, therefore, relates not to the utility of species but rather to their objective reality. Gregg's (1950) opinion is that if you do regard species as "real" there is no logical way this can be refuted completely, nor is there any way you can prove it universally. Much confusion has prevailed on what is meant by "reality." It seems clear that there are three kinds of reality with reference to species: (1) mental reality; (2) biological reality; and (3) evolutionary reality. Biological and evolutionary reality obviously also require acceptance of mental reality and evolutionary reality requires acceptance of the other two viewpoints.

Do species have mental reality? Insofar as any mental construct has any reality at all, it must be concluded that species certainly do pass this test. Virtually everyone agrees with this viewpoint (e.g., Bessey 1908; Shull 1923; Burma 1954; Davidson 1954; Cain 1962; Raven, Berlin, and Breedlove 1971; Levin 1979a). Some of the nominalists, however, view this as the *only* reality that species have. Bessey states that "species have no actual existence in nature. They are mental concepts, and nothing more. They are conceived in order to save ourselves the labor of thinking in terms of individuals, and they must be so framed that they do save us labor" (1908:218). Shull, a cytogeneticist, remarked that *"species are only quasi-natural entities* and that they are made so by the lack of agreement between external appearance and internal constitution and by the low visibility of many hereditary characteristics. Natural groups there certainly are, but these are the biotypes of the geneticist, not the species of the taxonomist. Only here and there is there a coincidence between biotype and species" (1923:227). Cain stresses that "in many micro-organisms, the species is not a valid concept. The practical unit is the strain, and this can be frankly recognized" (1962:227; cf. Cowan 1962). Burma, a paleontologist, calls species "highly abstract fictions" (1954:209), and Davidson, with his "dephlogisticated species concept," emphasizes that "they are mental units rather than biological units. The biological units are the individuals and these functioning individuals are interrelated through their phylogenetic lineages" (1954:250).

Beyond mental reality, we can ask if species have *biological* reality. Here the viewpoints begin to split with the nominalists, such as quoted above, saying no and others saying yes. Babcock, another cytogeneticist, said: "I shall assume that my audience accepts the evidence from the ever-increasing body of experience in classifying animals and plants, which certainly indicates that species do exist and that they are really natural groups of individual organisms (1931:5)." Clark commented that

> It is evident that species do exist. . . . This being so, we may reasonably question the justification for the discussion of the nature of a species. The answer is, of course, that the concept of a species is different at different levels of investigation and that the quibbles of the taxonomist, while proper at one level of inquiry, are irrelevant at another. This sounds as though the concept of a species should be akin to the physicist's concept of matter, as indeed it is, but this is not generally admitted by taxonomists, nor is it overtly recognized in systematic procedure. The difficulties of systematics spring from the fact that taxonomic procedure is based on a static view of the nature of a species, whereas a species is in fact a dynamic entity. (1956:1–2)

Workers who have studied primitive societies have generally concluded that species do indeed have biological reality. As Mayr (1969a) has pointed out, a primitive tribe of Papuans recognized 136 names for what he regarded as 137 species of birds. Although the anthropologist Brent Berlin, studying folk classifications in the Tzeltal Indian tribe of Chiapas, Mexico, initially viewed biological species as "spurious generalities" (Berlin, Breedlove, and Raven 1966; Raven, Berlin, and Breedlove 1971), he later (1973) reversed his position based on the similarity of correspondence between folk and scientific systems of classification. One might argue from these correspondences that people, whether culturally progressive or aboriginal, view the world in the same way and that this tells us much about the cognitive powers of the human population but nothing about the biological reality of species. However, the interest in species recognition in a behavioral sense (e.g., Roy 1980) further stresses that other species of organisms recognize each other in a consistent way, and that the resulting life forms are not completely continuous. It seems obvious that species do have biological reality based on finding phenetic gaps in the living world which correspond largely to our formal species designations. No biological concept is absolute; hence, imperfections in viewing species as real must be tolerated. Those who stress the reality of the individual (but reject that of the species) must consider the difficulty of deciding where the individual starts and the environment stops, especially when in the activities of breathing or eating. Hull points out: "If absolutely discrete boundaries are required for individuals, then there are no individuals in nature. It is only our relative size and duration which make the boundaries between organisms look so much sharper than those between species" (1976:185). Attempting to define the population is even more difficult. But these difficulties do not invalidate the biological reality of the individual nor of the population. As Grant has so aptly remarked: "The species as a unit of organization is probably no more and no less universal and well defined than the individual, cell, gene, atom, or any other unit with which we have to deal" (1963:342).

Do species have *evolutionary* reality? This also seems obvious in that the clear units of the biotic world have come about via evolution. If they are admitted to be biologically real, then they must also be judged to be evolutionarily real. They are clearly *products* of the evolutionary process, and even some of the nominalists agree with this point (e.g., Ehrlich and Raven 1969). It is important to distinguish between species being *passive products* or *active parts* of the evolutionary process. Whether species are or are not held together by gene flow among the populations (to be discussed more fully below) is irrelevant to the concept of their evolutionary reality. They are evolutionary units not in the sense of active formative units that give rise to new diversity but rather in the sense of passive products or results of the evolutionary process that have genetic and phylogenetic validity and evolutionary reality.

Confusing the issue of reality of species has been the very different and separate question, are species needed for evolutionary theory? This must be answered "probably not," at least not in most cases. To understand this question and answer, it is necessary to return to the biological species concept and recall that it stresses (1) the interbreeding of populations, and (2) the isolation of these populational systems from other such units. Recent studies and commentaries have suggested that gene flow within a species is much more limited than previously believed (e.g., Ehrlich and Raven 1969; Endler 1973; Levin 1979a; Ehrlich and White 1980; Grant 1980). Few studies actually have been done to really know the limits of gene flow within different animal and plant groups, but there is doubt that in plants gene exchange is extensive (Grant 1980). Similar selection regimes and descent from common ancestors seem to be more important in preserving species homogeneity. Previously, workers had stressed the presumed gene flow idea within the context of the biological species concept as evidence for the reality of species (e.g., Lehman 1967). Others took the reverse view and concluded that since gene flow is not as extensive and unifying as once believed, then the biological species concept is untenable. The issue is really whether the concept of species is needed to explain the origin of new organic diversity. If one considers that mutation, recombination and natural selection have their most pronounced effects on the local interbreeding population, then all known modes of speciation can be accounted for by dealing with this unit rather than the species. This is especially obvious in considering the origin of new distinct populations ("species") by geographic isolation in which several new units may result allopatrically from one diverse parental taxon. It is *convenient* to talk about the origin of new species when discussing the production of new diversity, especially in rapid modes such as allopolyploidy, but it is not essential for evolutionary theory. Even allopolyploidy can be discussed clearly by describing their origin via crosses between different populations and subsequent chromosomal doubling to yield a new, different, and reproductively isolated population. No species need be involved to explain this or other evolutionary phenomena.

NATURALNESS OF SPECIES

Another philosophical point regarding species is whether they are more natural than genera. Because of the many different definitions of "natural" (see chapter 6), this issue could be discussed at great length, but without great profit. It becomes of interest due to the survey of opinion completed by Edgar Anderson more than 49 years ago (1940; see also 1969). He asked the question: "Which in your opinion is the more natural unit among the flowering plants, the genus or the species?" (1940:364). He qualified his use of "natural" to mean: "which of the two more often reflects an actual discontinuity in organic nature." He sent this and other questions (regarding genera) to 50 practicing taxonomists and received 48 replies. Of the respondents 54 percent regarded the genus as more natural, 17 percent regarded the species as more natural, while 23 percent believed that the answer would depend upon the group under study. One respondent had no opinion and two believed that the question was meaningless.* I would agree with the 23 percent who believe it depends upon the taxonomic group in question. In older groups, such as the Magnoliaceae, both genera and species are well delimited from each other, whereas in recently evolved taxa, such as the Compositae and Gramineae, both limits are often subtle and delimited only with difficulty. Because extinction is an important factor in creating phenetic gaps between taxa, it seems reasonable that genera should be more distinct than species. But in some groups, such as the Orchidaceae, elaborate pollination mechanisms keep species well isolated and reasonably distinct whereas

---

*One of these was apparently Svenson, who explained that: "the *genus* and the *species*—at least in Linnaean taxonomy—are both natural by assumption, hence the limits of these entities are represented by discontinuities" (1945:286).

the generic limits are difficult to establish clearly. That genera are "natural" in the sense used by Anderson (1940) was further attested in his experimental study of the species concept (1957) in which he sent 16 specimens of *Uvularia perfoliata* and *U. grandiflora* (Liliaceae), both extremely well differentiated, to three New Zealand plant taxonomists (one a monographer of another family, one a biosystematist, and the third a phycologist) asking them to place the 16 sheets into genera, species, subspecies, or varieties. All three agreed on their being in the same genus and the first two recognized the same species. Only the phycologist partitioned each species into two additional species (i.e., he recognized four species instead of two). In this simple example the generic concept was not disputed but some difficulty existed with the recognition of species.

SPECIES AS INDIVIDUALS

Because of logical problems created by monotypic taxa (i.e., Gregg's Paradox), philosophical problems (the reality issue), and biological and evolutionary difficulties (such as gene flow), several studies have questioned whether species might not better be regarded as individuals rather than classes (see Bernier 1984, and Rieppel 1986, for the most recent commentaries). One could take the nominalist position of Rosen to the effect that "a species, in the diverse applications of this idea, is a unit of taxonomic convenience, and that the population, in the sense of a geographically constrained group of individuals with some unique apomorphous characters, is the unit of evolutionary significance. . . . If this view is accepted, it renders superfluous arguments about whether a 'biological species' is an individual or a class" (1978:176–177 and note). But most workers (including myself) prefer to regard species as real, placing the issue squarely back in front of us.

Part of the problem has been the inadvertent treatment in the literature of species in some instances as universals (classes) and at other times as proper names (individuals). Mayr (1976) reckons this to be easily resolved by clarifying that the species category is the class and that species taxa are the individuals. Ghiselin, who has advocated the "species-as-individuals" perspective most strongly, defines species as "firms" in an economic analogy to be "the most extensive units in the natural economy such that reproductive competition occurs among their parts" (1974:538). This has been called the *hypermodern species* concept by Platnick (1976; see also Ghiselin 1977). This concept has been extended even further by Reed, who points out that Ghiselin's definition places emphasis on a process (i.e., competition), in which in a mathematical sense the species can be viewed as a symmetry (i.e., an "invariant under a transformation, or a persistence in spite of a change") (1979:73). That is, the species becomes the only existing invariant within the total series of populations on earth undergoing reproductive competition.

Unquestionably, the regarding of species as individuals solves several difficulties with species concepts. The philosophical difficulty of reality of species becomes less problematical because obviously the reality of individuals has rarely been questioned (except in an ethereal philosophical context or in some plant groups with asexual modes of reproduction, e.g., ramets of an herbaceous clone as in *Podophyllum peltatum*, Berberidaceae). The biological and evolutionary difficulties with species concepts relate to the problems inherent in the widely accepted biological species concept in which potential interbreeding and gene flow are significant. If species are individuals, then it doesn't matter if they are interbreeding or not at a particular moment in time. Nor does it matter how loosely or tightly integrated an individual is by gene flow or any other criterion (Ghiselin 1974). The logical difficulty posed by monotypic taxa (Gregg's Paradox) does not disappear in the way that Ghiselin suggests: "No paradox arises when a single factory constitutes the entirety of a firm" (1974:539). The taxon *Ginkgo* still equals the taxon *Ginkgo biloba* (and hence the paradox, genus = species). Whether taxa are individuals or classes does not eliminate this logical problem.

Hull (1976, 1978) provided the clearest and most spirited presentation of the species-as-individuals position. He states: "Organisms remain individuals, but they are no longer members of their species. Instead an organism is part of a more inclusive individual, its species" (1976:174–175). He goes on to make the argument that because the concept of individual has some foggy boundaries, e.g., in terms of spatial and temporal unity, then "If organisms can count as individuals in the face of such difficulties, then so can species" (p. 177). He also lays out definitions for classes vs. individuals: "Classes have members not parts. These members are members of the same class because they are similar to each other in one or more respects" (p. 178). Individuals consist of parts unlike to each other (such as different organs being parts of a whole organism), whereas classes have very similar members. Hull's main point is the following: "Are organelles part of cells, cells part of organs, organs part of organisms, and possibly organisms part of kinship groups, but organisms are *members* of populations and/or species? I think not. The relation which an organ has to an organism is is the same as the relation which an organism has to its species" (p. 181).

But despite these attempts to regard species as individuals, and even in the face of being able to eliminate several problems with species concepts if they would be viewed in that fashion, in my opinion it is still clear that species (taxa) are classes and not individuals, at least not in the same sense as with organisms. There are numerous definitions of the term individual, but most stress not separable or divisible, which applies well to organisms but not as well for populations and even less well for species. Likewise, following Hull's definitions, species are made up of similar populations (fulfilling the membership criterion) in contrast to the very different organs of an organism. Species are *individual concepts* but they are not *individual organisms*. The species category is also defined by an *individual definition* (applicable only to this category; other categories have other individual definitions), but neither is it an individual organism because of this. Hull considers the evolutionary perspective and says: "if species are classes, it is difficult to see how they can evolve—but they do!" (1976:175). But there is a problem here with his interpretation. Species are populational systems that *result* from the evolutionary process. They do evolve, but only as a passive modification through time dependent upon the action of natural selection on the sources of genetic variation within the local interbreeding population. If one wished to regard these formative populations as "individuals," it would be more tolerable, but better still would be to regard them as classes in the sense of individual populations (i.e., local and interbreeding among the included organisms). The species, i.e., the collections of individual populations at the next hierarchical level, are also classes. Part of the difference in viewpoint here may derive from differing views on the process of evolution. Hull sketches his view of evolution: "A more precise description of evolutionary processes is complicated by the fact that the events operative in evolution occur at a variety of levels and these levels are integrated by the part-whole relation" (1976:181). Because of the "part-whole" viewpoint, the organism or "superorganism" status of species results. But this seems an inaccurate portrayal of the evolutionary process in which the mutations accrue to the individuals to be then reshuffled within the population by recombination and then acted upon by natural selection (again primarily at the populational level; the *action* of selection occurs at or even *within* the individual, but the *result* of selection is manifest evolutionarily at the populational level). Even though the individuals differ somewhat from each other, they still pass muster for class membership in the same population. Likewise, these similar populations qualify for class membership in the same species.

CURRENT SPECIES CONCEPTS

Morphological Species Concept

Having considered the history of species concepts and some of the logical and philosophical difficulties, it is appropriate to discuss the different species concepts now in use by practicing plant taxonomists. The one most frequently employed, especially by revisionary workers or museum taxonomists, is the *morphological species*, or morphospecies, concept (also called the *classical phenetic species* concept; Sokal 1973). This has also been called the *Linnaean* or *classical species* concept (Burger 1975), although this label might suggest that no variation is admitted. In the current usage of this concept, variation is regarded as the expected result of dealing with populations. Whether we like it or not, in practice we usually do not have sufficient information on reproductive behavior to allow the biological species concept to be applied successfully. As a result, workers have stressed the importance of recognizing species on morphological bases alone. The exact form this concept has taken has varied, but some examples are: "Species are the smallest groups that are consistently and persistently distinct, and distinguishable by ordinary means" (Cronquist 1978:15); "Species may be defined as the easily recognized kinds of organisms, and in the case of macroscopic plants and animals their recognition should rest on simple gross observation such as any intelligent person can make with the aid only, let us say, of a good hand-lens" (Shull 1923:221); or "A species is a community, or a number of related communities, whose distinctive morphological characters are, in the opinion of a competent systematist, sufficiently definite to entitle it, or them, to a specific name" (Regan 1926:75). An extension of this idea comes from the phenetic school, e.g., "the species level is that at which distinct phenetic clusters can be observed" (Sneath 1976a:437). The phenetic clusters could result from cytology, chemistry, anatomy, etc., although in practice they would tend to be based on morphology. From a practical standpoint in the preparation of floras, the circumscription of species based upon easily observable morphological features is the sensible approach. For example, in the *Short Guide for Contributors to Flora Europaea*, it is made clear that the species recognized in the flora "... *must be definable on morphology*" (Heywood 1958b:20). Even though it might be more desirable to examine species in reproductive terms, it is likely that the morphological discontinuities recognized formally do reflect biological limits of isolation, commonality of interbreeding, and genetic divergence (Stuessy 1972a) due principally to the "causal connection between interbreeding and character cohesion and dispersal" (Hull 1970b:281). Du Rietz has stressed that species are: "The smallest natural populations permanently separated from each other by a distinct discontinuity in the series of biotypes" (1930:357). Although we certainly tend to operate in this fashion, as Heywood has pointed out, "we delimit species in practice on the basis of the differences shown by populations irrespective of the evolutionary factors that contribute to the development of these differences" (1967:31). That is, it really doesn't matter how the discontinuities have arisen (i.e., whether they represent actual biotypes or not); if they exist, we will recognize taxonomic units accordingly. Despite the apparent narrowness or antique quality of the morphological species concept, it has served us well. As Burger points out, it has worked well even in difficult groups such as *Quercus* (Fagaceae) in which hybridization is commonplace: "The Linnaean or classical species-concept of readily recognized and morphologically defined species has served as a practical and efficient system for information retrieval in most flowering plants. There are very few groups where morphological correlations as a basis for taxonomy have failed to identify meaningful taxa. Even in those cases where intermediates and hybridization are known, the classical concepts have often continued to be useful and meaningful" (1975:45).

Biological Species Concept

The *biological species* concept is the one held conceptually by most systematists at the present time. This concept has two aspects (Mayr 1969a): (1) a group of interbreeding populations, (2) which are reproductively isolated from other such groups. Rarely does the practicing plant taxonomist have data about either of these biological aspects of the populations with which he or she is working. Nevertheless, most practicing workers would believe that the morphological differences used for species delimitation do indeed reflect similar degrees of interbreeding and reproductive isolation (Runemark 1961). Hence, although in practice the morphological species concept is emphasized by default (Sacarrão 1980), most workers adhere to the broader conceptual base of the biological species concept.

The obvious utility of the biological species concept has spawned many applications and added perspectives. Clearly one of the reasons for its utility is that it deals with reproductive isolation, which is admitted by nearly all workers to be important in evolutionary theory (e.g., Ehrlich 1961; see also the review by

Littlejohn 1981). The fact that biospecies also differ in their ecological contexts has been stressed by Baker (1952), Cain (1953), and Kruckeberg (1969), among others. Löve (1962, 1964) has been a strong advocate of the concept, but he has gone to extremes in plant groups in regarding reproductive isolation as an absolute criterion for species recognition (even morphologically indistinguishable cytotypes are accorded specific status; Löve 1954). Grant (1966a, b) used the biological species concept as the framework within which a new diploid species of *Gilia* (Polemoniaceae) was created experimentally through intense artificial selection over ten generations in sixteen years. This general concept emphasizing reproductive isolation has also been used for the description of a new diploid species of *Stephanomeria* (Compositae) occurring naturally (Gottlieb 1973, 1977b, 1978). Many difficulties prevail, however, with attempts to determine the degree of reproductive isolation between populational systems. Solbrig (1968) has emphasized caution in interpreting crossing results because of known genetic control of chromosomal pairing in some groups and other cytogenetic events. This point has also been made more recently by De Wet and Harlan (1972). In *Drosophila*, from which Dobzhansky originally laid the foundation for the biological species concept, the idea still basically holds although he pointed out (1972) that when one looks carefully, there are different types of biological species in the genus and different modes of speciation.

Mayr has recently suggested a modification of the definition of biological species to stress ecological aspects along with reproductive isolation. A species is viewed as "a reproductive community of populations (reproductively isolated from others) that occupies a specific niche in nature" (1982:273). The rationale for this alteration of viewpoint is to emphasize that not only are isolating mechanisms important in speciation but also adaptations to new environments in which the new populational systems arise. Hengeveld (1988) has criticized the use of niche in this definition on the grounds that it is too difficult to define accurately and also that it is too typological. I concur that this ecological criterion seems to cause more problems that it solves and recommend against it.

Many criticisms have been leveled at the original biological species concept. Some obvious problems are: the accurate determination of interbreeding among populations; the real extent of gene flow among populations; the common occurrence of interspecific hybridization between species of flowering plants; and the inapplicability of the concept to asexual species. It is clear that gene flow among plant populations is less than was previously believed (Levin and Kerster 1974; Grant 1980; Slatkin 1985), although it is yet unclear what level would be needed to maintain morphological and other character uniformity throughout a species range. Interspecific hybridization is common in many groups of plants (Heiser 1949, 1973), which might seem to vitiate the criterion of reproductive isolation. The point in angiosperms, however, is that due to the absence of ethological isolation, reproductive isolation is more complex with a series of prezygotic and postzygotic mechanisms (Levin 1971a, 1978). Hybrids are often formed naturally or can be produced in the garden and greenhouse, but problems of sterility or breakdown almost always occur in the F_1 or F_2 generations. The plant system is simply more open developmentally and therefore less sensitive to genomic disruption.

Ehrlich focuses on the difficulties of determining reproductive barriers and concludes: "I think that the biological species concept has outlived its useful-ness. The current revolution in data processing permits the relaxation of the rigid hierarchic system long employed to describe the products of evolution. We may now modify our system to permit more accurate and thus more useful description of the intricate relationships of living organisms. As a step in this direction I suggest that the genetic definition of species, never employed in practice, be discarded as an ideal. Relationships at the lower levels of the taxonomic hierarchy should be expressed numerically, in essentially the same way as relationships of higher categories are now expressed" (1961:175). He follows this with a similar perspective in 1964: "No species has ever really been defined 'biologically,' and it is unlikely that one ever will be: membership or nonmembership is determined primarily on phenetic grounds" (p. 119). He accepts the idea of reproductive isolation being important in evolution, but rejects basing the species concept primarily on reproductive criteria. Sokal and Crovello (1970) and Sokal (1973) also share this view and advocate use of what we might call the *numerical phenetic species* concept (i.e., the one used by numer-ical taxonomists). Sokal and Crovello (1970) logically analyze the various tenets of the biological species concept and conclude with a resounding "no" to a series of questions designed to evaluate its efficacy. Their questions deal with need of the concept for practical taxonomy, for evolutionary taxonomy, as a unique heuristic concept from which evolutionary hypotheses can be developed, or for evolutionary theory. While I agree that the concept of species of any type is not absolutely necessary for evolutionary theory (discussed earlier), I also believe that the biological species concept has helped and will continue to help in the development of new evolutionary hypotheses in some instances. Further, the concept is needed in evolutionary theory to explain the result or end product of populational phenomena. It obviously is not needed for practical taxonomy, although it is a stimulus for numerous workers who deal primarily with pre-served specimens to consider and discuss broader evolutionary implications of the relationships they see and document. This is, in fact, one of the most impor-tant benefits from classical revisionary studies (Stuessy 1975).

Additional comments have been offered in defense of the biological species concept. Ghiselin responds to Ehrlich's (1964) criticism (that no species has ever been defined biologically) by arguing that "the biological species definition is a definition of the word 'species' in abstract terms, not a collective term for species which have 'biological' definitions" (1966b:128). Hull (1970a) responds to Sokal and Crovello's (1970) "critical evaluation" of the biological species concept by acknowledging that certain difficulties exist with its use. However, he continues that the numerical phenetic species concept has even more problems, particu-larly on deciding which phenetic unit is the one to be called species. Some relationship of these phenetic units to biological, populational, or reproductive criteria must be used to establish a framework for application of the phenetic concepts.

Genetic Species Concept

Another idea closely related to the biological species concept is the *genetic species* concept. This assumes that the biological factors of gene flow and reproductive isolation are operative, but that the way to define species is by a measure of the genetic differences or distance among populations or groups of populations. In effect, this is really the numerical phenetic species concept using a quantitative measure of genetic, rather than morphological (or other), distance as the yardstick. This has its obvious difficulties in the simple fact that we rarely know the real genetic differences between populations. Newer techniques of measuring at least part of the genome via allozyme electrophoresis (e.g., Gottlieb 1977a, 1981a; Crawford 1983) are most helpful here, and the genetic divergence based on allelic frequencies can be measured by various statistics such as Nei's (1972) genetic distance. The data are not yet available to indicate general levels of genetic divergence for each of the levels in the taxonomic hierarchy in plants, and they may never be fully meaningful even when available. At the higher levels the approach may be severely limited by the likelihood of parallel point (and other) mutations yielding uninterpretable degrees of divergence. At the specific and infraspecific levels, however, they should prove to be most helpful. A modification of this genetic approach will be the direct measure of genetic distance from DNA sequences, which is now gaining momentum in plants particularly from chloroplast DNA studies (e.g., Palmer and Zamir 1982; see chapter 21). Even when complete DNA sequence data are available for entire genomes, however, which *will* be extremely helpful in taxonomic and phyletic studies, problems surely will arise in the interpretation of the data. Passive sites, feedback mechanisms, duplicated sites, and so on will all have to kept in mind. That is, the sequence alone will not provide the whole story; it will be the sequence plus how developmental considerations and intramolecular events affect the final coding from the sequence that will tell the tale. This level of understanding will be a long time in coming.

Bock has offered a modification of the biological species concept that puts it somewhat intermediate to that of the genetic concept. He recommends changing the words "which are reproductively isolated from other such groups" to "which are genetically isolated in nature from other such groups" (1986:33). This is not a genetic distance concept, but rather an emphasis on genetic, rather than reproductive factors which are responsible in nature for keeping populational systems isolated. The viewpoints are similar, however.

Paleontological Species Concept

Paleontologists, working with fossil materials, cannot deal directly with species concepts based on gene flow and reproductive isolation. Their material is often fragmentary, rarely shows populational variation even at the morphological level (for a rare exception, see Stuessy and Irving 1968), and few localities of a particular taxon are ordinarily known. While paleontologists can adhere philosophically to the biological species concept, in practice they must seek other means of definition. Further, they deal routinely with the time dimension in which species appear and later disappear in the fossil record, much different from the single-time reference afforded by extant taxa. Paleontologists, therefore, often speak of *paleospecies* (Simpson 1961) or *chronospecies* (George 1956) in which arbitrary time (and/or morphological) limits are used to delimit *paleontological species* (Cook 1899). These are essentially slices of time that allow workers to communicate about the ordered fossil diversity. A collection of pa-leospecies in a monophyletic succession has been termed a *gens* (Vaughan 1905). In practice, therefore, the paleospecies is usually a time-oriented morphospecies (Sylvester-Bradley 1956a). Sometimes distinct character state gaps occur be-tween forms at different time zones, thus according a good place for making a species break. But if evolution in a particular lineage is gradual, and if sampling is good, then no clear breaks may be discernable.

Evolutionary Species Concept

Although the biological species concept (as well as the genetic species concept) is useful in many ways, it does not by definition refer to evolution directly.

> It is the fact of evolution that has made genetical species separate and that keeps them from always being sharply, clearly separate. It is also evident that the genetical definition of species has evolutionary significance. Still it is striking that the definition does not actually involve any evolutionary criterion or say anything about evolution. It would apply equally well, or in fact a great deal better, to species that did not evolve. . . . Given the fact that the genetical definition of species is consistent with evolution, its lack of any direct and overt evolutionary element certainly does not invalidate it. Nevertheless it is desirable also to have a broader theoretical definition that relates the genetical species directly to the evolutionary processes that produce it. (Simpson 1961: 152–153)

As a result of this perspective, the *evolutionary species* concept was advocated by Simpson (1951) to read specifically: "An evolutionary species is a lineage (an ancestral-descendant sequence of populations) evolving separately from others and with its own unitary evolutionary role and tendencies" (1961:153). This definition is useful to give a time perspective to neontologists and a phyletic perspective to paleontologists (as opposed to a purely phenetic concept). Simp-son (1961) emphasizes that this concept avoids the difficulties with determining actual or potential levels of interbreeding and gene flow, and it allows some degree of interspecific hybridization (so common in plants), provided that it doesn't interface with the basic "evolutionary role" of each species. Determining what these "roles" are might be problematical, but Simpson (1961) suggests they are equivalent to niches taken broadly to mean the multidimensional rela-tionship of a taxon to its environment rather than just its microgeographic situation. This point has been extended by Van Valen in a more precise ecologi-cal definition: "A species is a lineage (or a closely related set of lineages) which occupies an adaptive zone minimally different from that of any other lineage in its range and which evolves separately from all lineages outside its range" (1976:233). He calls this the *ecological species* concept (cf. ecospecies, defined below).

Cladistic Species Concept

The evolutionary species concept might be especially appealing to the cladists, who would search for a concept to relate to dichotomous branches on a cladogram (i.e., their species). Hence, Wiley espoused the adoption of the evolutionary species concept with a few modifications: "A species is a single lineage of ancestral descendant populations of organisms which maintains its identity from other such lineages and which has its own evolutionary tendencies and historical fate" (1978:18). Although this definition is similar to Simpson's, several minor changes make it even more compatible with the cladistic viewpoint. The emphasis on "single" for the lineage more nearly equates this to a single branch on a cladogram. The use of "maintains its identity from other such lineages" rather than "evolves separately from others" opens the possibility of use of synapomorphies in detecting such lineages rather than the more general phrasing. And finally the stress on "historical fate" instead of "evolutionary role" (i.e., ecological context) is a significant shift from an ecological viewpoint to a historical context resulting from apomorphic changes within single branches of a cladogram. That is to say, Wiley's (1978) definition, although apparently embodying only minor alterations from that of Simpson (1961), is really different in substantial ways—so much so that it seems best to call this the *cladistic species* concept. Bremer and Wanntorp (1979), two other cladists, also favor this concept. So do Donoghue (1985b) and Mishler (1985b), although they call it the *phylogenetic species* concept. Lovtrup (1979), still another cladist, takes issue with Wiley's cladistic species definition and proposes in a more radical way to abandon the use of any species concept as "detrimental" in cladistic (his "phylogenetic," p. 391) classification. He admits that the species as a Linnaean category is probably "necessary in practical taxonomic work" (p. 391), but he stresses simple recognition of terminal taxa of cladograms as the meaningful units of diversity. Wiley (1980) rejects this as a largely artificial approach to the problem and emphasizes the need for an evolutionary view in which the termini of the branching points are cast as species resulting from the evolutionary process. Willis has gone even further and stated that "each species is an internally similar part of a phylogenetic tree" (1981:84). Wiley responded generally favorably to this suggestion, but regarded it "as a special case of the evolutionary species concept" (1981b:86).

Biosystematic Species Concepts

In addition to the principal types of species concepts in current use as discussed above, numerous other perspectives exist. It serves here to sketch some of these other concepts to indicate the breadth of viewpoints even beyond that already detailed. These additional species concepts reflect a desire to have units that more nearly reflect the diversity of reproductive relationships beyond the limitations allowed by the Linnaean hierarchy. Most of these have not received wide usage, but some have become helpful informal descriptors in specific situations. It is clear, however, that these experimental categories will not replace the conventional categories of the taxonomic hierarchy.

The experimental taxonomic studies of Turesson (1922a, 1923), Clausen, Keck, and Hiesey (1939, 1941a, b) and others led to special categories of taxa to express the variations encountered in their reciprocal transplant and hybridization studies (e.g., Valentine 1949). The most common ones are *ecotype, ecospecies,* and *coenospecies* (Cain 1953; Grant 1960). The ecotype refers to closely related but ecologically distinct populations that are largely interfertile. Ecospecies are similar but hybrids between them are of reduced viability, and coenospecies are not interfertile, even artificially. The *species aggregate* is used to describe a complex of species that simply will not sort out well taxonomically for a variety of reasons, but in which there is hope of eventual resolution; the components of a species aggregate have sometimes been called *microspecies* (Davis and Heywood 1963). Manton (1958) advocated use of the concept to refer to morphologically poorly defined cytological or genetical groups. Grant's (1957) *species group* is similar to the species aggregate concept, as is Mayr's (1931, 1969c) *superspecies*.

Numerous categories have been proposed to deal with the units resulting from biosystematic investigations in which much effort is placed on interpreting reproductive limits of taxa. The most extensive list is given in Camp and Gilly (1943a) in which twelve kinds are defined: *homogeneon, phenon, parageneon, dysploidion, euploidion, alloploidion, micton, rheogameon, cleistogameon, heterogameon, apogameon,* and *agameon.* It serves no purpose to indicate here all the definitions of these terms, but two are given as examples. The *homogeneon* is "a species which is genetically and morphologically homogeneous, all members being interfertile" (p. 334), and the *heterogameon* "a species made up of races which, if selfed, produce morphologically stable populations, but when crossed may produce several types of viable and fertile offspring" (p. 351). All of these concepts are based largely on morphological and interbreeding criteria. The apogameon and agameon apply to apomictic groups. Löve (1962) agrees with this biosystematic approach but he did not in practice use all of the categories. This did, however, lead him to recognize cytotypes as distinct species because of reproductive barriers even without morphological divergence. The *comparium* and *commiscuum* of Danser (1929) are similar to the coenospecies, but more stress is placed on the ability to hybridize and on geographic factors. The *coenogamodeme* and *syngamodeme* of Gilmour and Heslop-Harrison (1954) are equivalent to coenospecies and comparium, respectively (from Grant 1957). The *syngameon* of Lotsy (1925, 1931) is approximately the same as a breeding population or in some cases equating to biological species. Grant redefined it as "the sum total of species or semispecies linked by frequent or occasional hybridization in nature; a hybridizing group of species; the most inclusive interbreeding population" (1957:67). This is similar to Van Valen's *multispecies* concept: "A set of broadly sympatric species that exchange genes in nature" (1976:235). The recognition that some plant species often hybridize freely with neighboring taxa, especially the weedy relatives of cultivated crops, led Harlan and De Wet to propose the concept of the *compilospecies,* which "is genetically aggressive, plundering related species of their heredities, and in some cases . . . may completely assimilate a species, causing it to become extinct" (1963:499). The *semispecies* concept has been used in various ways to refer to an intermediate position between species and subspecies. Mayr (1940) regarded these as clear geographic segregates of a good species but so morphologically distinct as to be treated almost as distinct species. This viewpoint was followed by Valentine and Löve (1958). Baum (1972) stressed reproductive criteria and viewed semispecies as on the way to becoming species. This is similar to Legendre's: "a group of actually or potentially interbreeding populations, which are chromosomally somewhat distinct, but not effectively reproductively isolated from other such groups" (1972:402).

RECOMMENDED SPECIES CONCEPT FOR GENERAL USE

What approach, then, is recommended for species definition in sexually reproducing plants? The problem is that the determination of species is on the one hand one of the most important activities of the taxonomist, and on the other hand one of the most difficult. As Svenson has pointed out: "When a person asks for a definition of truth, he expects an all-embodying answer to this abstraction. This cannot be done, for the word 'truth' is referable only to an inherent circumstance or object" (1953:56–57). So it is in our attempts to define the category of species; there are no absolute answers which leaves a feeling of frustration. Species recognition has always been difficult in certain complex groups regardless of the concepts used. As Hitchcock commented many years ago: "We work over them for months, patiently noting differences and resemblances, assembling and segregating, seeming to have a scheme nicely worked out, only to have it upset by a new batch of specimens, going through all the stages of hopefulness, satisfaction, doubt, hopelessness, and finally tearing our hair and exclaiming 'Confound the things! What's the matter with them anyway?' " (1916:334). One approach is to fall back on experience and judgment without worrying about being explicit, as suggested by Hatch: "A species is, primarily, composed of those specimens which, upon examination, the taxonomist believes to be cospecific. The crux of the whole situation lies right there. If you could pass on to me or I could pass on to you the criteria that we employ at this stage of our study in such a way that I would always agree with you on your species and you would always agree with me on mine—that would be a taxonomists' Utopia indeed!" (1941:230). Griffiths has pointed out problems with the definition of physical objects, which one would probably regard as much easier to define than species: "We are accustomed to think of physical objects as occupying a constant amount of space and excluding other physical objects from that space. But this appearance is illusory. An atom consists mostly of empty space, through which high-energy particles can pass. So it is at higher levels of organization. Much biologically inert material passes into and out of the bodies of organisms. Therefore, I do not think that the concept of a physical object can be defined in terms of exclusive spatial boundaries" (1974a:94).

A further aggravation in defining species is that many different modes of speciation occur (e.g., Bush 1975; White 1978; Grant 1981; Templeton 1981; Barigozzi 1982; Rose and Doolittle 1983), and therefore the species that result from these different processes also will vary one from another. Consider the variations and reproductive barriers provided by rapid chromosomal divergence in *Oenothera* (Onagraceae; Cleland 1944) or in groups in which some apomixis is known (e.g., Löve 1960) such as *Taraxacum* (Compositae). As Lewis has nicely summarized: "The pattern of morphological differentiation may differ from one group of plants to another and is a reflection of the diversity of evolutionary processes. Consequently, species and subspecies are not necessarily equivalent in different genera or different sections of the same genus" (1955:18). Mishler and Donoghue (1982) have more recently summarized this point again.

Various attempts have been made to define the species (and other categories) in mathematical terms. Legendre and Vaillancourt defined the species as "the set of all the individuals that have the same genetic load, including the possible variation of alleles in each gene" (1969:248). This is essentially the numerical phenetic species concept based on genetic criteria assessed quantitatively. They offer another definition (p. 245) of the " 'species such as in nature' to be the union of certain vital neighborhoods in a multi-dimensional space, the said intervals obviously being those that correspond to the given species."

The species concept that still makes the most sense for most sexually reproducing flowering plants is the biological species concept. There is no question that species have objective reality and that they also have evolutionary reality. The reproductive barriers that keep species apart are most important for limiting gene exchange and for maintaining the integrity of each unit. Whether or not gene flow among the populations included within a species is limited or exten-

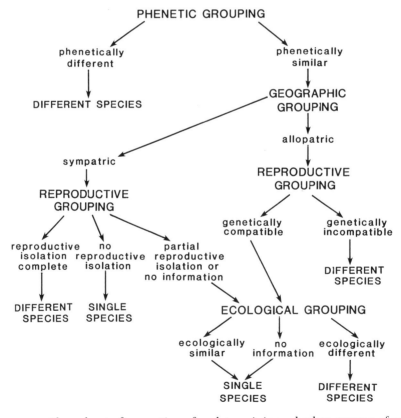

FIGURE 11.1 Flow-chart of operations for determining whether groups of populations represent distinct species. (From Doyen and Slobodchikoff 1974:241)

sive is irrelevant; the main point is that these included populations do form reproductively compatible units that are isolated from other such groups. Because of this genetic isolation and due to a similar genetic background, similar selection pressures, and some low degree of gene exchange, the included populations tend to resemble each other morphologically, at least more so to each other than to other populations of other species. Operationally, because we rarely have reproductive data for all the included populations of a presumed species, we rely heavily on morphological cohesiveness and distance from other units (the phenetic gap). It is assumed that these morphological relationships do reflect genetic and reproductive relationships of a similar degree. Doyen and Slobodchikoff (1974) have offered a useful flow-chart of operations toward the recognition of species (fig. 11.1). These involve first making phenetic groups. If these groups are different enough, they are regarded as good species (and also presumably as reproductively isolated). If they are similar, they need to be examined in more detail for their geographic, reproductive and ecological attributes. If they are sympatric, they will be good species if reproductively isolated. The degree of isolation necessary for species recognition obviously must be judged in each case on all available data. If no reproductive data are at hand, then ecological concerns come into play. Likewise, if the populations are allopatric, then reproductive and ecological data are brought to bear on the problem. This is a very useful operational perspective that is recommended for use (see Ehrendorfer 1984, for a similar view).

After all is said and done, I am afraid we all must still agree (to a greater or lesser extent) with Crum, who says: "A species cannot be fully defined, nor can it be intuitively sensed. Although subjectivity is involved in decision making, a species is only as good as the knowledge and insights used in its delimitation. Certainly methodologies help. So do good sense and good judgment based on meaningful experiences, and the more the better" (1985:221).

CHAPTER 12

Subspecies, Variety, and Form

Although in many cases the designation of species within a taxon is completely adequate to account for almost all of the meaningful patterns of morphological and other variation, in other cases it is not. In some situations, particularly those in which complex patterns of variation occur, there is a real need to circumscribe infraspecific taxa. This is perhaps more important now then ever with the renewed emphasis on the local population as the focus for evolutionary change (Ehrlich and Raven 1969).

The *International Code of Botanical Nomenclature* (Greuter et al. 1988) recommends use of no more than five infraspecific categories: subspecies, variety *(varietas)*, subvariety *(subvarietas)*, form *(forma)*, and subform *(subforma)*. Of these, three are used most commonly and will be discussed here: subspecies, variety, and form. Additional categories using *super-* as a prefix also can be used, but this approach has never been popular and certainly should not be encouraged. There is enough difficulty with attempting to apply consistently the three major infraspecific categories without complicating the problem with other units. Informal concepts and terms can be used if it is believed necessary to describe and name additional patterns of variation (to be discussed later in this chapter).

The usage of subspecies, variety, and form has changed over the years, which has confounded attempts to use the concepts in a consistent fashion. In fact, at the present time there is more confusion surrounding usage of these three categories, especially subspecies and variety, than with any other level in the taxonomic hierarchy. Some workers use only subspecies to describe initial patterns of infraspecific variation, others use only varieties, and still others use both categories. Some workers believe that subspecies and variety are unnecessary as separate categories and treat them more or less as synonyms (at least in a biological sense). Others advocate the use of forms to describe more minor morphological variations, and still others recommend never using this category. These different perspectives have led to enormous confusion in the way infraspecific taxa are circumscribed, the degree of variation that is believed useful to recognize in a formal way, and the resulting nomenclatural complexities. As Boivin had aptly quipped: "*Subspecies* is an almost sure-fire conversation gambit between botanists" (1962:328). These problems cannot easily be disregarded because many of the more difficult taxonomic situations occur at this level, and they also must be dealt with before one can really feel comfortable with specific delimitation in a particular group. To help the reader into and through these difficulties, therefore, this chapter will sketch the history in botany of usage of the three major infraspecific categories (subspecies, variety, and form), comment on some of the difficulties with usage of each of them, and offer recommendations for their consistent and effective employment.

HISTORY OF INFRASPECIFIC CATEGORIES

The variety *(varietas)* was the first category to be used below the species level for plants. Linnaeus in his *Species Plantarum* (1753) used the category commonly, and this was the beginning of its common use in plant systematics. In his *Philosophia Botanica* of 1751, Linnaeus stated clearly what he believed a variety should represent: "Plant changed by an accidental cause due to the climate, soil, heat, winds, etc. It is consequently reduced to its original form by a change of soil . . . Further, the kinds of varieties are size, abundance, crispation, colour, taste, smell . . . Species and genera are regarded as always the work of Nature, but varieties as more usually owing to culture" (Ramsbottom 1938:199). The variety to Linnaeus, therefore, was primarily an environmentally induced variation and in the terms of today, one that was not genetically controlled in a strong or rigid way. Although Linnaeus stated clearly his concept of the variety, the varieties he actually described in the *Species Plantarum* often did not coincide with these criteria (Clausen 1941).

The history of the use of subspecies in botany has not been examined until more recently, and several opinions have been offered. The subspecies apparently came by transfer from earlier zoological usage (Boivin 1962), and the variety is clearly the earliest used botanical category. Most workers in the past (Clausen 1941; Boivin 1962; Davis and Heywood 1963) have regarded the first usage of the subspecies to be in Persoon's *Synopsis Plantarum*, vol. 1 (1805). Weatherby (1942) believed that Link in his *Philosophiae Botanicae Novae Prodromus* (1798) was the first to use the subspecies, "defined as strains 'many of which are in cultivation and have become almost hereditary,' which commonly come true from seed, but originally arose from the progeny of a single individual. Varieties, in Link's view, did not come true" (translated in Weatherby 1942:160). But as pointed out by Boivin (1962) and Chater and Brummitt (1966a), although Link was apparently the first to define the concept, he did not actively use it in his work. Persoon did indeed use subspecies and clearly differentiated its use from the variety (Fuchs 1958; Chater and Brummitt 1966b). However, recently Chater and Brummitt (1966a) have shown that subspecies were first used clearly in a nomenclaturally distinct fashion by Ehrhart in a series of papers beginning in 1788 (see Manitz 1975, for bibliography). His concept of subspecies *(Scheinarten)* is as follows (from translation by Chater and Brummitt 1966a:98): "In this way I term plants which agree in essentials almost completely with each other, and are often so similar to each other that an inexperienced person has trouble in separating them, and about which one can conjecture, not without reason, that they have formerly had a common mother, notwithstanding that they now always reproduce their like from seed. They are, in a word, Varietates constantes, or an intermediate between species and *Spielarten* [= varieties]. They are separated from species in that they differ from one another in small particulars of little importance; and they differ from *Spielarten* in that they reproduce themselves unchangingly by seed and always beget their like." Subspecies in the view of both Ehrhart and Link, therefore, were variations hereditarily determined as opposed to the environmental modifications, or plasticities, which were more indicative of varieties.

From these beginnings, subspecies were not commonly used in botany in the nineteenth century. In the first *International Code of Botanical Nomenclature* of 1867 (Candolle 1867; see English translation by Weddell in Gray 1868) both subspecies and varieties were recommended but the emphasis on subspecies was clearly on the most striking variations of species in a horticultural sense rather than in a geographic series of natural populations. Emphasis in this period was clearly on use of the variety, e.g., it was the category of choice in the monumental *Prodromus* (A. P. de Candolle, 1824–1838; A. L. P. P. de Candolle, 1844–1873). In North America, Asa Gray, following European conventions, also used varieties almost exclusively in his works (Weatherby 1942). His view was: "Any considerable change in the ordinary state or appearance of a species is termed a *variety*. These arise for the most part from two causes, viz: the influence of external circumstances, and the crossing of races" (1836:289). During this time varieties came to be regarded more as geographical and morphological subdivisions of a species. This was forced in part by the large numbers of specimens accumulating from major exploring expeditions in the United States and elsewhere which revealed variations within wide-ranging species that had a geographical basis (Weatherby 1942).

The subspecies came once again into attention through the experimental taxonomic studies of H. M. Hall and associates of California (e.g., Hall and Clements 1923). Partly as an attempt to reflect the natural evolutionary units and to have these results correspond with those being obtained through cytology, cytogenetics, and genetic studies at this time with animals, and partly overreacting to the excessive splitting in taxa of the California flora by E. L. Greene of Berkeley, the subspecies was stressed as the category of choice (Weatherby 1942). These were the major "phy[lo]genetic lines from the great maze of connecting forms" (Hall 1929b:1573) encountered within species. He concluded that the term variety "has such a multiplicity of uses and so often applies only to races, ecologic responses, horticultural forms, or even to abnormalities that, in the opinion of the writer, its use in serious taxonomic work were better discontinued" (1929a:1461).

In the meantime, followers of Asa Gray in the eastern institutions continued to use the variety as the choice for infraspecific categories. A particularly strong advocate of that position was M. L. Fernald (e.g., 1936) of Harvard University who believed that the variety should be used for recognizing geographic variations of ordinary species. Subspecies, on the other hand, should be regarded as subdivisions of an aggregate species (Fosberg 1942), or what might also be called a species complex. This attitude has prevailed to the present day and many followers exist (e.g., Fosberg 1942; Weatherby 1942; Turner 1956; Northington 1976; Cronquist 1980; Keil and Stuessy 1981; Rollins 1981). The variety is also the category of choice among mycologists (Hudson 1970).

DIFFICULTIES IN APPLICATION OF CONCEPTS

As a result of these initial differences within the United States on the usage of subspecies and variety, two basic schools of thought have prevailed. The first can be regarded as the Californian School, represented earlier by H. M. Hall, and later by Clausen, Keck, and Hiesey (1940). These workers were attempting to understand the structure of plant species by means of reciprocal transplant studies and interpopulational hybridizations. Subspecies were used, varieties were not. Strong feelings prevailed on this point as expressed by Camp and Gilly: "In those species where an author wishes to express a great spread of intraspecific variability, the variety has often served as a useful vehicle for the multiplication of nomenclatural possibilities. It is certainly the happy hunting ground for those who wish to put their own 'authority names' after nomenclatural entities, but yet do not wish to expend any great amount of study on a group to determine the exact status of the population segments which they recognize" (1943a:370). The second viewpoint may be called the Eastern School, which emphasizes use of the term variety for the primary subdivisions of the species. Subspecies are used less often, if at all. Feelings here are also strong as evidenced by this quote from Weatherby: "All difficulty not wholly illusory would have been avoided by the simple, and one would suppose the obvious, expedient of following the rules and using variety as the term primarily to be employed for subdivisions of species. If the workers in experimental taxonomy have convinced themselves that only one infraspecific category is worth while, so be it; if they can prove it, well and good; variety would still better serve their turn and would meet with no opposition" (1942:167).

These two basic viewpoints with regard to usage of infraspecific categories result in two different operations in classification at this level in the hierarchy (fig. 12.1). If within a species the morphological diversity is great enough so that additional taxa need to be recognized beyond the primary subdivisions, the California workers will first delimit subspecies, and then recognize varieties within one or all of the subspecies. In approaching this same problem, the Eastern workers will first delimit varieties, and then, if some of the varieties seem more similar morphologically (or in other characteristics) to each other than to the rest of the varieties, the former will be grouped into a subspecies (Kapadia 1963). In the first approach, therefore, one can have subspecies without varieties at the infraspecific level, whereas in the second approach, one can have varieties without subspecies.

Obviously the existence of two principal schools of thought with reference to infraspecific categories causes confusion. Among the problems have been nomenclatural hardships by use of both variety and subspecies to refer to major subdivisions of species. Although they may be roughly equivalent in many (if not most) cases biologically, they are clearly not so nomenclaturally, and names applied to them have no priority outside of their own rank. In recognition of this problem, Raven commented: "I hope that botanists may ultimately adopt such a system [i.e., use only of subspecies] and consider it desirable meanwhile for names at both varietal and subspecific levels to be considered for purposes of priority at either one, even though it is not technically necessary to do so"

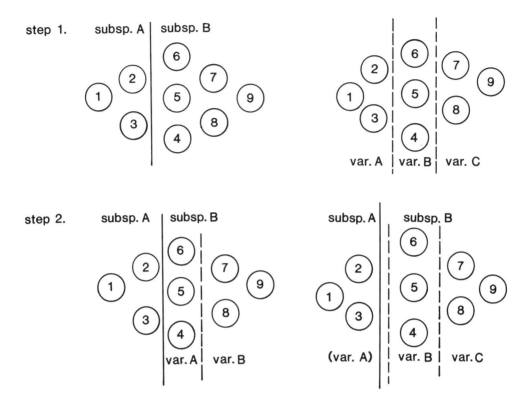

CALIFORNIAN SCHOOL   EASTERN SCHOOL

FIGURE 12.1 Two-step diagrammatic representation of the Californian and Eastern schools' approaches to the recognition of varieties and subspecies. Numbered circles refer to populations spatially distributed in a hypothetical geographical area.

(1969:168). This was followed by a formal proposal (Raven, Shetler, and Taylor 1974) to equate the two nomenclaturally, but this failed to be adopted during the nomenclatural sessions at the International Botanical Congress in Leningrad in 1975. Pennell (1949) earlier suggested the same route of nomenclaturally merging varieties into subspecies. This would be a difficult accomplishment, however, because the two categories each have been used for nearly 200 years independently, they have meant different things to different workers throughout all this time, and in some cases both have been used in the same group (e.g., Strother 1969).

FORMS

The previous discussions have indicated that the emphasis for the delimitation of subspecies or varieties is on morphological variation that is associated with

geography. We have not commented, however, upon variation that is not correlated geographically. It is not an uncommon occurrence within natural populations to find plants with unusual morphological features growing near individuals with more "typical" morphology. These variants usually represent small
genetic changes as a result of mutation and/or recombination, and they are not
much different in total genetic composition from the more normally appearing
types. The term *form (forma)* is used to apply to these unusual morphological
variants. The first usage of this term is uncertain, but Gray used it in the second
edition of his *Manual* (1856) to refer to "lesser varieties" (Weatherby 1942:158).
Some plant taxonomists never use this category in a formal scheme of classification on the grounds that trying to keep track of such minor morphological
variations is not a useful function of biological classification (Davis and Heywood 1963; for a practical example of rejecting the use of forms, see Camp and
Gilly 1943b). They would argue that one might as well attempt to provide a
category for almost every individual organism, because most every sexual organism is different genetically. Schaffner stressed this point: "As to the word, form,
which our European brothers are using so largely I haven't any use for it at all.
If there is a mere fluctuation as to ecological conditions then all well and good
but I would then prefer to call it a fluctuation. We simply don't want a system
that will make a local Indiana or Ohio plant list look like a Webster's Dictionary
in size" (1937). Proponents on the other side argue that recognizing forms serves
the useful function of highlighting unusual morphological variations which may
stimulate economic or evolutionary interest (e.g., Valentine 1975). The rebuttal
to this argument is that one can call attention to these variants by using an
informal label such as "race" without cluttering up the hierarchy with additional categories. And so the dialogue goes. This discussion could apply equally
well to *subvariety* as used by some workers (for usages see Farwell 1927; Heywood 1958a; and Lambinon 1959). Rosendahl has a nice perspective on this:
"We found relatively few cases where it seemed necessary or desirable to employ
forma and I believe this is the general experience of most taxonomists. However,
some authors when dealing with highly polymorphous groups seem to feel that
it is necessary to take account of all variants that can be distinguished and fit
them into the formal scheme. The trouble with this procedure is that in attempts
to set up a series of units of descending rank, a point of diminishing returns is
soon reached, beyond which confusion rather than clarification results. Such
schemes may have something to commend them in theory but not in practice"
(1949:27).

BIOSYSTEMATIC INFRASPECIFIC CATEGORIES

If controversy regarding these categories is not enough, some experimental workers
have advocated a large series of additional concepts and terms similar to what
has been done with species concepts (see p. 177). The most extensive list has
been compiled by Sylvester-Bradley (1952). The number of categories at the
subspecific level is 13 including such terms as *prole, ecotype, climatype, topotype*
(not to be confused with the nomenclatural term, which refers to a new collec-

tion from the original type locality), *Rassengrupp, geo-ecotype, tansient, waagenon,* and so on. At the level below the subspecies are 25 terms such as *natio, subnatio, ecovar, subecotype, topodeme, paganae, gamodeme, cline, eco-element, transitio,* etc. It serves no purpose to attempt to define these here; the reader is referred to the original manuscript for clear explanations and references for each of the terms. Although some of these concepts and labels have been used occasionally in the botanical literature, their occurrence is not frequent in a taxonomic sense and does not impact significantly on our discussion of infraspecific categories. They can be used in certain cases to describe biological conditions of populations, but they are a supplement to, rather than a replacement of, the three principal infraspecific levels of the taxonomic hierarchy.

RELATED ZOOLOGICAL CONCEPTS

Although zoologists have tended not to use varieties (Hale 1970) as commonly as botanists,* they have had numerous discussions and debates regarding the proper usage of subspecies (e.g., see symposium including Bogert, Burt, Clench, Hubbell, and Sibley, each 1954). A more recent example is the advocating of the use of the term *megasubspecies* to refer to "well-marked forms approaching the level of species, but nonetheless judged to be conspecific" (Amadon and Short 1976:161). In fact, they have expended far more effort on this question than botanists have on the proper employment of varieties vs. subspecies vs. forms (see review by Starrett 1958). Because of the complexity of character variation in plants, we would not welcome the "75 percent rule" which says that subspecies should be clearly differentiable in 75% of the individuals examined within a particular group (Amadon 1949). As Davis and Heywood (1963) point out, botanists would insist on a much higher degree of confidence for subspecific recognition. The zoologists have tended to fall into four camps with reference to subspecies (in part after Tilden 1961): (1) those who wish to dispense with the concept altogether (e.g., Wilson and Brown 1953; Brown and Wilson 1954; Gillham 1956); (2) those who defend the subspecies concept as valuable and useful (e.g., Mayr 1954; Durrant 1955; Fox 1955; Parkes 1955; Smith and White 1956); (3) those who would retain the subspecies as useful but suggest redefinition and limitations on its application (Edwards 1954, 1955; Van Son 1955; Pimentel 1958, 1959); and (4) those who have no quarrel with the *concept* but object to attendant nomenclatural difficulties (Gosline 1954; Moore 1954). Wilson and Brown stressed the abandonment of the subspecies and in its place would substitute "the simple vernacular locality citation or a brief statement of the range involved" (1953:110). It is doubtful that any botanist would subscribe to this radical departure from normal practice, especially considering the wide range of geographic variation commonly encountered in plants. As Fennah has expressed well: "faced with the thought of grappling with a *series* of ephemeral Tibetan vernaculars written by

---

*The principal animal group in which the situation is similar to higher plants is in the insects in which subspecies, race, form, variety, aberration, phase, and caste are all used, but somewhat inconsistently (Askew 1970).

Tibetan or Chinese taxonomists, all, at various times, indicating the same population (if we assume no doubtful synonymy or misidentified races), non-systematic zoologists [and botanists!] may be tempted to summarize their views about the proposed method of Wilson and Brown in 'frankly expressive' informal and flexible vernacular" (1955:140). One of the serious concerns is how to differentiate clinal variation from more definite and geographically partitioned subspecific variation (e.g., Hagmeier 1958). Recently further discussions have prevailed on the ecological bases of the subspecies concept (Böhme 1978, 1979; Botosaneanu 1979).

RECOMMENDED INFRASPECIFIC CONCEPTS

What then are useful definitions of subspecies, variety, and form for practicing plant taxonomists working with sexually reproducing angiosperm taxa? Several criteria need to be utilized and these include morphological distinctness, geographical cohesiveness, and where known, genetic divergence, natural reproductive isolation, and degrees of fertility or sterility of natural hybrids (table 12.1). Usually only morphological and geographical data are available, and decisions have to be made on these criteria alone. When genetic and reproductive data are at hand, however, they can provide a finer resolution of relationships.

Morphology obviously is important in the formal recognition of infraspecific taxa. If no morphological differences occur among populations within a species, then no formal designations should be provided even if cytological, genetic, chemical, or other differences prevail. Some of the comments in Meikle (1957), although largely from zoologists, suggest that the prime consideration should be

TABLE 12.1 Characteristics Useful for Distinguishing Subspecies, Varieties, and Forms in Sexually Reproducing Flowering Plants.

	Characteristic				
Category	*Morphological distinctions*	*Geographical patterns*	*Genetic divergence*	*Likelihood of natural hybridization*	*Fertility of hybrids*
Subspecies	several conspicuous differences	cohesive; largely allopatric or peripatric	usually markedly multigenic	possible along contact zones	markedly reduced fertility
Variety	one to few conspicuous differences	cohesive; largely allopatric with some overlap	multigenic or with some simple control	probable in overlap region	reduced fertility
Form	usually a single conspicuous difference	sporadic; sympatric	simple control (usually single gene)	always expected	complete fertility

reproductive divergence regardless of degree of morphological differentiation. For most purposes, however, this is not adequate in plant taxonomy and certainly it is not common to have reproductive isolation without morphological change. (It can occur, however; see Grant 1981.) Plants tend to have greater morphological differences and correspondingly fewer genetic (or reproductive) differences between subspecies than animals. If such nonmorphological infraspecific variation is encountered that has a geographical basis, then the designation *race* or *-type* is recommended, such as *cytotype, chemotype,* or cytological or chemical race. Others have used the *deme* terminology (Gilmour and Gregor 1939; Gilmour and Heslop-Harrison 1954) such as with flavonoid *chemo-demes* in *Thelesperma* (Compositae; Melchert 1966). For other applications of *-deme* terminology and its changing usage, see Briggs and Block (1981).

Because of the often complicated nature of infraspecific variation, especially in flowering plants, this is an ideal area for the application of multivariate statistics. In fact, of all the applications of phenetic-type methods at different levels in the hierarchy, it is at the infraspecific level that the greatest efficacy lies (Sokal 1965; Gilmartin 1974). Patterns of variation at this level can be unbelievably complex and computer algorithms are often used for patterns to emerge. Consider, for example, the large quantities of data generated in monoterpenoid (= essential oil) investigations in *Bursera* (Burseraceae; Mooney and Emboden 1968) and in *Juniperus* (Cupressaceae; Adams and Turner 1970; Adams 1975b, 1983; Von Rudloff and Lapp 1979; Adams, Zanoni, and Hogge 1984), in which both quantitative and qualitative variation in these components occur over a geographical area. Such trends as reflected by contour mapping can not only be useful for helping arrive at decisions on infraspecific classification but also in detecting hybridization and introgression (e.g., Flake, Von Rudloff, and Turner 1969), and the suggestion of evolutionary trends and tendencies. Many studies of this kind have been published for animal groups (e.g., Zimmerman and Ludwig 1975; see review by Thorpe 1976), and several exist for plants (for examples see Hickman and Johnson 1969; Arroyo 1973; Clayton 1974b; Jensen and Eshbaugh 1976). It is more challenging in plants, however, due in part to the difficulty of plasticity and variation so common in plants over broad geographic areas. Nonetheless, even in plants the power of numerical analyses of complex data is best revealed at the infraspecific level and these types of studies should be encouraged.

Geography is a most important component in the recognition of infraspecific taxa (Lewis 1955). The crux of the issue is the following: if morphologically distinct population systems are completely overlapping, they are probably reproductively isolated and hence best viewed as good species. Or, if the morphological difference is minor and with a simple genetic basis (such as petal color variants), then forms are probably indicated. Subspecies and varieties only are considered if the distributions are largely allopatric. In the following hypothetical example (fig. 12.2), individuals in populations 6, 11, and 15 are regarded only as forms because they do not form a cohesive populational system and they have only minor morphological differentiation from the other populations. Further, they even occur sporadically within the three populations and are intermixed with contrasting morphological types. If only population 6 had this unusual

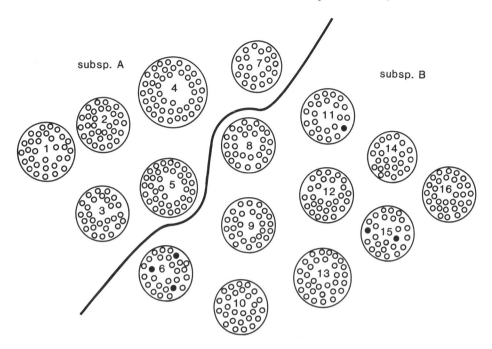

FIGURE 12.2 Diagram of hypothetical example of forms (dots representing individual plants) occurring in populations 6, 11, and 15 of a species with two morphologically and geographically distinct subspecies.

morphology, and if *all* of the individuals of this population were this way, then it might be useful to treat this population as a variety distinct from all the other populations within the same subspecies (i.e., nos. 8–16). An associated geographic concept is that subspecies have been regarded by some workers as a regional "facies" (or appearance) of a species whereas variety only as a local "facies" (Du Rietz 1930). This has merit if both categories are used together in the same group, but less merit if one or the other is used alone. In this case both are regional facies with greater or lesser morphological differences, respectively.

An important point nomenclaturally that bears strongly on these concepts is that if one subspecies, variety, or form is created within the next higher taxon, then automatically another taxon at the same rank must be established. This is simply another application of logical division discussed earlier. The subordinate taxon that contains the nomenclatural type of the next higher taxon carries the epithet of the higher taxa repeated without authority (called an *autonym*). The other subspecies, variety, or form not containing the type must have a new name with appended author. For example, if population no. 6 were treated as a distinct variety, and if the type of subspecies B did not fall within it morphologically (e.g., let's assume it belongs clearly within population no. 14), then it would need a new varietal name with attached publishing author. The remaining populational systems (nos. 8–16) would be treated as a second variety using the subspecific epithet and without author. For example, if the subspecies was *Mel-*

ampodium leucanthum subsp. *montanum*, then the two varieties would be *M. leucanthum* subsp. *montanum* var. *roseum* Stuessy [for population no. 6] and *M. leucanthum* subsp. *montanum* var. *montanum* for nos. 8–16. The creation of one infraspecific taxon always necessitates the automatic creation of another at the same rank.

The genetic criterion forms an important part of the evaluation of infraspecific patterns of variation. Runemark stresses that while species would be expected to have no gene exchange, subspecies would have gene exchange "restricted on genetic grounds or . . . limited or made impossible by external means" 1961:29). One would expect a "genetic yardstick" in which subspecies would be more genetically distinct from each other than would be varieties, and forms only differing by one or a few genes, such as was documented in *Dithyrea* (Cruciferae) by Rollins (1958); in this case the variations were originally treated as species and finally viewed by Rollins as forms in the sense used here, but without formal nomenclatural designation. Such a yardstick has been difficult, if not impossible, to apply in most cases due to the large quantities of crossing data needed for what is usually regarded as a relatively trivial taxonomic problem. With the advent of protein electrophoresis of isozymes, however, a technique is now at hand that can be used to reflect these degrees of genetic similarity and difference more easily. Allelic frequencies at particular genetic loci are compared via some measure such as Nei's (1972) genetic distance or measure of genetic similarity (see Gottlieb 1977a, 1981a, for reviews). A number of infraspecific populational systems have been examined in plants to date (although far fewer than in animals; e.g., see Avise 1974; Ayala 1975, 1982; Buth 1984), and these have been reviewed by Crawford (1983). Studies have been completed within *Chenopodium* (Chenopodiaceae; Crawford and Wilson 1977, 1979; Crawford 1979a); *Coreopsis* (Compositae; Crawford and Bayer 1981; Crawford and Smith 1982a, b, 1984), *Sullivantia* (Saxifragaceae; Soltis 1981, 1982); and *Helianthus* (Compositae; Wain 1982, 1983). Work has not been sufficiently extensive to determine whether varieties or subspecies are differentially genetically divergent. In fact, a complicating factor is: "In most instances, subspecies or varieties exhibit high genetic identities similar to conspecific populations rather than the lowered similarities characteristic of many species . . . , this being the case despite the fact that the subspecific entities are morphologically distinguishable. It may be that some subspecies or varieties represent recently diverged populations and thus changes have not occurred at isozyme loci due to the time factor" (Crawford 1983:267). Thus, one is sometimes faced with very reduced genetic divergence but with morphogeographic compartmentalization. In such cases, the low levels of genetic difference only serve to stress the taxonomic unity of the populations *within* the species and their probable common evolutionary origin, but these data will not be helpful for differentiating infraspecific taxa. In other instances they can be extremely helpful in such differentiation. A further complication is that many features in flowering plants are known to be controlled by single gene differences (Hilu 1983; Gottlieb 1984). The taxonomic import of this result must be evaluated carefully in each case, but it does raise the possibility of significant qualitative taxonomically valuable features, even at the specific level, being under relatively simple genetic control. Important in these considerations will be the consistency of the features, their presumed adaptive value, and the level of selection pressure on them. A related recent development is the assessment of mitochondrial DNA variation geographically within species (called *"intraspecific phylogeography"*; Avise et al. 1987), which although so far examined closely only in animals, may also be extremely useful at the infraspecific level in plants for suggesting geographic partitioning of populational systems.

The final recommendation for use of subspecies, variety, and form is, therefore, as follows. Forms should not be used in a formal nomenclatural sense except perhaps in groups with strong economic value such as wild relatives of crop plants, in which the formal highlighting of unusual variation might be a stimulus for incorporating these into an on-going breeding program of great potential for increased food production or horticultural improvements. If subspecies and/or variety has been used already within a group to describe the infraspecific patterns of variation, then this precedent should be followed insofar as possible. Changes of rank from subspecies to variety or vice versa without strong evidence should be resisted. If no infraspecific classification has ever been proposed within a group, and the patterns of morpho-geographic variation leave a question as to whether subspecies or varieties should be recognized, then I favor use of subspecies as the initial category of choice. However, both categories can and should be used if judged helpful in a particular group (Wilbur 1970; see as examples Cronquist 1947, in *Erigeron*, Compositae; Strother 1969, in *Dyssodia*, Compositae; Lowrey 1986, in *Tetramolopium*, Compositae). One point worth mentioning is that no complete indices exist to infraspecific epithets. The *Gray Card Index* has all such names from 1885 to the present of flowering plants and ferns from the New World, and since 1970 the *Kew Record of Taxonomic Literature* has all infraspecific names listed. *Index Kewensis*, until just recently, did not list these and hence there is a real burden in dealing with priority of such names when the indices are so inadequate. It is hoped that the future will bring resolutions to these bibliographic difficulties. I would not be as pessimistic as Burtt, who refers to "the muck-heap of two centuries of unindexed and inadequately described epithets. The best thing to do with a muck-heap is to leave it undisturbed so that it quietly rots down. In course of time the *Code of Nomenclature* will no doubt accept it as disposable refuse" (1970:238). A preferred perspective is that with further understanding of the genetic nature of infraspecific variation, with improved algorithms for their analysis, and with more comprehensive indexing, we will be even better able to deal with this level of the hierarchy. This in some ways is a significant challenge because it is at this level that the dynamics of the evolutionary process are active, which demands full understanding for proper interpretations of relationships. One suggestion toward uniformity would be to set a future start date, e.g., the year 2000, for use of only one infraspecific category (preferably the subspecies).

CHAPTER 13

Genus

The genus is the next principal category in the taxonomic hierarchy above the species. The *International Code of Botanical Nomenclature* (Greuter et al. 1988) allows several intervening categories also, viz., *series, section (sectio)*, and *subgenus*, but these are not fundamental to the hierarchy and are not always used in classification within a particular group. They can be very helpful, however, especially in providing an infrageneric structure within large genera (e.g., in *Senecio*, Compositae, with more than 1,000 species). Their use is to be encouraged in certain situations (to be discussed in more detail later in this chapter).

Although much less has been written about genera than species and infraspecific units, three past symposia have been held to discuss generic issues in detail, two with reference to plants (Bartlett 1940; Verdoorn 1953b), and one in paleontology (Amsden 1970). Interest continues, as evidenced by the more recent symposia dealing with the generic concept in Compositae (Lane and Turner 1985) and with more general issues (Young 1987). On a practical level there is a great need for good generic synopses (Just 1953; Verdoorn 1953a) so that the closely related species can be grouped together to allow for further in-depth evolutionary studies, for new floristic works, and so on. This point is stressed by Kendrick (1974) with reference to Hyphomycetes (fungi) as the "generic iceberg" in which many more inadequately understood names and taxa exist than those for which we have a good nomenclatural and taxonomic knowledge.

The genus is more difficult to define than the species. Considering the many problems and uncertainties at the specific level, this may appear a nearly hopeless situation. Robinson struggled with the problem and remarked: "we may roughly describe a genus (when plurityic) as a group of species which from likeness appear to be more nearly related to each other than they are to other species. But so varying are the degrees of similarity and so diverse is human judgment regarding them, that such a definition offers only an exceedingly vague basis for a uniform classification" (1906:81). Du Rietz is even more pessimistic: "The taxonomical units superior to the species can hardly be called 'fundamental' to the same degree as the species and its subordinate units, since the delimitation of the higher units is to much greater an extent a matter of taste and convenience. Every taxonomist knows how hopelessly opinions differ regarding the delimitation of genera and other units of higher rank" (1930:392). Legendre and Vaillancourt (1969) attempted to depart from this sea of hopelessness and place the concept of the genus on mathematical grounds, but the result was largely a cladistic definition based on holophyly and adaptive zones (Legendre 1971).

To help the reader appreciate the generic concept and its definition, therefore, this chapter attempts to sketch the history of the generic concept in the botanical literature, discuss the criteria which have been useful for the recognition of genera in flowering plants, cover related topics such as conservative vs. liberal approaches to generic circumscription, monotypic genera, etc., and finally to offer a recommendation on the definition of the category of genus and means for recognition of taxa.

HISTORY OF GENERIC CONCEPTS

The history of the generic concept in the botanical literature is much less tortured than that of the species concept. Still, considerable diversity of viewpoints has prevailed. Bartlett has remarked: "the generic concept is so useful in classifying knowledge and has been so logically and extensively applied in various parts of the world, that to trace its history would be to trace the history of language and thought itself" (1940:354). Some workers have suggested that the concept of the genus antedates that of the species:

> The concept of the genus is probably the oldest among all taxonomic categories and perhaps the oldest one recognized by mankind. In many languages there exists a terminology of plants and animals hardly different in meaning from modern scientific nomenclature on the generic level. Such names as pines, elms, poplars, willows, oaks, roses, palms are used by many peoples. The local kinds of these plants, equivalent to species, are also often recognized by natives. The generic concept could have been set up by the synthesizing of species or, conversely, the species concept may have been derived from differentiation from the generic level. However . . . , we have reason to believe that the generic concept antedates the species concept and that the latter was developed by differentiation from the former. (Li 1974:720)

I would not agree that the concept of the genus antedates that of the species. It seems clear that the basic units recognized in the natural world by primitive peoples are basically equivalent to our modern scientific species. A confusion has existed in equating modern species with the term "folk genera" (e.g., Berlin 1973). The latter are not genera in the modern sense but rather the third level in the folk taxonomic hierarchy (after folk varietal and specific levels). Varieties and species in folk classification refer usually to plant groups of high cultural value, such as those having food or construction importance. The basic units or organic diversity are the folk genera and some 250 to 800 usually prevail in primitive societies (Raven, Berlin, and Breedlove 1971), but these are equivalent to modern species rather than to modern genera (Berlin 1973). Hence, "generic" concepts are probably the older in primitive societies, but they are not the generic concepts discussed here in a modern sense. Atran regards this as a "dead issue," because of the problem of knowing what people mean when they talk about "generics", (1987:202). He suggests talking simply about the folk biological "generic-speciemes." It is also of interest to point out that the general limit for organization of information about a particular subject for the human mind ranges in the 250–800 item level. Beyond this level it becomes humanly difficult to remember all the variations involved in the classification system. It is not surprising, therefore, as pointed out well by Raven, Berlin, and Breedlove (1971) that as more species become known from the level of primitive peoples into the early classification systems, we see Tournefort (1700) using 698 genera of plants and Linnaeus (1737b) with 935. As the known diversity increased from futher exploration, the family concept became emphasized as a means of being able to hold all information about plants together more effectively, and thus Jussieu (1789) recognized 100 "families" which has now reached the level of 300 to 400 families of flowering plants in the modern schemes of Cronquist (1968, 1981), Takhtajan (1969, 1980), Thorne (1976, 1983), and Dahlgren (1975, 1983).

Theophrastus' generic concepts were essentially those of the primitive peoples modified only slightly (Greene 1909) and therefore are best regarded as folk generic rather than modern generic concepts. Another early usage of genus was in a logical sense by Aristotle as it applied to the concept of logical division. The genus was divisible into species and was simply a class to be subdivided into subclasses (Sinclair 1951).* However, as Lennox (1980) points out, Aristotle at times viewed the logical *differentiae* for specific delimitation as varying along a continuum and hence in less of a typological framework for both specific and generic concepts. The early herbalists, exemplified by Brunfels (1530–1536), used essentially folk generic concepts (Bartlett 1940). Bauhin (1623) did not improve the situation, and the use of generic names at that time was confused at best: "The name of a species by Bauhin's time has become something that need not indicate any genus and may even indicate a genus from which the species is excluded. A name is merely a name, not necessarily indicating generic affinity at all, and knowing where species belong has become merely a feat of memory. Truly simple generic grouping, as found in folk botany and reflected in language, had been lost, by the time of Bauhin's *Pinax*, in a maze of complexity and obscurity" (Bartlett 1940:358).

It remained for Tournefort (1700) to correct the situation and place the concept of the genus on sound footing. Although he certainly did not invent the generic category, nor was he the first to use it frequently, he did place all the plants in his *Institutiones Rei Herbariae* into genera. It is for this effort that he is regarded as the Father of the generic concept. He commented on his view of genera in the introduction (titled the *Isagoge in Rem Herbariam*) to his book (from Bartlett 1940), and he believed that of the six features of a plant (roots, stems, leaves, flowers, fruits, seeds) five should be considered for purposes of generic circumscription. He also stressed that usually features of the flowers,

---

*This usage continues in traditional formal logic. Consider this passage from Sinclair, which from a biological perspective seems somewhat amusing: "The usage of the words 'genus' and 'species' in Zoology (and Botany) is exceptional, and may be confusing unless distinguished as a special case. Though any zoological class is, logically considered, a genus in reference to inferior classes and a species in reference to superior classes, yet zoologists have in the main agreed to apply the name genus to one kind of class only, and the name species to its immediate sub-classes only" (1951:92 note)

and especially fruits, would give the best criteria upon which to found genera. In practice, however, he did use some single features of fruits, bark, underground stems, etc., for some generic distinctions. As Bartlett has aptly commented: "Tournefort's ideas of genera were clearly pragmatic in the extreme. If new generic names would be conducive to understanding the nature and affinities of plants, he had no scruples about establishing them. Nevertheless he did not do so thoughtlessly or without good reason. His criteria were generally well considered, and few of his generic propositions failed, in the long run, to win the approval of Linnaeus and his successors" (1940:361).

Linnaeus (1737a) based his generic concepts clearly on those of Tournefort (1700) and Plumier (1703). His approach to generic circumscription was outlined in detail in the *Philosophia Botanica* (1751) and it consisted of searching for three characters (after Svenson, 1945): (1) the *natural character* giving the complete description of all its features and upon which the classification system should be based; (2) the *factitious character* being a selection of features suitable for discrimination among genera in an artificial system of classification or even in a key; and (3) the *character essentialis* which equated to the features allowing for easiest description. Linnaeus' advice in establishing genera was to recognize species first and then to synthesize these into genera, thus essentially sidestepping the question of generic definition in a general sense. In practice he tended to emphasize characters of the fruit for generic delimitation (Larson 1971) which followed in part the tradition of Tournefort. As Larson (1971) suggests, in describing the genus Linnaeus first described each element of the fruit (and/or their features) by reference to number, figure, proportion, and situation, of the principal (or chief) species of the genus (i.e., the one selected as the basis for comparison), followed by comparison with similar species thought to be in the same genus, with the elimination of features missing in some of them. This left a residue of critical features, or the "natural characters," diagnostic for the genus. Sometimes Linnaeus apparently focused almost entirely on the features of one species to yield his generic characters, as pointed out by Pennell (1931) in the Scrophulariaceae.

From Linnaeus's time to the present, the concept of the genus has remained relatively stable. Genera have been added and others subtracted, obviously resulting in an increase of total genera as new material has been obtained through additional field exploration. As an example of this inflation, in the Gramineae, Linnaeus (1753) treated 38 genera, Bentham and Hooker, vol. 3 (1883) recognized 292 genera, and Airy Shaw (1966) listed 661 genera (all figures from Clayton 1972). But these increases in genera have not been due to great revolutions in applications of concepts of genera in flowering plants; they are due to increase in collections. The basic concept still is that a genus is an assemblage of species that has more significant features in common among its members than with any other species. A corollary is that there is a greater discontinuity, or phenetic gap, between groups of species than between species (called *"hiatus" taxonomy* by Singer 1986). What has changed, however, are the new comparative data available for making comparisons among taxa for purposes of generic circumscription.

TYPES OF DATA USED TO DELIMIT GENERA

Morphology

The traditionally used data for the recognition of genera have been morphology and anatomy. The importance of the former is obvious and hardly needs emphasizing (see Greenman 1940, for supportive comments). Genera have been, still are (e.g., Patterson 1977; Friis 1978; Jansen 1981), and will probably always in part be delimited by morphological features. As with all morphological features at any level in the taxonomic hierarchy, the variation in presumed generic characters must be assessed before their value can be determined (the same is true for microcharacters derived from SEM; Lane 1985). For example, detailed studies of infra- and interpopulational variation in species of *Cachrys* and *Prangos* (Umbelliferae) by Herrnstadt and Heyn (1975) showed continuous variation in presumably significant generic characters which caused the authors to combine all the populations into one genus (and even further into a single species). Sometimes the patterns of morphological variation between and among genera are complex such as indicated by Orchard in *Ixodia* (Compositae): "These examples show that there is no simple way to define genera in this group, and that the genera at present recognized are only semidiscrete assemblages of species from what is, in effect, almost a continuum. In defining genera in this context the most that can be reasonably expected is that the genera will be equivalent in their distinctiveness from each other. It is not possible to fully represent within a formal nomenclatural system the interwoven relationships of these taxa" (1981:187, 189).

Anatomy

Anatomical data have also been used for the recognition of genera. This approach became more common in the nineteenth century and has increased up to the present day (e.g., Heintzelman and Howard 1948; Carlquist 1958, 1967; Robinson 1969; Sherwin and Wilbur 1971; Schmid, 1972b; Calderon and Soderstrom 1973). Two practical difficulties prevail (Bailey 1953), however, and the first is the procurement of adequate comparative material. Herbaria do not always have satisfactory anatomical specimens, at least certainly not with the proper state of preservation (e.g., in FAA or other liquid fixative), nor are curators overly anxious to have material removed from regular herbarium material. As Bailey says facetiously: "Imagine the condition of an herbarium if a voracious swarm of anatomical beetles were allowed to digest the precious leaves, stems, flowers and fruits of type specimens" (1953:122). The second difficulty is the inordinate amount of time involved with the extraction of anatomical data, much slower than with morphological (exomorphic) features. As a result, anatomy tends to be used not with routine taxonomic work but to help resolve particular problems that have not yielded well to other modes of analysis. Floral, as well as vegetative, anatomy can be useful in generic delimitation (Eames 1953).

Geography

Geography has also been used traditionally to help delimit genera. For example, Stuessy (1969) referred to the improved geographic unity of the two species of *Unxia* (Compositae) confined to northern South America, as partial evidence (with other morphological and cytological data) for their recognition as a good genus apart from other presumed generic relatives. However, the circumscription of genera by reference to distribution is much more complex than with species and infraspecific taxa. In these latter situations, the geographical patterns are essential for arriving at a reasonable solution on the level of the hierarchy involved. Distributional considerations with regard to genera have more to do with estimating the age of the group, its center of origin, and its phytogeographic history. As Mason has stated: "As compared with the discontinuity evident in interbreeding populations, discontinuity in the genus is not so clear a concept. As most often used it implies a situation that calls for either an unusual dispersal mechanism or a long evolutionary history with intervening extinction" (1953:157). Much discussion has prevailed on the interpretation of these patterns, especially at the level of the genus. Willis (1922) stressed the importance of the area occupied by a group as an indication of its evolutionary age, the reasoning being that older groups have speciated more, have dispersed more widely, and have become more morphologically diverse. Cain (1944) gave many useful criteria for determining centers of origins in plants. There are general values to these perspectives, as more recently confirmed by analyses of geographic patterns in the Gramineae relative to age and area (Clayton 1975). Obviously, however, and as pointed out forcefully by Croizat, Nelson, and Rosen (1974), the size of a genus will not always be related to its age and a pattern of distribution may be due to many factors. Changing environmental conditions may promote speciation, and major earth events (such as mountain building or even continental drift) and long-distance dispersal may enlarge distributions within relatively short time periods and give the impression of much older age for a group as well as complicating significantly the interpretations of centers of origin.

The traditional approaches using morphology, anatomy and geography in the delimitation of genera have worked reasonably well over the past two centuries, but other data obviously have become available in more recent years. Lawrence calls for an "expanded outlook," to bring together as much data as possible to solve taxonomic problems at this level (1953:120). A good example of this viewpoint would be the study on generic limits in the tribe Cladothamneae (Ericaceae) in which morphology, anatomy, flavonoids, and pollen data are all utilized (Bohm et al. 1978).

Cytology

Two of the more recent types of data used for generic recognition beyond the traditional ones are cytology and cytogenetics. As for cytology, the features utilized of importance at the generic level have been largely basic chromosome number and chromosome size and shape. More recently extra-chromosomal features have been used at the generic level, such as crystalline inclusions in nuclei of the Scrophulariaceae (Speta 1977, 1979), but their use has been rare so far. A good example of the efficacy of base numbers for generic delimitation comes from Hassall (1976) in the Australian Euphorbieae (Euphorbiaceae), in which cytologically homogeneous groups correlated with those derived from phenetic cluster analysis. Jones (1985) summarizes the utility of chromosome number among genera of the Astereae (Compositae). Embryology is associated with cytological approaches and also has been useful in some instances in the delimitation of genera (Cave 1953). A surprisingly helpful perspective on use of cytology in dealing with genera comes from Löve (1963:49), who is well known for his emphasis on cytological differences and species recognition (e.g., Löve 1962, 1964): "But although we must admit that cytological evidence is an important auxiliary in studies aimed at natural classification of genera, we must also realize that it is no more the final answer to our problems than are other methods of study. I would like to emphasize that one ought to be reasonably conservative in splitting or uniting genera on basis of more or less insufficient evidence, be it cytological or morphological, and also that cytological differences which are clearly significant as generic characters in one group do not necessarily have to be so in another. To ignore cytological differences in generic revision is, however, equally objectionable as the over-estimation of their importance. Here as elsewhere the golden middle way is the best choice."

Cytogenetics

The data that have seemed to provide the most absolute criterion for generic delimitation are cytogenetic in that genera usually do not cross naturally and usually cannot be made to cross even artificially (Powell 1985). It stands to reason that if species are largely reproductively isolated, certainly genera should also be so, and even to a larger degree. This point has been mentioned by Anderson (1940), Löve (1963), and many others. The basic perspective of the crossibility of genera is summarized well by Rollins: "when two species properly placed in different genera will cross and produce a hybrid progeny of any sort, the validity of one of the two genera becomes suspect" (1953:135). Nonetheless, numerous intergeneric hybrids within the angiosperms are known involving many families (Knobloch 1972). The Gramineae are especially well known for the occurrence of artificial and natural intergeneric hybrids (e.g., Dewey and Holmgren 1962; Dewey 1967a, b, c, 1970, 1983, 1984; Prywer 1965; Pohl 1966; Sulinowski 1967; Knobloch 1968; Runemark and Heneen 1968; Sakamoto 1974; Rajendra et al. 1978; Wang, Dewey, and Hsiao 1985). Genera of the Orchidaceae cross easily artificially (e.g., Garay and Sweet 1966, 1969), but few natural intergeneric hybrids are known. Intergeneric hybrids in other families are not common, but some examples may be cited in the Caryophyllaceae (Kruckeberg 1962; Crang and Dean 1971); Compositae (Heiser 1963b; Anderson and Reveal 1966; Kyhos 1967; Yeo 1971); Cyperaceae (Fernald 1918); Hydrocharitaceae (Kaul 1969); Iridaceae (Chimphamba 1973); Leguminosae (McComb 1975); Rosaceae (Stutz and Thomas 1964; Byatt, Ferguson, and Murray 1977); Saxifragaceae (Soltis and Bohm 1985); Scrophulariaceae (Kruckeberg and Hedglin 1963);

Solanaceae (Menzel 1962); and Umbelliferae (Webb and Druce 1984). It is interesting that in families with elaborate pollination systems with high degrees of specificity, as in the orchids, few natural intergeneric hybrids are known, whereas in wind pollinated groups such as in the grasses, many more are known. The high levels of artificial intergeneric crosses in the Gramineae surely derives in part from the high economic value of the family and the considerable research activity in the group.

Use of crossability to delimit genera has many ramifications and the work is not simple. Because many genera will not interbreed, Rollins (1953) stresses the need for crossing studies *within* the genus (i.e., *among* the constituent species) to look for genetic structure and cohesiveness of the included taxa. Using this approach, Long (1973) discovered that in *Ruellia* (Acanthaceae) species groups began to emerge that cast doubt on the naturalness of the genus. But if a natural or artificial hybrid between two species of two separate genera is obtained, what does this say about the relationships of the genera? There are four alternatives. First, the two genera should be merged completely, as was done with *Franseria* into *Ambrosia* (Compositae; Payne 1964), or as with the suggestion to merge *Agropyron* and *Elymus* (Gramineae; Runemark and Heneen 1968; this perspective, however, is not shared by Dewey 1983). Second, one or the other of the species is misplaced and should be moved into the other genus, as with the studies of Heiser (1963b) in the Compositae in which *Viguiera porteri* was crossed successfully with four species of *Helianthus*, and was suggested as perhaps better placed into the latter genus. (He did not make the formal new combination, however.) Third, the crossability is regarded as showing that the genera are closely related, but no change in taxonomic position is made, as was the case in the artificial hybrid between *Boottia cordata* and *Ottelia alismoides* (Hydrocharitaceae; Kaul 1969). Fourth, the two species in question could be placed together in their own genus. To really understand the meaning of the crossability between two genera, *all* the crosses between *every* species pair should be attempted both within each genus and between them. Clearly this will rarely be attempted due to the extensive work involved. This would be the only way to understand clearly the implications of the observed crossability. Lacking this completeness in comparative data in most cases, we must assess all available data and make the best decision. Certainly conservatism in generic concepts is desirable and the remodeling of genera should be done only if the evidence is strong. The degree of fertility of the natural or synthetic intergeneric hybrid is also important to determine, the higher level of fertility suggestive of a closer genetic relationship.* I would agree fully with Heywood (1960) that an understanding of two crossable genera on a worldwide scope is needed before decisions can be made

---

* Some cladists have claimed that crossing data reveal nothing about evolutionary relationships because the ability to cross is a primitive trait, or plesiomorphy, and therefore useless for cladistic purposes (Funk 1985a; Platnick, pers. comm.). This misconception results from a confusion on the type of data that crossability represents. They are not comparative data such as leaf shape, stamen number, and so forth that exist in two or more states, but rather data on the *interaction* between two organisms and their genomes. Hence, to talk of apomorphies and plesiomorphies is not productive; one must use crossability data to indicate the *degree* of genetic cohesiveness of taxa which come from a single evolutionary line. Phenetic approaches to classification earlier had the same difficulty in attempting to integrate cytogenetic data with other types of comparative information with similar lack of success. Crossability data are different and should be used as a test of relationships based on structural information.

on uniting them. Rollins has a fitting closing to the use of cytogenetic data with genera: "Ordinarily, cytogenetic approaches will not provide the broad general outlines of classification within a genus, nor is it expected that given genera will always be sharply delimited by the use of cytogenetic criteria alone. But these data, added to the sum total built up by other means, often supply the needed critical facets to solve perplexing problems. Thus, cytogenetics may aid substantially in building the kind of generic classification that most accurately reflects the evolutionary relationships of the species which make them up, and the families to which they belong" (1953:139).

NUMERICAL DELIMITATION OF GENERA

Just as numerical approaches have been used effectively at the specific level, so also they have been applied efficaciously with genera. Most of the work has been done with phenetic algorithms. James (1953) prepared a very early "objective aid" to determining generic limits which was a data matrix of differences between species of different genera for each character. The values of differences between each pair of species was summed for a quantitative view of the relationships. In this fashion numerical gaps based on all characters of all taxa could be discerned. It is most important that all phenetic relationships be determined both within the genus and between all the included species and those of the related genera under question. Rowell (1970) has an excellent summary with emphasis on fossil groups. An early and effective phenetic study dealt with generic delimitation in the Chrysobalanaceae (Prance, Rogers, and White 1969). Other examples include Hassall's (1976) work in the Euphorbiaceae and Baum's (1978c) studies in the Triticeae. Only a few studies so far of plant groups have dealt with determining generic limits using cladistics, but one example is Bolick's attempt to divide *Iva* (Compositae) into several segregate genera.

NATURALNESS OF GENERA

The question of the naturalness of genera has been raised frequently by many workers, or as Booth puts it: "Do you believe in genera?" (1978:1). Some have responded in the fashion of Bisby and Ainsworth that: "Nature may make species, but man has made the genera" (1943:18). Cain remarked that "the genus cannot now be regarded as a naturally discrete group either in relation to its ancestors and descendants, or at any one time" (1956:108). Anderson, in his survey of "modern opinion" on the genus concept, showed that most respondents to his questions regarded genera as more natural than species (i.e., reflecting "an actual discontinuity in organic nature" (1940:364). A further interesting correlation is that in breaking the respondents down by background and experience, he found that the monographic workers regarded genera as more natural than species by a margin of 2 to 1, but that the "non-monographers" viewed it as just the opposite. It seems clear to me that genera are certainly less natural than species in terms of representing an actual discontinuity in the living world. They may be clearly delimited in some families, especially in older ones in which extinction has brought about definite phenetic gaps, but less so in others. But even if they are clearly delimited because of absence of intermediates, they are not as natural in the sense of being reproductive units of nature. Genera are the accumulations of groups of reproductive units (the species) rather than the direct result of their formation. The dogmatic cladistic view would regard genera as no more nor less natural than species or even higher taxa so long as they are treated as holophyletic groups (based on synapomorphies) in the reconstructed phylogeny.

REMODELING OF GENERA

A number of suggestions have been offered as guides to the remodeling of genera, especially in the recognition of generic segregates. The most comprehensive are the recommendations of McVaugh (1945:15–17; also rephrased by Gillis 1971, and used again in that form by Grashoff 1975), herewith presented condensed: (1) special consideration should be given to qualitative morphological characters; (2) the recognition of segregate genera based on minor or single characters should only be allowed in particular instances to preserve usage; (3) the biological unity of a genus is more important than the "gap" between it and its close relatives; (4) changes made in generic limits should be done only after a full study of variation within the complete range of the group; (5) decisions on whether to establish segregate genera should be based on the relationship of the segregate to its core genus and not on relationships of the core group to other established segregates; (6) segregate genera should be sharply delimited (any intermediate species should be included in the larger genus); (7) the strength of the argument to recognize segregate genera varies proportionally to the number of differentiating characters; and (8) the decision to recognize a generic segregate is strengthened if the group has a distinctive geographical range.

These recommendations basically form a good framework within which to approach the remodeling of genera, although they represent a conservative approach. Much difference of opinion has prevailed over the past several decades on conservative vs. liberal views of generic limits. Camp has advocated an aggressive posture to latch hold of any new data and remodel genera wherever necessary. As only he could put it: "Perhaps we should adopt as our motto, not 'Back to Linnaeus' but, 'Forward to the truth.' Perhaps, if we were not afraid of the puling croaking of certain of our confreres every time we broaden and particularize our concepts, we could put new life into old taxonomic bones, long interred in the musty vault of nomenclatural conservatism" (1940:389).

Generic splitters have surfaced from time to time, e.g., Rydberg (1922, 1924a, b), King and Robinson (1970, 1987), Robinson and King (1977, 1985), and Nordenstam (1977, 1978), and these approaches always have caused a furor among other workers. The recent generic fragmentation of the tribe Eupatorieae of the Compositae by Robinson and King (1977, 1987; also King and Robinson, 1987) is an instructive example in this regard. Over the past two decades their work has resulted in a 211 percent generic inflation (Turner 1977b) in the tribe, which is nearly double that of any other tribe and nearly five times the rate for the family as a whole (45 percent). Especially irritating in this instance has been the touting of the more liberal view of genera in the Eupatorieae as the "New Synantherology": "It is frustrating to see phanerogamic taxonomists ignore a wealth of microscopic characters. It is inexcusable when the alternative characters prove to be as unreliable as those now widely used in the study of the family Compositae. The new Synantherology is many things, including a rejection of bad characters, elucidation of numerous new characters at an anatomical level, and the application of techniques that make these characters available in the routine taxonomic study of Compositae" (King and Robinson 1970:6). The presumed efficacy of the use of floral microcharacters in generic delimitation in the family has not been accepted by most workers, however. Strong rebuttals have been offered by Grashoff and Turner (1970), and McVaugh gave a penetrating critique of the approach. "No other authors known to me in modern times have proposed such sweeping reorganization of a major plant group in such a short time, and asked the public to take so much of it on faith, without much documentation" (1982:189). This is still true even after the publication of a book bringing together in a useful fashion all the previous work on the tribe (King and Robinson 1987). The clear difficulty is that many small genera now exist in small coordinate subtribes which provide no internal predictive power for the tribe. Small deviating groups have been recognized, but these have not been brought together again in a meaningful synthesis of relationships at the subtribal level. Wetter (1983) has shown that floral microcharacters can vary within wide limits in the Senecioneae (Compositae) and stresses caution in their general use. As B. L. Robinson said more than eighty years ago; "The burden of proof should always rest upon the writer suggesting the change. It is rather surprising to notice how lightly this matter is taken by some, who attempt sweeping changes. It is by no means rare to see a few habitually similar species of a large genus split off and set up as a new genus with scarcely any attempt to give accurate definition to the new group or tell just what traits are of diagnostic value in separating it. The authors of such work indolently and carelessly shift the burden of proof upon others" (1906:90). An example of a balanced approach to remodeling generic limits is the combining of *Notoptera* and *Otopappus* (Compositae) by Hartman and Stuessy (1983) after detailed study of all relevant characters of all species in both genera.

An important point to consider is the availability of creating new genera with the resulting necessary new combinations vs. the establishment of subgenera, sections and series to reflect relationships without having to make any new combinations that must be indexed by the *Index Kewensis* and elsewhere (pointed out by Grant 1959, and McVaugh 1982). Greenman said it well: "Unless some very definite object is attained by segregation of relatively homogeneous groups of plants, such for example as *Aster, Erigeron, Conyza, Baccharis, Senecio, Euphorbia*, and *Cassia*, I am personally inclined to think that it is more practical to retain these groups in their traditional sense. Certainly such a treatment is less disconcerting to botany in general than to make numerous possible changes. Generic segregation almost invariably means the introduction of new combinations and new names" (1940:373). The alternative recognition of subgenera, sections and series avoids many nomenclatural difficulties. It also allows a better reflection of evolutionary relationships among the species, due to the greater number of available subdivisions than if the segregates are all treated as coordinate genera. See Philipson (1987) for agreement on this point. (Of course one could erect new supergenera, subtribes, etc. to achieve some of this same objective, but with even more nomenclatural burden.) Reveal (1969) has used infrageneric units to useful effect in *Eriogonum* (Polygonaceae) and so also has Stuessy (1972a) in *Melampodium* (Compositae).

Another related point is the need to understand genera on a world-wide basis before remodeling part of it. Sherff stresses that: "The entire earth must be taken as the source-book of our generic concepts" (1940:376). Taylor (1955) showed clearly the need to take such a world view of *Anagallis* (Primulaceae) in order to revise just the species for the flora of Tropical East Africa.

PALEONTOLOGICAL GENERA

The genus for paleontologists is somewhat different than for neontologists (Amsden 1970). The obvious problem is with the fragmentary fossil record in which only single organs or even parts of organs become preserved (especially acute with dispersed pollen and spores; Hughes 1972). This leads to the inevitable establishment of *form genera* or *organ genera* until such time as organic connections are found between isolated parts to yield a whole organism. A classical example is the Devonian gymnospermous form genus *Callixylon*, common as stem and trunk sections, and *Archaeopteris*, known from leaf and fertile material only. Beck (1960) showed organic connection between these two organ genera which subsequently yielded a better interpretation of its overall evolutionary relationships (Beck 1962, 1970). The same problems apply to recognizing genera from isolated pollen and spores (Hughes 1976) so common in the fossil record and so helpful for understanding the early evolution of the angiosperms (e.g., Hickey and Doyle 1977; Doyle 1978).

MONOTYPIC GENERA

We return once again to the problem of monotypic genera, considered earlier with regard to the logical difficulties called Gregg's Paradox (see p. 159). Whatever one may think of the desirability of having monotypic genera, Clayton (1972) has shown that they are recognized more or less consistently in different flowering plant families. The most recent challenge leveled at the concept of monotypic genera comes from the cladists. Platnick (1976) argues that if one accepts only dichotomous speciation in phylogeny (a most dubious assumption), then monotypic genera (and their sister groups) can only be paraphyletic because they do not contain all of the species descended from the common ancestor of the entire group. In my opinion this is an insignificant problem because genera should be recognized based on all features of phylogeny and not just cladistic results.

CLADISTICS AND GENERIC DELIMITATION

Most cladistic workers would argue that all genera should be defined holophylet-ically, and that paraphyletic genera should be rejected (e.g., Funk 1985b; Young 1987). These same workers would suggest that the "general purpose" aspect of generic delimitation should be subordinated in favor of a more precise cladistic evaluation (e.g., Stevens 1985). Needless to say, I do not share this view (see general comments in chapter 9); the most predictive and useful delimitation of genera will be by phyletic, rather than cladistic, means. From a theoretical standpoint it is of interest that genera may be paraphyletic but also conservative in size and shape of certain features reflecting clear morphological discontinui-ties (Leman and Freeman 1984), and therefore deserving of formal taxonomic recognition. For a balanced attempt to determine generic relationships within the tribe Andromedeae (Ericaceae), in which both phenetic and cladistic ap-proaches were used to suggest major lines of evolution, see Judd (1979).

In conclusion, a genus is a group of species held together by several to many character states and distinct from other such groups, and between which natural or artificial hybridization is usually not possible. I agree with Boivin that "from a practical point of view, genera should be easily recognizable groups, in such a way that once a number of species of a group are known, most other species will at once be recognized as members of the same genus, although the species themselves may be unknown" (1950:39–40). Some general perspectives need to be kept in mind (after Davis and Heywood 1963), such as whether the genus is natural or not (i.e., a good evolutionary unit), or how to make it so, where to draw the line(s) between closely related genera, whether to recognize a group as an independent genus or include it in another, and so forth. Certainly to be considered are the degrees of phenetic distance between the groups, the concepts used traditionally in other parts of the same family, the size and homogeneity of the constructed groups, the numbers of intermediates to be dealt with, tradi-tional usage in the group and close relatives, subgeneric vs. separate generic status for the groups, etc. Defining the generic category is difficult and recogniz-ing generic taxa in practice is even harder.

CHAPTER 14
Family and Higher Categories

To rise beyond the generic level in classification is to enter a world of much greater uncertainty. Families, orders, classes, divisions and even kingdoms depart significantly from biological concerns at the populational level and force a treatment based almost entirely on comparative data that are often incomplete. Taxa at higher levels will be well-defined or ill-defined depending upon the group in question.

Kingdoms of life used to be extremely well-defined, traditionally as plants and animals, but in recent years much change of viewpoint has prevailed. Whittaker (1969) has suggested the now reasonably well-accepted concept of five kingdoms of organisms: Monera, Protista, Animalia, Fungi, and Plantae. Even more recently the discovery of the very primitive methane-bacteria (Woese and Fox 1977; Fox et al. 1977) suggests that they and their relatives may belong in a separate kingdom (Yang, Kaine, and Woese 1985).

The perspective on divisions of plants has changed in the past two decades from the traditional Algae, Fungi, Bryophyta, etc. (essentially the scheme of Eichler 1883), to a much more dissected approach stressed by Bold (1967) and others. In particular the algae now are viewed as containing many divisions (Bold and Wynne 1985), such as Chlorophyta, Rhodophyta, Phaeophyta, etc., and the blue-green algae are now treated usually as bacteria (as cyanobacteria) and placed in the kingdom Monera. Lewin (1981) has also suggested the possibility of a new prokaryotic division of algae, the Prochlorophyta (see also Chapman and Trench 1982).

In the angiosperms, if we treat the flowering plants as a division (Magnoliophyta; Cronquist, Takhtajan, and Zimmermann 1966), then in general the classes (monocots and dicots) are well defined. An exception to this viewpoint has been offered by Bremer and Wanntorp, in which, based on a cladistic reinterpretation of Takhtajan's (1969) phyletic classification, they suggest recognition of six "major groups" (1978:323), instead of the traditional monocot and dicot lines. This approach suffers from the inadvisability of reinterpreting intuitively generated phyletic relationships in cladistic terms and a failure to accept the evolutionary value of paraphyletic groups.

The orders of angiosperms have been very ill-defined, except for a few such as the Caryophyllales. This level of the hierarchy in the angiosperms is still very much in flux as evidenced by the different views in the systems of Cronquist (1968, 1981, 1983), Thorne (1968, 1976, 1983), Takhtajan (1969, 1980, 1986, 1987), and Dahlgren (1975, 1983). The concepts of families have varied less widely, and many, such as the Umbelliferae, Compositae, and Melastomataceae, are well-defined, in contrast to such nebulous groups as the Rosaceae and Ranunculaceae (Walters 1961).

In terms of the concepts to be discussed in this chapter, the focus will be on the family, but the issues relating to this level in the hierarchy apply equally well to orders, classes, and divisions. We shall begin with a brief historical summary of the development of higher categories, consider some of the philosophical, logical, and practical problems with them, discuss data and methods that have been used in their recognition, and briefly touch on the interesting question of the evolutionary origin of higher taxa.

HISTORY OF CONCEPTS OF HIGHER CATEGORIES

Although strikingly similar groups of plants that we now call families, such as the mints (Labiatae) or carrots (Umbelliferae), were recognized many centuries ago by Theophrastus (370–285 B.C.), the formal taxonomic recognition of the concept, obviously, did not appear until after genera were recognized. Although Tournefort (1700) used informal groups of genera, he made no attempt to formally characterize these units or to provide them with descriptions. Nor did Linnaeus (1753) use a family concept consistently. Parts of Linnaeus' "classes" were in fact more or less equivalent to some of our modern families, but his artificial system did not allow for this to be developed well. The use of family in the modern sense did not appear until the publication in 1789 of the *Genera Plantarum* by Antoine-Laurent de Jussieu. Jussieu's units (he called them "orders") correspond closely with many of the families of today, and he deliberately grouped the genera and indicated the characteristics that were used to hold them together. He could appropriately be called, therefore, the "Father of the Familial Concept" in flowering plants. However, the term "family" was not used for these units until the mid-nineteenth century, and it appears as such in the system of Bentham and Hooker (1862–1883) and elsewhere. The concept of order in flowering plants apparently originated with Lindley (1833), in which a collection of related families was called a *"nixus"* (Davis 1978).

NATURALNESS OF HIGHER CATEGORIES

In general, the higher categories would be regarded as less real and less natural than at the generic level and below. But the terms reality and natural relate to groups that are easily recognizable in a number of features, which can occur at any level in the hierarchy. The family Compositae, for example, is surely more real and natural than the genus *Aster,* in which numerous segregates have been recognized. Likewise, the Caryophyllaceae are surely more natural and real than even the species of *Crataegus,* well known for difficulties due to hybridization and apomixis (Phipps 1984).

HIGHER CATEGORIES AS INDIVIDUALS

Brothers (1983) has raised the question of whether higher taxa cannot be regarded as individuals, rather than classes, in a manner similar to the perspectives for species (Ghiselin 1966a, 1974, 1977; Hull 1976, 1978; see p. 169 of this book for discussion). Wiley (1980) suggests that any natural (i.e., holophyletic, in his sense) taxon at any level may be regarded as having characteristics of both classes and individuals and calls them "historical entities." As with species, my viewpoint is that higher taxa are definitely classes and not individuals. Each family, order, etc., has its members which are the genera, families, etc., respectively. These higher taxa do not at all qualify for the "part of the whole" concept of Hull (1976) such as with different organs of a single individual, which would be necessary in order to regard higher taxa as individuals. A family consists of a collection of coordinate genera, all more or less similar in their features (the family features), and so on for families of orders, orders with classes, etc. These higher taxa are best regarded as classes and not individuals.

SIZE OF HIGHER TAXA

Reasons for the size of higher taxa are obscure, but worthy of discussion. Perhaps most surprising is the large number (37 percent) of monogeneric angiosperm families (Clayton 1974b), which seems far higher than one would initially expect. This is also true for monofamilial orders, especially in the monocots (Van Valen 1973). Strathmann and Slatkin (1983) consider the problem of the improbability of animal phyla with few species, using a model of speciation and extinction rates over geological time. They conclude that the persistence of small phyla for long periods can occur when speciation rates exceed extinction rates early in the adaptive radiation of a group, but when shortly thereafter (before the clade becomes large) speciation rates become equal to extinction rates and both are low. Clayton speculates on the reasons for so many small families and suggests that "taxonomists have an obvious predilection for the excision of solitary outliers, thus exposing further outliers which they are tempted to chip off next" (1974a:278). This must be due in part to our conceptual inability to deal well with large families which contain much diversity. Another point is that more modern taxonomists tend to be trained in critical data analysis and place emphasis on finding differences between taxa rather than looking for similarities for syntheses which was the main thrust when the basic structure of the system of classification of the angiosperms was erected. A type of "domino effect" occurs when critical studies in one group lead to splitting of taxa and elevation of groups to higher rank, which makes other neglected taxa appear in need of similar attention and dismemberment (whether they actually do or not).

VERTICAL VS. HORIZONTAL CLASSIFICATION

Some discussion in the literature has prevailed on the desirability of *vertical* vs. *horizontal classification* of higher taxa. These terms are of special importance in paleontology because the time dimension must be taken into account for purposes of classification. Horizontal classification is based on relationships among taxa in the same time zone, determined by similarities and differences, and is hence largely phenetic (e.g., Bigelow 1961). It can certainly also reflect evolutionary relationships to a large degree, but only insofar as phenetic relationships are able to do this. Vertical classification is based on determining the ancestral-descendent lineages and is largely cladistic in perspective (e.g., Nelson and Platnick 1981). That both have been used in the construction of higher taxa should be obvious, and often a compromise approach is taken (Simpson 1961) to yield a more broadly based evolutionary classification (also advocated in this book; see chapter 9). Davis (1978) points out that more phenetic overlap occurs among taxa when going up the hierarchy to the familial and higher levels. He also uses the concepts of vertical and horizontal but in a different context to mean *changes* (rather than approaches) in classification, and he emphasizes that further study of a group will often result in a vertical change or elevation in rank (= *hierarchical inflation*).

PRACTICAL DIFFICULTIES

In addition to logical and philosophical problems with higher categories and taxa, there are also practical difficulties. These come in different forms, but one factor that must be contended with more than at other levels in the hierarchy is the history of classification within a particular group. Awareness of history is always important in classification to help maintain stability insofar as possible, but because the families, orders, and classes are more subjectively defined than at other levels of the hierarchy, this factor looms more importantly. For example, Rydberg (1922, and elsewhere) viewed the family Compositae as consisting of several families, instead of the traditional single natural family, but this view departed so violently from past approaches, and also seemed to provide no positive gain, that it has been followed by few other workers. Another practical difficulty with working with the higher levels of the hierarchy is the large size of some of the taxa. As Jacobs has clearly pointed out, it takes a tremendous commitment (ten to twenty years), or even a lifetime, for a worker to address the larger families such as the Orchidaceae, Compositae, Gramineae, etc., as well as the medium-sized families, even for making contributions to regional floristic projects. He stresses cooperation and teamwork: "the time has come that three, four, five taxonomists set out to make discoveries in one large group of plants— together" (1969:262). Even if the higher taxon is manageable in size, the extensive literature and herbarium resources needed to deal with relationships at that broad level are practical discouragements for people in most institutions. A final concern is the importance of floral vs. vegetative features at the higher taxonomic levels. Almost all the families and orders are defined by ensembles of floral or fruit features, with a few exceptions, such as the parallel veins of the leaves of the Melastomataceae, the punctate oil glands of the Rutaceae, or the obvious growth form of most Cactaceae. The P_{III}-type sieve-tube element plastids of the Caryophyllales is another conspicuous exception (Behnke 1976a). In large measure in the angiosperms the focus is clearly on reproductive features for delimitation of higher taxa, and sometimes these are not fully represented in herbarium material.

TYPES OF DATA USED WITH HIGHER CATEGORIES

The types of data used for the recognition of families and higher taxa are essentially the same as with genera, but the families (and above) should not be able to cross either naturally or experimentally. New tissue culture techniques may well change this picture (Levin 1975), but for the present it certainly holds true. Morphology, obviously, is the cornerstone of data for recognition of higher taxa. As Greenman expressed many years ago: "there is a definite philosophical principle underlying the system, namely, the arrangement of the large categories in such a manner as to indicate, through comparative morphology, their genetic relationships and to some extent their probable phylogeny" (1940:371). Extensions into micromorphology and ultrastructure have more recently expanded the role of morphology and anatomy (Stuessy 1979b; Walker 1979), such as exemplified in the works on types of sieve-tube element plastids (Behnke 1975a, 1976a, 1977a, 1981a; Behnke and Barthlott 1983) and pollen (Walker 1979; Blackmore, 1984).

Of the newer approaches, cytology and cytogenetics have been less useful at the higher levels due to the inability of taxa to cross, and because chromosome number varies widely across the angiosperms and has evolved in parallel within virtually every family (Grant 1982a, b). Micromolecular approaches have been used with various compounds, but the most useful have been shown to be benzylisoquinoline alkaloids, iridoids, and betalains, with glucosinolates, polyacetylenes and other types of alkaloids also helpful (Gershenzon and Mabry 1983). Macromolecular approaches have yet to fulfill their tremendous potential, but proteins have been helpful, especially with serology and amino acid sequences, followed by DNA hybridization, and Fraction I proteins with isoelectric focusing (Fairbrothers 1983; Fairbrothers and Petersen 1983; Jensen and Fairbrothers 1983). More recently DNA restriction endonuclease fragment comparisons offer hope of real gains in being able to compare directly the genomes of higher taxa (e.g., Palmer 1986; see chapter 21). In fact, the breakthroughs in higher level classification will certainly come about through use of these data. Exciting times lie ahead in this area.

NUMERICAL APPROACHES

Numerical approaches have also been used in the delimitation of higher taxa. These methods, principally phenetics and cladistics, do not offer new data but can often help structure the existing data more clearly and can reveal insights not obtainable intuitively. A most serious problem with use of these approaches at the higher levels is homology (Meeuse 1966; Patterson 1982a; Sattler 1984; Stevens 1984a; Tomlinson 1984a; Lammers, Stuessy, and Silva 1986). The further one goes up the hierarchy in the use of character state comparisons, the higher the probability that the states originated in parallel rather than from common ancestry. There is no real solution to this problem other than to spend literally years determining the developmental patterns for each character and state (e.g., as was done for degrees of dissection of leaves in *Acacia* by Kaplan 1984), which is obviously impossible in most cases. Hence, one must choose the characters and states as carefully and thoughtfully as possible, look for their structural and developmental (if available) correspondences, and hope for the best. For this reason it is not surprising that few phenetic studies have been done at this level in the angiosperms. An early study by Young and Watson (1970) used phenetics to assess relationships of families of dicots. A clear result was the lack of separation between the subclasses Dilleniidae and Rosidae in the Cronquist-Takhtajan systems, which substantiates problems encountered in attempting to teach these units and concepts (they simply do not separate well). Another attempt, somewhat less positive, resulted from a phenetic study of 85 families of monocots (Clifford 1977) in which little agreement with existing higher taxa occurred except for the Zingiberales and Alismatidae. Few cladistic studies have been done at the higher taxonomic levels in plants, and some of the attempts have been flawed by serious problems (e.g., Bremer and Wanntorp 1978; Parenti 1980). Rodman et al. (1984) used both phenetics and cladistics to useful effect in helping delimit the Centrospermae and show relationships among the included families. Lammers, Stuessy, and Silva (1986) used both (with patristics and chronistics also) in determining the relationships of the Lactoridaceae within the Magnoliales. The recent cladistic treatment of the families of monocots is also noteworthy (Dahlgren, Clifford, and Yeo 1985).

EVOLUTION OF HIGHER TAXA

Although this book attempts to deal with the classification of *patterns* that result from the evolutionary process, an interesting question that has been addressed repeatedly in the literature (e.g., Rensch 1959; Stebbins 1974) is the evolutionary *origin* of higher taxa. Because of this previous interest, a few comments are offered here. Recall the responses to the questions of Edgar Anderson (1940) in which most respondents (70 percent) believed that genera originate via the same processes that produce species. The evidence to date suggests that there are no additional mechanisms, other than those already known, to explain the existence of higher taxa (Stebbins 1975; Stebbins and Ayala 1981), with the exception of *extinction*. The reason for the phenetic gap among higher taxa is due largely to extinction. As Bigelow commented: "The taxonomic gaps between the higher categories [taxa] exist because the extinct, ancestral, intermediate forms no longer exist. These gaps are not only useful in classification, they are also meaningful in terms of evolution; they should be recognized and used, both for taxonomic convenience and as food for evolutionary thought. Any system of classification that tends to obscure these gaps, or even to obliterate them, should be re-examined closely" (1961:86). That higher taxa may come about via developmental shifts of relatively simple means has been stressed by Stebbins (1973, 1974). Developmental studies such as those of Tucker (e.g., 1984) on the origin of symmetry in flowers, are very helpful in providing insights on the types of characters and character state shifts reflected in differences among taxa at higher levels. Sachs stresses that the developmental and structural modifications are accompanied by "specific adaptive specializations which give them competitive advantages for part of the environmental resources," and that "physiological adaptations to particular conditions, symbioses which aid in obtaining nutrients and, especially, chemical defense mechanisms could be major components of these specializations and thereby the raison d'etre of plant families" (1978:1).

In conclusion, we can say that the criteria for the recognition of families and higher taxa relate to internal homogeneity of features (primarily morphological but from all available data) plus external discontinuity from other such groups. They also should not cross under any circumstances. They should be monophyletic (i.e., natural) in the general sense that includes both paraphyly and holophyly. Historical usage plays a most important role at these levels, and a constructive approach is recommended with changes being made only when required after careful study and documentation. In the final analysis both within higher taxa and at all levels of the hierarchy, "so long . . . as we have to use judgement at all, the accuracy and soundness of the application of any taxonomic category, definition or no definition, will be in direct proportion to the accuracy and soundness of the judgment of the individuals who apply it" (Weatherby 1942:160).

PART TWO
Taxonomic Data

SECTION FOUR
Types of Data

T O PROVIDE a classification of organisms involves having comparative data for analysis. Data come in many forms, and the plant taxonomist must be able and ready to handle many different types. In addition to using data available in the literature, a taxonomist usually finds it necessary to generate additional original data, so that the relationships among the considered plants can be understood more clearly. To be able to utilize data fully for taxonomic purposes involves being aware not only of the taxonomic implications in the problem at hand, but also of the suitability of sampling, measurement, and display of the gathered information. To provide a background for the eventual analysis of data, therefore, this section discusses the major types of data commonly used in plant taxonomy.

Any piece of information about a plant or series of plants is potentially useful for determining and understanding systematic relationships. This gathered information may be available in small or large quantities, and it may come from only one or from many parts of the plant. From whatever source, all the different kinds of data can be regarded as being of three basic types: (1) those that come from the organism itself, such as morphology, cytology, genetics, and chemistry; (2) those that result from organism-organism interactions such as cytogenetics and some reproductive biology studies (e.g., pollination, animal-mediated dispersal, etc.); and (3) those that come from the organism-environment interactions in a broader sense, such as distribution and ecology. All three types of information are useful in taxonomic studies, but the major sources of data have come from the plants themselves. Within this category, two different kinds of data can be obtained: (1) those that reflect the structural composition or architecture of the plant, such as anatomy, morphology, and chemistry; and (2) those that refer to the dynamic interactions among the structures, i.e., the processes of development and physiology. Both kinds of data have been used in solving taxonomic problems, but most have tended to be structural.

Many different kinds of data will be discussed in the following chapters with comments on history, equipment and other investments needed, problems encountered, and suitability for helping solve particular taxonomic problems. Much of the historical information comes from Sachs (1890), Greene (1909, 1983), Reed (1942), Steere (1958), Humphrey (1961), Ewan (1969), Raven (1974), Stuckey and Rudolph (1974), and Morton (1981). It is obviously not possible (nor even desirable) to discuss every aspect of all data, but enough will be offered with sufficient selected examples to serve as an appreciation of the vastness of the data absorbed and utilized by taxonomic botanists.

Morphology

It is no exaggeration to state that morphology, or the external form of an organism, has been and still is the type of data used most in plant classification. Morphological features have the advantage of being easily seen, and hence their variability has been much more appreciated than for other kinds of features. This is especially true with herbarium material, on which most taxonomic work must be based (Davis and Heywood 1963). The early plant taxonomists relied almost exclusively upon morphology to classify the plants being sent to them from many parts of the world. As a result of these factors, the system of classification among flowering plants developed by these and subsequent workers was based primarily upon morphological data. It is still the foundation for most of our classification today.

HISTORY OF MORPHOLOGY IN PLANT TAXONOMY

To outline the history of use of morphology in plant taxonomy is to describe the development of the entire field. From the earliest recorded observations of the ancient Greeks (e.g., Theophrastus, 370–285 B.C.), through the age of the herbalists (1470–1670), into the early classifiers such as Ray (1686–1704) and Tournefort (1700), then Linnaeus (1753), Jussieu (1789), the Candolles (1824–1873), Bentham and Hooker (1862–1883), and so on to the systems of today (e.g., Dahlgren, Rosendal-Jensen, and Nielson 1981; Dahlgren and Clifford 1982; Dahlgren and Rasmussen 1983; Dahlgren, Clifford, and Yeo 1985; see also Smets 1984), morphology has been dominant. It serves little use, therefore, to exhaustively detail the various contributions of morphology in taxonomy, but rather to examine different kinds of morphological data and examples of their efficacy in recent times. Suffice it to say that the bulk of the data that we have used, are still using, and likely will continue to use, is morphological.

GENERAL MORPHOLOGICAL TEXTS AND REFERENCES

There are no general texts that deal solely with the significance of morphology in taxonomy. There are many that deal with systems of classification based largely upon morphological data (e.g., Cronquist 1981). There also are taxonomy textbooks that have discussions of morphological features with illustrated glossaries (e.g., Johnson 1931; Lawrence 1951; Core 1955; Porter 1967; Radford et al. 1974; Benson 1979; Jones and Luchsinger 1986; Shukla and Misra 1979), and books that deal with the terminology of plant structures, or *phytography* (e.g., Featherly 1954; Harrington 1957; Stearn 1983). General morphological texts also exist (e.g., Troll 1937–1939; Bold 1967; Bierhorst 1971; Foster and Gifford 1974; Sporne 1975; Guédès 1979; Bold, Alexopoulos, and Delevoryas 1987). Books, symposia, and review articles also exist on the use of micromorphology in systematics (also including ultrastructure, to be discussed in this book under anatomy; see chapter 16), such as Heywood (1971b); Cole and Behnke (1975), Stuessy (1979b), Tyler (1979), and Behnke (1981c). In this chapter emphasis will be placed on selected morphological features of interest, with more attention given to newer types of data.

TYPES OF MORPHOLOGICAL DATA

Morphological data, or *exomorphic* features of a plant, can be viewed as being of two basic types: *macromorphological* and *micromorphological*. The former are those features seen with the unaided eye or with the resolving power of a hand lens or binocular microscope. These are the most easily observed features, and most of the taxonomic characters tend to be of this type (fig. 15.1). They are also used most commonly in keys because of ease and speed of observation and documentation. Micromorphological features are those seen only with compound light microscopes or with the scanning electron microscope (SEM). A useful adjunct to the regular transmitted light compound microscope is the instrument with the same optics but with a direct-reflecting light source, which allows observation (but with shallow depth of field) of surface features, such as on seeds or leaves, at medium magnifications (e.g., × 450).

Vegetative Morphological Characters

Although in a general sense the floral features have been most useful in angiosperm taxonomy, many vegetative characters have also been used to good effect. Why vegetative morphology should have been "neglected" has been discussed by Davis and Heywood (1963) and Tomlinson (1984b), and there are some, at least partially satisfactory, reasons. The vegetative part of the plant is of a more "modular construction" (Tomlinson 1984b:51; see also White 1979), in which there are repeating units of structure without fixed numbers of parts, in contrast to floral features which are more definite in number. This makes adaptive sense in that the vegetative parts of the plant have many varied functions such as support, food production, water transport, etc., in contrast to the more narrow (but obviously important) role of floral features in reproduction. Because of their more numerous functions, vegetative features tend to be more plastic and/or variable and hence more difficult to use for taxonomic purposes. Despite these problems, many vegetative features dealing with leaves, stems, and roots, have been used with most attention being given to leaves.

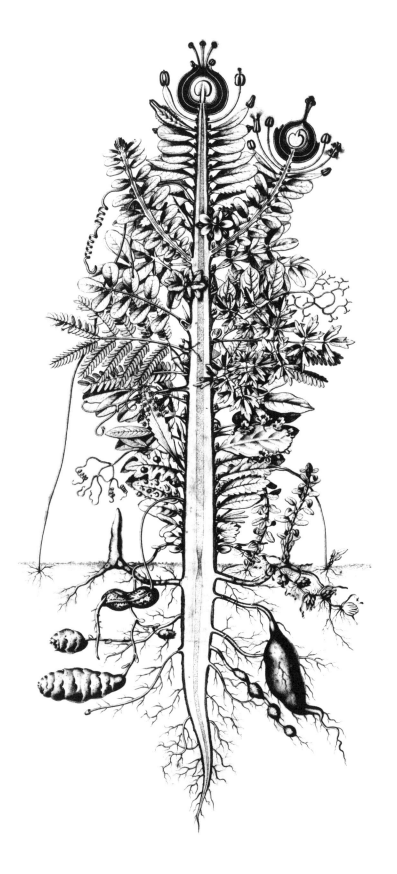

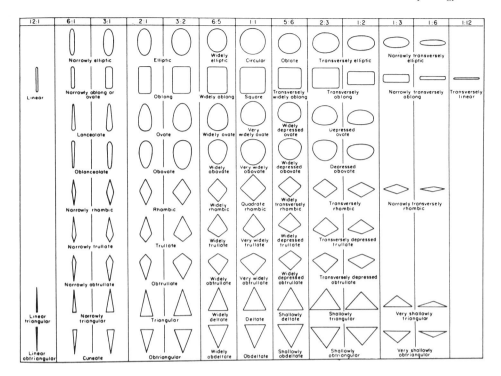

FIGURE 15.2 Standardized shapes for descriptions of outlines of leaf blades (and other symmetric plane features). (From Radford et al. 1974:131), after Systematics Association Committee 1962)

Leaf Blade. Because of its conspicuousness and ease of observation plus obvious differences in size and shape, the leaf blade has been examined extensively in taxonomic studies. Two aspects are of particular importance: the blade outline, and the internal architecture. Because of the wealth of terminology dealing with leaf blade outline extending back centuries (e.g., Linnaeus 1751; Lindley 1832; see list of references in Systematics Association Committee 1960), and due to the profusion of terms even now in use (e.g., Stearn 1983), an attempt has been made to stabilize them as shown in figure 15.2. More detailed efforts have also been made to describe the blade outline features in more precise quantitative ways (Dale et al. 1971; Dickinson, Parker, and Strauss 1987), but the acknowledged gain in precision of information must be balanced against the increased time needed for such approaches. The terms for internal architecture of leaf blades have also been addressed recently (Hickey 1973, 1979; Dilcher 1974; Hickey and Wolfe 1975; Melville 1976) with the result being classifications of terms and features for venation patterns, margin types, and so on. Both aspects of blade features have been used extensively in taxonomic studies. Hill (1980) has shown

FIGURE 15.1 A visual summary of the wealth of vegetative and floral morphological features available in angiosperms as comparative data. (By P. F. F. Turpin, from Eyde 1975b:432)

a successful application of numerical taxonomy to these architectural leaf features.

Epidermis and Cuticle. Another important feature of leaves for taxonomic purposes is the epidermis and the cuticle (Metcalfe and Chalk 1979; Barthlott 1981; Juniper and Jeffree 1983; Stace 1984). Many studies have been published utilizing different aspects of the epidermis (e.g., Stace 1965, 1966, 1969; Gray, Quinn, and Fairbrothers 1969; Palmer, Gerbeth-Jones, and Hutchinson 1985). The epicuticular waxes have been examined early by transmission electron microscopy (TEM) and carbon-replica techniques (e.g., Mueller, Carr, and Loomis 1954; Schieferstein and Loomis 1956; Juniper and Bradley 1958; Eglinton and Hamilton 1967; Hallam and Chambers 1970; Martin and Juniper 1970), and more recently by SEM (e.g., Wells and Franich 1977). These can show useful patterns of variation (fig. 15.3); especially infraspecific levels, but the waxes can be adaptations to different moisture stresses, and hence may be most useful in suggesting ecotypic differentiation rather than formal infraspecific taxa.

Variations in the epidermal cells also have been used taxonomically to include features of the generalized epidermal cells, the stomata, and the trichomes. A survey of stomata reveals different patterns that have importance at many levels in the hierarchy, although more at higher levels (Stace 1969; Payne 1970, 1979; Baranova 1972, 1987; Fryns-Claessens and van Cotthem 1973; Raju and Rao 1977; Eggli 1984; Inamdar, Mohan, and Bagavathi Subramanian 1986). Environmental conditions, especially variations of moisture, can also cause some alterations of stomatal (and other epidermal) features (e.g., Sharma and Dunn 1968, 1969), but some are clearly under strong genetic control (e.g., Cutler and Brandham 1977). Generalized epidermal cells also have been utilized taxonomically, such as in the Liliaceae (fig. 15.4; Newton 1972; Brandham and Cutler 1978, 1981; Cutler 1978b, 1979; Cutler et al. 1980) or cacti (Schill et al 1973). Trichomes have been useful for centuries, and even more so now that they can be revealed more clearly and dramatically in SEM (fig. 15.5; e.g., Mulligan 1971a, b; Roe 1971; Knobloch, Rasmussen, and Johnson 1975; Rollins and Banerjee 1975; Hardin 1976; Gangadhara and Inamdar 1977; and Ladiges 1984). For description and a new classification of trichomes, see Payne (1978) and Theobald, Krahulik, and Rollins (1979). See also the classification of trichomes for the Scrophulariaceae (Raman 1987). Studies on the function of trichomes have also been completed to help reveal the adaptive value of these leaf features (Heslop-Harrison 1970; Schnepf and Klasova 1972; Levin 1973; Johnson 1975; Parkhurst 1976; Lersten and Curtis 1977; Rodriguez, Healey, and Mehta 1984; Fahn 1986). Another interesting series of papers deals with the adaptive value of trichomes in the epiphytic Bromeliaceae (Benzing et al. 1976, 1978) and Orchidaceae (Pridgeon 1981; Benzing and Pridgeon 1983).

Other Vegetative Morphological Characters. Tomlinson (1984b) emphasizes the importance of vegetative morphology in helping solve taxonomic problems, and he gives good examples of use of shoot morphology, seedling morphology, and stem morphology/anatomy in different groups. The interest in stem architecture, or the overall growth patterns, is increasing and can be helpful taxonomically (e.g., Donoghue 1981); moreover it can lead to adaptive insights (Horn 1971; Bell and Tomlinson 1980; Tomlinson 1982), especially with the newer applications of computer modeling (e.g., Honda and Fisher 1978). Seedling morphology also has been shown to be helpful in some cases as in *Calophyllum* (Guttiferae; Stevens

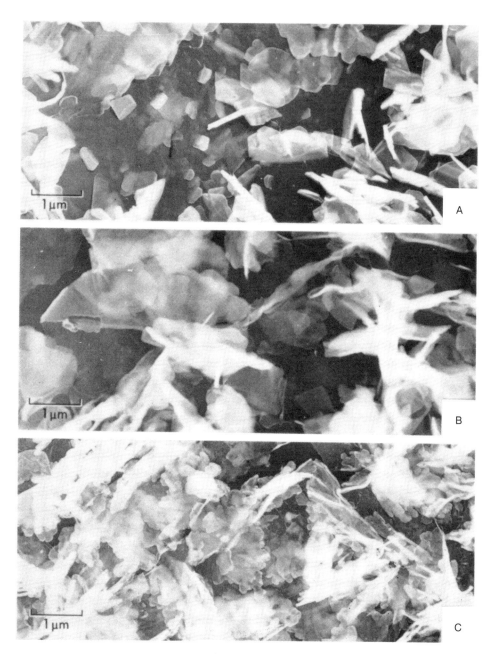

FIGURE 15.3 Epicuticular waxes among species of *Eucalyptus* (Myrtaceae) as revealed by carbon-replicas and TEM. A, *E. terminalis;* B, *E. planchoniana;* C, *E. citriodora.* x15,000. (From Hallam and Chambers 1970:342)

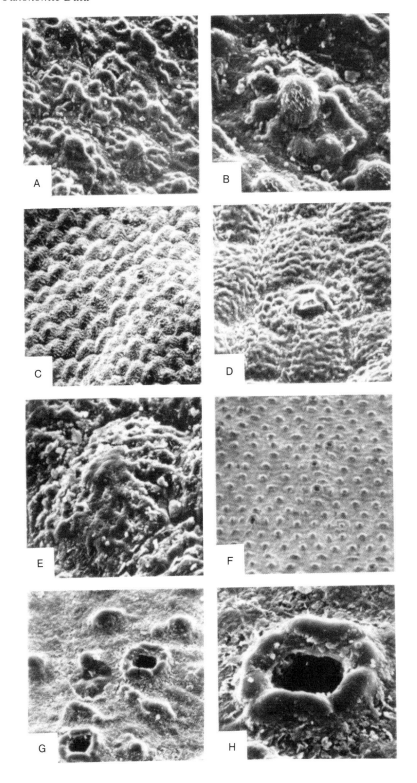

1980b). Morphological features of roots are useful, but they have been employed mostly in a broad sense as fibrous vs. tap roots, general size, and so forth. However, a remarkable degree of variation does exist in root systems (e.g., fig. 15.6), and more attention should be placed on these features. For root structures of representatives of many families, see Kutschera (1960) and Köstler, Brückner, and Biebelriether (1968). Refer to Jenik (1978) for models of basic root types.

Floral Morphological Characters

Despite the impressive contributions of vegetative morphology to angiosperm taxonomy, the data from external reproductive morphology have had far more impact. The diversity of floral features of the angiosperms is enormous, and they have been used extensively in the development of systems of classification at all levels. The various features have come primarily from the flowers, fruits, seeds, and variations in symmetry. Recent use of the SEM has allowed more detailed characters to be obtained.

Epidermal Cells. Epidermal cell patterns on corollas have shown a wide diversity of types (fig. 15.7) that correlate with plant pigments (flavonoids) to result in certain UV reflectance and absorbance patterns (Strickland 1974; Brehm and Krell 1975; Baagøe 1977a, b, 1980). The epidermis of inflorescence bracts has also been examined taxonomically (e.g., Vignal 1984).

Seeds and Fruits. Similar studies with SEM have been done on dry indehiscent fruits (Fig. 15.8) to which the term *"carposphere"* has been applied (Heywood 1969; see also Heywood 1968; Heywood and Dakshini 1971; Schuyler 1971; Theobald and Cannon 1973; Walter 1975). Likewise seeds, because of their small size, have been ideal materials for investigation with the SEM for additional useful taxonomic characters (fig. 15.9). One advantage of seed coats is the existence of elaborate terminology for the reticulations and other variations that are commonly found (Isely 1947; Martin and Barkley 1961; Gunn 1972; Stearn 1983); even so, Hill (1976) has suggested the need for additional descriptors. Several studies have already been completed in various families with considerable success (e.g., Echlin 1968; Niehaus 1971; Chuang and Heckard 1972; Whiffin and Tomb 1972; Skvortsov and Rusanovitch 1974; Ehler 1976; Hill 1976; Musselman and Mann 1976; Heyn and Herrnstadt 1977; Seavey, Magill, and Raven 1977; Sharma et al. 1977; Crawford and Evans 1978; Kujat and Rafinski 1978; Newell and Hymowitz 1978; Canne 1979, 1980; Celebioglu, Favarger, and Huynh 1983; Barthlott 1984; Chance and Bacon 1984; Matthews and Levins 1986). As with all micromorphology, there exists a real need to do accompanying anatomical studies (or ultrastructural studies) to correlate the observed surface sculpturing with the internal architecture. Good examples in this regard are in *Cor-*

FIGURE 15.4 Epidermal cell features on abaxial leaf surfaces in species of *Aloe* (Liliaceae) by SEM. A, B, *A. keayi;* C–E, *A. macrocarpa* var. *major;* F–H, *A. schweinfurthii.* C, x65; F, x170; A, D, G, x310; B, E, x760; H, x950. (From Newton 1972, plate 2)

FIGURE 15.5 Trichomes on leaves of selected species of *Quercus* (Fagaceae) by SEM. A, *Q. margaretta* (x100), B, (x400); C, *Q. arkansana* (x200); D, *Q. lyrata* (x500); E, *Q. rolfsii* (x1500); F, *Q. chapmanii* (x700). (From Hardin 1976:157)

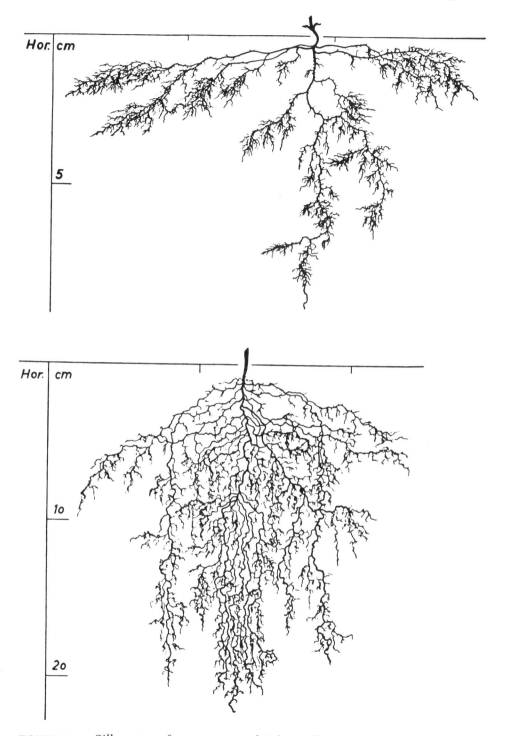

FIGURE 15.6 Silhouettes of root systems of *Valerianella carinata* (top) and *V. rimosa* (bottom) of the Valerianaceae. (From Kutschera 1960:495, 496)

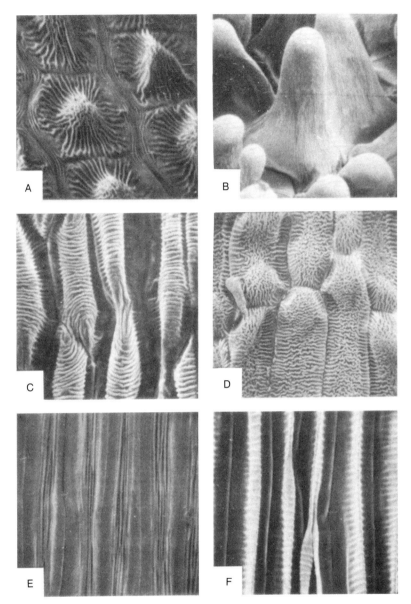

FIGURE 15.7 Epidermal patterns on the ray corollas of selected genera of Compositae as viewed by SEM. A, *Lasthenia;* B, *Rudbeckia;* C, *Grindelia;* D, *Doronicum;* E, *Calendula;* F, *Perezia.* All x530–780. (From Baagøe 1977a:122)

dylanthus and *Orthocarpus* (Scrophulariaceae; Chuang and Heckard 1972, 1983), and in *Mentzelia* (Loasaceae; Hill 1976). Symmetry in flowers also has been used taxonomically but primarily at the higher levels of the hierarchy and especially in the broad interpretation of evolutionary trends (e.g., Leppik 1956, 1968a, b, 1972, 1977).

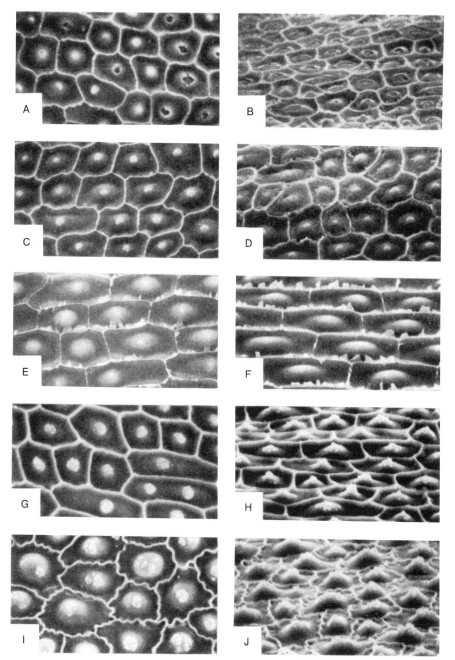

FIGURE 15.8 Internal structure of epidermal cells on the achene surface in species of *Eriophorum* (Cyperaceae). A, B, *E. latifolium;* C, D, *E. microstachyum;* E, F, *E. comosum;* G, H, *E. crinigerum;* I, J, *E. japonicum.* All x1000 (From Schuyler 1971:45).

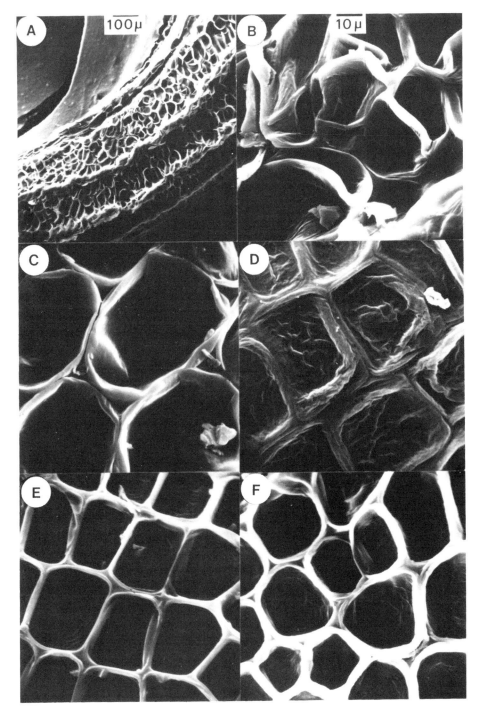

FIGURE 15.9 Seed coats of species of *Hypericum* (Guttiferae). A, B, *H. lissophloeus, Adams 685;* C, *H. lloydii, Ahles 54918;* D, *H. reductum, Adams 725;* E, *H. fasciculatum, Adams 364;* F, *H. chapmanii, Adams 340.* B–F same scale. Original.

Morphogenetic Characters

In addition to structural morphological data, a few comments need to be made on morphogenetic or developmental characters. Here one considers the interactions among the structures in an ontogenetic framework. The importance this work has for cladistic analysis has already been stressed (chapter 8), particularly as it applies to revealing homologies among the mature structures (Kaplan 1984). These types of studies have been limited no doubt due to the amount of time and effort needed for data collection. Numerous separate studies of developmental morphology (and anatomy) exist that offer insights on the origin of a particular structure or structures but do not make comparisons between taxa for taxonomic purposes (e.g., F. J. F. Fisher 1960; J. B. Fisher 1974; Pandey and Singh 1978; Pandey and Chopra 1979). Others use the developmental data to assess relationships but few of these types of works exist. Some positive examples include Maze, Bohm, and Beil (1972) in species of *Stipa* (Gramineae) and Kam and Maze (1974) among species of *Oryzopsis* (Gramineae). More frequently the data are used at the familial and ordinal levels such as with the relationships of the Scheuchzeriaceae (Posluszny 1983), in the Piperales (Tucker 1980, 1982a, b), or in the woody Ranales (Benzing 1967a, b). Stebbins (1973, 1974) has rightly pointed out the great value of developmental data for interpreting evolutionary relationships and trends at the higher levels of the hierarchy in the angiosperms, but this is still largely an untapped area.

INVESTMENTS FOR GATHERING MORPHOLOGICAL DATA

The investments needed to gather useful morphological data for taxonomic purposes are less than with any other type of data, and in most cases probably more effective (more on this in chapter 24). For sound morphological studies one needs only a hand lens, a binocular dissecting microscope, a compound light microscope (with transmitted and/or reflected light), and access to an SEM. As Tomlinson has stressed: "there is enormous scope for research which requires the simplest apparatus: an axe, a knife and a scalpel, a clear eye, and the patience and willingness to settle down and observe elemental things" (1962:43). The expertise needed depends upon the morphological features involved; some are more complex than others and some have had controversial histories (e.g., problems of homologies of some features such as stipules in different angiosperm families). But because the technical aspects of observing morphology are relatively minor, these features are more easily examined than with almost any other type of data. But a caution is needed here: the level of critical evaluation and comparison of morphological features varies greatly from worker to worker, and careful scrutiny of these data is required. The time commitment varies with the particular morphological features and the scope of the taxonomic problem, the size of the sample at hand, and whether or not SEM is involved (much more time-intensive). Costs are minimal except for SEM beam time ($25–50 per hour) in which a few to many samples can be examined (depending upon the size of the SEM stub and the total samples to be observed, etc.). A good way to reduce

SEM costs is to do most of the routine sampling of morphology under a compound microscope with direct-reflecting light. Hundreds of populations can be sampled inexpensively this way, with SEM used for clarifying three-dimensional relationships and for documentation for publication.

EFFICACY OF MORPHOLOGICAL DATA IN THE TAXONOMIC HIERARCHY

Morphological data are helpful at all levels of the taxonomic hierarchy from the variety to the division. Some families are clearly delimited even vegetatively (e.g., the Melastomataceae). Tomlinson (1984b) has listed many families for which vegetative features are diagnostic for major taxa (e.g., the tendril morphology of the subfamilies of the Cucurbitaceae). The bottom line is that both vegetative and floral morphology are the most useful data for nearly all taxonomic problems in almost all groups. If one were forced to choose only one set of data with which to do taxonomy, the choice would have to be morphology. In most instances it provides the best mirror of genetic and evolutionary relationships and gives clues to the way in which the plants have adapted to their environment. Morphology may be regarded as old-fashioned by some, but it is still the foundation for solving any taxonomic problem.

CHAPTER 16

Anatomy

Anatomy, or the internal form and structure of plant organs, is another classical source of data used in plant taxonomy. Anatomical data are often extremely useful in solving problems of relationships because they can often suggest with greater confidence the homologies of morphological character states, and they can help in the interpretation of evolutionary directionality (= polarity). It is sometimes scoffed that comparative anatomy has outlived its usefulness and that it is a sterile discipline, bereft of new advances. Nothing could be further from the truth. A glance at the papers in recent symposia volumes (e.g., Robson, Cutler, and Gregory 1970; Baas 1982b; White and Dickison 1984) plus the stimulating ecological hypotheses of Carlquist (1975) and the architectural form and function analyses of Tomlinson, Fisher, Honda, and coworkers (e.g., Tomlinson and Zimmermann 1978; Halle, Oldeman, and Tomlinson 1978; Honda, Tomlinson, and Fisher 1981, 1982; Fisher 1984) reveal many new challenges. In fact, interest is now focused clearly on the function and adaptive value of anatomical features, even in fossil plants (e.g., Taylor 1981), which will help reveal more clearly the homologies of structure for purposes of classification and the reconstruction of phylogeny.

HISTORY OF ANATOMY IN PLANT TAXONOMY

The history of use of anatomical data in systematics is long and follows in parallel fashion the use of more explicit morphological data. Good summaries are provided by Eames and MacDaniels (1947), Carlquist (1969a), and Metcalfe (1979). The beginnings can be traced back to the English worker, Nehemiah Grew, especially as revealed in his book *The Anatomy of Plants* (1682), and to the Italian Marcello Malpighi (*Anatome Plantarum*, 1675–1679). John Hill (1770) followed with his book on *The Construction of Timber*. The real development of plant anatomy came in the middle of the 19th century, beginning with Hugo von Mohl's *Gründzuge der Anatomie und Physiologie der Vegetabilischen Zelle* (1851). Among the many anatomical works that followed was Radlkofer's anatomical treatment (1895) of the Sapindaceae for Engler and Prantl's *Die natürlichen Pflanzenfamilien*. His student, Solereder, wrote the important *Systematische Anatomie der Dicotyledonen* (1908) and set the stage for continued presentations of comparative anatomical data ending with the most recent works of Metcalfe and

Chalk (1950, 1979). In the United States comparative plant anatomy developed primarily under E. C. Jeffrey at Harvard as reflected in his book *The Anatomy of Woody Plants* (1917). His students, such as I. W. Bailey, A. J. Eames, E. W. Sinnott, and R. H. Wetmore, all have made significant contributions. Other important figures and their works in more recent years include A. S. Foster (*Practical Plant Anatomy*, 1942), K. Esau (*Plant Anatomy*, 1953; *Anatomy of Seed Plants*, 1960; *Vascular Differentiation in Plants*, 1965), and S. Carlquist (*Comparative Plant Anatomy*, 1961).

GENERAL ANATOMICAL TEXTS AND REFERENCES

Many books contain anatomical data about plants. Basic anatomical textbooks include Eames and MacDaniels (1947), Esau (1953, 1960), Carlquist (1961), Molisch and Hofler (1961), Cutter (1971, 1978), Fahn (1982), and Mauseth (1988). Carlquist (1961), who deals exclusively with comparative anatomy, offers useful perspectives for the taxonomist. Books of interest that deal with general aspects of developmental anatomy are Torrey (1967), Steward (1968), and O'Brien and McCully (1969). Wood anatomy is covered well by Desch (1968), Jane (1970), Panshin and de Zeeuw (1970), Zimmermann and Brown (1971), Core, Côté and Day (1979), and Zimmermann (1983). Other specific aspects of vegetative anatomy include works on shoots and meristems (Clowes 1961; Romberger 1963; Dormer 1972; Williams 1974), leaf development (Maksymowych 1973; Dale and Milthorpe 1983), phloem (Esau 1969), cambium (Philipson, Ward, and Butterfield 1971), general features of vascular differentiation (Esau 1965), and roots (Torrey and Clarkson 1975). Ultrastructural topics are covered in Côté (1965), Robards (1974), and Gunning and Steer (1975). Information on applied and economic aspects of plant anatomy can be found in Hayward (1938) and Cutler (1978a). Standard texts for microtechnique used in anatomical research are Johansen (1940), Sass (1958), Jensen (1962), Berlyn and Miksche (1976), and O'Brien and McCully (1981); for SEM and TEM techniques see Hayat (1970, 1978, 1981), and Gabriel (1982a, b). The most valuable sources for comparative anatomical data about angiosperms are Metcalfe and Chalk (1950, 1979, 1983), which are organized by family in the Bentham and Hooker system of classification, and which also have many references to the included genera. A useful source of data on leaf structure of tropical trees is Roth (1984). More detailed modern anatomical monographs are available for some groups, especially in the monocots (e.g., Gramineae, Metcalfe 1960; Palmae, Tomlinson 1961; Juncales, Cutler 1969; Commelinales-Zingiberales, Tomlinson 1969; Cyperaceae, Metcalfe 1971; Dioscoreales, Ayensu 1972). In short, there is a wealth of anatomical data available for potential use as taxonomic characters.

TYPES OF ANATOMICAL DATA

Anatomical data can be viewed as consisting of two types: endomorphic (as contrasted with exomorphic, or morphological, data), and ultrastructural. The former are observable largely with the light microscope and the latter by use of the transmission electron microscope (TEM). As one approaches comparative cellular structure with TEM, the transition with cytology is reached. In this book cytological data are mostly confined to information about chromosomes.

Numerous applications of anatomical data for solving systematic problems exist in the literature, and it is not our task here to chronicle all of them. The purpose of the following discussion is to give examples of kinds of anatomical and ultrastructural data that can be useful in solving different kinds of taxonomic problems. For a good list of potential anatomical characters and states from different plant organs, see Radford et al. (1974).

Vegetative Anatomical Characters

In contrast to vegetative morphological features, vegetative anatomical characters have been used with more regularity than floral ones. This is probably due to the viewpoint that if additional data are believed desirable to solve a taxonomic problem, then looking inside the leaves, stems and roots could potentially yield different information than that from reproductive organs. Data from floral and fruit anatomy *usually* correlate well with observed reproductive morphological features, and hence serve to refine the relationships already documented instead of offering totally new insights.

Leaves. Leaves provide many anatomical characters as shown by the following examples: in the Bromeliaceae (Robinson 1969); Leguminosae (Lackey 1978); Palmae (Glassman 1972); and Myrtales (Keating 1984). Within leaves, data can be taken from either the petiole, blade or cotyledons. An example of the former comes from Schofield (1968; fig. 16.1) in the Guttiferae, and the latter by Philipson (1970) in *Rhododendron* (Ericaceae).

Most leaf features of taxonomic significance derive from the blade. The epidermis (and hypodermis, when present) provide many useful characters as shown by Stace (1966), Baranova (1972), and Cutler, Carter, and Harris (1980), but this topic (as micromorphology) has already been discussed in the previous chapter. There are ultrastructural studies of epidermal cell features, such as the guard cell variations in the Gramineae (Brown and Johnson, 1962), phytoglyphs in *Eucalyptus* species (Carr, Milkovits, and Carr 1971), and crystals in nuclei of epidermal cells (Speta 1977). The mesophyll offers some useful features, including also the presence of crystals (Heintzelman and Howard 1948, Icacinaceae; Mathew and Shah 1984, Verbenaceae). The structure of the bundles can also vary (e.g., figs. 16.2, 16.3; Brittan 1970; D'Arcy and Keating 1979), especially in Gramineae with C_4 photosynthesis (Kranz anatomy) in which bundle/sheath cells have chloroplasts centrifugally localized and without grana (Johnson and Brown 1973; Brown 1977). Patterns of venation are also useful as has been shown on numerous occasions (e.g., fig. 16.4; Dickison 1969). Sclereids in leaves are taxonomically valuable too, as indicated by Rao, Bhattacharya, and Das (1978; Rhizophoraceae), Tucker (1977; Magnoliaceae). See also Rao and Das (1979) for the distribution of leaf sclereids within Dahlgren's system of classification of the angiosperms. Leaf anatomical data, as with most all other types of data, have

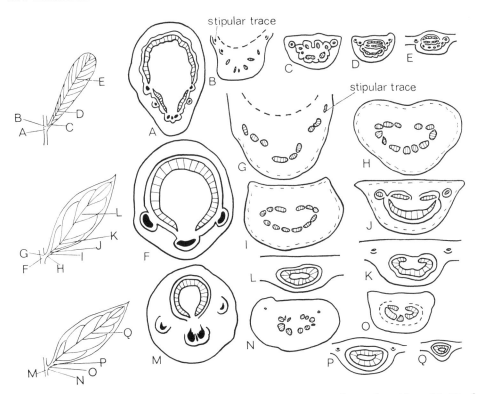

FIGURE 16.1 Petiolar anatomy in *Blastemanthus* (A–E) and *Brackenridgea* (F–Q) of the Guttiferae, including comparisons with nodal and midrib anatomy. A–E, *B. gemmiflorus;* F–L, *B. australiana;* M–Q, *B. nitida*. x15. (From Schofield 1968:20)

also been treated phenetically for a more quantitative view of relationships (e.g., Lubke and Phipps 1973).

Stems

Tissues and cells of stems also provide many helpful lines of taxonomic evidence. Throughout the stems of both herbaceous and woody plants are found starch grains that are of some limited taxonomic utility (Czaja 1978). In plants that have laticifers (or latex-containing ducts, as in the Euphorbiaceae), the anatomy of these structures, their starch grains, as well as the chemistry of the latex itself is taxonomically important (Mahlberg 1975; Biesboer and Mahlberg 1981; Mahlberg and Pleszczynska 1983; Mahlberg, Rauh, and Schnepf 1983). At the ultrastructural level, it has been shown that members of the closely related Cruciferae and Capparaceae (Capparales) have dilated cisternae of the endoplasmic reticulum (ER) in phloem parenchyma cells filled with protein granules, filaments, or tubules (Hoefert 1975). These features are also known from root cap and epidermal cells and in leaves (Iversen 1970a; Behnke 1977a, c), as well as in the inner integuments of ovules (Ponzi, Pizzolongo, and Caputo 1978). A related feature is

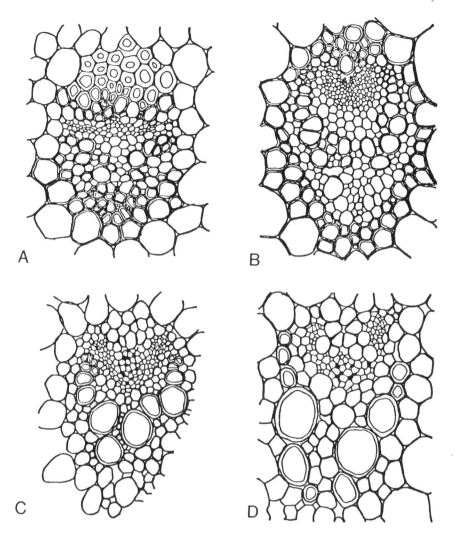

FIGURE 16.2 Vascular bundles of leaves of species of *Thysanotus* (Liliaceae). A, *T. arenarius;* B, *T. multiflorus;* C, *T. tuberosus;* D, *T. formosus.* A–C, x175; D, x300. (From Brittan 1970:62)

ER-dependent vacuoles that also contain proteinaceous material. More recent studies by Behnke (1977c) suggest that these may be structurally closely related to the dilated cisternae. The function of the protein-containing ER segments is not yet clear (Behnke 1977c), but there exists the possibility that they contain myrosinase (Iversen 1970b), an enzyme hydrolyzing glucosinolates (mustard oil glucosides) found in families of the Capparales.

The most interesting and useful taxonomic feature of stems to appear recently has been data on sieve-tube element plastids. In a series of papers (Behnke 1967, 1968, 1969, 1971, 1972, 1973, 1974a, b, c, 1975a, b, c, d, 1976a, b, c, 1977a, b,

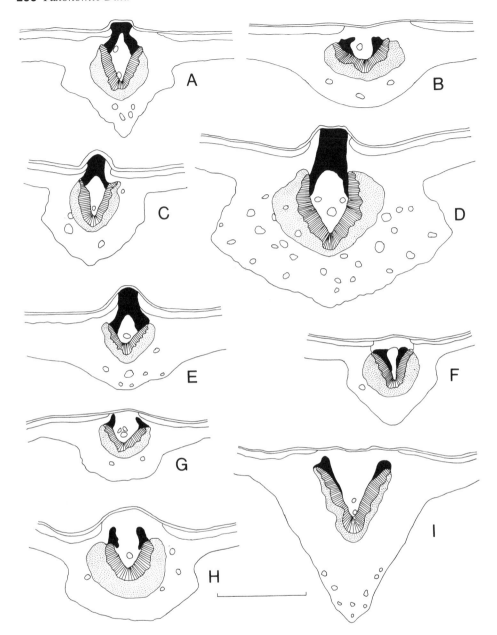

FIGURE 16.3 Leaf midrib cross-sections in *Calophyllum* (Guttiferae). A, *C. angulare;* B, *C. brasiliense;* C, D, *C. longifolium;* E, *C. nubicola;* F, G, *C. rekoi;* H, *C. soulattri;* I, *C. inophyllum.* Scale line = 1mm. Hatching = xylem; stippling = phloem; solid black = sclerenchyma. (From D'Arcy and Keating 1979:562)

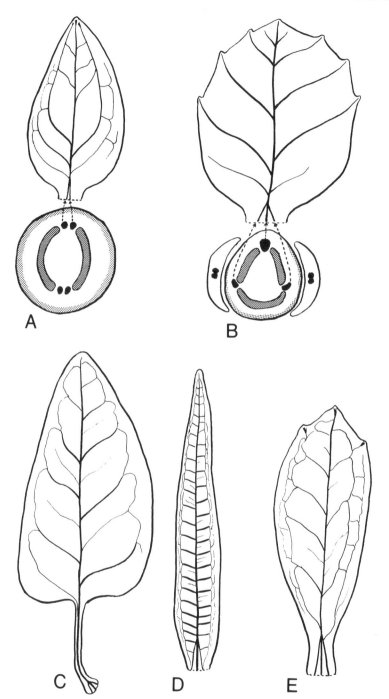

FIGURE 16.4 Major leaf venation in *Dillenia indica* (A, cotyledonary node and vascularization of cotyledon; B, vascularization of first foliage node and leaf; C, *Hibbertia dentata;* D, *H. tontoutensia;* and E, *H. cuneiformis*) of the Dilleniaceae. (From Dickison 1969, plate ix)

1978, 1981a, b, 1982a, b, c, d, 1984, 1986a, b; Behnke and Turner 1971; Behnke et al. 1974; Hunziker et al. 1974; Behnke and Dahlgren 1976; Mabry and Behnke 1976; Mabry, Behnke, and Eifert 1976; Behnke and Mabry 1977; Behnke, Pop, and Sivarajan 1983; Eleftheriou 1984), Behnke and colleagues have shown the occurrence of several basic plastid types (fig. 16.5), with differences relating to the shape and occurrence of starch and/or protein inclusions. The most significant result is that the P_{III}-type plastid is restricted to the Caryophyllales, which in addition to morphological features, is also characterized by betalains (discussed in chapter 21). This is the best example of the efficacy of an ultrastructural feature at the ordinal level. The different character states have also been treated in transformation series to reflect their presumed phylogeny (fig. 16.6), with the starch type probably being most primitive in the angiosperms (Walker 1979). The function of these specialized plastids is unclear, but the P-type plastid at least may have to do indirectly with sealing the sieve-plate pores of injured sieve-tubes with P-protein. That the angiosperms alone have this type of protein may relate to the larger size and greater penetrability of their sieve-area pores compared to gymnosperms and lower vascular plants (Evert 1984).

Many significant taxonomic features have been derived from stem tissue. Leaf gaps and nodal anatomy are two of these useful features, and they can be helpful in both herbaceous and woody plants (Dickison 1969, 1975; Howard 1970; Keating 1970; see also fig. 16.1). The general utility of wood anatomy has been documented extensively (e.g., Brazier 1968) and hundreds of studies exist (e.g., Carlquist 1966, 1969c, 1970, 1971, 1978, 1981b, 1982a, b, c, d, e, 1984a, b, 1985a, b; Stern 1967; Keefe and Moseley, 1978; Schmid and Baas 1984; Vliet and Baas 1984). Generally transverse (cross-), radial, and tangential sections are made of the secondary xylem (fig. 16.7) and the arrangement of tissues compared and contrasted for taxonomic utility. Detailed microscopic study of the individual cell types may also prove significant (fig. 16.8; see Findlay and Levy 1970, for good SEM photos of details of wood structure).

Roots. Roots are neglected vegetative features for taxonomic characters perhaps due in part to the difficulty of obtaining materials, especially in large woody species. Also, the data available so far reveal fewer variations in comparative structure than with the other organs of the plant (fig. 16.9); hence, roots are regarded as less valuable taxonomically. No doubt variations in mycorhizal associations and in rootcap and meristem organization might provide more data, but so far these have been little explored. One positive example is the use of root anatomy to identify eight genera of the Caprifoliaceae found in the British Isles (Gasson 1979).

Reproductive Anatomical Characters

Reproductive anatomical features have also been used in taxonomic studies to good effect (see general review by Puri 1951; see also Eyde 1975a). The flower contains many useful patterns of vascularization that are helpful in showing relationships in specific instances (e.g., in the Hamamelidaceae, Bogle 1970). Features of the ovarian disc (Rao 1971), placentation (Rao 1968), gynoecium

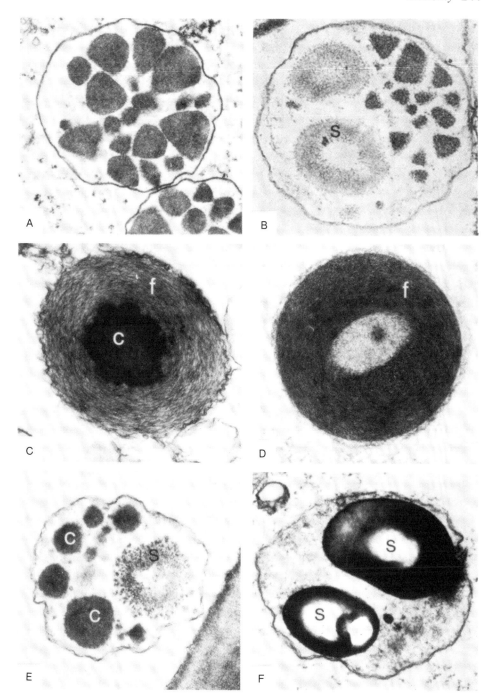

FIGURE 16.5 Sieve-element plastids of the angiosperms. A, B, monocotyledon sub-type, with cuneate protein crystalloids; C, D, Caryophyllidae subtype, with ring-like bundle of filaments (f) and crystalloids (c); E, F, Fabales subtype, with polygonal crystals (c) and starch inclusions (s). (From Behnke and Dahlgren 1976:292)

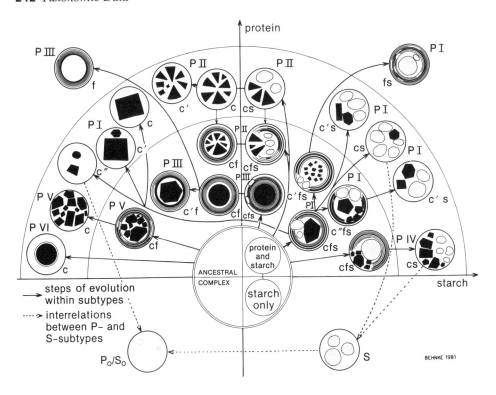

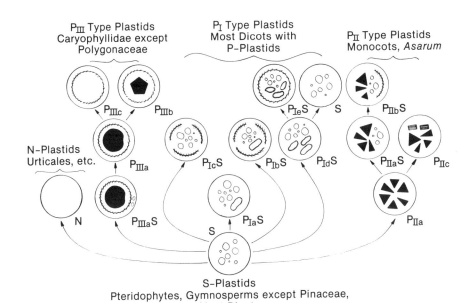

FIGURE 16.6 Occurrence and evolution of sieve-element plastids in the angiosperms. (From Behnke 1981b:389, top; and Walker 1979:614, bottom)

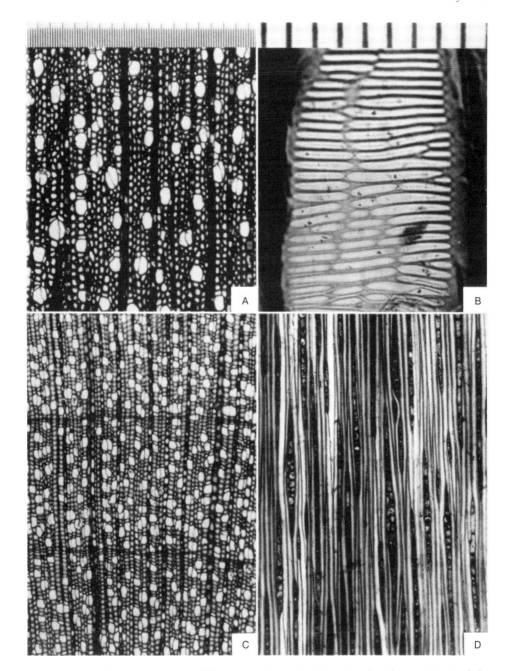

FIGURE 16.7 Wood anatomy of *Illicium cubense* (A, B) and *I. floridanum* (C, D) of the Illiciaceae. A, C, transections; B, aberrant perforation plate from radial section; D, tangential section. Finest divisions = 10μm. (From Carlquist 1982d:1590)

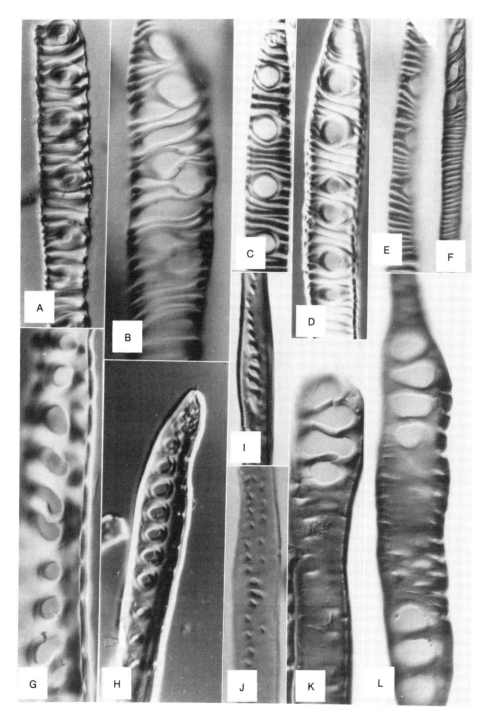

FIGURE 16.8 Portions of a tracheid of *Gnetum montanum* (A) and vessels of *G. montanum* (B–G, K–L), *G. gnemon* (H), all Gnetaceae, and *Comptonia perigrina* (I, J), of Myricaceae. Nomarski interference contrast, x235–590. (From Muhammad and Sattler 1982:1009)

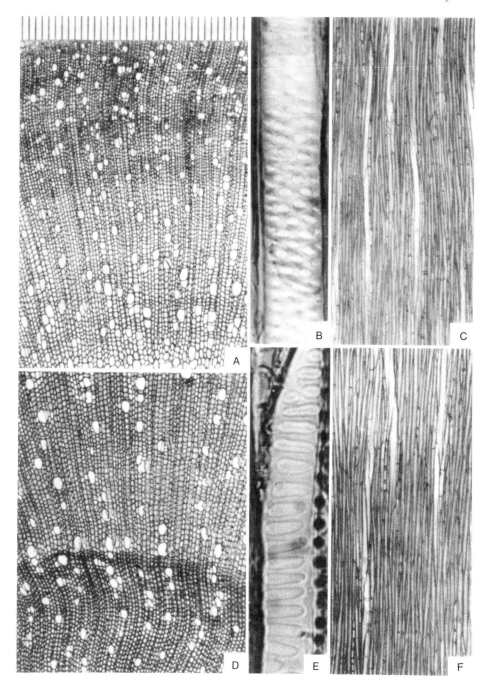

FIGURE 16.9 Woody root structure in transections (A, D; pith below), vessels from tangential sections (B, E), and tangential sections (C, F) in *Circaea lutetiana* (A–C) and *Lopezia suffrutescens* (D–F) of the Onagraceae. Divisions = 10 μm. (From Carlquist (1982e:763)

(Eyde 1967), and more specifically the carpels (Endress 1980a, b; Endress and Lorence 1983, fig. 16.10) are also important. The anatomy of fruits also can be examined (e.g., in achenes, Chute 1930; Pandey, Chopra, and Singh 1982; Stuessy and Liu 1983), as well as seeds (e.g., Tobe, Wagner, and Chin 1987; fig. 16.11)

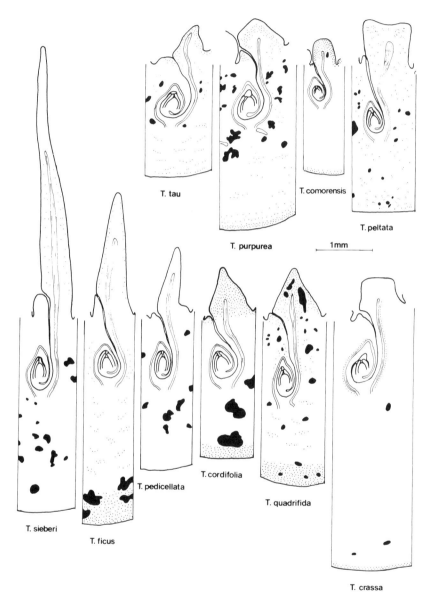

FIGURE 16.10 Median longitudinal sections of carpel and segment of floral cup at anthesis in species of *Tambourissa* (Monimiaceae). Hatched = dorsal and ventral vascular bundle of carpel (bundles in the floral cup not drawn); black = stone cells and stone cell groups; dotted = tanniferous tissue. (From Endress and Lorence 1983:62)

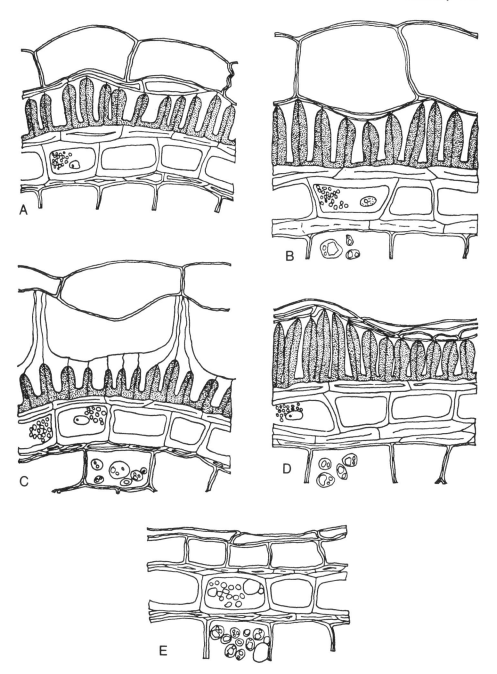

FIGURE 16.11 Transections of testa, endosperm and outer portion of cotyledon of seeds of Cruciferae. A, *Brassica carinata;* B, *B. juncea* (European); C, *B. nigra;* D, *B. juncea* (Indian); E, *Crambe hispanica.* x600. (From Vaughan 1970:36)

including protein body inclusions (Lott 1981). A good example of a detailed seed anatomical study is by Elisens (1985) in the tribe Antirrhineae of the Scrophulariaceae. Inflorescences provide a wealth of anatomical data too, such as revealed in the studies of the subtribe Madiinae of the Compositae (Carlquist 1959), and in *Uncinia* of the Cyperaceae (Kukkonen 1967).

INVESTMENTS FOR GATHERING ANATOMICAL DATA

The equipment needed for anatomical work at the light microscopic level differs from that of ultrastructural studies involving the TEM. For light microscopy one might need microtomes, usually rotary, which can yield tissue slices several microns thick. Finely honed steel knives or glass knives are used for the cutting. Tissues are usually embedded in paraffin or plastic and the cut sections mounted on glass slides, dehydrated, stained and counter-stained to reveal the different tissues. For TEM an ultramicrotome is used to cut 70–100 nm thick sections of very small pieces of tissue (ca. 1 mm square), embedded in hard resins. These sections are then cut with glass or diamond knives, mounted on grids (usually of copper), and stained to increase electron density. Along with these techniques is the very great need for voucher specimens so that the anatomical sections can always be referable to an herbarium sheet and hence to the macroscopic features of the taxon (Stern and Chambers 1960).

The expertise, time and cost needed to do anatomical work requires more of a commitment than with morphological studies. To be able to cut good sections with microtomes requires a definite skill and patience not found in all taxonomists. Further, the numerous other preparative steps of embedding, dehydration, staining, etc., all require considerable skill and experience. A good five-week course in plant microtechnique should be adequate to begin light anatomical studies with an additional ten-week course for TEM. The cost for such efforts in terms of equipment is small for light microscopic studies, assuming a good compound microscope is already available. Several thousand dollars will set up a rudimentary bench for beginning anatomical work. With TEM, however, the costs escalate tremendously with the cost of a good TEM itself being in the range of $100,000–$200,000. Ultramicrotomes, darkroom equipment, and other expensive tools all require many thousands of dollars for just the initial laboratory setting. Fortunately, at most major institutions, a well-outfitted TEM lab is already available and the services can be purchased for an hourly fee or by access through training courses. The costs for TEM work, however, are high. SEM is also expensive, as mentioned in the previous chapter.

EFFICACY OF ANATOMICAL DATA IN THE TAXONOMIC HIERARCHY

Anatomical data have been used to good effect at all levels in the taxonomic hierarchy. At the familial level and above, examples would include Baranova's (1972) use of leaf anatomical features to determine relationships in the Magnoliaceae and related families. Armstrong (1985) helped resolve the delimitation of the Bignoniaceae and Scrophulariaceae based on floral anatomy. The spectacular results of sieve-tube plastids in delimitation of the Caryophyllales by Behnke and coworkers (mentioned earlier in this chapter) is the best example of efficacy of ultrastructural data at the ordinal level. Many anatomical studies have dealt with problems of generic delimitation and one of the most interesting is the *Eugenia-Syzygium* problem in the Myrtaceae resolved by Schmid (1972b). Carlquist and co-workers have also contributed considerably to the solution of generic problems via floral and wood anatomy (1961). Sherwin and Wilbur (1971) give another good example in the Crassulaceae, as does Theobald (1967) in placement of *Uldinia* (Umbelliferae). At the specific level the contributions of anatomy tend to be less helpful than at the generic level. As Metcalfe has mentioned: "Even when the identity of the genus has been established by traditional methods the anatomical separation of species is often by no means easy, especially where one is dealing with a large genus in which the species are rather alike" (1954:434). Anatomy can be useful at this level in some instances, however, and even in the documentation of hybridization (Hillson 1963; Webb and Carlquist 1964).

Of all the anatomical characters available for employment in taxonomic studies, do some of them *tend* to be more efficacious than others? The answer, of course, is that it will depend on the type of question being asked and the nature of the group under study. A useful perspective is offered by Metcalfe:

> The investigator has to take the plants on which he is working as he finds them and to make the best use of the characters which they exhibit. Provided the plants under investigation have at least one, or preferably more, of the following attributes there are reasonable prospects that the systematic anatomist will be able to make something of their taxonomy. With plants that do not fulfill these conditions he is less likely to be successful. The desirable attributes are as follows: (a) well developed secondary xylem; (b) distinctive trichomes or other dermal appendages; (c) a characteristic distribution pattern of sclerenchyma; (d) ergastic substances such as crystals and siliceous bodies which are deposited in the plant body in distinctive morphological forms. (1968:48)

SPECIAL CONCERNS WITH ANATOMICAL DATA

Perhaps the most successful contributions of anatomy to plant taxonomy so far have been in the suggestion of hypotheses on evolutionary trends of character states. As Eames has well summarized: "These internal characters are as valuable as the external; and some of them are perhaps even more valuable because of the frequent persistence of the vascular supply of lost organs after all external evidence of the organs has disappeared. These buried vascular vestiges give information about ancestral forms and so provide evidence of relationship to other groups" (1953:126). Few would argue with the broad-scale trends of xylem evolution in the angiosperms, such as shown in the evolution of vessels from tracheids (see Baas 1982a; basic ideas originally from Frost 1930b, 1931, and

Cheadle 1943a, b), but within certain genera or families the situation must be interpreted with more caution. Carlquist (1969b) strongly criticized the overly zealous attitudes of Eames and followers and essentially concluded that floral anatomy could be terribly misleading for the unraveling of phylogeny, despite the fact that he himself had used these data to suggest such trends (e.g., Carlquist 1957, 1959). This somewhat harsh and equally exaggerated view in the opposite direction from that of Eames has been called into question by Kaplan (1971) and Schmid (1972a). The truth lies between these extremes (as is often the case): floral anatomy can be extremely helpful in suggesting phyletic sequences but only within reasonable limits. A good example of a most helpful application is the detection of the origin of hypogyny from epigyny in *Tetraplasandra* (Araliaceae; Eyde and Tseng 1969), which is the reverse of the common trend in the angiosperms (Grant 1950). Here geography provided useful ancillary data.

More recently interest has turned to the interpretation of floral and especially wood anatomy in a functional and ecological way. An important book by Carlquist (1975) hypothesizes many adaptations of xylem structure with ecological conditions. These hypotheses have been followed by several detailed studies (Baas 1976, 1982a; Carlquist and Bissing 1976; Carlquist 1977, 1978, 1982a, b, 1984a, b, 1985a, b; Carlquist and Debuhr 1977; Dickison, Rury, and Stebbins 1978; Rury and Dickison 1984). The results so far confirm some of the ecological correlations with wood structure, but other aspects seem more questionable. The full results are not yet in on this issue (for a recent review see Baas 1986). What is certain is that many of the structural features of secondary xylem surely must be ecologically (and/or physiologically) adaptive, and more tests and correlations are needed to determine these tolerances and relationships. This new information, in turn, will allow for better understanding of homology of anatomical character states for taxonomic use.

In conclusion, what realistically can be said about the use of anatomical data in plant taxonomy? Bailey long ago summarized the proper perspective which is every bit as appropriate today as it was over thirty years ago:

> It should now be clearly recognized and freely admitted that internal or endomorphic characters are inherently no more conservative or reliable than are exomorphic ones. Extensive comparative investigations of a wide range of angiosperms demonstrate that each morphological character tends to be relatively stable in certain groups of plants and highly plastic and variable in others. From a taxonomic point of view, there is no fundamental difference between the utilization of endomorphic, as contrasted with exomorphic, characters, except for differences in the methodologies of obtaining data. Thus, anatomical evidence merely adds another string to the taxonomist's bow, and strengthens the summation of evidence that may be essential in the solution of difficult problems. In other words, anatomical characters are inherently no more reliable than exomorphic ones, but are susceptible to equivalent and equally valid uses. (1953:121)

CHAPTER 17
Embryology

Features relating to the origin and development of the embryo in angiosperms have been used successfully to help determine taxonomic relationships. It is believed helpful to view embryology in the broad sense to encompass features of three generations: the old sporophyte, the gametophyte, and the new sporophyte, and hence involving "the development of the entire ovule and anther, including micro- and macrosporogenesis, gametophytes, gametogenesis, and growth of the embryo, endosperm, nucellus, and integuments" (Cave 1948: 344; see also Cave 1953). As Davis puts it clearly, this includes "all processes and structures associated with sporogenesis, gametogenesis, and embryogeny" (1966:7). The close relationship of embryology to other disciplines such as morphology, anatomy, cytology, physiology, and morphogenesis (e.g., Johansen 1950) makes the borders of this data-gathering approach somewhat vague, but there does exist a core of literature that can rightly be regarded as forming a separate and valid area of comparative data.

The use of embryological characters in plant taxonomy has not yet reached its potential in part due to the great number of embryological features that potentially can be documented and of which only a few usually are described in any one study. Further, "The analysis of embryological characters in comparative studies, while sufficiently broad, often lacks in depth and in sufficient attention to detail" (Herr 1984:647). The potential utility of embryological data for determining the early evolution of the angiosperms is obvious if the immediate progenitors of the flowering plants can ever be determined. Another interesting contribution that has already been recorded is to give one of the best examples of *neoteny* in the plant world in the change from the multinucleated mature embryo sac* condition in the gymnosperms to the mature 8- (or 16-) nucleate condition of the angiosperms (Herr 1984). Embryological data of different types have already been used successfully at different levels in the taxonomic hierarchy. As Herr has fetchingly put it: "To discover true phylogenetic relationships remains an obsession among botanists in spite of the recent exciting advancements in other aspects of plant sciences [e.g., in plant molecular biology], and new contributions from embryology toward this end are replete" (1984:647).

---

*Herr points out *(in litt.)* that the term "embryo sac" would best be replaced by "female gametophyte" or "megagametophyte" because it "does not accord the true nature and full character of the female (sexual) gametophyte." Because "embryo sac" has been used extensively in the literature, however, it is still used here in this chapter.

HISTORY OF EMBRYOLOGY IN PLANT TAXONOMY

Discussing the early history of embryology is challenging inasmuch as the definition of the discipline dictates the starting point. The discussion here comes primarily from Maheshwari (1950, 1963). One could go back to the ancient Arabs and Assyrians and point out that date palms were hand-pollinated to insure good crops (Maheshwari 1950), but this seems better regarded as the beginnings of pollination biology rather than embryology. Nehemiah Grew (1682) first clearly mentioned that the stamens were the male parts of the flower, and Rudolph Jakob Camerarius (1694) was the first to experimentally demonstrate sex in plants using the mulberry, castor bean, and corn. Further pollination studies that dealt with sex in plants were by Joseph Gottlieb Kölreuter (1761–1766). Perhaps the first observations that might truly be called embryological were made by Giovanni Battista Amici (1824) on the germination of a pollen grain on a receptive stigma of *Portulaca oleracea* (Portulacaceae) and the penetration of the pollen tube into the style. He later with further observations concluded that the pollen tubes grew down into the style and entered the ovary, coming finally in contact with the ovules (Amici 1830). There ensued a spirited debate on whether the embryo sac developed from the tip of the pollen tube or whether it developed independently inside the ovule prior to the arrival of the tube. The former incorrect view was strongly advocated by Matthias Jakob Schleiden (1837) and followers, but Amici favored the latter correct view, finally convincingly elaborated upon by Wilhelm Hofmeister (1849). In a series of papers, Hofmeister (1847, 1848, 1849, 1859, 1861) also greatly extended our knowledge of the structure of the embryo sac. Edward Strasburger (1877) showed clearly the binucleate nature of many pollen grains, and made many other useful observations on embryo sac development and function (1879), and discovered fertilization (1884). Double fertilization was revealed by Sergius G. Nawaschin (1898) and Leon Guignard (1899). The early text by Coulter and Chamberlain on *Morphology of Angiosperms* (1903) summarized the basic understanding of the reproductive processes in plants and signalled the readiness of the discipline for comparative studies for purposes of assessing relationships. The earliest compendium of comparative data was given by Karl Schnarf (1929, 1931) and the most recent by Gwenda Davis (1966). Obviously numerous additional data from various taxa have been steadily accumulating over the past fifty years, and the most significant insights in structure and function of embryology have come more recently with the use of EM techniques (e.g., Russell 1979). Much still remains to be known at the detailed level.

GENERAL EMBRYOLOGICAL TEXTS AND REFERENCES

The literature of comparative embryology is not as extensive as that of morphology or anatomy, but it is substantial nonetheless. The basic general texts are those of Maheshwari (1950), Johansen (1950), and Bhojwani and Bhatnagar (1983). Good general chapters are found in the edited volumes of Maheshwari

(1963) and Johri (1984). Compendia that list known embryological data for particular taxa are those of Schnarf (1929, 1931), Johansen (1950) and Davis (1966), the last of which is the most complete and is arranged according to Hutchinson's system of classification. For related topics on embryogenesis, one may consult chapter 2 of Steeves and Sussex (1972), the numerous papers by Jensen and collaborators on cotton (e.g., 1968a, b, c), and the book by Raghavan (1986). Works dealing with experimental embryology, which cover manipulations from the egg to the seedling, include those of Cutter (1966), Steeves and Sussex (1972, their chapter 3), Raghavan (1976), Johri (1982), and Johri and Rao (1984). The terminology used to describe the embryological structures and processes come from several sources (following Davis 1966): embryo sac development (Maheshwari 1948, 1950), embryogeny (Johansen 1950), embryo types (Crété 1963, and Johansen 1950), endosperm (Chopra and Sachar 1963; Swamy and Parameswaran 1963), and embryological processes (Maheshwari 1950). Updates from these contributions may be found in Johri (1984). A helpful glossary of embryological terms is found in Johansen (1950).

TYPES OF EMBRYOLOGICAL DATA

Many different types of embryological data are available for use in solving particular taxonomic problems. The most recent and complete list of potential features can be found in Herr (1984), although older but still useful lists also occur in Maheshwari (1950), Cave (1962), Johri (1963), Davis (1966), Palser (1975), and Bhojwani and Bhatnagar (1983). It is not helpful to consider every type of embryological data in detail but rather to emphasize those features of most value. The data come from several main areas of structure and processes (from Herr 1984): anther, male gametophyte, ovule, archesporium, megasporogenesis, female gametophyte, fertilization, endosperm, embryogeny, seed-coat, and special features of apomixis and polyembryony. The following discussion is structured on the more detailed outlines of data in Herr (1984) with additional cited references.

Anther

The anther contains many taxonomic features of value, such as the number of microsporangia, number of vertical rows of cells in each archesporium (= cells inside the anther which differentiate into parietal and sporogenous tissues), type of wall development, number of cell layers and level of differentiation in the endothecium, persistence of the epidermis, glandular or amoeboid tapetum, number of ploidy of nuclei in the tapetal cells (e.g., Buss and Lersten 1975), mode of delimitation of microspores, shape of the tetrad, and degree of aggregation of microspores (solitary, tetrad, etc.). These features of the anthers are utilized frequently in studies that often treat both micro- and megagametogenesis together, such as in Pullaiah (1978, 1979a, b, 1982a, b, 1983), Davis (1962a, b), Maheswari Devi (1975), Maheswari Devi and Pullaiah (1976), and Maheswari Devi and Lakshminarayana (1977). Variations in endothecial cell wall patterns that have systematic import in the Compositae have been shown by Dormer (1962; fig. 17.1) and Vincent and Getliffe (1988) and in the Araceae by French (1985a, b). These could also be regarded as micromorphological data (see chapter 15).

Male Gametophyte

The male gametophyte and its development have been utilized far less than the corresponding female structures. Pollen grains have both structural and developmental features of taxonomic value, but these will be treated in detail in chapter 18. Prósperi and Cocucci (1979) and Cocucci (1983) have shown variations in presence or absence of callose in pollen tubes (viewed under fluorescence microscopy) which seems fairly diagnostic for the Polemoniales. This needs to be examined further. Ultrastructural details of sperm cell formation have been examined by Russell and Cass (1981, 1983), but comparative TEM studies for systematic purposes await to be done. For a detailed summary of recent work on all aspects of microsporangial development, see Bhandari (1984).

Female Gametophyte

Many detailed studies have examined megagametogenesis and the megagametophyte (e.g., Harling 1960; Anderson 1970; Howe 1975; Ahlstrand 1978, 1979a, b, c; Anton and Cocucci 1984) to include aspects of the ovules, archesporium, megasporogenesis, and early and mature gametophytes. The form of the ovule has long been used to help sort out taxonomic problems, with the major types being orthotropous, campylotropous, amphitropous, and anatropous (Bocquet 1959; Bocquet and Bersier 1960; fig. 17.2). The number of integuments, type of nucellus (crassinucellate, pseudocrassinucellate, or tenuinucellate), presence or absence of endothelium, associated ovular structures such as an aril, and extent of vascular supply (just in the funiculus or extending also into the integuments) are other ovular features of taxonomic value. The situation in the parasitic mistletoes (Loranthaceae) is especially fascinating in that the ovules are sometimes without integuments and even disappear as distinct units (Maheshwari, Johri and Dixit 1957; Cocucci 1983).

The ovular archesporium can also provide helpful comparative data. These are the one to several hypodermal cells of the ovule primordium detectable by their large size and large nuclei and dense cytoplasm (Radford et al. 1974). The number of archesporial cells separately differentiated, their position in the nucellus, the extent of parietal tissue (if present), the origin and number of sporogenous cells (from archesporial cells), and the persistence or degeneration of the defunct sporogenous cells are all potentially useful features. However, these have been stressed less than other megagametophyte characteristics.

Megasporogenesis and megagametophyte development have received the most attention for characters of taxonomic value in the angiosperms. Herr (1984) lists features of significance during megasporogenesis: changes in the megasporocyte prior to meiosis (nuclear movement, degree of vacuolation, etc.); location and

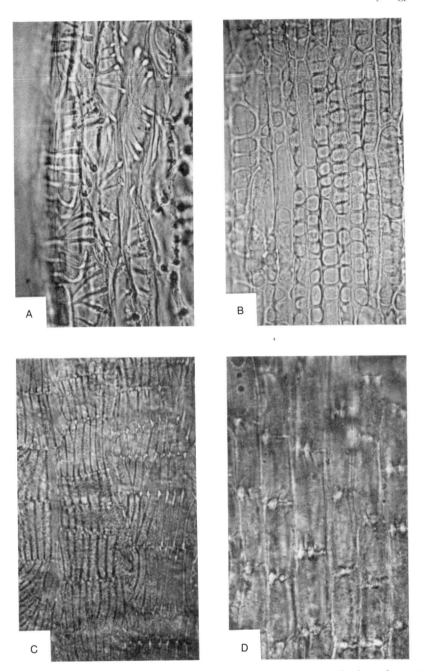

FIGURE 17.1 Variations in endothecial cell patterns in anthers of selected species of Compositae. A, *Tanacetum vulgare*, radial tissue with bars; B, *Hypochaeris radicata*, transitional tissue; C, *Dahlia* sp., polarized tissue with several bars per cell; D, *Onopordon tauricum*, polarized tissue with 1–2 ribs per end wall, with facial bars incomplete. x700. (From Dormer 1962, plate 4)

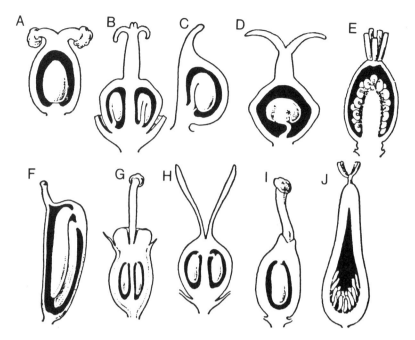

FIGURE 17.2 Types of ovules and their attachment in the ovary (=placentation) in angiosperms. A, basal, orthotropous (*Rheum;* Polygonaceae); B, basal, anatropous (*Rhamnus;* Rhamnaceae); C, basal on ventral suture, anatropous (*Ranunculus;* Ranunculaceae); D, basal, campylotropous (*Chenopodium;* Chenopodiaceae); E, on basal central columnar placenta, anatropous (*Lychnis;* Caryophyllaceae); F, pendulous, anatropous (*Spiraea;* Rosaceae); G, suspended, anatropous (*Cornus;* Cornaceae); H, pendulous from an axile placenta, anatropous (*Linum;* Linaceae); I, suspended or pendulous (*Prunus;* Rosaceae); J, basal, parietal (*Salix;* Salicaceae). (From Johnson 1931:50)

orientation of spindle during meiosis I; cytokinesis in relation to the nuclear division; meiosis II synchronous, nonsynchronous, or restricted to one dyad nucleus; number of nuclei in the megaspore; in monospory, the arrangement and relative size of megaspores; position of functional megaspore; and the persistence of the nonfunctional megaspore. Figure 17.3 illustrates many of these features, and shows the major types of embryo sacs in the angiosperms. Another type, the Poaceae variant of the basic *Polygonum* type, has been proposed by Anton and Cocucci (1984) in which the antipodals continue to divide to form 6–300 cells. The features of taxonomic value in the mature female gametophyte include (from Herr 1984): differentiation of egg apparatus and antipodals (synchronous or nonsynchronous); structure of antipodals, synergids, and egg; form and behavior of synergids and antipodals; position of polar nuclei; and timing of fusion of polar nuclei. Fish (1970) discovered two different embryo types in *Clematis* (Ranunculaceae) in fertile (monosporic) and sterile (tetrasporic) ovules

FIGURE 17.3 Diagram of major types of embryo sacs and their development in angiosperms. (From Willemse and van Went 1984:160)

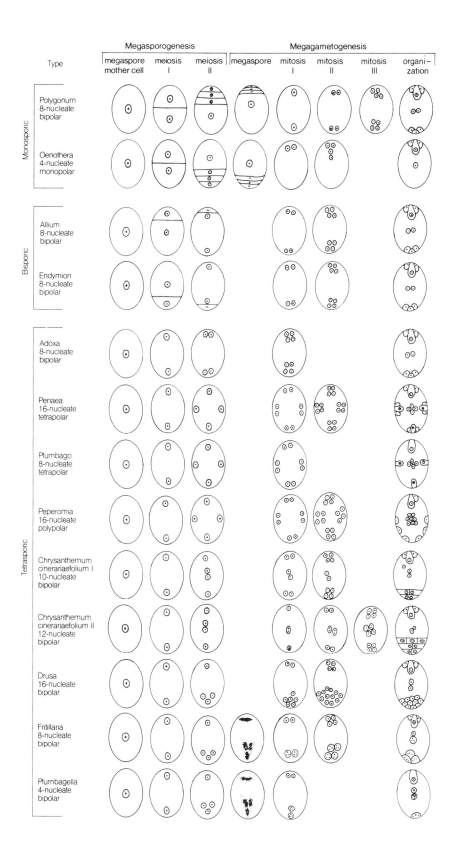

which led to inferences on evolutionary directionality within the genus. Cave (1955) used embryo sac features (in conjunction with other embryological data) to suggest that *Phormium* deserves an isolated position from Hemerocallideae of Liliaceae and also from Agavaceae. Anderson (1970) has shown variations among species of *Chrysothamnus* (Compositae), especially in the number of antipodal cells present (fig. 17.4). Many more details of these embryo sac developmental patterns are being revealed by TEM studies such as those of Newcomb (1973a, b) in sunflower (*Helianthus*, Compositae) and Russell (1979) in corn (*Zea*, Gramineae), but again, detailed ultrastructural comparative studies await completion. The best recent review can be found in Willemse and van Went (1984).

Fertilization

Much recent work has taken place on the details of fertilization in angiosperms with EM techniques, but data of a helpful comparative nature have not yet been developed. The series of papers by Jensen (1965a, b, c, 1968a, b c; Jensen and Fisher, 1968) on events prior to, during, and after fertilization in cotton (*Gossypium*, Malvaceae) provide a much more detailed view of the process and again offer the potential of additional comparative data to be used for taxonomic purposes. The papers by Russell (1980, 1982, 1983) and Russell and Cass (1981) on the ultrastructure of fertilization in *Plumbago zeylanica* (Plumbaginaceae) are worth noting in this same regard. See also the excellent review by van Went and Willemse (1984). The most useful taxonomic features so far used are the location of entrance of the pollen tube (porogamous, chalazogamous, mesogamous) and the time interval between pollination and fertilization (Herr 1984). Another feature of some value at the higher levels of the hierarchy is the type of nuclear fusion in angiosperms (fig. 17.5), of which the intermediate type occurs in a number of Liliaceae.

Endosperm

Differences in endosperm are also taxonomically significant (fig. 17.6). Important here are the patterns of development (nuclear type, cellular, or helobial, although there is no clear-cut demarcation between these types; Vijayaraghavan and Prabhakar 1984), the origin and type of haustoria, the type and amount of food reserves, and the persistence in the mature seed. Swamy and Parameswaran (1963) describe different types of endosperm and give details of the helobial variation. For extreme intrafamilial variation the Acanthaceae reveal a diversity, especially in the haustoria (Mohan Ram and Wadhi 1964). Trela-Sawicka (1974) has shown how endosperm formation can be related to the compatibility system. In *Anemone ranunculoides* (Ranunculaceae) from Poland, due to an incompatibility system, the second sperm nucleus fails to fuse with the coalesced polar nuclei ($2n$ secondary fusion nucleus) yielding a diploid endosperm with inhibited development and seeds which will not germinate, despite the normal external appearance of the achene.

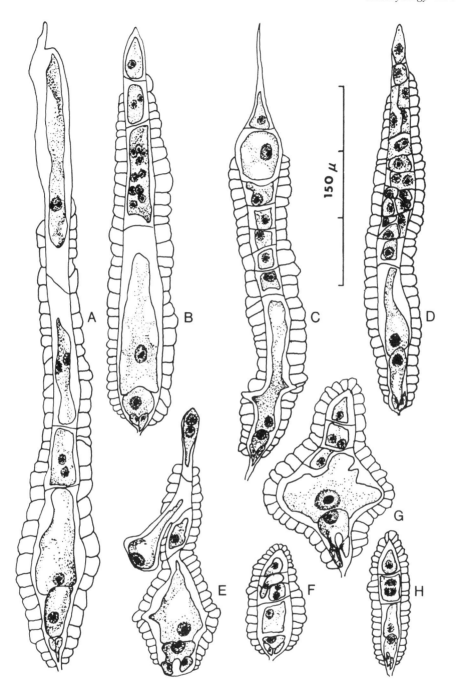

FIGURE 17.4 Mature embryo sacs and endothelium in *Chrysothamnus* (Compositae), especially showing variation in number of antipodals. A, *C. viscidiflorus* ssp. *lanceolatus;* B, *C. nauseosus* ssp. *leiospermus;* C, *C. pulchellus;* D, *C. parryi* affin. ssp. *nevadensis;* E, *C. viscidiflorus* ssp. *humilis;* F, *C. linifolius;* G, *C. nauseous* ssp. *nauseosus;* H, *C. viscidiflorus* ssp. *viscidiflorus.* (From Anderson 1970:338)

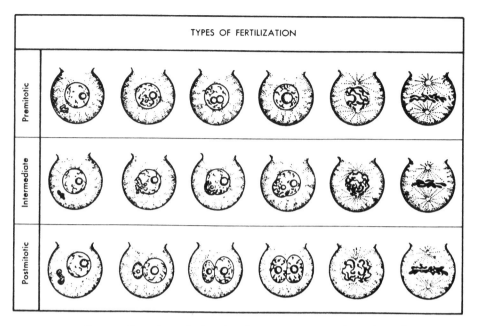

FIGURE 17.5 Types of nuclear fusion in the angiosperms, based on first nuclear contact and completion of mixing of chromatin. In the Premitotic type the nuclear fusion is completed before the zygotic division. In the Intermediate type the chromatin of male and female nuclei remain partly separated until the first mitotic division. In the Postmitotic type the chromatin of male and female nuclei remain separated completely until the mitotic division of zygote. (From van Went and Willemse 1984:305; after Gerassimova-Navashina 1960)

Embryogeny

The variations of embryogenesis (= embryogeny) yield different types of patterns (Johansen 1950; Maheshwari 1950; Crété 1963; Bhojwani and Bhatnagar 1983; Yamazaki 1982), of which the most common are: Asterad, Caryophyllad, Chenopodiad, Onagrad, Piperad, and Solanad (fig. 17.7). These different patterns are based primarily on the occurrence and direction of cell division of the terminal and basal cells of the embryo. Recent EM work again has clarified the process (Newcomb 1973b; Schulz and Jensen 1968a, b, c, 1969, 1971, 1973, 1974). One of the most dramatic examples of the systematic utility of the embryo is in the Gramineae (Reeder 1957, 1962; fig. 17.8) in which six major patterns are found: festucoid, bambusoid, centothecoid, arundinoid-danthonioid, chloridoid-eragrostoid, and panicoid. These are based on the course of the vascular system, the presence of an epiblast (= a small scale-like structure opposite the scutellum; Maheshwari 1950:289), fusion of the scutellum to the coleorhiza, and the cross-sectional appearance of the embryonic leaves. The different embryo types correlate with other features to help define tribal limits and suggest further taxonomic investigations on confusing complexes (e.g., the Festucoideae).

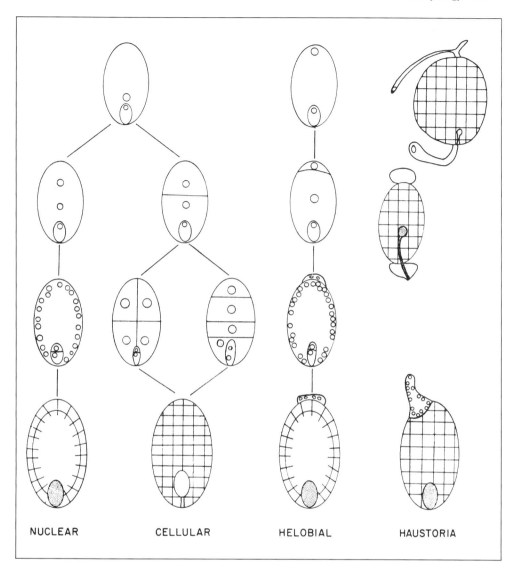

NUCLEAR   CELLULAR   HELOBIAL   HAUSTORIA

FIGURE 17.6 Types of endosperm developmental patterns. (From B. Palser in Radford et al. 1974:227)

Seeds

Seeds also are extremely helpful in providing taxonomic characters, but the import of seed coats has already been considered in the chapters on morphology and anatomy and will not be repeated here. The nature of the vasculariziation of the seed, however, can also provide useful taxonomic information and is worth mentioning (fig. 17.9).

Apomixis and Polyembryony

Other special embryological features of taxonomic interest include apomixis and polyembryony. The former is mostly important for determining the nature of the breeding systems to understand the biology of the plants in question, rather than

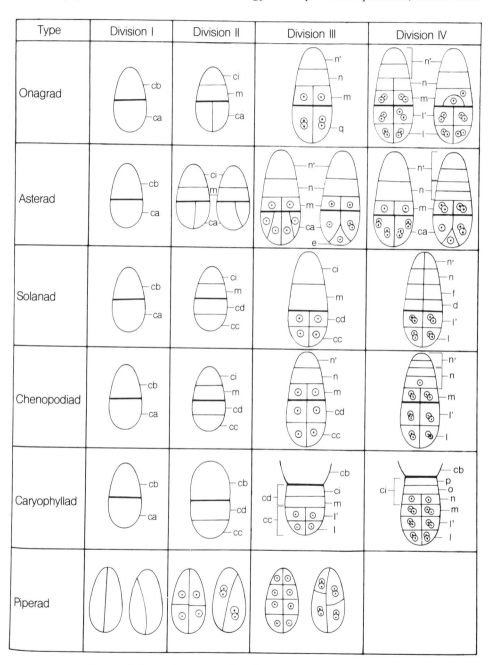

FIGURE 17.7 Some of the major types of embryogeny in angiosperms. (From Natesh and Rau 1984:389)

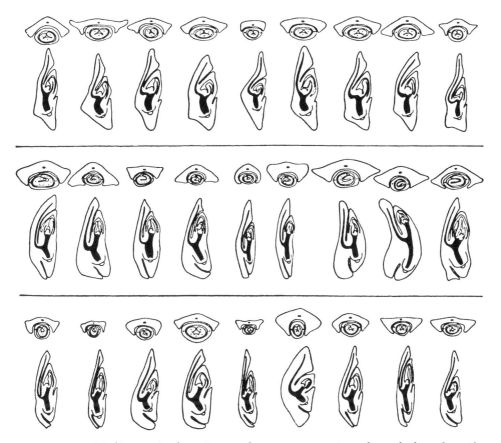

FIGURE 17.8 Median sagittal sections and transverse sections through the coleoptile region of embryos of various species of Gramineae, showing three principal types. Top row, festucoids; middle row, panicoids; bottom row, chloridoid or eragrostoid type. Not all to same scale. (From Reeder 1957:763)

to show comparative differences for taxonomic purposes. The broad differences (Nogler 1984) between diplospory (unreduced embryo sac coming from a generative cell by mitosis or modified meiosis to yield an unreduced restitution nucleus) and apospory (unreduced embryo sac coming from somatic cells of the ovule), however, could well be taxonomically significant within a particular group. It is likely, though, that apomixis has evolved many times in parallel in the angiosperms. (For examples of developmental studies with taxonomic import, see Davis 1967; Malecka 1971a, b, 1973; and Spooner 1984). Polyembryony is mostly an abnormal feature in which two or more embryos or proembryos occur in a developing ovule. For a classification of types see Yakovlev (1967) and Lakshmanan and Ambegaokar (1984).

INVESTMENTS FOR GATHERING EMBRYOLOGICAL DATA

The equipment for embryological studies is much the same as for anatomical investigations. Regular light microscopy is the standard tool of the trade with

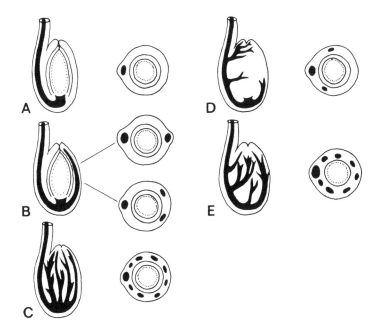

FIGURE 17.9 Types of seed vasculature in angiosperms. A, seed with well-developed raphal bundle; B, one postchalazal bundle in the median plane, or two postchalazal bundles; C, many postchalazal bundles; D, pre-raphe bundles; E, pachychalazal bundles. Left diagrams are side views; the right are transverse sections. (From Boesewinkel and Bouman 1984:572)

options added to enhance its vision (Johri and Ambegaokar, 1984). For example, polarizing optics are useful for revealing starch grains, cell walls, and crystalline inclusions in cells. Phase-contrast is good for living cells, and fluorescence microscopy is helpful for showing presence of callose during microsporogenesis and pollen-stigma interactions. Nomarski optics (differential interference contrast microscopy) gives good surface relief to thick specimens. EM techniques are obviously pertinent here too, both SEM and TEM. New techniques of cyto- and histochemistry (also with autoradiography) are very useful, as are now whole-mount clearing techniques (e.g., of ovules; Herr 1971, 1982; Smith 1973). The minimal equipment needed would be a good quality compound light microscope with rotary microtome, warming trays, paraffin ovens, staining jars, solvents, stains, etc., going up to the fully equipped EM laboratory at much higher cost.

The expertise, time, and costs needed for gathering of embryological data are again much the same as with anatomical information. The sectioning, embedding, staining, and microscopic observations all take considerable knowledge of these techniques and a time investment of some magnitude is essential. As with anatomy, a good one-quater course in plant microtechnique (now becoming rare in most university curricula) is beneficial as a beginning. The time needed to gather embryological data is at least as consuming as with anatomy, but probably even more so, depending upon the type of observations desired. To do a complete embryological examination of even one taxon takes much effort to find all the proper stages and document them unequivocally. The most important ingredient is patience in making the numerous sections.

EFFICACY OF EMBRYOLOGICAL DATA IN THE TAXONOMIC HIERARCHY

Embryology certainly has no claim to supremacy with regard to ranking of taxa at different levels in the hierarchy. As Davis and Heywood put it: "What embryology *cannot* do is indicate how different groups must be before they should be recognised as genera, subgenera or sections, any more than other lines of evidence can do" (1963:191). Embryological data have been successfully used at all levels of the hierarchy, but most success has prevailed in solving problems at the generic level and higher (Cave 1953). Many studies are largely descriptive and document the pattern of development in particular taxa (e.g., Davis 1962a, b; Eliasson 1972; Maheswari Devi 1975; Maheswari Devi and Pullaiah 1976). Czapik (1974) showed only quantitative embryological variations among five species of *Arabis* (Cruciferae), although Fish (1970) did find useful variations (especially for determining evolutionary directionality) among 26 taxa of *Clematis* (Ranunculaceae), as did Smith (1975) among five species of *Cornus* (Cornaceae). As might be expected, between cultivars of *Nerium indicum* (Apocynaceae) little differences were noted except for ovarian degenerations in one of them which often led to failure of seed set, of some horticultural and economic importance in this case (Maheswari Devi and Narayana 1975).

Many excellent studies using embryology at the generic level exist. Cave's many papers (1955, 1968, 1974, 1975; see early summary in 1953) on Liliaceae in this regard are worth noting. Other examples include Harling (1960; *Anthemis*, Compositae), Kapil and Vani (1966; *Nyctanthes*, Oleaceae), Howe (1975; *Grindelia*, Compositae), and Tobe and Raven (1984a; *Alzatea*, Alzateaceae). An especially good modern example is the work on *Rhynchocalyx* of Myrtales by Tobe and Raven (1984b), which was variously placed in Lythraceae, Crypteroniaceae, or treated as a monotypic family. Based on a complete examination of the embryological development patterns and comparisons with data from other families, the decision to place the genus in its own family, Rhynchocalycaceae, was substantiated.

Equally good examples of the efficacy of embryological data exist at the subfamilial and familial levels and above. The positive utility of embryo features in the Gramineae at the tribal level has been noted already (Reeder, 1957, 1962; fig. 17.8). A study at the same level of the hierarchy by Pullaiah (1981) confirmed the retention of *Lagascea* (Compositae) in the tribe Heliantheae as recommended earlier by Stuessy (1976, 1978) based on morphological and cytological data. The subfamilial classification of the Aizoaceae has been improved with embryological data (Prakash 1967). The conclusion that the Cercidiphyllaceae are not close to Magnoliaceae of Magnoliales but closer to Hamamelidaceae of Hamamelidales was confirmed by the embryological studies of Bhandari (1971). A good overview of results as they impact family relationships is found in Herr (1984), dealing with Basellaceae, Centrolepidaceae, Frankeniaceae, Podostemaceae, Salvadoraceae, Stachyuraceae, and Tropaeolaceae. Tobe and Raven (1983) help define Myrtales based on embryological data.

SPECIAL CONCERNS WITH EMBRYOLOGICAL DATA

A number of concerns should be kept in mind with embryological information. First, the importance of vouchers must be emphasized. The more complex the data-gathering technique, i.e., the further one becomes removed from the whole specimens themselves, the greater the tendency to forget about making proper vouchers and depositing them in a recognized herbarium. Davis and Heywood concur: "Clearly, a strict check of the taxonomic identification of material used for embryological studies is vital, as indeed it is for any other systematic studies. It is exceedingly difficult to assess to what extent early results may be suspect due to such reasons, in the same way that many of the earlier chromosome counts have had to be rejected" (1963:191). Second, sample sizes tend to be small, due naturally to the complexity and time-consuming nature of the techniques, but quantitative data are much needed to account for the range of variation within taxa before making final taxonomic judgments. The large sample sizes used by Czapik (1974) to investigate the embryology of five species of the *Arabis hirsuta* complex (Cruciferae) are laudatory (e.g., 134 ovaries in 15 plants of 5 populations of *A. planisiliqua* examined). The quantitative survey (Anderson 1970) of antipodal variation in 13 species of *Chrysothamnus* (Compositae) is also positive, as also is the quantitative work of Smith (1975) on the megagametophyte of five species of *Cornus* (Cornaceae). Third, homology again becomes an issue when comparing taxa at the higher levels of the hierarchy, which is the most suitable arena for use of embryological data, and cautions are needed. A good review of this issue is provided by Favre-Duchartre (1984).

With these caveats, one can turn the picture around in a positive way, however, and point to comparative embryological data as really ontogenetic data, especially if the entire male and female and embryo developmental sequences are resolved. From this viewpoint these data can help profoundly in the resolution of difficulties in determining homologies of mature structures such as seeds. Further, they can provide independent (and quite different) evaluations of relationships hypothesized via use of other characters.

CHAPTER 18

Palynology

Palynology, or the study of pollen grains and spores, could be regarded as simply one aspect of embryology instead of being treated separately. Because so much work has been done with pollen grains in taxonomy, far more in fact than with all other aspects of embryology combined, it makes more sense to regard this area as a separate source of comparative data. The field is relatively youthful, however, with the label palynology not having been coined until 1945 (Hyde and Williams 1945). Palynology also obviously intergrades with pollination biology and reproductive biology, particularly in the area of pollen-stigma interactions and compatibility systems. For taxonomic purposes most emphasis has been placed so far on the comparative features of the pollen grains themselves, especially those of apertures and wall structure, and a wealth of detail has been found. Thousands of studies have now been completed that show the proven efficacy of use of pollen grains at all levels of the hierarchy. Although pollen grains are small and the features only observable with compound light and electron microscopes, Keating is correct when he says: "The usefulness of palynology has become so obvious that it is now routinely incorporated into most systematic and evolutionary studies" (1979:592).

HISTORY OF PALYNOLOGY IN PLANT TAXONOMY

The history of palynology is tangled with that of embryology, cytology, and pollination biology. The comments that follow have been taken principally from Wodehouse (1935) and Nair (1970). The understanding of pollen grain structure and function clearly has depended to some extent on advances in microscopy, especially in the earliest years. As Keating (1979) points out, however, many subsequent conceptual advances can be shown to be independent of instrumental improvements. Both Marcello Malpighi (1628–1694) in Italy and Nehemiah Grew (1641–1712) in England observed and reported pollen grains in the seventeenth century using microscopes developed prior to 1665 by Robert Hooke, but their biological function in reproduction was unknown.

From these early beginnings, serious study of pollen grains did not rejuvenate until the nineteenth century. The Englishman Francis Bauer, working as a botanical artist at Kew Gardens from 1790 to 1840, compiled a series of drawings of pollen grains that included 175 species in 120 genera and in 57 families, but which were never published. These are now bound in one volume and reside in the British Museum. They are remarkable for their clarity and accuracy, and certainly Bauer was well aware of the taxonomic value of these comparative data. He also drew plates for Robert Brown of the British Museum and the latter noted pollen differences in some of his taxonomic works, such as in the Proteaceae (1811; quoted in Wodehouse 1935:38): "I am inclined to think, not only from its consideration in this family, but in many others, that it [the pollen] may be consulted with advantage in fixing our notions of limits of genera." During this period several major advances were made. Johannes Evangelista Purkinje of Moravia (now part of Czechoslovakia) published (1830) the results of many observations on different forms of pollen grains and made an attempt to develop a system of terminology to describe the detail he saw.

The first deliberate use of features of pollen grains for classification was by the Englishman John Lindley (1799–1865) from 1830–1840. He showed that some features of pollen correlated with others in establishing tribes of the Orchidaceae. Hugo von Mohl (1805–1872), the illustrious German botanist, made many important palynological discoveries published mainly in 1835. He produced a classification of pollen grains of angiosperms and gymnosperms (211 families) with emphasis on variations in number of pores, furrows, and surface features. Carl Julius Fritzsche (1808–1871), also a German, published several works on pollen, some antedating those of von Mohl, but his most notable was *Ueber den Pollen* in 1837. He coined the terms *intine* and *exine*, investigated the nature of spines, and developed classifications of pollen grains in the flowering plants. Many other workers were interested in pollen during this time, but Sergius Rosanoff deserves mention because of his detailed study of the pollen of the Mimosoideae (1866). Carl Albert Hugo Fischer produced an outstanding doctoral thesis in 1890 on *Beiträge zur vergleichenden Morphologie der Pollenkorner* in which he examined over 2,200 species. Among his conclusions, he indicated that the pollen grains of related species are usually similar and that often pollen of close relatives cannot be distinguished from each other.

In the twentieth century the study of palynology has mushroomed enormously. One very important book is Wodehouse's *Pollen Grains* (1935) which was a complete manual of history, technique, and known features of pollen grains within the gymnosperms and angiosperms. He must be credited with placing palynology in its modern context. Erdtman's *Pollen Morphology and Plant Taxonomy* (1952) was another landmark which was followed by numerous publications from his laboratory in Stockholm, Sweden, including the *Handbook of Palynology* (1969). He set forth more detailed methods of pollen analysis and provided a more complete set of terms for pollen grain description. Many students and visitors were trained in his laboratory which had a strong effect on the development of the field.

No history of palynology could possibly overlook the enormous impact the electron microscope has had on obtaining a better view of features of pollen grains, especially of wall structure (fig. 18.1). The excellent work at the light microscopic level of earlier workers, such as Stix (1960) in the Compositae, was overshadowed by greater wealth of detail with TEM data (such as seen in Skvarla and Larson 1965, and Skvarla and Turner 1966a, also in the Composi-

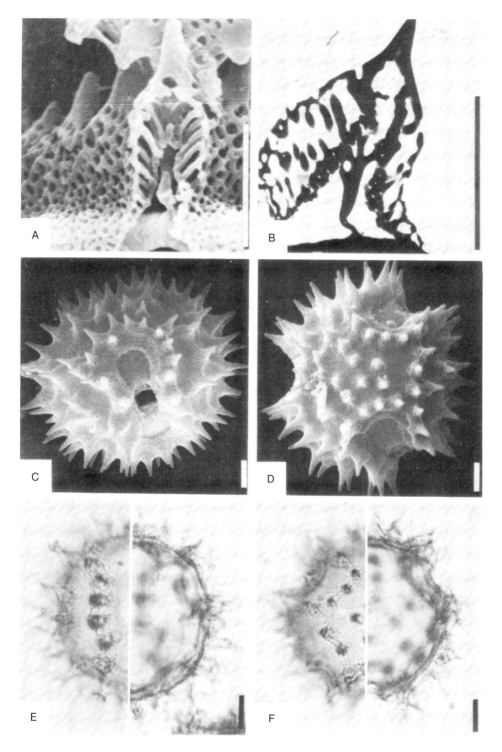

FIGURE 18.1 Light, SEM and TEM of pollen grains of Scorzonerinae (Compositae). A, *Scorzonera graminifolia;* B, *S. laciniata;* C–F, *S. humulis.* C, E, apertural; D, F, polar views. E and F show two focal planes under light microscopy. Scale lines = 5 μm. (From Blackmore 1982a:153)

tae), replica techniques with TEM (e.g., Barth 1965; Tsukada 1967; Graham, Graham, and Geer 1968), and later SEM data (e.g., Skvarla et al. 1977; see also the earlier general surveys of Martin 1969, and Martin and Drew 1969, 1970). Both TEM and SEM approaches are now considered routine for pollen analysis and both should be used in conjunction to provide the best understanding of outer (sculptural) and inner (structural or architectural) features and their interrelationships.

GENERAL PALYNOLOGICAL TEXTS AND REFERENCES

Many texts and references are available for learning the terminology that relates to pollen grains, their development, and their patterns of variation throughout the angiosperms. General texts would include Wodehouse (1935), Erdtman (1952, 1969), Nair (1966, 1970), Kapp (1969), Tschudy and Scott (1969), Stanley and Linskens (1974), and Shivanna and Johri (1985). For pollen analysis, with reference to use in systematics and paleoecology, consult Faegri and Iversen (1950, 1964) and the excellent newer book by Moore and Webb (1978). A good symposium on palynology and systematics can be found introduced by Keating (1979); for a series of papers on form and function of pollen, see Blackmore and Ferguson (1986). The nature of the very hard pollen wall material, sporopollenin, has been examined by Brooks et al. (1971) and Brooks and Shaw (1978). General reviews on the systematic and phylogenetic value of palynology are Walker and Doyle (1975), Sivarajan (1980), and Blackmore (1984), and on the development of pollen grains by Heslop-Harrison (1971), Dickinson (1982) and the excellent overview by Knox (1984). Many pollen floras have been completed or are in progress, such as for South Africa (Zinderen Bakker 1953), the western Himalayas (Nair 1965), Chile (Heusser 1971; Parra and Marticorena 1972), Argentina (Markgraf and D'Antoni 1978), and northwest Europe (the most completely understood part of the world palynologically, Punt 1976; see also Huntley and Birks 1983). The ambitious *World Pollen Flora* (Erdtman 1970) has been superceded by the *World Pollen and Spore Flora* (Nilsson 1973 and continuing). Airborne pollen, of great importance to hayfever sufferers, has been chronicled by Hyland et al. (1953) for Maine, and by Lewis, Vinay, and Zenger (1983) for North America. This latter volume is especially useful for physicians and of high general interest for the educated layperson. For insights on possible functional and evolutionary import of pollen grains, see Ferguson and Muller (1976), and for reproductive biology involving gamete competition, breeding systems, compatabilities, recognition systems, and so forth, consult Mulcahy (1975), Clarke and Gleeson (1981), and Mulcahy and Ottaviano (1983). There is also an extensive literature dealing with fossil pollen and spores, which is obviously important in classification of extant angiosperms which have fossil pollen histories (e.g., Graham and Graham 1971), and some further mention will be made later in this chapter. The most important indices of palynological data both by subject and taxon (for fossil and extant taxa) are the *Bibliographie Palynologie*, published as a supplement to the journal *Pollen et Spores* (e.g., Van Campo and Millerand 1985), and the five volumes of Thanikaimoni (1972–1986).

TYPES OF PALYNOLOGICAL DATA

The data of palynology derive from features of the pollen grain involving the aggregations and shape of grains, aperture number, shape and position, external wall layers (primarily the exine), and internal protoplasm. Because the terminology for describing pollen grains is complex, fig. 18.2 shows some of the general features of importance to help in the discussion to follow. Good glossaries can be found in Wodehouse (1935), Erdtman (1952, 1969), Kremp (1965), Kapp (1969), and Moore and Webb (1978). Different schools of terminology have evolved, especially those of Erdtman and of Faegri, and the literature is filled with descriptors which adequately label the same structure. A computer-based numerical coding system has even been developed to deal with the problem of adequate description of pollen grains (Germeraad and Muller 1970). Hideux and Ferguson (1975, 1976) have suggested a numerical approach to pollen grain description involving geometrical location in three-dimensional space, similarity, and set theory, as applied to examples in the Saxifragaceae. Factor and cluster analysis have also been used to show similarity among pollen grains quantitatively in the Saxifragaceae (Hideux and Mahe 1977) which can bypass the need for qualitative descriptors (see also Hideux 1979). The proliferation of terms has been so extreme that it occasioned Davis and Heywood to state directly: "The results are impressive but, we feel, self-defeating due to their very complexity" (1963:187). Progress is being made in standardization (e.g., Reitsma 1970; Nilsson and Muller 1978) and "Hopefully it will eventually prove possible to achieve a state where the descriptive palynological works can be understood much more readily by the nonspecialist, thus making pollen characters more accessible to the plant taxonomist" (Blackmore 1984:136). Even Erdtman himself (1966) has made a plea for reduction in terminology.

External Characters

Pollen usually occurs as single grains at maturity, but in some cases they can be aggregated into clusters (fig. 18.3) of two (a dyad), four (tetrad) or many (polyad) grains. These arrangements can have taxonomic utility as shown in the Annonaceae (Canright 1963; Walker 1971a). The type of binding between the grains in multiple arrangements can vary also, such as seen in the Epilobieae of the Onagraceae (Skvarla, Raven, and Praglowski, 1975). The shapes of the grains in outline also vary (fig. 18.4), in a similar fashion as do shapes of leaf blade outlines (see chapter 15, p. 221). Descriptors have been provided to help verbally distinguish these variations. One of the most unusually shaped grains (and also one of the largest) has to be the "foot-long hot dog" in *Crossandra stenostachya* (Acanthaceae) which measures 520 X 19 μm (Brummitt, Ferguson, and Poole 1980). Another extremely large, but spherical, grain is found in *Cymbopetalum odoratissimum* (Annonaceae) measuring nearly 350 μm in diameter (Walker 1971b). Aperture number also varies remarkably within the angiosperms (fig. 18.5), and many terms have been used to describe these variations. In fact, the number of pores and furrows (= colpi; singular, colpus) and their configuration form the

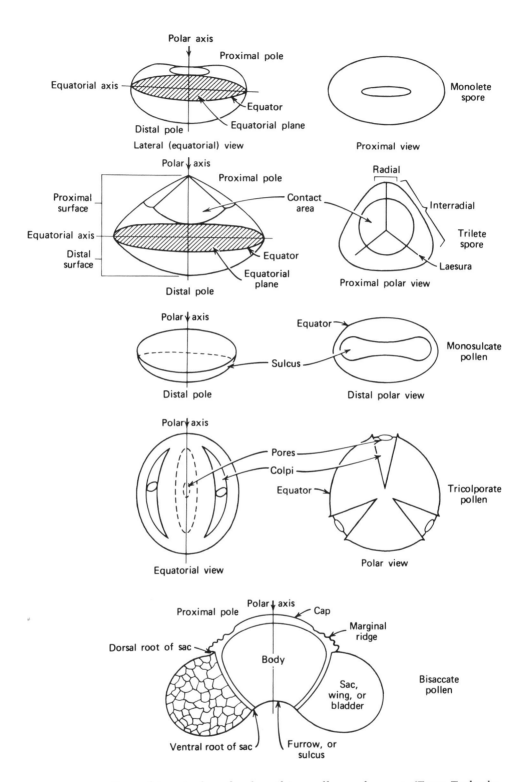

FIGURE 18.2 General terminology for describing pollen and spores. (From Tschudy 1969:19)

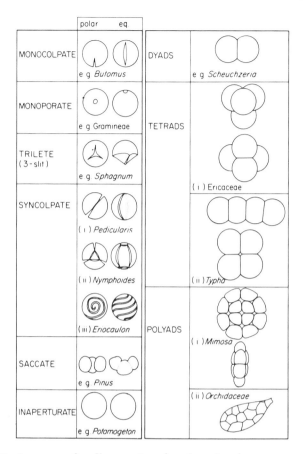

FIGURE 18.3 Basic types of pollen grains showing simple aperture variations and multiple associations. (From Moore and Webb 1978:36)

basis for the description of most grains with additional descriptors added for other details. With TEM and SEM techniques, further details of apertures can be discerned and these have taxonomic and evolutionary significance (e.g., Blackmore 1982b). Next to the number and position of apertures, the details of surface sculpturing provide an amazing wealth of detail of taxonomic value (fig. 18.6). Just as with surface features of seed coats and vesture of leaf surfaces (see chapter 15), a plethora of descriptors exists here also.

Internal Wall Characters

A very important point in working with pollen grains, however, is that the exine structure or stratification be studied as a correlate of the variations seen in surface features. The wall of the pollen grain is divided into different layers structurally and/or histochemically, and different sets of terminology have been developed. Figure 18.6 shows terms from Erdtman (left) and Faegri (right). The

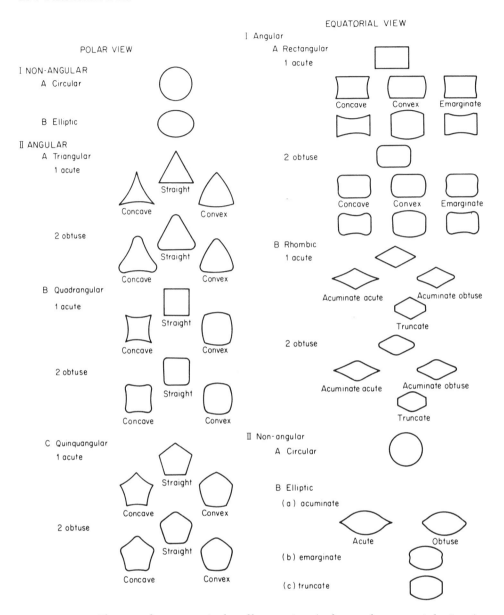

FIGURE 18.4 Shapes of symmetrical pollen grains (polar and equatorial views). (From Moore and Webb 1978:40; after Reitsma 1970)

general point is that there is an inner wall called the intine which is degraded completely under acetolysis, a severe method of treating pollen grains with boiling acids to completely remove all protoplasmic contents and intine (Erdtman 1960) and leave the walls clean as well as looking comparable to fossilized material (useful for comparative purposes in paleoecological studies). Using the Faegri terminology, outward from the intine is the endexine, and then the ektex-

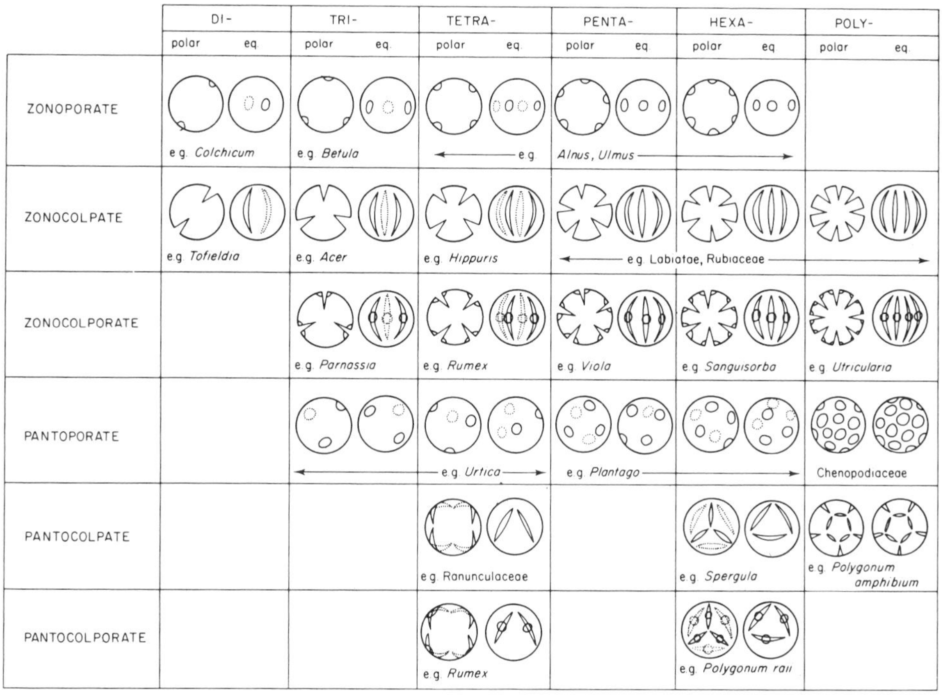

	DI- polar	DI- eq	TRI- polar	TRI- eq	TETRA- polar	TETRA- eq	PENTA- polar	PENTA- eq	HEXA- polar	HEXA- eq	POLY- polar	POLY- eq
ZONOPORATE	e.g. *Colchicum*		e.g. *Betula*		← e.g.		*Alnus, Ulmus*		→			
ZONOCOLPATE	e.g. *Tofieldia*		e.g. *Acer*		e.g. *Hippuris*		← e.g. *Labiatae, Rubiaceae*				→	
ZONOCOLPORATE			e.g. *Parnassia*		e.g. *Rumex*		e.g. *Viola*		e.g. *Sanguisorba*		e.g. *Utricularia*	
PANTOPORATE			← e.g. *Urtica* →				e.g. *Plantago*		→		Chenopodiaceae	
PANTOCOLPATE			e.g. Ranunculaceae						e.g. *Spergula*		e.g. *Polygonum amphibium*	
PANTOCOLPORATE			e.g. *Rumex*						e.g. *Polygonum raii*			

FIGURE 18.5 More complex aperture variation showing classification of pollen grains based upon number and arrangement of apertures (shown in polar and equatorial views). (From Moore and Webb 1978:37)

ine which may be simple or elaborated into a roof-like structure (tectum) with supporting columns (columellae) and an inner layer (foot layer).

One of the serious problems with detecting homologies of ektexine structure has been the need to have both external sculpturing data as well as internal structural information (seen most clearly with TEM or fractured surfaces with SEM). Virtually identical sculpturing can derive from very different wall structures (fig. 18.7) and hence analogous for taxonomic purposes. To help detect this, in earlier studies with light microscopy Erdtman (1952) stressed the importance of making optical sections through the grain to reveal internal structure. Because the patterns that result have light and dark contrasting areas, this was called "LO analysis" (*lux-obscuritas*; = light-darkness; fig. 18.8). This is extremely useful in light microscopic work but now less valuable due to the many EM techniques. Also, different refractive indices of wall material and internal holes can give misleading patterns. Recently, even the internal surface of the endexine (= endosculpture) has been examined for taxonomic utility (Van Campo 1978). The intine, usually removed with acetolysis, is ordinarily not examined, although in some studies done with fresh pollen material it does show clearly (e.g., El-Ghazaly 1980). A most remarkable situation occurs in the banana relative *Heliconia* (Heliconiaceae) in which only a very thin layer of exine is present with also a few spinules. The wall is almost entirely intine (Kress, Stone, and

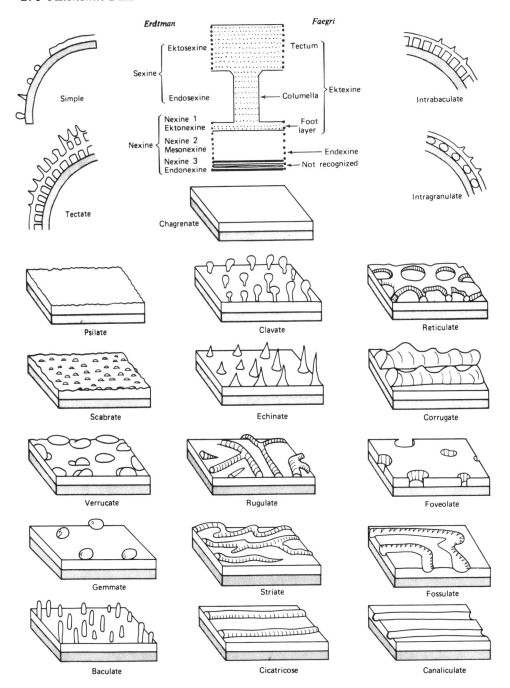

FIGURE 18.6 Exine layers and different types of surface sculpturing. (From Tschudy 1969:24)

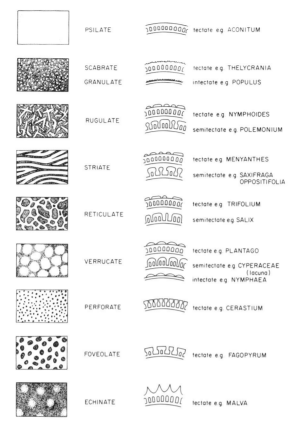

PSILATE	)0000000000(	tectate e.g. ACONITUM
SCABRATE	)0000000000(	tectate e.g. THELYCRANIA
GRANULATE		intectate e.g. POPULUS
RUGULATE	)000000000(	tectate e.g. NYMPHOIDES
	ЛоЛооЛЛоо	semitectate e.g. POLEMONIUM
STRIATE	)0000000000	tectate e.g. MENYANTHES
	ЛоЛЛоЛЛ	semitectate e.g. SAXIFRAGA OPPOSITIFOLIA
RETICULATE	)000000000(	tectate e.g. TRIFOLIUM
	ЛоЛоЛоо	semitectate e.g. SALIX
VERRUCATE	)000000000	tectate e.g. PLANTAGO
	ЛоЛооЛооЛоо	semitectate e.g. CYPERACEAE (lacuna)
		intectate e.g. NYMPHAEA
PERFORATE	)000000000(	tectate e.g. CERASTIUM
FOVEOLATE	)оЛ)оЛЛ)оЛ	tectate e.g. FAGOPYRUM
ECHINATE	)0000000(	tectate e.g. MALVA

FIGURE 18.7 Selected sculpturing types of pollen grains in surface view (left) and in optical section (right), showing how a single sculpturing type can result from different exine structures. (From Moore and Webb 1978:42)

Sellars 1978; Stone, Sellars, and Kress 1979). A similar situation in known in *Canna* (Cannaceae; Skvarla and Rowley 1970; Rowley and Skvarla 1986) and in most other Zingiberales (Kress 1986). The adaptive value of this lack of exine is unknown, but it may relate to the absence of sporophytic incompatibility mediated by enzymes usually found in the exine.

Protoplasmic Characters

Although most features of pollen grains deal with the wall, there are also some protoplasmic variations of taxonomic value. One of the most well surveyed is the presence of two or three nuclei in the mature pollen grains, shown to vary in the angiosperms (Brewbaker 1967). The binucleate condition appears primitive in the angiosperms and this was used by Dahlgren and Clifford (1982) as one of the characters in classification of monocot families. Most genera have either the bi- or trinucleate condition with only six genera of angiosperms showing both (Grayum 1986). Within the Araceae both conditions occur, and the trinucleate

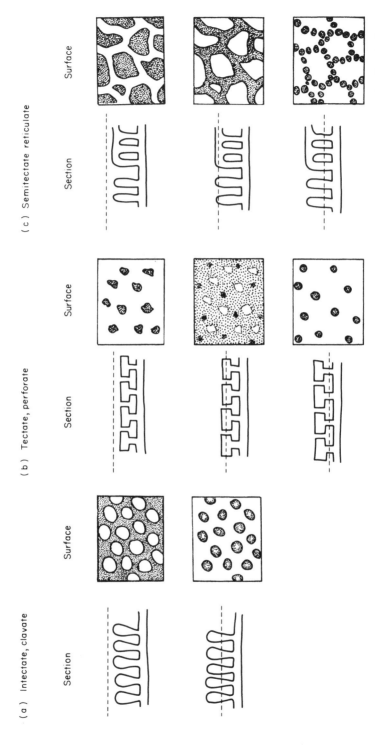

(a) Intectate, clavate Section Surface

(b) Tectate, perforate Section Surface

(c) Semitectate reticulate Section Surface

FIGURE 18.8 Types of sculpturing at different focal planes, also referred to as the LO method (Erdtman 1952). The focal plane is indicated by the dashed line on the sections. (From Moore and Webb 1978:44)

state appears to have evolved independently many times within the family (Grayum 1986). Both conditions also are known to occur in the Rubiaceae (Mathew and Philip 1986). The New Zealand flora has also been surveyed for these same data with similar distributional results (Gardner 1975). The original position of the generative nucleus in pollen tetrads has also been shown to vary in the angiosperms (Huynh 1972), but more surveying is needed before the taxonomic efficacy of these features is known. A very interesting study of the nature of stacks of the endoplasmic reticulum (ER) in mature pollen grains in subtribe Castillejinae of the Scrophulariaceae has been done by Jensen, Ashton, and Heckard (1974). Among the four genera examined *(Castilleja, Orthocarpus, Ophi ocephalus,* and *Cordylanthus),* the intracisternal width varied from 880 to 300 A° and showed average values for each genus that correlated with existing generic boundaries.

INVESTMENTS FOR GATHERING PALYNOLOGICAL DATA

The investments needed for successful collection of palynological data are substantial. One can do rapid light microscopic observations with relative speed and ease to obtain an idea of the potential of variation in the grains of the taxa under study, but to develop the data fully requires SEM and TEM approaches. These methods are costly (described in detail in chapters 15 and 16) involving ancillary equipment such as shadow-coaters, ultramicrotomes, etc., and one does not go this direction without serious commitment. The time investment is substantial as with all technically complex data-gathering methods, although rapid preliminary surveys can be done with the light microscope to good purpose. This also helps set the stage for subsequent EM work if it seems warranted.

EFFICACY OF PALYNOLOGICAL DATA IN THE TAXONOMIC HIERARCHY

Data from pollen grains are known to be useful at all levels of the taxonomic hierarchy. Of all the ultrastructural data obtained for taxonomic purposes, a general correlation can be noted with use of SEM for external features that have value mostly at the lower levels of the taxonomic hierarchy, and TEM for internal features useful at the higher levels (Stuessy 1979b). The exceptions are pollen grains which are useful at all levels of the hierarchy, due surely to the strong selection forces at work in different ways in dispersal, water-stress, pollination, germination, and stigmatic interactions. Some taxa are *stenopalynous,* i.e., with very little variation, such as in *Ambrosia* (Compositae; Payne and Skvarla 1970) in which no significant differences were found among 37 species of the genus. Others, however, are *eurypalynous,* i.e., containing much variation, such as is well known in the Acanthaceae (Erdtman 1952), Nyctaginaceae (Nowicke 1970), or at the generic level in *Cuphea* (Lythraceae; Graham and Graham 1971).

At the specific level pollen can often be helpful in suggesting relationships (e.g., in *Sonneratia,* Sonneratiaceae, Muller 1969, 1978; *Tournefortia,* Boraginaceae, Nowicke and Skvarla 1974; and *Myriophyllum,* Haloragaceae, Aiken 1978).

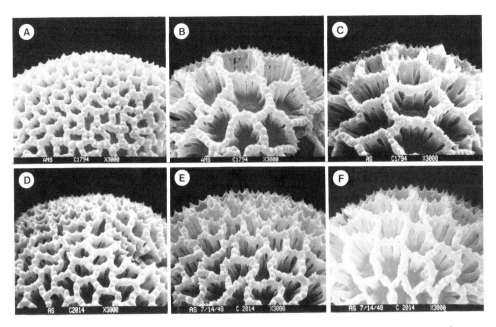

FIGURE 18.9 SEM of variations of the tectum in *Armeria maritima* var. *sibirica* (Plumbaginaceae). A–C, all from one inflorescence; D–F, all from another inflorescence. x2010. (From Nowicke and Skvarla 1979:666)

Variation within a species is also known, such as seen in *Leiphaimos* and *Voyria* of the Gentianaceae (Nilsson and Skvarla 1969). Somewhat startling infraspecific variation has been documented in *Armenia maritima* var. *sibirica* of the Plumbaginaceae (fig. 18.9; Nowicke and Skvarla 1979), even within the same inflorescence. Sometimes, however, the variation is "subtle and quantitative" among species as found in *Erythrina* (Leguminosae; Graham and Tomb 1974), although in this case there were some small features of value at the sectional level. In other instances the variation is minor when seen at some distance (fig. 18.10), but upon close examination helpful details can be discerned (fig. 18.11). And in some genera remarkable variations of great and obvious taxonomic value occur among species as in *Polygonum* (fig. 18.12).

Pollen data also have been used to good effect at the generic and subgeneric (or sectional) levels on numerous occasions, e.g., in subtribe Hyoseridinae (Compositae; Blackmore 1981), with subsections in *Vernonia* (Compositae; Jones 1979), among genera of tribe Lactuceae (Compositae; Tomb 1975), within subtribe Stephanomeriinae of the same tribe (Tomb, Larson, and Skvarla 1974), and in Polemoniaceae (Taylor and Levin 1975). Two good examples of the referral of genera to other positions within families come from the Compositae. *Blenno-*

FIGURE 18.10 Pollen grains of species of *Virola* (Myristicaceae). A, *V. minutiflora;* B, *V. weberbaueri;* C, *V. malmei;* D, *V. surinamensis;* E, *V. calophylloidea;* F, *V. multinervia.* x2000–3400. (From Walker and Walker 1979:740)

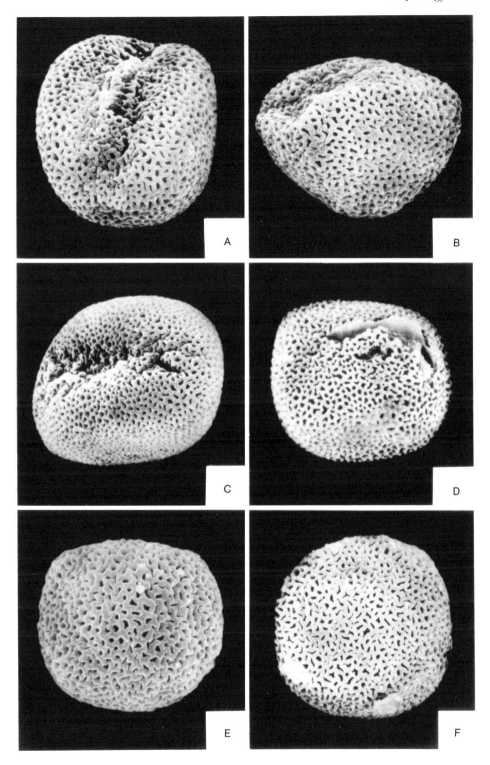

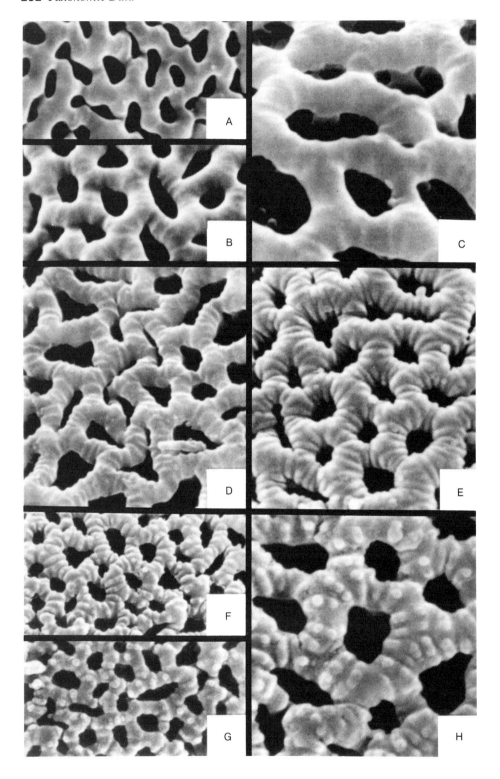

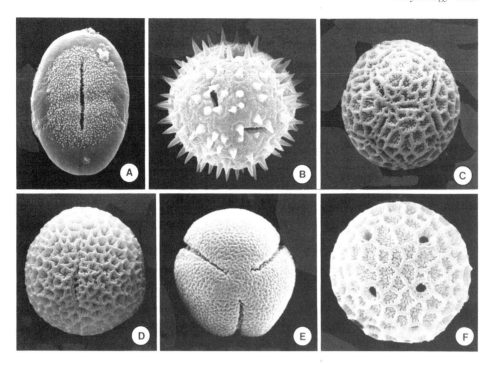

FIGURE 18.12 Pollen grains of species of *Polygonum* (Polygonaceae). A, *P. convolvulus;* B, *P. forrestii;* C, *P. amphibium;* D, *P. glaciale;* E, *P. cilinode;* F, *P. orientale.* x1160–2370. (From Nowicke and Skvarla 1979:674)

sperma was placed traditionally in the tribe Helenieae but it clearly had morphological ties with *Crocidium* of the Senecioneae. Pollen data showed the close relationship of these two genera, as well as the ill fit of the former in the Helenieae, and provided the critical information for a transfer of the former to the Senecioneae (Skvarla and Turner 1966b). Subsequently, flavonoid data corroborated this placement (Ornduff, Saleh, and Bohm 1973). Marticorena and Parra (1974) using light microscopic data showed how three genera of the tribe Mutisieae of the Compositae belonged in two other tribes: *Anisochaeta* and *Feddea* to the Inuleae, and *Chionopappus* to the Liabeae. Here again, uniformity can prevail within particular groups of genera such as in the Casuarinaceae in which neither species nor "genera" were distinct among 34 species examined (Kershaw 1970). Variation is slight within groups of species of *Virola* (Myristicaceae; fig. 18.10), but more variation is detected at higher levels of magnification (fig. 18.11; Walker and Walker 1979).

Among families and higher taxa, pollen data have been most helpful in suggesting relationships. A good review with many examples is given by Nowicke

FIGURE 18.11 Exine surface features of species of *Virola* (Myristicaceae). A, *V. glaziovii;* B, *V. malmei;* C, *V. carinata;* D, *V. multinervia;* E, *V. calophylloidea;* F, H, *V. weberbaueri;* G, *V.pavonis.* x12000–24000. (From Walker and Walker 1979:743)

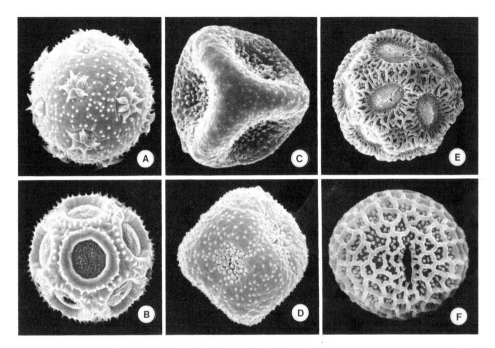

FIGURE 18.13 Pollen grains of selected families of the Centrospermae. Amarantha-
ceae, A, *Psilotrichum amplum;* Caryophyllaceae, B, *Siphonychia americana,* C,
Herniaria glabra, D, *Cardionema ramosissima;* Cactaceae, E, *Opuntia lindheimeri;*
Nyctaginaceae, F, *Abronia angustifolia.* x840–6000. (From Nowicke and Skvarla 1977:25)

and Skvarla (1979). Between some families striking differences prevail such as
in the order Centrospermae (fig. 18.13). In this group the pollen of the betalain-
containing families (see chapter 21) show a stronger similarity to each other
than any one does to the non-betalain containing Caryophyllaceae, which correlates with the chemical and morphological evidence. Pollen have been helpful in
showing relationships of the Austrobaileyaceae between the Laurales and Magnoliales (Endress and Honegger 1980). Other pollen data also have been helpful
in improving the tribal classification of the Caesalpinioideae (Leguminosae;
Graham and Barker 1981, showing relationships among subtribes of the Inuleae
(Compositae; Leins 1971), and in conjunction with types of viscin threads helping delimit tribes of the Onagraceae (Skvarla et al. 1978). The delimitation of the
order Myrtales has also been aided by pollen information (Patel, Skvarla, and
Raven 1984).

SPECIAL CONCERNS WITH PALYNOLOGICAL DATA

Because pollen grains have such important roles biologically and because the
walls are so resistant during fossilizaiton, a few related topics should be mentioned briefly. One area of active research is in the development of pollen (e.g.,
Heslop-Harrison 1971; Rowley 1981; Dickinson 1982; El-Ghazaly 1982; Rowley

and Skvarla 1987; Skvarla and Rowley 1987), and an excellent review has been provided by Knox (1984). Specific studies have focused on wall development (e.g., in Austrobaileyaceae; Zavada 1984b), synchronous vs. asynchronous formation of tetrads (in Winteraceae; Sampson 1981), pore development (in *Helianthus*, Compositae; Horner and Pearson 1978), spatial orientation in the anther (in *Sorghum*, Gramineae; Christensen and Horner 1974), and germination and early tube development (in *Lycopersicon*, Solanaceae; Cresti et al. 1977). Again, these events may be considered as embryological data, but they are worth mentioning here also.

Work is also continuing on the chemical nature of sporopollenin, "the most resistant organic material known" (Brooks and Shaw 1978:91; see also Shaw 1971). The economic potential of understanding this extremely durable substance (which resists boiling acids) is substantial and merits strong investigation. Sporopollenin is apparently a complex carotenoid polymer that is soluble in fused potassium hydroxide (Southworth 1974). Despite its resistant qualities, the pollen grains literally collapse within seconds after landing on a receptive stigma due presumably to the action of enzymes from the papillae interacting with those of the exine itself (Heslop-Harrison 1976, 1979).

Such pollen-stigma interactions are also of interest in pollination and reproductive biology. The nature of the "recognition" of the grain to the receptive surface is important (Vithanage and Knox 1977) especially at the molecular level (Clarke and Gleeson 1981). Some pollen grains are held together by viscin threads probably for pollination efficiency (Cruden and Jensen 1979). The relationships of pollen grains to their roles in breeding and compatability systems (Cresti, Ciampolini, and Sarfatti 1980; Williams, Knox, and Rouse 1981) are also important.

The adaptive significance of the pollen exine has attracted considerable attention due to the desire to explain in evolutionary terms the bewildering variations that exist among all the flowering plants. As Blackmore (1982c) has pointed out, these variations must be due in large measure to the following not necessarily mutually exclusive points: (1) accomodation of volume changes due to water stress (harmomegathy); (2) pollination biology; and (3) exine-held substances. Payne has emphasized that: "It is hypothesized that evolution of structural features that meet harmomegathal stress requirements provides the principal explanation for pollen wall form, composition, organization, and architecture" (1981:39). This seems too much stress on this one selective regime, but it is no doubt important in pollen grain evolution. That pollinators have also affected exine adaptations has been suggested by Ferguson and Skvarla (1982). Other perspectives on form and function can be found in Bolick (1978, 1981), Muller (1979), and Melville (1981).

Because exines are so resistant, they persist through millions of years of geological time and are useful in paleoecology, archeology, and the unraveling of angiosperm origins and phylogeny. In paleoecology they have been used for nearly fifty years to help reconstruct paleoclimate (e.g., cf. figs. 23.9 and 23.10). By assuming that extinct relatives of extant species grew in similar habitats to those living today, it is possible to reconstruct past climates by determining the presence and quantity of indicator species. Concerns here must deal with dispersal distances (Andersen 1974a, b), especially in relation to moisture in the air (Solomon and Hayes 1972), sampling (Funkhouser 1969), and the necessity of making still other assumptions before fossil communities can be reconstructed (Janssen 1970). Computer techniques have been developed to deal with the numerous data available (e.g., Gordon and Birks 1972) in seeking fossil pollen zones. Pollen likewise can be helpful in archeology (Gray and Smith 1962) as nicely shown in the interpretation of the events surrounding a Neanderthal burial in northern Iraq (Leroi-Gourhan 1975).

The preservation of exines in fossil sediments makes pollen grains ideal for helping understand the origin and early evolution of the angiosperms. This seemingly intractable problem has not yielded final understanding despite the considerable efforts being directed toward this end (for an overview, see symposium introduced by Dilcher and Crepet 1984). The fossil pollen record for angiosperms has been summarized well by Muller (1981, 1984), and other perspectives on the origin of the group are found in Walker and Walker (1984) and Zavada (1984a). The papers by Walker (1974a, b, 1976a, b) and Walker and Doyle (1975) deal with the evolution of different pollen grain features and give a good idea of the kind of interpretations possible with these data. Sporne (1972) shows how low numbers of apertures correlate with other primitive features in the angiosperms. Stern (1970) examines aperture evolution among species of *Dicentra* (Fumairaceae) by superimposing drawings of pollen grains on a Wagner Groundplan/divergence cladistic analysis of the same taxa.

Because palynology is so dependent upon technique, a few words about new developments seem appropriate. Innovations for more versatile use of SEM provide more data for those workers not skilled in TEM or without convenient access to such equipment. Fracturing or sectioning grains for SEM can often help substitute for lack of TEM (Hideux 1972; Hideux and Marceau 1972). Freeze-drying and critical-point drying can be used with SEM to keep grains from collapsing under vacuum, as shown successfully by Pacque (1980) in using the latter for observations on pollen of Passifloraceae. Newer TEM approaches involve freezing microtomy (= cryomicrotomy) of pollen (Muller 1973) and sectioning in gelatin (Blackmore 1981, 1982a; Blackmore and Dickinson 1981). Ferguson (1978) has suggested the technique of combining SEM and TEM observations by embedding pollen in resin with microsectioning and observation in TEM, followed by mounting on SEM stubs and partly dissolving away the resin containing the sections with sodium methoxide, and finally shadow-coating and examining with SEM. In this way the same sections can be examined with both TEM and SEM. Electron microscopes that allow both TEM and SEM to be done with the same machine (STEM or TSEM) also help provide this capability. Other newer approaches include "thick" section deplasticization, "coverslip" acetolysis (Skvarla, Rowley, and Chissoe 1988), ion-beam etching of the exine (Barthlott, Ehler, and Schill 1976; Blackmore and Claugher 1984), and plasma ashing (Nowicke, Bittner, and Skvarla 1986; Claugher and Rowley 1987).

Other miscellaneous topics worth mentioning include the use of pollen features in detecting hybridization and introgression, and nomenclature. Pollen is often aborted in hybrids, as shown in *Erythrina* (Leguminosae; Graham and Tomb 1974), and different stains can be used to test this viability (Hauser and

Morrison 1964) in addition to relying upon distorted appearance. Patterns of introgression have also been detected palynologically in *Salvia* (Labiatae; Emboden 1969). For nomenclatural details, consult Schopf (1969). This is complicated, because there are many isolated fossil palynomorphs which are treated as taxa and must go under the rules for fossil plants, as given in the *International Code of Botanical Nomenclature* (Greuter et al. 1988).

Despite the impressive power of palynology in solving taxonomic problems, certain cautions must be kept in mind. The chemical processing of the pollen grains can alter their size (Reitsma 1969) and different treatments bring about different size alterations. Nutritional factors can also affect size of grains (Bell 1959), which is important when trying to correlate ploidy level with grain size. A positive correlation has been seen in many taxa with the higher polyploids having larger grains and often more pores (e.g., in *Carya*, Juglandaceae, Stone 1961, 1963; *Bouteloua*, Gramineae, Kapadia and Gould 1964; *Campanula* sect. *Heterophylla*, Campanulaceae, Geslot and Medus 1971). The proper identification of the voucher specimens is another source of error, as shown by the example of *Parthenice mollis* (Compositae; Bolick and Skvarla 1976). Perhaps the most difficult problem with use of palynological data is the numerous parallelisms in grain features (Kuprianova 1969). This is not surprising, considering the strong selection pressures on grains in every population, but it does lead to problems of interpretation. Gentry and Tomb (1979) show that areolate pollen occur in unrelated groups in two tribes of the Bignoniaceae. Rogers and Xavier (1972) also showed parallelisms in *Linum* (Linaceae), and Small et al. (1971) also indicated them in the well-known genus *Clarkia* (Onagraceae), for which a wealth of biosystematic data is available and for which a convincing phylogeny already has been constructed. This is an especially telling example.

With these cautions in mind, we can conclude that palynology can contribute useful types of data for taxonomic purposes at all levels of the hierarchy. The major contributions have tended to be at the generic level and above and in angiosperm origins and phylogeny. In this latter context the contributions have been impressive, and were it not for pollen grains, we would have more difficulty in continuing to accept the Besseyan principles that suggest the Magnolialean complex as the most primitive of all the flowering plants.

CHAPTER 19

Cytology

Although cytology in a broad sense deals with all aspects of cells, in practice in taxonomic work the focus has been on "chromosomes and their various attributes" (Lewis 1969:523). Other features of cells, e.g., sieve-tube plastid variation, in this book have been treated in chapter 16 as ultrastructural data. The chromosomes (fig. 19.1) play a special role as a source of comparative data in taxonomy, because these structures contain the genetic material which is responsible for maintaining reproductive barriers and the integrity of species and other taxa. As Lewis has stressed: "Chromosomes derive their prominence as a tool in taxonomy from their direct relation to the genetic system of which they are an integral part" (1957:42). Despite their importance, there is often broad variation in chromosome numbers within taxa and, as with other data, the limits of this variation need to be understood clearly before sound taxonomic decisions can be made.

HISTORY OF CYTOLOGY IN PLANT TAXONOMY

The history of cytology in a broad sense goes back to the discovery of cells by Robert Hooke in 1665. This was followed by the formation of the cell theory in plants by M. J. Schleiden (1804–1881) in 1838. More significant for our interests, however, was the detection of chromosomes and the discovery by W. Flemming (1843–1915) in 1882 that they split, with the longitudinal halves going to the daughter cells during division. They were subsequently named by Waldeyer (1836–1921) in 1888. These and other events concluded the "descriptive period" of cytology (Swanson 1957) and opened the door for the "experimental period" with its focus on integration with embryology, genetics and evolution. The rediscovery of Mendelism by A. de Vries (1900a, b), E. Tschermak (1900) and C. Correns (1900) at the turn of the century set the stage for developments in cytogenetics, which will be discussed in chapter 20. Plant cytology, with a direct bearing on taxonomic issues, gained momentum with the work by East (1928)

FIGURE 19.1 Photomicrographs of chromosomes of species of Commelinaceae in mitotic metaphase. A, *Commelina* sp., $2n = 90$; B, *Callisia fragrans*, $2n = 12$ (note tandem satellite); C, *Phyodina navicularis*, $2n = 32$. (From Jones and Jopling 1972, plate 1)

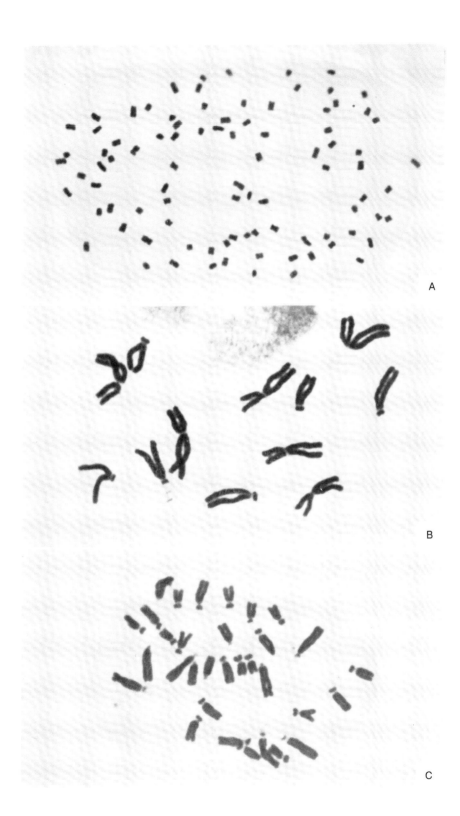

A

B

C

289

on *Nicotiana* (Solanaceae), by Cleland (1923, 1936) on *Oenothera* (Onagraceae), and by the long series of papers on *Crepis* (Compositae) by Babcock and numerous collaborators (e.g., Hollingshead and Babcock 1930; Babcock and Cameron 1934; Babcock, Stebbins, and Jenkins 1937; Babcock and Jenkins 1943). This last set of papers, in particular, was most influential in showing the positive value of cytological data in plant systematics.

GENERAL CYTOLOGICAL TEXTS AND REFERENCES

From these beginnings in plant cytology have come explosions of research on many plant groups too numerous to mention. Today exist many excellent books that will introduce the student to different cytological topics. Some of the general texts are Sharp (1943), Gresson (1948), Swanson (1957), Cohn (1969), and De Robertis and De Robertis (1987). Special topics include chromosome structure (Bostock and Sumner 1978); genome organization (Nagl, Hamleben, and Ehrendorfer 1979; Davies and Hopwood 1980; Leaver, 1980; Dover and Flavell 1982; Grierson and Covey 1984); meiosis (Sybenga 1975); polyploidy (Lewis 1980b); chromosomal evolution in plants (Stebbins 1971b; see also the review paper by Jones 1978); cytology of cultivated plants (Darlington 1973); B-chromosomes (Jones and Rees 1982); techniques for making chromosomal preparations (Haskell and Wills 1968; Darlington and La Cour 1975; Löve and Löve 1975b; Dyer 1979; Sharma and Sharma 1980); and chromosomal identification (Caspersson and Zech 1973). Two excellent recent symposium volumes are Jones and Brandham (1976) and Brandham and Bennett (1983). A good glossary of terms is found in Smith (1974). General summaries of the import of cytology in plant taxonomy, from which many of the perspectives and references in this chapter were derived, are in Davis and Heywood (1963:193–213), Meyer (1964), Lewis (1969), Moore (1978), Grant (1984), Greilhuber (1984), and Jones (1984).

   As the observational data on plant chromosome numbers have increased over the decades, the need has arisen for indices to this information so that it will be more accessible for purposes of helping make taxonomic decisions. A good starting point is the index of Darlington and Wylie (1955), followed by the yearly coverage provided by Cave (1958–1965), Ornduff (1967–1969), Moore (1970–1977), and Goldblatt (1981, 1984, 1985, 1988). The index of Fedorov (1969) is also important because of coverage of the Russian literature not found in the other indices. New isolated chromosomal reports are appearing in the literature constantly, but one place to watch is the International Organization of Plant Biosystematists (IOPB) chromosome number reports section of the journal *Taxon*. Chromosome number observations without much text are also sometimes published in *Sida* and the *Annals of the Missouri Botanical Garden*.

GENOMIC ORGANIZATION

To appreciate the significance of chromosomes in taxonomy, it is necessary to have some understanding of recent ideas of genomic organization. The main focus here will be on the structure of the chromosomes. Most of these data are taken from Flavell, O'Dell, and Thompson (1983), Haapala (1983), Hutchinson (1983), Schweizer (1983), Sharma (1983), Sumner (1983), Risley (1986), and Lewin (1987).

The DNA of eukaryotic organisms is much too long to fit into chromosomes unless it can be packaged tightly. The condensation of the nucleic acids depends upon the presence of basic proteins with which it binds. The DNA exists as double-helix strands closely associated with the protein into fine strands called *chromatin.* Less densely packed regions in the cell are the *euchromatin* whereas densely packed regions are called *heterochromatin.* The latter is transcriptionally inactive. Regions that are never expressed are called *constitutive heterochromatin,* such as in satellite DNA sequences, and *facultative heterochromatin* exists as entire chromosomes that are inactive. A chromosome can be viewed, therefore, as a single long DNA duplex, but the DNA is packaged in several stages of organization. First, approximately 200 base pairs of DNA are wound onto an octamer of small, basic proteins (histones) into a beadlike structure called a *nucleosome,* with the DNA on the outside and the proteins on the inside. Second, these beads are connected by single DNA strands and are coiled in groups into an helical array to form a fiber that corresponds to interphase chromatin. The third and final packing is with the chromatin strands, which is especially noticeable during condensation into chromosomes during mitotic and meiotic divisions. The important point about chromosomes, therefore, is that they represent a stage of packing of the genetic material rather than a direct reflection of it. This will be important to keep in mind as we consider the question of variation in chromosome number and its taxonomic significance.

TYPES OF CYTOLOGICAL DATA

Different types of chromosomal data have been used taxonomically, including number, size and shape, behavior in meiosis, and DNA content. Because chromosomes are only visible with the light microscope during division of the nucleus, the sources in the plant for chromosomal data of value to the taxonomist are most commonly mitosis and meiosis. These processes take place and can be observed with proper techniques in two regions of the plant: (1) the shoot, root, and lateral meristems (mitotic divisions); and (2) the sporogenous tissue. In practice, most mitotic data come from root tips, and almost all meiotic observations are made of microsporogenesis in young anthers (meiosis in the megasporocyte can be examined also, but it is difficult with few chances for seeing the desired chromosomal configurations).

Chromosome Number

Chromosome number has been used most often in taxonomic work due to ease of observation and its discrete nature (i.e., it can be used conveniently as a character along with other features). That chromosome numbers have varied during evolution of the angiosperms is obvious from the wide numbers ranging from $n=2$ in *Haplopappus gracilis* (Jackson 1957) and *Brachyscome* [= *Brachycome*] *lineariloba* (Smith-White 1968; Watanabe and Smith-White 1987), both of the Compositae, and *Colpodium versicolor* of the Gramineae (Sokolovskaya and Probatova 1977), up to $n=$ca. 250 in *Kalanchoe* sp. (Baldwin 1938) of the Crassulaceae. Variation in numbers has been especially characteristic of the herbaceous dicots with woody dicots showing much less diversity. As a further comparison, the conifers and cycads have even less variation (Levin and Wilson 1976; see also an example from the Taxodiaceae, Schlarbaum and Tsuchiya 1984), but the ferns have broad ranges (Stebbins 1966) with the highest numbers known in plants, viz. $n=620$ in *Ophioglossum reticulatum* (Löve and Kapoor 1967b). Some families have remarkable variation, such as the Compositae, with $n=2$ mentioned above and also $n=110$–120 in *Montanoa guatemalensis* (Funk and Raven 1980), and others tend to be more uniform, such as the Phytolaccaceae with only $n=9$ or its multiple (Raven 1975). Many genera have completely uniform chromosome numbers, such as *Simsia* of the Compositae (all $n=17$; D. M. Spooner, unpubl.), whereas others have extreme ranges of variation such as *Stylidium* (Stylidiaceae; James 1979) with $n=5, 6, 7, 8, 9, 10, 11, 12, 13, 14, 15,$ 16, 26, 28, and 30, and *Melampodium* (Compositae; Stuessy 1970, 1971a) with $n=9, 10, 11, 12, 18, 20, 23, 25\pm1, 27, 30, 33$. Within species the variation in number tends to be less, but one can cite *Chaenactis douglasii* (Compositae; Mooring 1965) as an extreme example of variation ($2n=12, 13, 14, 15, 18, 24, 25,$ 26, 27, 28, 36, and ca. 38) as well as the bizarre diversity in *Claytonia virginica* (Portulacaceae; with $2n=12, 14, 16, 17, 18, 19, 20, 22, 24, 26, 28, 30, 31, 32, 34,$ 36, 41, 48, and 72 (W. Lewis 1962, 1970a, b; W. Lewis and Semple 1977). Many other species are completely cytologically uniform even with ecological diversity, as in *Danthonia sericea* (Gramineae; Fairbrothers and Quinn 1970). Another problem, however, is that chromosome number varies within an individual plant in different tissues (D'Amato 1984), although this in practice does not cause much difficulty because almost all material for taxonomic and evolutionary studies comes from cytologically conservative root-tip or flower-bud regions. In the face of such variation at the infraspecific level, it may seem overwhelming to even consider use of these types of data for purposes of classification. Fortunately, such broad variation in numbers is rare. A humorous perspective from the taxonomic standpoint is expressed by H. Lewis: "I would argue that if an organism does not take its chromosome number seriously, there is no reason why the systematist should" (1969:526).

The utility of chromosome numbers in some families has spawned collaborative projects to help accumulate these data. Notable efforts have been in the Umbelliferae (Bell and Constance 1957, 1960, 1966; Constance, Chuang, and Bell 1971, 1976; Crawford and Hartman 1972; Constance, Chuang, and Bye 1976; and Constance and Chuang 1982), Hydrophyllaceae (Cave and Constance 1942, 1944, 1947, 1950, 1959; Constance 1963) and Compositae. In this last family there have been two large series, one by B. L. Turner and collaborators (e.g., Turner and Ellison 1960; Turner and Irwin 1960; Turner and Johnston 1961; Turner, Beaman, and Rock 1961; Turner, Ellison, and King 1961; Turner, Powell, and King 1962; Powell and Turner 1963; Turner and King 1964; Turner and Lewis 1965; Turner and Flyr 1966; Turner, Powell, and Cuatrecasas 1967; Turner 1970a;

Turner, Powell, and Watson 1973; Turner et al. 1979), and the other by P. H. Raven, O. T. Solbrig, and associates (e.g., Raven et al. 1960; Raven and Kyhos 1961; Ornduff et al. 1963, 1967; Payne, Raven and Kyhos 1964; Solbrig et al. 1964, 1969, 1972; Anderson et al. 1974; Powell, Kyhos, and Raven 1974; King et al. 1976), as well as many smaller projects (e.g., Keil and Stuessy 1975, 1977; Jansen and Stuessy 1980; Jansen et al. 1984).

Further interest in chromosome numbers has led to numerous cytotaxonomic studies over the past three decades on particular groups, often at the generic level. Literally thousands of papers exist, but only a few will be cited here to provide a sample of this type of approach. For the monocots a good series of studies on genera of the Commelinaceae has been very fruitful (e.g., W. Lewis 1964; Jones and Jopling 1972; Faden and Suda 1980; Jones, Kenton, and Hunt 1981) and has revealed taxonomically invaluable data that have brought into question the existing classification of the family based primarily on morphological data. A good study of the cytology of the morphologically cryptic Lemnaceae (Urbanska-Worytkiewicz 1980) is also worth noting. The excellent studies on the Agavaceae, with their variation in size and shape of chromosomes should be mentioned (Cave 1964; Gomez-Pompa, Villalobos-Pietrini, and Chimal 1971), as well as efforts in the Araceae (*Cryptocoryne;* Jacobsen 1977; Arends, Bastmiejer, and Jacobsen 1982), and *Eleocharis* of the Cyperaceae (Harms 1968). A few studies in the dicots are helpful to cite also, such as Malvaceae (Bates and Blanchard 1970), the "chromosome atlas" of *Lotus* (Leguminosae; Grant 1965) and *Rumex* (Polygonaceae; Löve and Kapoor 1967a), cacti of western North America (Pinkava and McLeod 1971; Pinkava et al. 1973), *Centaurium* (Gentianaceae; Broome 1978), and the Caribbean species of *Lantana* (Verbenaceae; Sanders 1987).

Aneuploidy. The documented kinds of chromosome number variation within taxa are of different types. Most of the variation encountered deals with the gain or loss of regular chromosomes either by changes of individual chromosomes *(aneuploidy)* or by duplication of complete sets (*euploidy;* also sometimes called *polyploidy* in a general sense). For proper taxonomic evaluation of these euploid and aneuploid patterns of chromosomal variation, some discussion on their evolutionary origin is required. As is the case with use of all types of comparative data, the more the taxonomist knows about the biological basis of the characters being used, the better will be the likelihood of sound judgments for classification. This is especially true with cytological data.

Aneuploid chromosomal changes, i.e., the gain or loss of one or more regular chromosomes, is less problematical taxonomically than with euploid alterations. Frequently a species has one chromosome number throughout its range, and sometimes this number differs by one from that of its closest known relative. The difference is obviously a useful taxonomic character and may suggest a chromosomal basis for reproductive isolation of the two taxa, especially if they are sympatric. Such a close cytological relationship also can be suggestive of an evolutionary directionality between the two taxa, and the cytogenetic evidence accumulated so far in most groups shows the loss of a single chromosome to be a more common event than a single gain. (More about this in chapter 20.) In

FIGURE 19.2 Variation of diploid chromosome numbers in populations throughout the range (enclosed area) of *Claytonia virginica* (Portulacaceae). (Redrawn from Lewis, Oliver, and Suda 1967:154)

some groups, however, widespread aneuploid changes occur, and the most spectacular example is in *Claytonia virginica* (Portulacaceae; Lewis 1962, 1967, 1970a, b; Lewis, Suda, and MacBryde 1967). Here in the single species, variation occurs on a broad geographic scale throughout its range (fig. 19.2), on a more local area around St. Louis (fig. 19.3), and even within single populations. These observations in *Claytonia* have led W. Lewis to conclude:

> Clearly, I cannot support the dogma of one chromosome number: one taxon! We know that for *C. virginica* an individual may have one chromo-

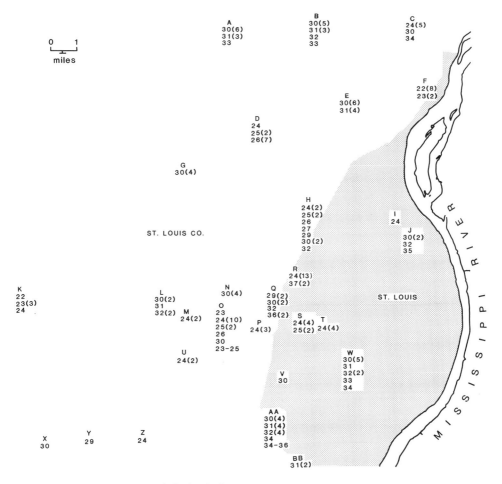

FIGURE 19.3 Variation of diploid chromosome numbers in populations (A–BB) of *Claytonia virginica* (Portulacaceae) in the St. Louis, Missouri, area. (Redrawn from Lewis, Suda, and MacBryde 1967:150)

some number in cells above ground and still another in those below ground; that individuals may have certain chromosome numbers at one topography, yet others at another topography; that individuals flowering during the early part of the season may average lower in chromosome number than those flowering later; that highest polyploids, here 8x, may be found late in the flowering season, but only lower polyploids early in the season; and finally that individuals may vary markedly in chromosome number from one season to another—in this case, downward, illustrating that aneuploidy may be increased or decreased depending on the circumstances. These data stress the absolute necessity of obtaining chromosomal information from numerous plants under different conditions before the characteristic number of a taxon is clear. (1970:181)

Fortunately for taxonomic purposes, this type of rampant aneuploidy (and euploidy) is rare, but the example does point toward the need for understanding the existence and biological basis of aneuploid changes before taxonomic decisions can be made effectively.

Euploidy. Euploid chromosomal relationships are even more important to understand biologically before reaching taxonomic conclusions (Jackson 1976). Here taxa, or populations or individuals, differ in their number of whole chromosome sets, such as diploid vs. tetraploid conditions. The documentation of the cytological patterns and determining what they suggest taxonomically demand understanding their evolutionary origins. There are two main ways in which polyploids arise from diploids. (The reverse process, or polyhaploidy, is known in higher plants, but it is uncommon; for some examples see De Wet 1968; Anderson 1972; Kasha 1974; Savidan and Pernès 1982). The first is the failure of cell division after the chromosomes have reproduced themselves at mitosis in roots or shoots. The cell which results can proliferate itself giving rise to tetraploid tissues or organs. The second is a failure of the reduction process during meiosis resulting in gametes with the diploid chromosome number. If these participate in the formation of zygotes, polyploidy will result and will be tetraploidy when two unreduced gametes fuse together. Mitotic origins will give rise to tetraploids in which the chromosome sets exactly replicate those of the plants in which they arose. Fusion of unreduced gametes produces tetraploids in which the sets are homologous but not necessarily identical because of crossing over and the randomness of gametic fusion. Despite this difference, tetraploids from both types of origin are known as *autopolyploids*. These are distinct from *allopolyploids*, which arise from diploid hybrids whose parents have substantial differences in the genetic content and/or the stuctural organization of their sets. The resulting polyploid progeny will have incorporated into the genome the genetic material of both parents. Fertility may often be restored because the doubled chromosomes from each parent can now pair with themselves in the new hybrid cells.

Such offspring of either autopolyploid or allopolyploid origin are reproductively isolated to a large degree from their diploid progenitors and hence raise the possibility that they are new biological, and good taxonomic, species. The problem, however, is that although polyploids are reproductively isolated in large measure from the diploids (some backcrossing in autoploids via a $4n$ bridge to the diploids cannot be excluded, e.g., Zohary and Nur 1959; Carroll and Borrill 1965; Marks 1966; Ladizinsky and Zohary 1968; Wagenaar 1968), in the case of the autoploids they are extremely similar genetically to the parents. With the new alloploids, however, they are genetically quite different from either diploid parent. From a taxonomic perspective, therefore, one is compelled to treat the new alloploid as a new and distinct species (e.g., in *Tragopogon*, Ownbey 1950, Roose and Gottlieb 1976; and in *Picradeniopsis*, Stuessy, Irving, and Ellison 1973; both Compositae), but the autoploid is best regarded as nothing more than a cytological form, or if it becomes more geographically widespread, as a cytological race (H. Lewis 1967; for examples, see in *Melampodium*, Compositae, Stuessy 1971b; in *Galax*, Diapensiaceae, Nesom 1983; and in *Tolmiea*, Saxifragaceae, Soltis 1984a). Some workers emphasize that new species can originate via autopolyploidy (e.g., Soltis and Rieseberg 1986), but, until morphological distinctions accrue, these are *taxonomically* best treated as cytological forms or races rather than as distinct species. Tetraploids can also originate between plants whose chromosome sets show partial structural homology (such as between varieties or subspecies), as reflected in their meiotic pairing (usually some multivalents), and these have been called *segmental allopolyploids* (Stebbins 1971b). Other categories also exist, such as *interracial autoploids* and *autoallopolyploids* (Grant 1981), all of which stress that there is a continuum of conditions of polyploids ranging from strict autopolyploidy to strict or "genomic" allopolyploidy.

Further complications occur in groups with excessive polyploidization and hybridization yielding evolutionary relationships of the most complex nature, and in which taxonomic disposition of the cytotypes is challenging, to say the least. These are called *polyploid pillar complexes*, as documented recently in *Antennaria* (Compositae; Bayer and Stebbins 1982; Bayer 1984) and *Acmella* (Compositae; Jansen 1985). Further difficulties arise with the addition of asexual (agamic) modes of reproduction to yield an unbelievably complex picture of relationships as seen in *Crataegus* (Rosaceae; Muniyamma and Phipps 1979) and other groups. Clearly the only way to make taxonomic sense of these patterns of variation is to study the group thoroughly and then base judgments primarily on genetic and phenetic grounds. There are no simple answers here—only years of diligent efforts provide helpful answers.

B-Chromosomes. Variation also occurs in the presence and number of *supernumerary chromosomes* or *B-chromosomes*. These largely consist of genetically inactive heterochromatin and tend to be inconstant in number, and smaller (Li and Jackson 1961). Although of lesser significance than regular chromosomes, they are common and have been documented for at least 250 angiosperm families (Stebbins 1971b; for specific studies see Tothill and Love 1964; Mendelson and Zohary 1972; Greilhuber and Ehrendorfer 1975). These supernumeraries can have an effect on quantitative characters which may result in their showing a nonrandom distribution in relation to ecological factors. Nonetheless, they are usually of limited taxonomic signficance, even more so because they often vary during the ontogeny of the plants containing them. In groups with small regular chromosomes, B-chromosomes may resemble the active ones and cause misinterpretations of the actual chromosome numbers (for example and discussion, see Soltis 1983). An excellent overview of the occurrence and biological significance of B-chromosomes is found in Jones and Rees (1982).

Base Numbers. To deal adequately with relationships between taxa having different stable chromosome numbers, it is necessary to make comparisons with *base numbers*. A base number is the lowest detectable haploid number within a group of related taxa. It may also be regarded in some circumstances as the lowest common denominator in a series. For example, if in a genus the haploid numbers are $n = 5$, 10, 15, and 20, $x = 5$ would be regarded as the base number. Even if $n = 5$ were not known within the genus, $x = 5$ would still be inferred as basic. The importance of base numbers relates to the existence of numerous polyploid taxa in the angiosperms; at least 70 percent of all species have had a polyploid origin (Goldblatt 1980; Lewis 1980a). Hence, the closest evolutionary relationships of angiospermous taxa relate more often up and down polyploid lines rather than between them. It is most important, therefore, that these euploid relationships be understood before assessing the meaning of chromosome numbers for purposes of classification. For example, in *Melampodium* (Compositae), consisting of 37 species, there are known haploid numbers of $n = $ 9, 10, 11, 12, 18, 20, 23, 25 ± 1, 27, 30, and 33 (Stuessy 1970, 1971a). These numbers relate to each other in polyploid lines of evolution to form the $x = 9$ group ($n = 9, 18$), $x = 10$ group ($n = 10, 20, 30$), $x = 11$ groups ($n = 11, 33$), and $x = 12$ group ($n = 12, 23, 25 \pm 1$). Morphological and limited cytogenetic data substantiate the recognition of these chromosomal lines as evolutionary units within the genus, and their treatment formally as taxonomic sections. When the species of a genus have clear euploid relationships, such as above, or in *Celosia* (Amaranthaceae; Grant 1961), the determination of a base number is straightforward. Sometimes errors of judgment can occur if the numbers are wide-ranging (Favarger 1978), and sometimes dubious reports have caused problems (e.g., in *Neptunia*, Leguminosae; Turner and Fearing 1960). It is more problematical, however, when aneuploid numbers must be related to a single base number for the group. In the *Melampodium* example above, $x = 10$ was determined to be the base number for the genus due to consideration of morphology, and size and ecological diversity of the $x = 10$ group. The related genera *Acanthospermum* and *Lecocarpus* are both $n = 11$, but it seems hard to believe that *Melampodium* was on this base because the single species with this number, *M. montanum*, occurs in high elevation forests in Mexico in zones different from the rest of the genus and very different from the ecological tolerances in the other two genera. (This is another example of misleading interpretations that can derive by sole reliance on out-group relationships for the understanding of evolutionary directionality; see chapter 8.) Similar problems with base number determination occurred in *Lotus* (Leguminosae; Grant and Sidhu 1967) with $n = 6$ and $n = 7$ species in the genus. Chemical data and degree of polyploidy off of each diploid number helped suggest that $x = 7$ was the base for the genus.

It is often useful to distinguish between the *immediate base number* of a group and its *ancestral base number*. This is obviously important when dealing with larger groups or those with long evolutionary histories, such as the primitive angiosperms (Ehrendorfer et al. 1968; Ehrendorfer 1976). For example, in a genus with haploid numbers of $n = 10, 20$, and 30, the base number of the genus is clearly $x = 10$. The ancestral base number, however, could well be $x = 5$. Only detailed cytogenetic studies and comparisons with other genera can help unravel these problems.

The importance of cytogenetics (in conjunction with isozyme electrophoresis) to help resolve base number difficulties is shown in the tribe Astereae of the Compositae in which common numbers are $n = 4, 5$ and 9. One view was that the base number was $x = 9$ followed by aneuploid reduction to $n = 5$ and $n = 4$ (Raven et al. 1960). The alternative view was that both $x = 4$ and $x = 5$ were base numbers for the tribe and the $n = 9$ taxa arose via allopolyploidy (Turner, Ellison, and King 1961). Detailed studies involving DNA content (Stucky and Jackson 1975) and enzymes (Gottlieb 1981b) suggested that the $n = 9$ taxa did not have alloploid origins (or at least there was no evidence for it) because they lacked the high DNA amounts and enzyme complementarities expected.

Karyotype

In addition to chromosome number, the size and shape of chromosomes, and the position of the centromere have been used taxonomically. (This latter feature is the bead-like body in the chromosome, often seen as a constriction, to which the fibers of the spindle apparatus appear to be attached.) These features of chromosome morphology as seen in mitotic metaphase are called the *karyotype* (fig. 19.1). Good reviews of this vast topic can be found in Moore (1968), Jackson (1971) and Bennett (1984). Although no uniform system of nomenclature exists for describing karyotypic variation (Stace 1980; Bennett 1984), some standardization has been attempted (e.g., Hamerton 1973), and there is enough similarity of approach to allow discussion here. The metaphase chromosomes seen in fig. 19.1 can be cut out (or redrawn) and arranged in sequence by size to yield comparative karyotypic data (fig. 19.4). To make the similarities and differences more obvious, it is sometimes useful to diagram the karyotypes, in what are called *idiograms* (fig. 19.5). These features can be further described by following

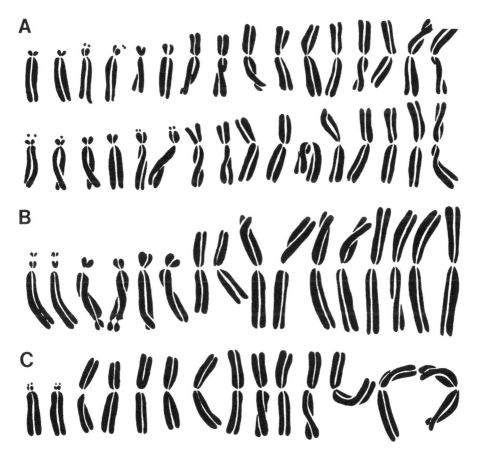

FIGURE 19.4 Drawings of metaphase chromosomes of *Anemone* (Ranunculaceae). A, *A. quinquefolia*, 2n=32; B, *A. rivularis*, 2n=16; C, *A. richardsonii*, 2n=14. x2500. (From Heimburger 1959:592)

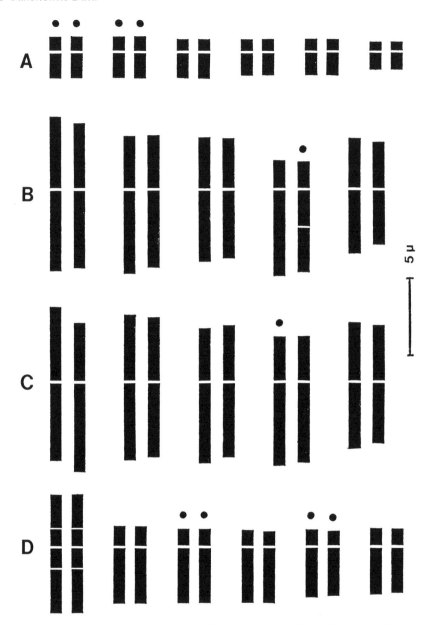

FIGURE 19.5 Chromosomal idiograms of *Claytonia* spp. (Portulacaceae). A, *C. perfoliata*, 2*n* = 12; B, *C. cordifolia*, 2*n* = 10; C, *C. sarmentosa*, 2*n* = 10; D, *C. virginica*, 2*n* = 12. (From Lewis and Suda 1968:67)

some convention involving the position of the centromere and the ratios of the arm lengths (fig. 19.6). Bentzer et al. (1971) emphasize the need for care in determining these ratios, and they point to errors of interpretation based upon different techniques of analysis (e.g., comparing drawings with photos, different

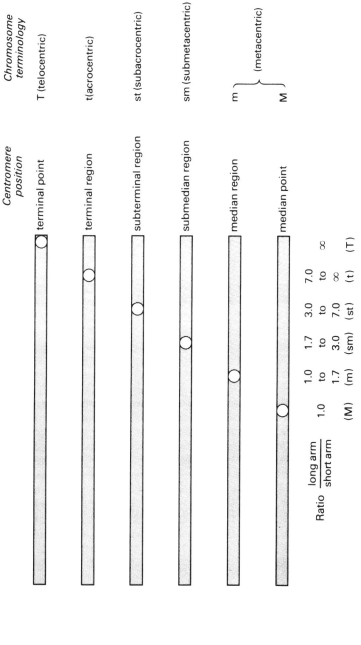

FIGURE 19.6 Suggested terminology for describing chromosome morphology based on position of centromere and ratios of arm lengths. (From Stace 1980:124)

people making measurements, etc.). As with chromosome numbers, variation in karyotypes can also occur within species, as shown in variations in satellited chromosomes in *Elymus striatulus* (Gramineae; Heneen and Runemark 1972). This again stresses the need for broad sampling before reaching final conclusions. Another important point emphasized by Bennett (1984) is that to simply order the chromosomes in a linear size sequence from large to small (as is typically done) is not based upon biological considerations of the number of possibly included haploid genomes if the taxon is obviously polyploid (usually judged above the $n = 13$ level if evidence does not suggest that it is lower). Hence the evolutionary relationships within the karyotype cannot be assessed properly. Recent studies by Bennett and colleagues (summarized in Bennett 1984) have shown that the spatial distribution of haploid genomes in mitosis is non-random and can give clues (with other evidence) to the subunits involved. The order of chromosomes that results is called the *natural karyotype*, and it is certainly worth seeking. A further point is the new relationship between the karyotype and DNA patterns that will make even more sophisticated judgments possible on the utility of karyotypic data for classification. Finally, some taxa have bold karyotypic patterns that are very helpful taxonomically, as in the striking size differences in chromosomes in *Agave* (Agavaceae; Cave 1964; Gómez-Pompa, Villalobos-Pietrini and Chimal 1971), whereas other taxa have chromosomes so small that this approach is not very productive (e.g., in *Carex*, Cyperaceae; Löve, Löve, and Raymond 1957; Kjellqvist and Löve 1963; or in the Crassulaceae; Uhl 1972).

Banding Patterns. More recently the information contained within karyotypes has been refined by the use of banding patterns on the chromosomes (see reviews by Fukuda 1984, and Greilhuber 1984). Although some early work was done with banding techniques (e.g., Darlington and La Cour 1938), the interest has developed more strongly in the past decade (e.g., Linde-Laurson and Bothmer 1986). For yet unknown reasons, different dyes (such as Giemsa) combine with heterochromatic regions of the chromosome forming dark bands (fig. 19.7). Another technique newly applied to plants is that of *in situ hybridization* of satellite DNA from one taxon to another using radioactively labelled DNA of one species and autoradiography (fig. 19.8). These give additional distinguishing markers to the chromosomes and allow more detailed comparisons between complements of different taxa. Caution in the uncritical use of these new data is stressed, however, especially in view of documented infraspecific patterns of variation (fig. 19.9; see also Serota and Smith 1967). Broad sampling is again recommended before drawing final conclusions.

The total DNA content of the genome varies incredibly from species to species even within the same genus. Further, there are significant differences of DNA content between chromosomes within the same genome. The observed variations of DNA are so far of limited taxonomic value (Rees 1984), because the variation is measured quantitatively and does not deal with sequences of base pairs. These types of studies can be helpful in detecting the polyploid nature of a taxon in consort with isozyme analysis (see chapter 20). For a few additional studies using DNA see Grant (1969), Bennett (1972), Greilhuber (1977), Roose and Gottlieb (1978), and Lawrence (1985).

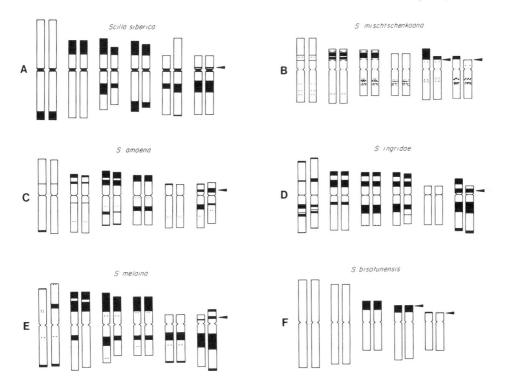

FIGURE 19.7 Giemsa C-banded karyograms in species of the *Scilla siberica*-alliance (A–E) and *S. bisotunensis*, a species of the *S. hohenackeri*-group s.1. (Liliaceae). C-value proportional presentation; arrows mark nucleolar organizing regions. (From Greilhuber 1984:164)

Chromosomal Behavior

Meiotic chromosomes are also useful for providing data of taxonomic import (fig. 19.10). In fact, much of the chromosomal data gathered in the course of routine taxonomic investigations is from meiotic material due to the ease of fixing of buds in the field in vials with appropriate solutions (normally 3 ethanol: 1 glacial acetic acid, or some variation thereon). Karyotypic information is difficult to assess with any accuracy here because of the different degrees of condensation of the chromatin during meiosis. Degrees of chiasma frequency and terminalization can be used to suggest relationships (e.g., Patil and Deodikar 1972), but the greatest power of meiosis comes from cytogenetic analysis of pairing relationships of genomes in hybrids. (This will be covered in detail in the next chapter.) Some chromosomes in meiosis are simply hard to observe, sometimes being "sticky," sometimes of vastly different sizes and shapes, and sometimes staining poorly (e.g., in *Polygala*, Polygalaceae; Lewis and Davis 1962). Once again, infraspecific variation in meiotic configurations is known (fig. 19.11), and this can cause problems in interpretation.

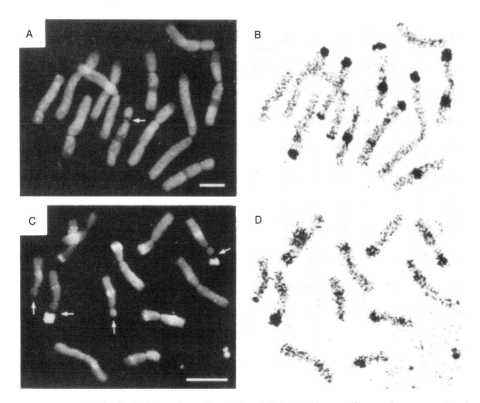

FIGURE 19.8 *In situ* hybridization of satellite-DNA-cRNA onto fluorochrome stained chromosomes. A, B, *Scilla siberica* "Spring Beauty" (A, quinacrine staining); C, D, *S. mischtschenkoana* (C, DAPI staining; Liliaceae). Arrows mark nucleolar-organizing regions; bar represents 10 μm. (After Deumling and Greilhuber 1982; from Greilhuber 1984:168)

INVESTMENTS FOR GATHERING CYTOLOGICAL DATA

The equipment needed for cytotaxonomic investigation is basically a good compound microscope, preferably with phase-contrast and with camera lucida and/or photographic attachments, a freezer-refrigerator for storage of fixed materials, a warming tray, and miscellaneous other glassware, slides, cover slips, stains, etc. Apparatus for fluorescence microscopy is more elaborate.

The techniques for cytological work can be simple or complex depending upon the type of approach. Most data now are obtained by squash preparation of microsporocytes or of root tips, which has greatly accelerated the accumulation of data. The cells are stained, separated from each other, and squashed so that the chromosomes are visible and spread out. Cell wall digestion by enzymes is especially useful in mitotic preparations. Much trial and error is needed to obtain good results. Good descriptions of techniques may be found in Darlington and La Cour (1975), Dyer (1979), and Sharma and Sharma (1980). Most all material must be collected and fixed in the fresh condition, although one case

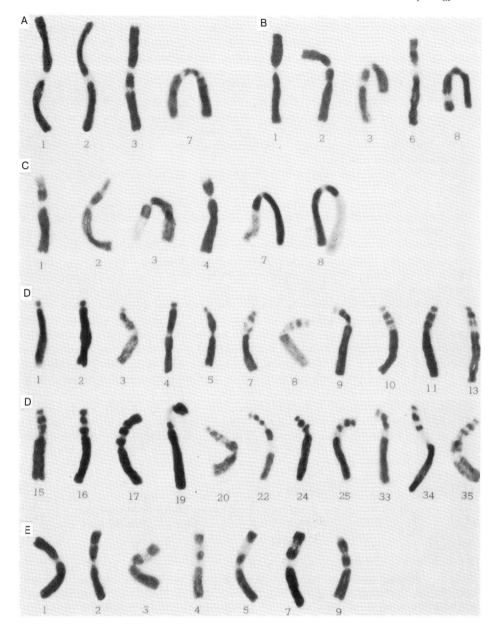

FIGURE 19.9 Photomicrographs of representative cold-induced banding patterns in the five chromosomes (A to E) in populations of *Trillium grandiflorum* (Liliaceae); heterochromatic segments are shown as clear bands. Chromosome D shows the most conspicuous variation. (From Fukada and Grant 1980:85)

exists (probably the only one) in *Impatiens* (Balsaminaceae) in which chromosome counts have been obtained from the generative nucleus of mature pollen grains on herbarium sheets. The chromosomes apparently are arrested in mitotic metaphase and stay that way for an indefinite period (Gill and Chinnappa

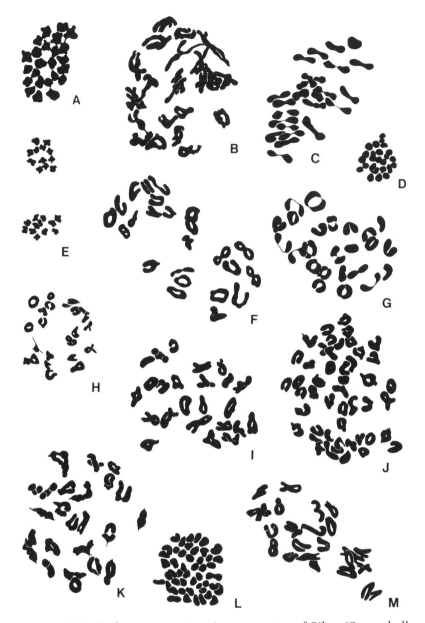

FIGURE 19.10 Meiotic chromosomes in microsporocytes of *Silene* (Caryophyllaceae), ca. x825. A, *S. bridgesii*, met. I; B, *S. campanulata*, diak.; C, *S. douglasii*, met. I; D, *S. menziesii*, met. I, tetraploid; E, *S. menziesii*, tel. I, diploid; F, *S. nuda*, diak.; G, *S. oraria*, diak.; H, *S. oregana*, diak.; I, *S. parryi*, diak., tetraploid; J, *S. parryi*, diak., octoploid; K, *S. parishii*, diak.; L, *S. scouleri*, met. I, octoploid; M, *S. scouleri* subsp. *grandis*, diak., tetraploid. (From Kruckeberg 1954:241)

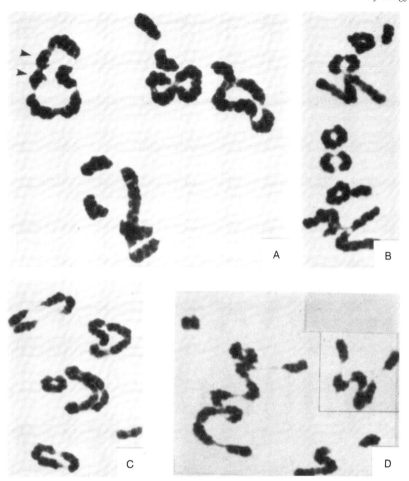

FIGURE 19.11 Variation in meiotic configurations (met. I) in *Tradescantia commelinoides* (Commelinaceae). A, $2_{VI}1_V1_{IV}1_{II}$, note two telocentrics in one hexavalent; B, $1_{VI}1_{IV}1_{III}4_{II}1_I$, note telocentric in hexavalent; C, $1_{VIII}1_{VI}1_{IV}2_{II}$; D, $1_{VIII}1_{VI}1_{IV}1_{II}$. I = univalent, II = bivalent, III = trivalent, etc. (From Jones, Kenton, and Hunt 1981:170)

1977). Broad sampling of populations and taxa is absolutely essential as the preceeding comments have attempted to stress. Grant puts it well: "It is only when a cytological analysis of the majority of the representatives of a taxon has been carried out that a complete or nearly complete picture of the relationships of its members can be revealed with some certainty" (1961:49). The costs for this type of work vary, with the main expenditure being for the compound microscope at approximately $5,000–$25,000.

EFFICACY OF CYTOLOGICAL DATA IN THE TAXONOMIC HIERARCHY

Cytological data have been useful at different levels in the taxonomic hierarchy, but they have been especially pertinent at the specific level due to their close relationship with reproductive factors. Species closely related morphologically that differ chromosomally include *Digitaria adscendens* and *D. sanguinalis* (Gramineae; Gould 1963), *Picradeniopsis woodhousei* and *P. oppositifolia* (Compositae; Stuessy, Irving, and Ellison 1973), and *Vaccinium boreale* and *V. angustifolium* (Hall and Aalders 1961). Thousands of other examples exist, however. At the same time, other species closely related morphologically have the same chromosome numbers, as in *Lupinus texensis* and *L. subcarnosis* (Leguminosae; Turner 1957). The close connection of chromosomes to reproductive isolation has caused some workers, especially Askell Löve and collaborators (e.g., Löve, Löve, and Raymond 1957; Löve 1964) to insist that any difference in chromosome number between populations must be recognized taxonomically at the specific level. This is difficult to accept in the face of abundant infraspecific aneuploidy and euploidy discussed earlier. A better perspective is offered by Mosquin: "I have found that, if the criterion of reproductive isolation were applied with any measure of consistency, the taxonomic units so created would not only have less biological meaning and consequently be less useful but would often be far more arbitrary than the present groupings based principally on morphology" (1966:213).

Dealing with chromosomal variation at the infraspecific level is more complex, especially because morphological variation often does not partition geographically in the same way or at all within a particular species. The range of infraspecific patterns of cytological variation is shown in fig. 19.12. How to handle taxonomically these different patterns is suggested in table 19.1. Implicit here is that taxonomic varieties or subspecies (or even species) should not be recognized formally without corresponding morphological variation. The greater the morphological differences between *cytotypes*, especially of a qualitative nature, the higher the probability of their formal recognition at some level in the hierarchy. If no other observable differences exist, and the cytological variation is sporadic within one or more populations, a *cytoform* label is recommended. If this variation is more widespread, then a *cytorace* designation is appropriate, as in the case with autoploid races (e.g., Raven et al. 1968; Barlow 1971; Stuessy 1971b; Miller 1976; Miller, Chambers, and Fellows 1984). If ecological separations are also evident, but still without morphological difference (e.g., as seen in the cytoraces of *Atriplex canescens*, Chenopodiaceae; Dunford 1984), the term *cytoecorace* is recommended. As regional morphological variations occur that correlate with the chromosomal differences, varieties or subspecies are in order. If significant (conspicuous) *qualitative* morphological differences are present between the cytotypes with a sporadic distributional pattern, then sympatric species are suggested and the whole structure of the group needs to be reevaluated.

Chromosome numbers and karyotypes are often useful at the generic level. For example, Soltis (1984b) found karyotypic differences between the two monotypic genera, *Leptarrhena* and *Tanakaea*, of the Saxifragaceae. He also used karyotypic data to help suggest two natural groups of genera, the *Heuchera* group and *Boykinia* group, in the tribe Saxifrageae (Soltis 1988). Stuessy (1969) used chromosome numbers to help segregate a species of *Melampodium* to a bitypic and unrelated genus, *Unxia* (Compositae). Chuang and Heckard (1982) used chromosome numbers successfully to help suggest relationships among genera of the subtribe Castillejinae (Scrophulariaceae). A useful concluding perspective on cytology and generic delimitation is given by Löve:

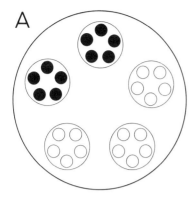

**Uniform cytological differences
between sets of populations
(major geographic correlation)**

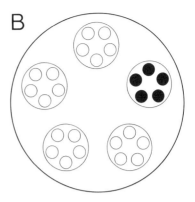

**Uniform cytological differences
between populations
(some geographic correlation)**

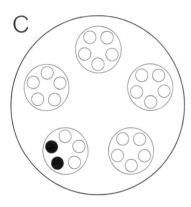

**Uniform cytological differences
between subsets of populations
(minor geographic correlation)**

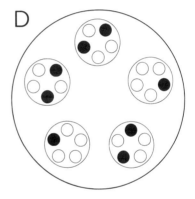

**Non-uniform cytological differences
(no geographic correlation)**

FIGURE 19.12 Models of geographic distributions of cytologically different individuals (closed circles) within and among populations in species of flowering plants. The smallest circles (open or closed) represent individual plants; the medium circles, populations; the all-inclusive circle, the species.

When using cytological observations in order to aid natural groupings at the generic level, three characteristics are known to be of greatest importance: the basic number of chromosomes, their size, and their morphology. Dissimilarity in all these characters is a reasonably safe indication of heterogeneity that has been caused by clear evolutionary divergence, whereas the significance of dissimilarity in some of these cytological characters and similarity in others has to be carefully evaluated in context with morphological and other characteristics. It should, however, also be born in mind

TABLE 19.1 Guide to the Taxonomic Treatment of Infraspecific Variation in Chromosome Numbers in Sexually Reproducing Angiosperms.

Patterns of chromosomal variation	No other observable differences	Ecological differences	Morphological differences	
			quantitative	qualitative
Uniform				
between sets of populations (fig. 19.12A)	cytorace	cytoecorace	variety (or subspecies)[a]	subspecies (or variety)[a]
between populations (fig. 19.12B)	cytorace	cytoecorace	variety	variety (or subspecies)[a]
between subsets of populations (fig. 19.12C)	cytoform	cytoecoform	cytoform	form
Non-uniform (fig. 19.12D)	cytoform	cytoecoform	cytoform	sympatric species

[a] Use of varietal or subspecific distinctions here will depend to some extent on one's view of their use in general, e.g., some workers use only variety, others only subspecies (see chapter 12). The more major the morphological and geographic differences, however, the stronger is the argument for subspecies.

that similarity in some or even all these characters is no indication of close relationship: in such cases only genetical experiments can give conclusive evidence as to the kindred qualities of the groups in question. (1963:47)

Chromosomal data tend to be less useful at the subfamilial, familial, and higher levels. This is due to the many changes in chromosomal diversity during speciation in the angiosperms so that the same number in different evolutionary lines is a most common occurrence. As pointed out well by Kowal, Mori, and Kallunki for the Lecythidaceae: "We find that chromosome numbers are enormously useful for indicating a natural subdivision of the family but are useless for indicating its relationship with other families" (1977:408). Turner (1966) found chromosome numbers to be of no particular aid in determining familial relationships of the Stackhousiaceae. However, he did find these data helpful in suggesting familial status for the Krameriaceae (1958). A good review of utility of cytology at the higher levels of the angiosperms is provided by Raven (1975). The important role of botanical gardens should be stressed here, as they often harbor representatives of rare families not yet counted chromosomally and for which material is easy to obtain (Solbrig 1973).

SPECIAL CONCERNS WITH CYTOLOGICAL DATA

Even further interest in chromosome numbers has led to projects to document the cytological diversity within geographic areas. These have often been completed as a part of other floristic investigations, because the techniques of field collecting of seeds or buds for mitotic or meiotic chromosomal studies, respectively, are relatively simple. A very nice project was the flora of Queen Charlotte Islands (Taylor and Mulligan 1968) in which all of volume 2 was devoted to cytological data. Other books include cytotaxonomical atlases of the Arctic (Löve and Löve 1975a) and Slovenian (Yugoslavian) floras (Löve and Löve 1974). Papers with such regional cyto-floristic scopes have considered the Hawaiian Islands (Carr 1978, 1985), the Canary Islands (Borgen 1970, 1974, 1975, 1977, 1979, 1980), the Juan Fernandez Islands (Sanders, Stuessy, and Rodriguez 1983; Spooner et al. 1987), Taiwan (Hsu 1967, 1968); Portugal (Fernandes and Queirós 1971; Fernandes and Santos 1971; Fernandes and Leitão 1971, 1972; Fernandes and Franca 1972), the Canadian Arctic (Mosquin and Hayley 1966), Poland (Skalinska and Pogan 1973), and southern France (Kliphuis and Wieffering 1972). Still other projects have attempted to take inventory of the chromosomal diversity within ecological and/or phytogeographic regions, such as in alpine and subalpine zones (Stoutamire and Beaman 1960; Beaman, De Jong, and Stoutamire 1962; Hedberg and Hedberg 1977). Inventories of regions are productive because species are often counted again from new geographical areas, and this can begin to provide clues to cytological diversity within taxa for taxonomic and evolutionary considerations.

A point worth reemphasizing is the power of polyploidy for determining evolutionary directionality, especially for cladistic analysis. All evidence to date strongly supports the notion that higher level polyploids have evolved from those at lower levels, and especially from diploids. Polyhaploidy is known in angiosperms (e.g., De Wet 1965, 1968, 1971), but it is uncommon and occurs at an extremely low level in contrast to the appearance of polyploids. In groups with polyploid relationships, therefore, these data can offer significant insights on the primitive or derived status of taxa (Stuessy and Crisci 1984b).

A number of studies have attempted to show the relationship of polyploidy to the environment. Clearly within particular taxa, polyploids often have different ecological tolerances than closely related diploids (Stebbins 1971b). On a broader scale, however, such as the correlation of polyploidy with latitude or elevation, studies have suggested increased levels of ploidy with increased latitude or elevation, i.e., as the climate becomes more severe, the levels of polyploidy also increase due to increased survivorship of the polyploids (Löve and Löve 1949). The situation is not simple, however, and it is important to distinguish between neo- and paleo-polyploids in addressing these issues (Ehrendorfer 1980). Powell and Sloan (1975) showed a lower percentage of polyploidy (17.8 percent) in species of gypsum floras of the Chihuahuan Desert in contrast to 32.7 percent for non-gypsum vegetation. They suggest that the ecological stability of gypsum areas might be responsible for the low level, but it seems also possible that the high salt concentrations might select against the survival of new variations. Pojar (1973) also related higher levels of polyploidy in certain vegetation types in British Columbia to more changing environments. Stebbins (1984b) has pointed out that in Pacific North America there seems to be no correlation of higher levels of polyploidy with increasing latitude, but there does seem to be an increase of polyploidy in areas showing the greatest degrees of Pleistocene glaciation. The main point is that polyploidy does seem to confer additional ecological adaptability for many taxa, and they can sometimes colonize unstable habitats as these develop.

In conclusion, cytological data are extremely important as comparative information for plant classification. They are complex, however, and much sampling and study are needed before their signficance can be fully appreciated. As Harlan Lewis expressed some time ago: "chromosomal traits are useful in systematics only to the extent that one can reasonably conclude that the observed characteristics reflect common ancestry rather than convergence. Such a conclusion can be reached only on the basis of correlation with other phenotypic traits. Obviously, therefore, cytological traits so evaluated cannot possibly have greater significance than the phenotypic traits used to evaluate them" (1969:524–525).

CHAPTER 20

Genetics and Cytogenetics

Going beyond the chromosomes to the hereditary material contained within them, we enter the realm of genetics and cytogenetics. Genetics is the science of heredity and variation, and cytogenetics is the study of chromosomes and their genetic implication (Gardner and Snustad 1984). Hence, they are logically and frequently considered together as a source of comparative data for taxonomic purposes. Repeatedly, systematists have been attracted by the promise of assessing relationships of organisms directly by comparing genetic similarities and differences and by their organization into chromosomes. Looking at chromosomes is one important dimension, but understanding what the chromosomes contain genetically and how they behave in meiosis, particularly in artifically generated hybrids, is a deeper level of insight. Such work is often complex, time-consuming, and hence expensive, but the long-term payoffs are immense as has been ably demonstrated decades ago by the work of Clausen, Keck, and Hiesey (1940, 1945, 1948) on plants in the Sierra Nevada of California. These landmark studies, in fact, laid the foundation for future work in this country on the nature of plant species.

HISTORY OF GENETICS AND CYTOGENETICS IN PLANT TAXONOMY

The origins of genetics can be traced from the pioneering work of Gregor Mendel (1866), although earlier work could be cited and discussed going back to the ancient Greeks (Sturtevant 1965). The fact remains that Mendel's paper, published in an obscure journal, when rediscovered independently by de Vries (1900a, b), Correns (1900), and Tschermak (1900), became the foundation for all subsequent investigations. The full understanding of the relationships of the hereditary material to their occurrence on chromosomes was documented by Sutton (1903), and linkage was explained by Morgan (1911). That DNA was the specific hereditary material was first reported on by Miescher (1871) as "nuclein" (Sturtevant 1965), followed by Wilson's (1896) conviction that it was indeed responsible for heredity. Feulgen showed its localization in the nucleus (Feulgen and Rossenbeck 1924), and Watson and Crick (1953) suggested its structure.

Population genetics developed early after rediscovery of Mendel's work as a natural desire of Darwinian evolutionary biologists to explain in genetic terms the adaptation of organisms by means of natural selection. Early papers by Yule

(1902), Castle (1903), Pearson (1904a, b), Hardy (1908), and Weinburg (1908) can be cited to illustrate developments. More sophisticated statistical insights were provided by Haldane (1924a, b; 1925–1932; 1932), Fisher (1930), and Wright (1931) as they attempted to synthesize ideas and data for a quantitative view of the genetic basis of evolution.

Although plant breeding can be traced back to early Egyptian date palm pollination (Zirkle 1935), in a more modern sense it began with the plant hybridization experiments of Kölreuter (1761–1766) and others of this period. Cytogenetics *per se* began after the cytological basis of heredity was recognized in the beginning of the twentieth century. Winge (1917) stressed the importance of hybridization in polyploidy. Subsequently, there came a series of investigations of different plant groups which showed the utility of in-depth cytogenetic approaches. Further, these studies also offered the potential of improving crop plants through improved breeding programs. The work of Goodspeed (e.g., 1934) on *Nicotiana* (Solanaceae), Cleland (1936, 1972) on *Oenothera* (Onagraceae), Babcock (e.g., with Stebbins 1938) on *Crepis* (Compositae), and Clausen, Keck, and Hiesey (1940, 1945, 1948) on the Madiinae (Compositae) are examples. This helped give impetus to the concept of the New Systematics (Huxley 1940) in which more experimental approaches (frequently cytogenetic) would be used to understand the biological bases of relationships among taxa. The importance of this was stressed in an understated fashion by Turrill: "it is becoming more and more obvious that recent discoveries in cytology, ecology, and genetics have often a bearing on taxonomy" (1940:47).

GENERAL GENETIC AND CYTOGENETIC TEXTS AND REFERENCES

The literature of genetics and cytogenetics is vast and no attempt will be made here to summarize this adequately. A few texts, however, may be useful as references. For general cytogenetics, one may consult Burnham (1962), Brown (1972), or Swanson, Merz, and Young (1981). Good general genetics texts are by Gardner and Snustad (1984) and Suzuki et al. (1986). For molecular evolutionary genetics, see Nei (1987). For the plant taxonomist, the book by Grant (1975) on genetics of flowering plants is highly recommended. For an introduction to population genetics, see Roughgarden (1979) and Hartl (1987); for quantitative genetics, Falconer (1981) is a good start, plus the recent symposium volume of Weir et al. (1988). Population cytogenetics is reviewed by John (1976). Other topics include plant breeding (e.g., Sinha and Sinha 1980), chromosomal evolution (Stebbins, 1971b), and the detailed analyses of meiotic configurations (Sybenga 1975). For a handy compilation of pertinent classic papers on the cytogenetic aspects of polyploids, refer to Jackson and Hauber (1983).

TYPES OF GENETIC AND CYTOGENETIC DATA

The great attraction of genetic and cytogenetic data for the plant taxonomist is that they probe the real hereditary bases of evolutionary divergence and hence offer the possibility of a more refined yardstick for classification. No panaceas exist, of course, but the idea is compelling nonetheless. The genetic similarities and differences among taxa can be determined in two different ways: (1) crossing studies to determine the genetic basis of selected (usually diagnostic) taxonomic characters; and (2) isozyme analyses to yield an estimate of genetic distance between taxa. Direct comparison of DNA similarities and differences could be included here, but this seems best treated under chemical data (chapter 21). Isozyme analyses could be regarded as chemical data also (as done by Crawford and Giannasi 1982), but I prefer to include them here, because they attempt to determine the genetic variation within and between taxa; these data really are presented in genetic terms.

Genetic Data

Genetic Analysis. Actually, not many studies have been done to determine the genetic basis of taxonomic characters for purposes of classification. Numerous investigations have been completed on the genetic basis of features of crop plants, but few have been published in a purely taxonomic context. One classical example is the work by Rollins (1958) on the genetic basis of pubescence on the fruits of *Dithyrea wislizenii* (Cruciferae) and *D. griffithsii* (sometimes treated as a variety of the former). The former was regarded as typically pubescent and the latter glabrous. Suspecting simple genetic control of this feature due to observed intra-populational variation, Rollins performed the necessary artificial crossings to obtain F_1 ratios that showed convincingly that a single gene was responsible. Rollins' conclusion: "it is safe to reject the phenotypic characteristic of glabrous siliques as having no significance for taxonomic purposes" (1958:150). The variation in fruit pubescence encountered within populations of these taxa made one suspect the taxonomic validity of the feature, but the genetic studies confirmed it. Another study showed single gene control of fruit characters in *Valerianella ozarkana* (Valerianaceae; fig. 20.1 and table 20.1; Eggers Ware 1983). Another recent study by Jackson and Dimas (1981) revealed that one of the distinguishing features separating the *Haplopappus phyllocephalus* group (Compositae) from sect. *Isocoma* of the same genus, the presence vs. absence of ray florets, was controlled by a single gene. This information, plus other considerations, led to the transfer of the *H. phyllocephalus* group from sect. *Blepharodon* (placed there by Hall, 1928) to sect. *Isocoma*. The series by Bachmann, Price, and collaborators is giving insight on the genetic basis of several morphological features of taxonomic value in *Microseris* (Compositae; e.g., Bachmann et al. 1983, 1987). Two good recent reviews of the genetic bases of various morphological features in flowering plants are Hilu (1983) and Gottlieb (1984). Both of these authors stress that many of the taxonomically useful features of structure, shape, and arrangement of parts that have been used at different levels in the hierarchy are governed by one or only a few genes. For taxonomic decisions, therefore, the important point is not to know only the genetic basis of a taxonomically useful character, but more significantly to understand its *consistency* within and between groups and its *correlation* with other features.

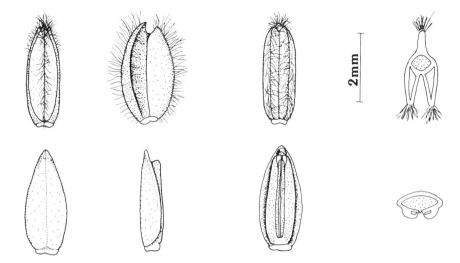

FIGURE 20.1 Different views of fruits of *Valerianella ozarkana* forma *ozarkana* (top) and forma *bushii* (bottom; Valerianaceae). (From Eggers Ware 1983:34)

Isozyme Analysis. Isozyme analysis offers a more rapid means of estimating genetic distance between taxa (Gottlieb 1971, 1977a). It has the advantage of conveniently assessing overall genetic differences and similarities between taxa in contrast to the laborious genetic trials which deal only with a few traits of taxonomic significance. Isozymes are genetically controlled allelic variants of enzymes which have various metabolic functions in the plants. The number of different isozymes that can be reasonably extracted and dealt with electrophoretically is small (typically 20–30) in comparison with the total number contained within a higher plant. Hence, only a small portion of the genotype is

TABLE 20.1 Genetic Results of Crosses Within and Between *Valerianella ozarkana* and *V. bushii.* Superscripts (a, b, c, d, and e) on Fruit-Type Symbols (OZ/BU) Indicate Individual Plants. (From Eggers Ware 1983: 40)

		Progeny	
	Parents	ozarkana	bushii
bushii			
Self 1	BU^a × BU^a	0	43
Self 2	BU^b × BU^b	0	17
Cross 1	BU^c × BU^a	0	39
Cross 2	BU^d × BU^b	0	25
ozarkana			
Self 1	OZ^a × OZ^a	35	12
Self 2	OZ^b × OZ^b	16	7
Cross 1	OZ^c × OZ^a	1	0
bushii × *ozarkana*	BU^e × OZ^b	19	16

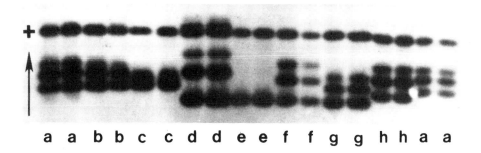

a a b b c c d d e e f f g g h h a a

FIGURE 20.2 Gel phenotypes in tetraploid populations of South American *Chenopodium* (Chenopodiaceae). (From Wilson 1981:383)

being sampled in any one instance. Nevertheless, the data are extremely useful and have been shown to be reliable indicators of overall genetic affinities among taxa (Hunziker 1968; Gottlieb 1977a; Crawford, Stuessy, and Silva 1987). The enzymes are extracted from the plant, spotted on gels, and separated via electrophoresis to yield (with appropriate stains) bands of proteins that reflect the isozyme phenotype (fig. 20.2). These are interpreted in genetic terms and the frequencies of their occurrence within taxa are calculated. The degrees of genetic polymorphism, including number of alleles per locus, as well as the proportion of the loci which are heterozygous, can be calculated (table 20.2). The genetic distance between taxa can also be determined by various statistics (table 20.3) such as Nei's (1972) or Rogers' (1972). Additional crossing studies can (and should) be done to determine the actual genetic basis for each of the isozymes separated and used, as has been advocated by Gottlieb (1977a) and Crawford (1983).

Cytogenetic Data

Cytogenetics attempts to reveal degrees of relationship among taxa by chromosome homologies, reproductive compatibilities, and reproductive capacities via natural and/or artificial hybridization. Obviously, some elements here grade into reproductive biology, others into cytology, and still others into genetics. The basic idea is to find naturally, or to produce artificially, intertaxon hybrids and analyze the resulting progeny in terms of a spectrum of interrelationships ranging from F_1 seeds to advanced hybrid generations. The assumption is that the more robust and fertile the hybrids, the closer will be the genetic relationships between the taxa involved. Sometimes hybrids simply cannot be obtained between two taxa no matter what is attempted. In this case no cytogenetic analysis of the hybrid progeny is possible. Frequently in closely related flowering plants, however, intertaxon hybrids are obtained, but they must survive to maturity to be very helpful in assessing relationships. Just obtaining seed set between the two parents (representing the F_1 generation) gives some information, but it is so much more useful to allow the seeds to germinate, examine their survival as seedlings, and test their fertility upon maturation. A commonly used and quick

TABLE 20.2 Proportion of Loci Polymorphic, Mean Number of Alleles per Polymorphic Locus, and Proportion of Loci Heterozygous for Populations of the Varieties of *Coreopsis grandiflora*. (From Crawford and Smith 1984: 222)

Taxon	Population designation	Proportion of loci polymorphic	Mean number of alleles per polymorphic locus	Proportion of loci heterozygous
var. *grandiflora*	1.	0.42	2.55	0.149
	2.	0.35	2.11	0.144
	3.	0.43	2.33	0.112
var. *harveyana*	4.	0.40	2.20	0.099
	5.	0.11	2.00	0.077
	6.	0.20	2.00	0.071
	7.	0.35	2.63	0.095
	8.	0.35	2.57	0.077
	9.	0.31	2.88	0.160
	10.	0.46	2.90	0.160
	11.	0.31	2.67	0.122
	12.	0.35	3.00	0.192
	13.	0.33	3.13	0.163
	14.	0.31	2.38	0.088
var. *longipes*	15.	0.46	2.45	0.117
	16.	0.31	2.88	0.148
	17.	0.39	2.20	0.126
var. *saxicola*	18.	0.42	2.55	0.156
	19.	0.39	2.50	0.130
	20.	0.39	2.60	0.087

TABLE 20.3 Nei's Coefficients of Genetic Identity (Upper Right Triangle) and Modified Rogers' Distances (Lower Left Triangle) Among Taxa of *Zea* (Gramineae). Numbers in the Diagonal are the Genetic Identities Among Populations of the Same Taxon. (From Doebley et al. 1984: 213)

	1	*2*	*3*
1. *Z. diploperennis*	0.916	0.844	0.812
2. *Z. perennis*	0.337	0.920	0.818
3. *Z. luxurians*	0.385	0.382	0.948
4. *Z. mays* var. *huehuetenangensis*	0.425	0.424	0.415
5. *Z. mays* var. *parviglumis* Balsas	0.380	0.314	0.399
6. *Z. mays* var. *parviglumis* Jalisco	0.392	0.334	0.446
7. *Z. mays* subsp. *mexicana* Chalco	0.444	0.400	0.468
8. *Z. mays* subsp. *mexicana* Central Plateau	0.404	0.376	0.437
9. *Z. mays* subsp. *mexicana* Nobogame	0.475	0.444	0.457
10. *Z. mays* subsp. *mays*	0.404	0.342	0.442

test is of the pollen viabilities in the F_1 hybrids as determined with various stains, such as lactophenol cotton blue, tetrazolium dyes (e.g., Hauser and Morrison, 1964), etc. These data are frequently presented as a crossing diagram (or polygon) and sometimes arranged geographically (fig. 20.3). The degree of seed set in the F_1s can also be measured (table 20.4). Even more useful in analyzing the nature of the hybrids is to examine the degree of chromosomal pairing. This analysis begins with prophase in an early stage called pachytene. Pachytene analysis attempts to examine the way in which the homologous chromosomes form pairs. At the initial stage of meiosis, each chromosome is long and relatively uncondensed and the degree of pairing between the chromatids often can be seen (fig. 20.4). The smaller the number of chromosomes, the better the configurations can be analyzed. One of the most effective series of studies of this type has been done by R. C. Jackson and colleagues on *Haplopappus* (Compositae, see Jackson 1962, and Jackson and Dimas 1981, for examples; see also Whittingham and Stebbins 1969). Meiosis as revealed in Metaphase I is often instructive as the degree of bivalent and multivalent formation can often be observed (table 20.5). The fewer number of bivalents and the greater number of multivalents and abnormalities, the less closely related the taxa are believed to be. These meiotic irregularities, in fact, are known to be positively correlated with low levels of pollen fertility (fig. 20.5). Abnormalities such as bridges and lagging chromosomes can also be detected in anaphase and on into telophase. If F_1 hybrids are sufficiently fertile to allow for development of an F_2 generation, then these same types of data can be gathered and often breakdown of some type is noted (e.g., in *Mimulus*, Scrophulariaceae; Vickery 1974).

A caution needs to be introduced with reference to the above types of analysis. In 1958 Riley and Chapman showed that bivalent formation in meiosis in wheat is genetically controlled by only one gene. This has caused some workers to question seriously the validity of some types of crossing data (e.g., De Wet and Harlan 1972), especially among polyploid taxa. While such concern is certainly warranted, the evidence to date shows overwhelmingly that pairing data are extremely powerful indicators of genetic relationship. Care must be exercised in their gathering and interpretation, and they provide no panaceas, but the same is true with all other types of comparative data.

4	5	6	7	8	9	10
0.767	0.793	0.795	0.724	0.762	0.692	0.774
0.772	0.863	0.853	0.781	0.799	0.736	0.842
0.794	0.797	0.756	0.722	0.752	0.740	0.755
0.981	0.808	0.770	0.753	0.776	0.766	0.803
0.385	0.927	0.979	0.913	0.929	0.858	0.975
0.430	0.131	0.886	0.877	0.912	0.820	0.953
0.438	0.248	0.305	0.933	0.960	0.889	0.889
0.412	0.219	0.256	0.168	0.848	0.933	0.888
0.430	0.321	0.372	0.286	0.221	0.959	0.802
0.394	0.135	0.189	0.271	0.281	0.385	0.946

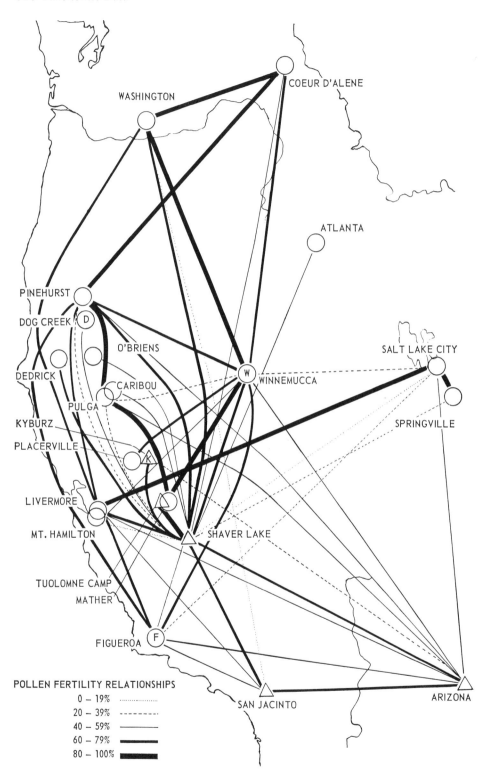

WASHINGTON

COEUR D'ALENE

ATLANTA

PINEHURST

DOG CREEK

O'BRIENS

SALT LAKE CITY

DEDRICK

CARIBOU

WINNEMUCCA

PULGA

KYBURZ

SPRINGVILLE

PLACERVILLE

LIVERMORE

MT. HAMILTON

SHAVER LAKE

TUOLOMNE CAMP

MATHER

FIGUEROA

POLLEN FERTILITY RELATIONSHIPS

0 – 19%	....
20 – 39%	- - - - -
40 – 59%	————
60 – 79%	▬▬▬
80 – 100%	▬▬▬

SAN JACINTO

ARIZONA

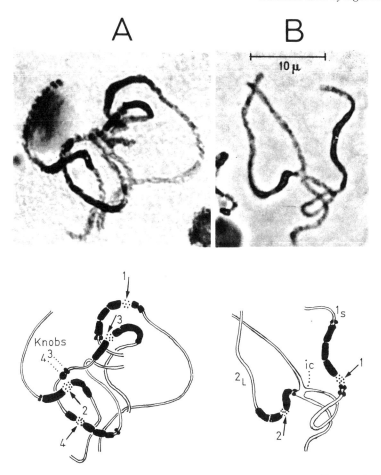

FIGURE 20.4 Photomicrographs (top) and interpretative tracings (bottom) of pachytene chromosomes of standard (A) and rearranged (B; translocation heterozygote) karyotypes in *Plantago insularis* (Plantaginaceae). (From Whittingham and Stebbins 1969:454, 455)

INVESTMENTS FOR GATHERING GENETIC AND CYTOGENETIC DATA

The investments needed for data-gathering in genetics and cytogenetics are somewhat similar to those for cytology. Certainly for cytogenetic studies, it is necessary to see chromosomes clearly after the buds or root tips have been collected from plants in the field or greenhouse. For these observations is needed a good phase-contrast microscope with the material being appropriately stained.

FIGURE 20.3 Fertility relationships among populations of *Clarkia rhomboidea* (Onagraceae) as shown by percentages of good pollen. Circles designate populations with the Northern chromosomal end arrangement; triangles indicate populations with the Southern arrangement. Circles with letters designate populations with unique chromosomal end arrangements. (From Mosquin 1964:20)

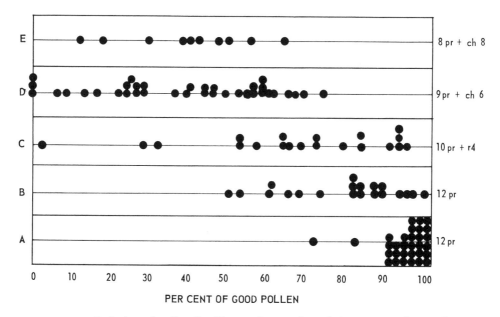

FIGURE 20.5 Relation of pollen fertility to the number of chromosomal interchanges in *Clarkia rhomboidea* (Onagraceae). A, garden progeny from wild seed; B, structurally homozygous interpopulational hybrids; C–E, interpopulational hybrids heterozygous for one, two, and three translocations, respectively. (From Mosquin 1964:21)

Generally meiotic material is used because here the pairing behavior of chromosomes in hybrid genomes can be seen better than in mitosis. The kind of data needed depends upon the type of systematic question being posed. For cytogenetic crosses and subsequent genetic analyses, space in field plots or greenhouses

TABLE 20.4 Survival and Fertility in the F_1 Generation in Crosses Among Three Species of *Melampodium* (Compositae). (From Stuessy and Brunken 1979: 58)

Hybrid ($♀ \times ♂$)	Survival Data		Pollen Stainability		Seed Set (# hybrids/ % seed set)	
	# seeds produced	% survival to maturity	# plants examined	% stained (range)	selfed	crossed
microcephalum × *paniculatum*	64	36	36	12 (6–25)	7/0	14/0
paniculatum × *microcephalum*	41	23	13	10 (5–20)	5/0	10/0
paniculatum × *gracile*	36	19	5	6 (2–10)	7/0	7/0
gracile × *paniculatum*	27	10	9	6 (4–10)	4/0	4/0
gracile × *microcephalum*	56	17	13	11 (6–25)	7/0	5/3.6
microcephalum × *gracile*	12	0	—	—	—	—

TABLE 20.5 Chromosomal Associations in Populations of Tetraploid Species of *Phlox* (Polemoniaceae). (From Levin 1968: 620)

Species		No. of PMC's	14II %	13II + 2I %	12II + 1IV %	12II + 4I %	11II + 1IV + 2I %	10II + 2IV %	9II + 2IV + 2I %	8II + 3IV %	6II + 4IV %
P. floridana	1	128	26	32			5	17	2	18	
"	2	35	29	9	30		9	14		9	5
"	3	30	10	35			5	38		7	
"	4	55	32	15	28		4	21			
"	5	50	32	8	30		6	18		6	
"	6	71	30	6	34		10	17	2	1	
"	7	34	23	6	31		3	31		5	
"	8	100	35	8	30		7	18		2	
"	9	30	39		33		10	18			
"	10	58	45	3	39		3	7		3	
"	11	70	21	2	33		9	23		10	2
"	12	92	46	3	30		4	16		1	
"	13	79	41	1	40		3	14		1	
"	14	73	41	9	31		6	12		1	
P. villosissima	1	22	45	9	36		5	5			
"	2	119	40	8	32		11	9			
"	3	61	50	5	33		3	9			
P. aspera	1	76	37	23	29	6		5	3		
"	2	150	33	14	28		11	11			
"	3	92	44	8	26	11	6	5			
"	4	21	62	4	30		4				
"	5	86	40	11	31	4	10	4			
"	6	79	68	9	14		4	5			
"	7	42	60	14	24			2			
"	8	151	60	4	30		3	3			

Chromosome Associations

is needed. For electrophoretic surveys, one must have a power source, equipment for pouring gels, and reagents for staining the protein bands on the gels after they have been run. The time involved with each of the specific steps of these analyses is not great, depending upon the particular step involved. Electrophoretic surveys can be done very rapidly with large numbers of individuals which facilitates the assessment of intra- and interpopulational genetic variation. For detailed cytogenetic studies, however, the amount of time required for a comprehensive program of crosses can be enormous. Such an effort takes literally years to complete, and it should not be initiated frivolously. What really consumes the time is the nearly daily attention to the crosses to insure pollen transfer at the proper moment and the careful tabulation of results. This necessitates a much greater time commitment than is obvious at first glance. Costs for such crossing studies are low by themselves, but measured in terms of time, they can be extremely expensive.

EFFICACY OF GENETIC AND CYTOGENETIC DATA IN THE TAXONOMIC HIERARCHY

Genetic and cytogenetic data are most valuable at the lower levels of the taxonomic hierarchy. There is no recorded successful interfamilial cross between families of angiosperms. The only way to overcome this substantial genetic barrier to such attempts is to go the route of somatic cell hybridization as proposed by Levin (1975). This has potential but has so far been little used taxonomically. Likewise, genera of angiosperms rarely cross. There are many recorded artificial intergeneric crosses in several families, especially in the Gramineae and Orchidaceae (e.g., Knobloch (1972), but this is a very low percentage of successful possibilities. In fact, if genera *are* successfully crossed in some way, it invariably raises the taxonomic question of whether they should be better merged into one genus (e.g., Heiser 1963b; Anderson and Reveal 1966).

Within the genus at the specific, series, sectional, and subgeneric levels, however, artificial crossability is commonly achieved, and it is here that the most useful cytogenetic data are obtained and applied. This gets at the real genetic and cytological basis of the partitioning of diversity into natural populations. Caution obviously must be counseled, especially taking into account differences in breeding systems which can affect crossing results (Mulcahy 1965). The numerous papers by Verne Grant over many years (see Grant 1981, for a list) in *Gilia* (Polemoniaceae) provided experimental data for delimitation of species and subspecies. These studies bear directly on the biological species concept and its application to reproductive isolation. Numerous examples can be cited of the positive value at this level including Stort (1984) in *Cattleya* (Orchidaceae) and Bayer and Crawford (1986) in *Antennaria* (Compositae). Likewise, isozymes are most useful at the specific and infraspecific levels. Isozyme data can be compared at higher levels in the hierarchy, but the problem of homology intensifies in such a way as to render the comparisons of dubious value. The probability of parallel mutations giving rise to identical isozymes is just too high for this be a fruitful approach at this taxonomic level.

SPECIAL CONCERNS WITH GENETIC AND CYTOGENETIC DATA

As with all types of data, it is important to stress once again the need for extensive sampling to understand the background variation that exists within each taxon being compared. Figure 20.6, for example, shows variations of pollen stainability within one subspecies of *Atriplex longipes* subsp. *praecox* in Scandinavia.

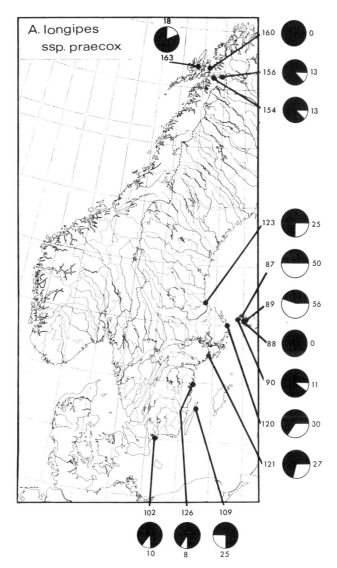

FIGURE 20.6 Percentage of plants with reduced fertility (less than 90% stainable pollen) in populations of *Atriplex longipes* ssp. *praecox* (Chenopodiaceae). (From Gustafsson 1973:363)

A point that needs to be emphasized is that crossing programs must be as complete as possible to be effective. For example, to successfully cross one species of one genus with another species of another genus is not sufficient to resolve the issue of the generic relationships. Four hypotheses exist to deal with these new data: (1) leave the generic limits intact with no change; (2) merge the two genera together; (3) transfer one of the species (either one) into the other genus; or (4) treat the two crossable species together as forming their own genus. To decide on the proper taxonomic decision requires having much more cytogenetic (and other) data. For the data to be truly comparative, crosses should be made among all the other taxa of both genera, both within and between them. This can be a signficant expenditure of resources to achieve the desired results. This should not be surprising, for who would consider remodeling genera based on morphology, for example, without having looked morphologically at all the other species of both genera? Cytogenetic data are no different in this respect from other comparative information.

A most important contribution of cytogenetic data, although more in the broad context of systematics, is in the unraveling and documentation of suspected pathways of evolution. Gaining such understanding, however, can have useful consequences for constructing classifications. The origins of species by interspecific hybridization at the diploid level is known, and useful comments are provided by Gottlieb (1972) and Levin (1979b). This mode of evolution in plants is not nearly as common as with polyploid hybrid evolution of which many examples are known (e.g., Ownbey 1950, in *Tragopogon*, Compositae; Stuessy, Irving, and Ellison 1973, in *Picradeniopsis*, Compositae). These types of origins can yield unbelievably complex patterns that are difficult to resolve taxonomically, such as seen in the *Acmella oppositifolia* polyploid complex (fig. 20.7). Attempting to distinguish the origin of taxa via autopolyploidy or allo-

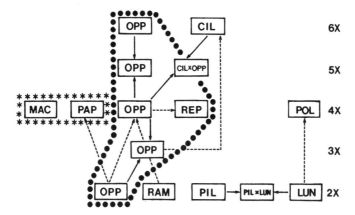

FIGURE 20.7 Hypothesized relationships in the *Acmella oppositifolia* (Compositae) polyploid complex. Asterisks and dots enclose *A. papposa* (with two varieties) and *A. oppositifolia* (with two varieties and five ploidy levels), respectively. Solid arrows indicate known natural hybridizations; dashed arrows suggest hypotheses of origin of the polyploids. (From Jansen 1985:9)

polyploidy is a challenging business especially in view of the recent questions regarding the use of quadrivalent vs. bivalent pairing, respectively, to help sort them out (e.g., Jackson and Casey 1980, 1982; Jackson and Hauber 1982; Jackson 1982, 1984b; Stebbins 1984a). Autopolyploids, even artificially produced ones, can become "diploidized" and show increased bivalent formation (e.g., Lavania 1986). Aneuploid evolution is also common in angiosperms, and the time-tested examples of Togby (1943) in *Crepis* (Compositae), Lewis and Roberts (1956) in *Clarkia* (Onagraceae), and Kyhos (1965) in *Chaenactis* (Compositae) serve to show the indispensible value of cytogenetic data for revealing this mode. The recent work by Jones (1974) on chromosomal evolution in the Commelinaceae via Robertsonian translocation (the fusion of two chromosomes together to form a descending series) is worth noting also.

An interesting point relevant to this book is the use of cytogenetics in relation to cladistic analysis. Some cladists have remarked on the lack of utility of cytogenetic data for determing evolutionary trends because they serve only to give plesiomorphic information about two taxa being crossed (Funk 1985a). Nothing could be further from the truth. The ability to cross does not just deal with a primitive genetic background; it deals with the degree of genetic compatibility developed in a particular evolutionary line (Stuessy 1985), as well as the evolutionary divergence *within* each line. A higher degree of crossability reflects a close relationship in the same fashion as does a morphological synapomorphy between the two. The difference is that the data are not manageable in the same way as with characters and states that can be obtained from each taxon. The cytogenetic data are obtained for each *pair* of taxa and really are similar to phenetic coefficients of similarity or cladistic distance measures.

That chromosomal inversion data can be useful in determining evolutionary directionality can hardly be disputed, but the *degree* to which it can be helpful has been questioned by some. The determination of patterns of homology between chromosomes of different taxa is to obtain comparative data similar to those of other types. This is not pairwise data as obtained with crossing information (discussed above) but point-data for each taxon. Hence, they can be used effectively as with any other type of data, and this has been openly acknowledged (e.g., Farris 1978; Wülker, Lörincz, and Dévai 1984). I personally believe that these data may be even more powerful if sufficient cytogenetic data are available to suggest modes of chromosomal speciation. The inversion data themselves will allow the formation of a network among the taxa, but this will lack directionality. Rooting this network has the same inherent difficulties as with any other network (see, e.g., a specific recent approach in *Robinsonia*, Compositae; Pacheco et al. 1985). Computer algorithms can also be used to assess these relationships more precisely (e.g., Wülker, Lörincz, and Dévai 1984), but they are complex and no optimal algorithms are available (Day and Sankoff 1987). The difference is that certain cytogenetic transformations are more common than others. For example, clearly descending aneuploidy is more common than ascending aneuploidy (Grant 1981; Goldblatt 1988). Hence, the probability of evolutionary directionality at the diploid level is strongly suggested simply on the knowledge of understood cytological mechanisms. This can be an extremely powerful help in unraveling evolutionary trends (provided no cytogenetic evidence exists to the contrary). Polyploidy is another such circumstance (see Stuessy and Crisci, 1984b, for an example of this approach in *Acmella*, Compositae, from Jansen, 1985).

This brief discussion does not do complete justice to the wealth of cytogenetic data obtainable and useful for taxonomic purposes. To show more fully the depth of this approach, the following is a list of different types of information that can be gathered (from Jackson 1984a:68): (1) number of meiocytes analyzed at each stage; (2) synapsis as observed at early pachytene; (3) number or frequency of chiasmata per cell, and any differences among meiotic configurations; (4) univalent frequency; (5) diakinesis or metaphase I configurations and their frequencies; (6) early anaphase I disjunction patterns, especially for multivalents and heteromorphic bivalents; (7) chromosome behavior at anaphase I and II; and (8) pollen stainability.

CHAPTER 21

Chemistry

Chemical data in taxonomy have inherent appeal by offering a look at relationships of plants via internal characters and at still another level of structural organization. Extended to the ultimate hope, chemical data should be able to go beyond cytology and genetics to direct comparisons of DNA sequences, the hereditary material itself. This goal has been in the back of people's minds since the early 1960s (e.g., Zuckerkandl and Pauling 1965), but technical difficulties prohibited serious advances in this area until recently. Also, it is now clearly recognized that many of the plant secondary metabolites (e.g., alkaloids and terpenoids) have major ecological roles in the way plants relate to their environment. Therefore, certain kinds of chemical data are likely to be of great value in determining phylogeny whereas others are of equally great value in understanding predator-prey relationships. Through it all, chemical characters have continued to prove valuable in helping solve different kinds of taxonomic problems, and it is now clear that chemistry is here to stay as a major source of comparative data for understanding relationships.

HISTORY OF CHEMISTRY IN PLANT TAXONOMY

One might suggest that the history of chemistry in plant classification dates back to the earliest *Materia Medica* of Dioscorides (ca. 300 B.C.). Here and into the later age of the herbalists (1470–1670), plants were grouped in part on their medicinal properties, which obviously derived from chemical substances (Gibbs 1963). For serious use of chemistry in plant classification, however, one of the earliest studies must be that of Abbott (1886). She examined the distribution of saponins (triterpenoids or steroids) in plants and made several useful (although naive by our standards) general assertions about the role of chemical data in evolutionary studies. The early work of Reichert (1916, 1919) on starches must be mentioned as also should that on terpenoids of *Eucalyptus* by Baker and Smith (1920). Mez and coworkers (1922) in Königsberg, Germany, carried out a series of innovative serological studies on plant relationships, but the affinities so demonstrated among the angiosperms were so controversial and criticized by workers that few people took the results seriously. By modern standards, we view the work as technically flawed. This actually hindered legitimate serological efforts in plant taxonomy until more recently. McNair (1934, 1935, 1945) published a remarkable series of insightful papers on the taxonomic and ecological roles of alkaloids, and cyanogenetic and sulphur compounds. Mirov (e.g., 1948) contributed many papers on terpenes in *Pinus* and their taxonomic significance. The early review on taxonomy and chemistry of plants by Weevers (1943) also should be noted.

Despite these and other early suggestive studies, the direct and more general application of chemical data to taxonomic problems did not appear until the mid-1950s. Hegnauer (1954, 1958) from the Netherlands was an early contributor, and he used the term "chemotaxonomy" in his writings. Bate-Smith from England pointed out (1958) the potential of phenolics in classification. Many genetic and phytochemical studies were published in the latter part of this decade, but the most influential with a direct taxonomic focus were those from R. E. Alston and B. L. Turner of the University of Texas at Austin. They began with examination of flavonoids in *Baptisia* (Leguminosae) especially to help in interpreting complex patterns of hybridization. This proved a good system to analyze because the morphological differences among the species were as striking as the flavonoid markers. The publication in 1963 of their book, *Biochemical Systematics*, was an important step in drawing attention to the field and helping to consolidate it. Establishment of the Phytochemical Society of North America (preceded by the Plant Phenolics Group of North America) and the Phytochemical Section of the Botanical Society of America were also important steps at this time. In England, Tony Swain of Cambridge was very active as illustrated by his edited books *Chemical Plant Taxonomy* (1963) and *Comparative Phytochemistry* (1966) as was Jeffrey Harborne of the University of Reading with his *Comparative Biochemistry of the Flavonoids* (1967) and *Phytochemical Phylogeny* (1970). Serology also made an early comeback as evidenced by the books of Leone, *Taxonomic Biochemistry and Serology* (1964) and Hawkes, *Chemotaxonomy and Serotaxonomy* (1968).

From these beginnings has come a wide variety of books and papers on uses of chemistry in plant classification, all falling under the labels of *chemotaxonomy, chemosystematics, biochemical systematics,* or *taxonomic biochemistry*. Once again it is impossible here to review all the literature and types of compounds potentially useful in solving taxonomic problems. Fortunately, the recent excellent review by Giannasi and Crawford (1986) eases our task considerably, and this is recommended reading for those wishing a more in-depth review.

GENERAL CHEMICAL AND CHEMOTAXONOMIC TEXTS AND REFERENCES

A few books and general reviews might be cited here to give an indication of the resources available on selected topics of chemotaxonomic interest. Relatively recent good general texts include Smith (1976), Ferguson (1980), and Harborne and Turner (1984). Symposia volumes of a general chemosystematic scope include Swain (1963), Bendz and Santesson (1974), Averett (1977), Bisby, Vaughan, and Wright (1980), and Goldstein and Etzler (1983). General literature reviews include Alston, Mabry, and Turner (1963), Alston (1965, 1967), Turner (1967, 1969, 1974, 1977a), Erdtman (1968), Throckmorton (1968), Fairbrothers et al.

(1975), Harborne (1984a), and Kubitzki (1984). Phytochemical phylogeny was addressed directly in the symposium volumes edited by Harborne (1970) and Young and Seigler (1981). Molecular evolution has also received attention from Anfinsen (1959), Jukes (1966), Calvin (1969), Hochachka and Somero (1973), Ayala (1976), and Terzaghi, Wilson, and Penny (1984), among others. There now exists the *Journal of Molecular Evolution.* Ecological aspects of phytochemistry, especially in plant-animal interactions and coevolution, are covered in Chambers (1970), and Harborne (1977, 1978). Books dealing with general surveys of plant secondary metabolites are Geissman and Crout (1969), Bell and Charlwood (1980), Robinson (1980), Goodwin and Mercer (1983), and Harborne (1984b.) Medicinal plants are treated by Frohne and Jensen (1973) as well as Tétényi (1970), the latter of which also deals with chemistry of infraspecific taxa. Compendia of what compounds have been reported previously in which taxa are Hegnauer (1962-present) and Gibbs (1974). Chemical surveys of selected large plant families have been included in symposia on the Compositae (Heywood, Harborne, and Turner 1977), Leguminosae (Harborne, Boulter, and Turner 1971), Rutaceae (Waterman and Grundon, 1983), Solanaceae (D'Arcy 1986), and Umbelliferae (Heywood 1971a). Serological techniques have been discussed by Leone (1964), Hawkes (1968), and Jensen and Fairbrothers (1983). Volumes also have been published on different types of compounds, such as flavonoids (Geissman 1962; Harborne 1967; Mabry, Markham, and Thomas 1970; Ribéreau-Gayon 1972, this more generally on plant phenolics; Harborne, Mabry, and Mabry 1975; Harborne and Mabry 1982; Markham 1982; Cody, Middleton, and Harborne 1986; and Farkas, Gabor, and Kallay 1986), sesquiterpene lactones (Yoshioka, Mabry, and Timmermann 1973; Fischer, Olivier, and Fischer 1979), terpenoids in general (Pridham 1967; Goodwin 1970; Newman 1972), alkaloids (Raffauf 1970; Robinson 1981), general pigments (Goodwin 1965), and proteins (Bryson and Vogel 1965; Goodman 1982; Oxford and Rollinson 1983). Specific reviews also deal with chemical races (Turner 1970b; Hegnauer 1975) and chemical aspects of disjunctions (Turner 1972; Mabry 1974). Additional journals that contain much chemical data of systematic import include *Phytochemistry, Journal of Natural Products Chemistry* (formerly *Lloydia*), *Biochemical Systematics and Ecology,* and the annual *Recent Advances in Phytochemistry* and *Fortschritte der Chemie organischer Naturstoffe* (Progress in the Chemistry of Organic Natural Products).

TYPES OF CHEMICAL DATA

Space permits only a brief discussion of different types of chemical data potentially useful in plant taxonomy. The emphasis here will be on the most common types of information employed in flowering plants. A convenient and biologically meaningful distinction exists between micromolecules and macromolecules. The former are low molecular weight compounds and include flavonoids, terpenoids (mono- and sesquiterpenes), alkaloids, betalains, and glucosinolates. These are products of plant secondary metabolism. Macromolecules include proteins and nucleic acids (DNA and RNA) found in the nucleus and in mitochondria and chloroplasts. These high molecular weight compounds are part of the basic metabolic machinery of plant cells and therefore quite different in their characteristics and informational content from secondary plant products. In addition to these specific types of micromolecular and macromolecular compounds, one could also talk about polysaccharides, or various lipids, polyacetylenes, xanthones, carotenoids, lichen acids, and iridoids, among other topics, but this is beyond the scope of this book. See Giannasi and Crawford (1986) for more in-depth discussion of these issues.

Micromolecular Data

Flavonoids. The most frequently used compounds in plant taxonomic studies to date have been flavonoids. Reasons for this utility include ease of isolation, separation and identification, stability of compounds (being preserved well even in herbarium material nearly a hundred years old, although some alterations can occur in drying, especially in dihydroflavonols; Bate-Smith and Harborne, 1971; see also Cooper-Driver and Balick 1978; as well as in fossils, Giannasi and Niklas 1977); and existence of many different kinds of compounds which often show taxon-specific patterns. Flavonoids are phenolic compounds which occur generally in a three-ring system derived through cyclization of an intermediate from a cinnamic acid derivative and three malonyl CoA molecules. They may function in defense (Levin, 1971b), but recent studies suggest that at least some of them may play a role in regulating (inhibiting) auxin transport (Jacobs and Rubery, 1988). More than 500 different flavonoids exist and these are distributed into several major structural classes (fig. 21.1). Flavonoids are found in all plants except for most of the algae; they are known, however, in *Nitella* (Markham and Porter 1969). From bryophytes (e.g., Markham and Porter 1978) on through the flowering plants they are found abundantly in most taxa in vegetative and reproductive organs. Some studies have dealt with compounds from either vegetative (e.g., Whalen 1978; Denton and Kerwin 1980) or floral structures (e.g., Bloom 1976), or both (e.g., Giannasi 1975). Flavonoid data can be handled in the same fashion as other comparative data, and numerical analyses can be done. Here much discussion has prevailed on the correct algorithm to use, with a special emphasis that the absence of a compound is much less meaningful than its presence (see Runemark 1968; Adams 1972, 1974a; Weimarck 1972; Crawford 1979b). The spectrum of types of analyses has gone from the simple polygonal graphs (e.g., Ellison, Alston, and Turner 1962; see fig. 25.7) to factor analysis (e.g., Parker and Bohm 1979). Although recent interest in DNA has taken some of the lustre from flavonoids in systematic work, these remain a valuable source of comparative data especially at the lower levels of the hierarchy (e.g., Bain and Denford 1985) and in helping to document interspecific hybridization. Much infraspecific variation exists (Bohm 1987), however, and caution must again be exercised in the interpretation of relationships.

Terpenoids. Terpenoids are another important class of secondary plant compounds that have been used in plant taxonomic studies (fig. 21.2). Different types of terpenoids exist, all derived from the geranyl pyrophosphate pathways, including mono-, sesqui-, di-, and triterpenoids leading eventually to the steroids.

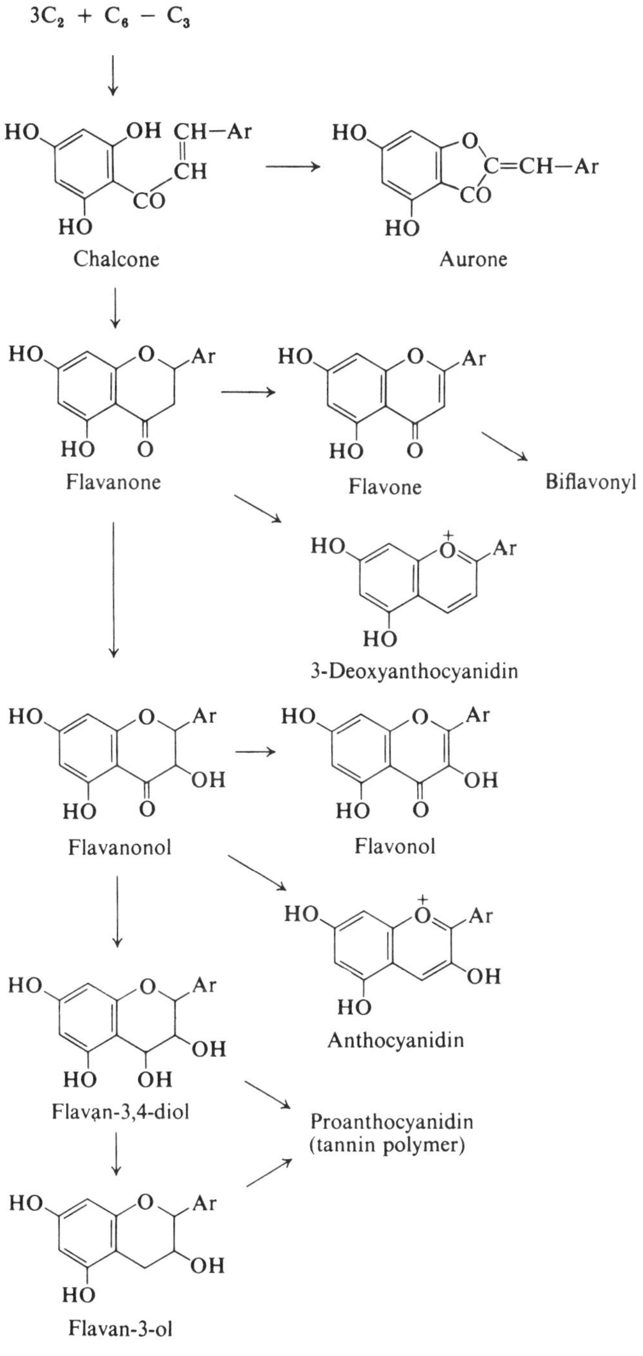

FIGURE 21.1 Representative flavonoids and the biosynthetic pathways of the major classes. (From Harborne and Turner 1984:134)

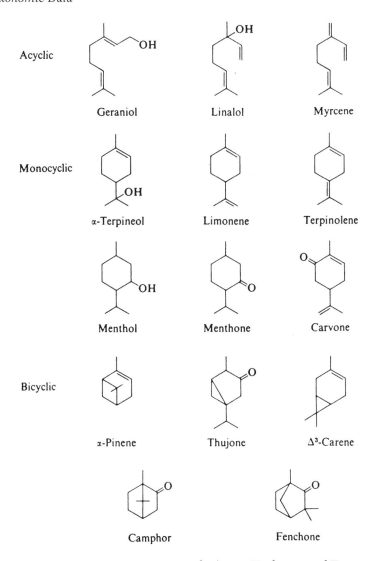

Acyclic — Geraniol, Linalol, Myrcene

Monocyclic — α-Terpineol, Limonene, Terpinolene

Menthol, Menthone, Carvone

Bicyclic — α-Pinene, Thujone, Δ³-Carene

Camphor, Fenchone

FIGURE 21.2 Representative monoterpenoids. (From Harborne and Turner 1984:52)

The most commonly used compounds are the mono- and sesquiterpenoids, and comments will be restricted to these.

The monoterpenoids, also known as "essential oils" from their "essences" or odors, are found in many plant families, but they are especially common in the Labiatae, Rutaceae, Umbelliferae, and also in the gymnosperms. These are low molecular weight compounds that are extremely volatile; hence their ease of detection as odors. The great advantage of these compounds is that they can be analyzed *quantitatively* by gas chromatography by means of integration of areas under the peaks for each compound as well as qualitatively, which gives great sensitivity to the determination of relationships. For example, in *Juniperus*

(Cupressaceae), many studies have been done to determine the limits of intra-individual, ontogenetic, seasonal, and populational variation (Adams 1975b, 1979; Adams and Hagerman 1976, 1977; Adams and Powell 1976). Mints have also been examined in similar ways, e.g., in *Hedeoma* (Firmage and Irving 1979; Firmage 1981) and in *Satureja* (Gershenzon, Lincoln, and Langenheim 1978; Lincoln and Langenheim 1978), as have taxa of the Rutaceae (e.g., Hopfinger, Kumamoto, and Scora 1979), which have important economic implications for the citrus-oil industry. The results do show variation, but not enough to vitiate use for taxonomic purposes. The wealth of data obtained in some groups demands careful statistical analyses, and a good visual presentation is by contour-mapping which shows rates of change of character states from one geographic area to another (fig. 21.3). Because of this sensitivity, monoterpenoid data are

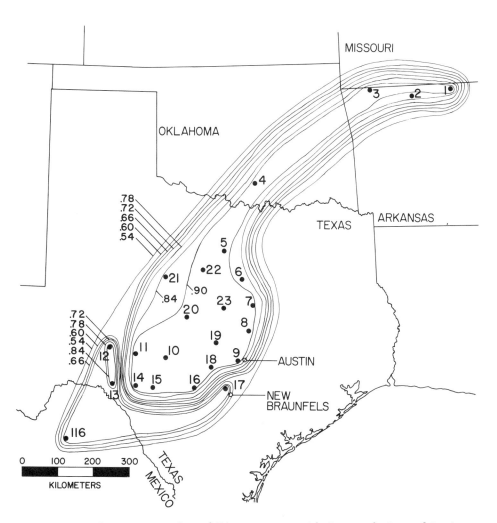

FIGURE 21.3 Contour mapping of 54 monoterpenoids in populations of *Juniperus ashei* (Cupressaceae). (From Adams 1977:187)

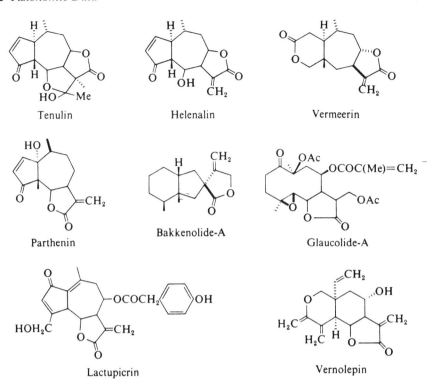

FIGURE 21.4 Structures of typical sesquiterpene lactones. (From Harborne and Turner 1984:111)

especially well-suited for detecting patterns of hybridization and clinal variation (e.g., Flake, von Rudloff, and Turner 1969; Zavarin, Critchfield, and Snajberk 1969).

Sesquiterpene lactones have been used extensively in studies of some families, particularly the Compositae (Seaman 1982), where they can occur in as much as 2 percent of the dry weight of the plants. Here again, many different types of compounds exist with major classes being divided into more minor structural variants (fig. 21.4). One advantage in dealing with sesquiterpene lactones over the monoterpenoids is that plants may be air-dried without loss of compounds during the field work. With monoterpenoids care must be taken to cool or freeze the material until it can be analyzed in the laboratory. The most comprehensive recent work is by Seaman (1982) who surveyed these compounds in the Compositae (in 474 pages!). So much is known that anyone working on the taxonomy of virtually any genus of the family should investigate the existing phytochemical reports for additional potentially useful characters. These compounds are principally useful at the lower levels of the taxonomic hierarchy, such as shown in populations of *Ambrosia psilostachya* (Miller et al. 1968). In another case, in *Ambrosia confertiflora* (fig. 21.5) there was no clear correlation of morphological and chemical variation with known infraspecific cytological variation. This pop-

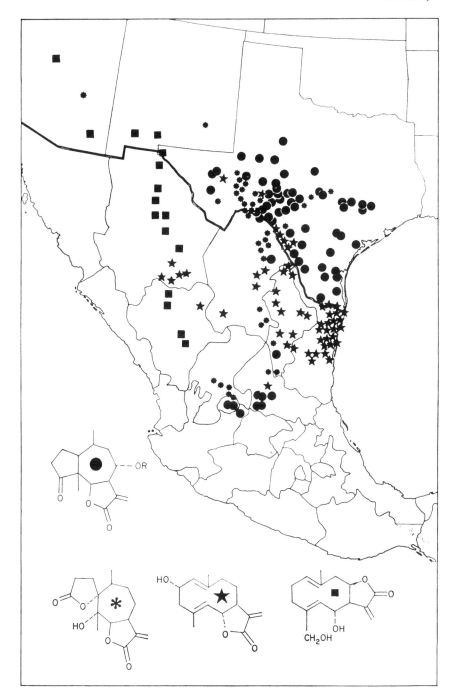

FIGURE 21.5 Populational variation of sesquiterpene lactones in *Ambrosia confertiflora* (Compositae). (From Seaman 1982:166; after Renold 1970)

ulational system represents complicated and still confusing non-correlated patterns of differentiation over a broad geographical area.

Alkaloids. Alkaloids comprise a structurally diverse group of compounds based on one or more nitrogen-containing rings (fig. 21.6). They often show dramatic physiological effects on people and higher animals. Because of this, they have received much attention for pharmaceutical uses and to lesser degrees for their systematic potential. One problem is that alkaloids may be produced in one organ of the plant and then transported for storage to another (Giannasi and Crawford 1986). Careful sampling, therefore, becomes extremely important. Another difficulty is that much of the data in the literature has been gathered by pharmaceutical and natural product chemists (for chemical and/or pharmaceutical studies) and not within the context of systematic surveys. One of the excellent classical studies was by Kupchan, Zimmerman, and Afonso (1961) on *Veratrum*

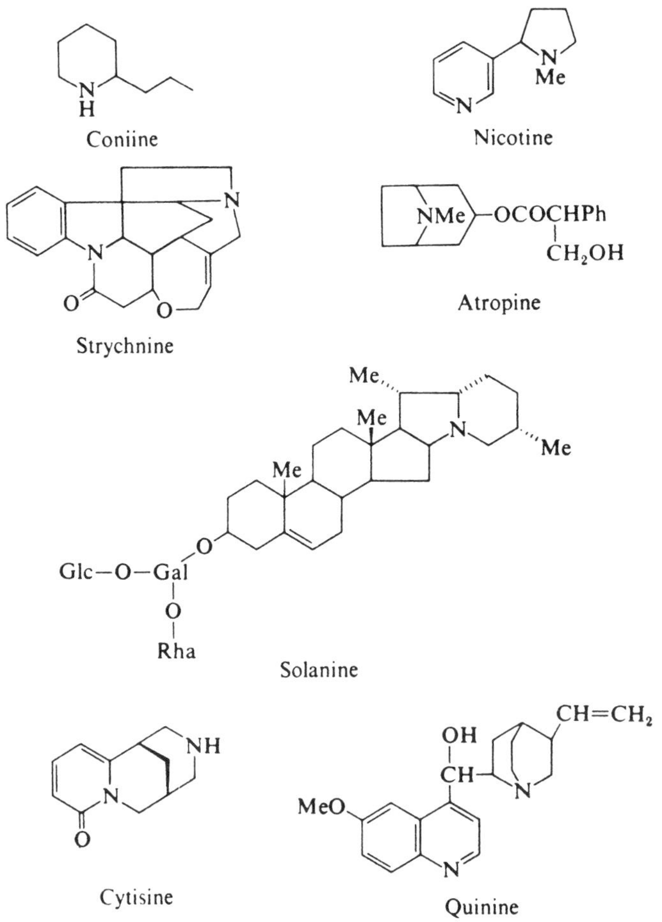

FIGURE 21.6 Structures of some common plant alkaloids. (From Harborne and Turner 1984:79)

(Liliaceae) and relatives, in which the alkaloid data correlated well with morphology at the generic level and gave insights on affinities.

Betalains. Betalains are a fascinating group of compounds that, along with the mustard oils (glucosinolates) to be discussed next, are the best examples of correlations of micromolecular data with other evidence at the ordinal level in flowering plants. Betalains could be regarded as alkaloids because they contain heterocyclic nitrogen (fig. 21.7), but they will be treated separately here for special taxonomic emphasis. The great interest in these compounds among angiosperms comes from their sole occurrence in the order Caryophyllales (Centrospermae; Mabry 1977). These are reddish (betacyanins) or yellowish (betaxanthins) pigments found in petals and sometimes other organs of plants such as cacti, beets, portulacas, bougainvilleas, etc. These compounds were originally called "nitrogeneous anthocyanins" until their true chemical nature (and biosynthesis) was revealed in the 1960s by Dreiding, Mabry and coworkers (e.g., Mabry and Dreiding 1968). Betalains are never found with anthocyanins (a type of flavonoid) in the same plant (or taxon)—only one or the other is present. They often have similar colors in ordinary light and almost identical UV reflectance spectra. Clearly, selection has opted for either one or the other chemical alternative in the production of pigments in petals of flowers of the angiosperms. Some recent ideas suggest that blockage of one pathway can lead to the devel-

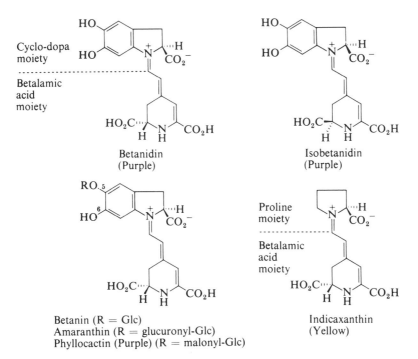

FIGURE 21.7 Structures of betacyanins and betaxanthins. (From Harborne and Turner 1984:159)

opment of the other (Giannasi 1979), and despite the marked biosynthetic differences between the two, they may be more correlated than previously believed. In any event, taxonomically the betalains are restricted to families of the Caryophyllales. Based on phenetic and cladistic analyses, Rodman et al. (1984) placed these betalain-containing families in the "suborder Chenopodiineae," viz., Aizoaceae, Amaranthaceae, Basellaceae, Cactaceae, Chenopodiaceae, Didiereaceae, Nyctaginaceae, Phytolaccaceae, and Portulacaceae. The only two remaining families of the order not containing betalains, the Caryophyllaceae and Molluginaceae, are placed in the separate suborder Caryophyllineae. Recall that these same families also have the P_{III}-type sieve-tube plastids of phloem tissue (see chapter 16).

Glucosinolates. The final type of micromolecular data to be mentioned is the glucosinolates or mustard oil glucosides (fig. 21.8). These are relatively simple sulfur-containing compounds of which about 85 different structural types occur and which, when hydrolyzed with acid or enzyme (myrosinase), give an isothiocyanate and glucose. A review of this group of compounds can be found in Ettlinger and Kjaer (1968). The glucosinolates are largely restricted to the Capparales, and they are most structurally diverse in the Cruciferae. These are useful principally at the lower levels of the hierarchy and gas and paper chromatography are the techniques for identification of the compounds. For example, Rodman, Kruckeberg, and Al-Shehbaz (1981) found species-specific profiles in both *Caulanthus* and *Streptanthus* (Cruciferae) as well as considerable infraspecific variability in some species. These data can be sensitive indicators of relationships as shown by the work on *Cakile* by Rodman (1976) in which patterns of introduction of particular isolated populations from parental source areas were suggested. These data are of limited application to angiosperms in general, however, simply due to their restricted occurrence in the Capparales.

Macromolecular Data

Macromolecular data have for several decades held promise for solving taxonomic problems, especially at the higher levels of the hierarchy. Workers have

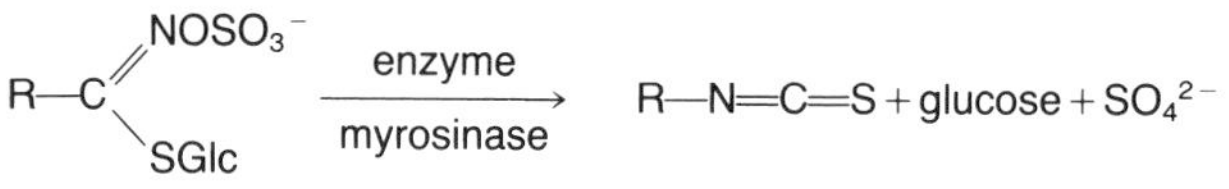

$$R-C\overset{\displaystyle NOSO_3^{-}}{\underset{\displaystyle SGlc}{\diagup}} \quad \xrightarrow[\text{myrosinase}]{\text{enzyme}} \quad R-N=C=S + \text{glucose} + SO_4^{2-}$$

Glucosinolate   An Isothiocyanate

Typical glucosinolates:

Glucocapparin:	$R = CH_3$
Sinigrin:	$R = CH^2{=}CH-CH_2$
Glucoibervin:	$R = MeS(CH_2)_3$
Glucotropaeolin:	$R = PhCH_2$
Sinapine:	$R = p\text{-}OH-C_6H_4$

FIGURE 21.8 Structure of plant glucosinolates. (From Harborne and Turner 1984:69)

hungered for the day when nuclear DNA base-pair sequences could be read directly and compared with each other. To some, this would be the ultimate source of comparative data. The full efficacy of these data still is unknown, but there have been useful studies in recent years. The types of data to be discussed are proteins and nucleic acids. Of the former, seed storage proteins, amino acid sequences, and serology will be mentioned. Electrophoresis of isozymes has already been dealt with in chapter 20 (Genetics and Cytogenetics), as these data are most useful in revealing genetic variation within and between populations, origin of closely related taxa, hybridization, etc., rather than in broad taxonomic affinites within and between genera (and higher). Nucleic acid data include total DNA content, DNA-DNA hybridization, and restriction endonuclease analyses of nuclear and chloroplast DNA (cpDNA).

Proteins. Electrophoresis of total seed storage proteins yields a series of bands as the proteins separate along an electrical gradient based on the polarities of their constituent amino acids (fig. 21.9). These bands can be compared from one taxon to another for an estimate of relationships, and the data can be treated phenetically as with other types of information (e.g., Crawford and Julian 1976). These studies have been particularly effective in helping understand cases of complex hybridization and/or reticulate evolution because they are usually inherited in an additive fashion (e.g., Levin and Schaal 1970). One difficulty, however, is that the bands being compared may not always be the same proteins (Giannasi and Crawford 1986). Another is that if the profiles are complex, the interpretation of relationships becomes correspondingly difficult. And finally, the exact techniques used can have a significant effect on the isolation and separation of the proteins making comparisons of data from one study to the next problematical.

Amino acid sequences of the respiratory enzyme cytochrome c in animal systematics have yielded phylogenetic trees comparable to those based on morphological data (e.g., Wilson, Carlson, and White 1977; Baba et al. 1981). As a result, plant taxonomists hoped to generate equally valuable data for plants to give a much better view of familial and ordinal relationships. This was especially hopeful due to the difficulties of determining ordinal limits based on structural and other micromolecular chemical data which have, without a doubt, evolved in parallel many times in the angiosperms. In fact, the evolution of this group of plants is more characterized by the conspicuous parallelisms of almost all characters than by anything else! It was hoped that protein evolution would be more conservative and hence more reliable as an indicator of relationships. The available data have been generated largely by the pioneering efforts of Donald Boulter and associates, and amino acid sequences of cytochrome c, plastocyanin, and ribulose bisphosphate carboxylase have been investigated (see overview by Boulter 1980; also Martin and Stone 1983; Martin and Jennings 1983; Martin, Boulter, and Penny 1985; Martin and Dowd 1986). These are technically difficult biochemical studies not within the reach of the ordinary taxonomist. The results have been mixed and are basically disappointing for higher level classification. For example, the data for the Compositae (fig. 21.10) give some relationships that are clearly not compatible with other data (e.g., the

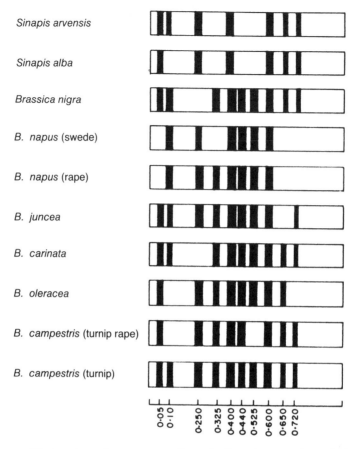

FIGURE 21.9 Electrophoretic patterns of seed albumin proteins of the *Brassica-Sinapis* complex (Cruciferae). (From Harborne and Turner 1984:412; after Vaughan and Denford 1968)

separation of *Taraxacum* from the rest of the Lactuceae such as *Sonchus* and *Lactuca*).

Plant systematic serology has been beset with numerous technical difficulties over the past fifty years; only in the last twenty years has this approach regained the confidence of taxonomists. Extracts of proteins from particular taxa (the antigens) are injected into rabbits or other laboratory animals to produce antibodies (called antisera when drawn out of the animals) which are then cross-reacted on gels to antigens from a series of related taxa. The degree of precipitation observed gives clues to the affinities of the taxa. For more refined comparison of relationships, the antigens are first electrophoresed to separate the consituent proteins and then treated with antiserum. This results in a series of bands for each pair of taxa (fig. 21.11). The pattern of bands is then compared to yield statements of overall relationships. One must be careful to remember the limitation of this approach, especially if crude protein extracts are employed. However, many useful insights have been obtained, especially at the higher

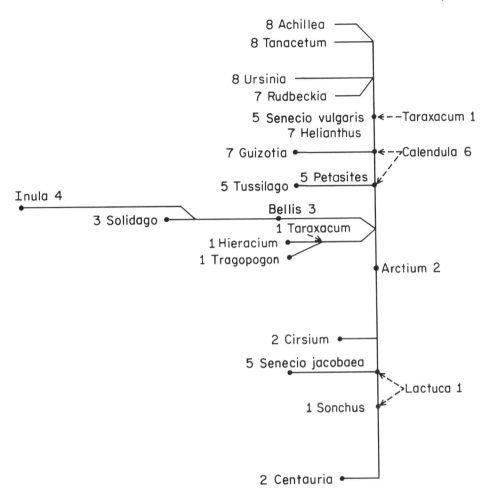

FIGURE 21.10 Trees generated by ancestral sequence method from partial sequences of plastocyanin in species of Compositae. Dashed lines indicate alternative positions of *Taraxacum, Calendula* and *Lactuca* in an equally parsimonious tree. Numbers refer to tribes (1) Cichorieae, (2) Cynareae, (3) Astereae, (4) Inuleae, (5) Senecioneae, (6) Calenduleae, (7) Heliantheae, (8) Anthemideae. (From Giannasi and Crawford 1986:164; redrawn from Boulter et al. 1978)

levels of the hierarchy (e.g., in the circumscription and major division of the "Amentiferae" by Petersen and Fairbrothers 1985; or in the relationships of the Magnoliidae proposed by Jensen and Greven 1984).

Nucleic Acids. Nucleic acids have received much attention in recent years as the hope for the near and distant future in the assessment of plant relationships. Here the actual genetic material is compared which should offer significant phyletic insights. Especially promising is the report that good quality DNA can be obtained from dried herbarium material a few months old (Doyle and Dickson 1987). DNA has also been obtained from mummified plant tissues up to 45,000

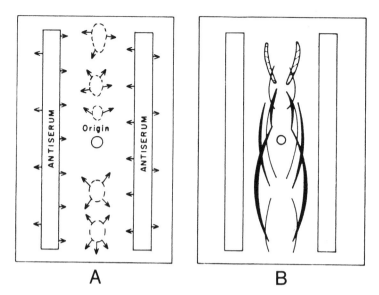

FIGURE 21.11 Immunoelectrophoresis of plant proteins. A, after gel electrophoretic separation, antisera are placed in troughs at side. B, when antigens and antibodies diffuse towards each other, they meet and precipitation occurs along an arc of optimal antigen-antibody proportion. (From Harborne and Turner 1984:430; after Smith 1976)

years old (Rogers and Bendich 1985), but it has been significantly degraded and of minimal systematic value. The work of Price (e.g., Price et al. 1986) and others on total DNA content is of limited taxonomic import but of great interest for helping to explore the adaptive reasons for large-scale total DNA fluctuations within and between populations even within a single species. So far, however, few correlations have been found (e.g.,. Sims and Price 1985; see additional comments in chapter 19). DNA-DNA hybridization was early believed to have great potential for plant systematics (e.g., Bendich and Bolton 1967), but the technical difficulties in dealing with plants made this dream vanish. Especially problematical was that when DNA was reannealed with that from the same species, often low (around 20 percent) values were obtained. The degree of reannealing was tested by chemical and thermal gradients, but the overall low level of reannealing made it difficult to evaluate confidently the inter-taxon comparisons. A few studies were completed, such as that by Chang and Mabry (1973) on the rRNA-DNA relationships of families within the Centrospermae, but the reannealing differences observed were small, so much so that greater differences were seen between some betalain-containing families than between betalain- and anthocyanin-containing taxa. Recent studies in animal systems using DNA-DNA hybridization have been more successful (e.g., Sibley and Ahlquist 1983, 1984; Caccone and Powell 1987), as has newer work on plants (e.g., Bendich and Anderson 1983; Okamura and Goldberg 1985). King and Ingrouille (1987) have used high-resolution thermal degradation of total DNA to yield base composition comparisons in the Gramineae, but the results are of uncertain value due to the mixed taxonomic results.

The difficulties with DNA-DNA hybridization and base composition studies have caused workers to turn to restriction endonuclease fragmentation of nuclear and chloroplast DNA (cpDNA). Mitochondrial DNA studies have also been done (e.g., Hasegawa, Kishino, and Yano 1985), but rarely in plants (for overview, see Sederoff 1987). Current progress in genetic engineering techniques has made new and more effective approaches possible, so that at the present time much promise is held for using these data at different levels of the hierarchy. Basically, one separates the circular cpDNA molecule from plant chloroplasts and then breaks this molecule at numerous points by use of different restriction enzymes that are specific for short (usually six) sequences of base pairs. These fragments are then separated on gels electrophoretically and a "map" of the sequence of the fragments and their size is developed. These circular maps are then presented for each taxon (fig. 21.12) and compared by synapomorphic cladistic techniques to yield cladograms of the DNA fragment data (fig. 21.13). The results of the few studies of this type to date are most encouraging. Direct sequencing is also being done, and these data are best analyzed phenetically or with maximum likelihood methods (Ritland and Clegg 1987). General reviews of the methods and potentials of these new data are found in Palmer (1986, 1987). Most impressive have been the results of cpDNA in attempting to determine the primitive tribe of the Compositae (Jansen and Palmer 1987). A unique inversion sequence has been found in nearly all members of the family that is unknown in related families such as the Rubiaceae, Campanulaceae, etc. This same sequence is also absent in the Mutisieae, a tribe abundant and diverse in South America, which gives support for the Mutisieae being the primitive tribe of the family. This idea was first proposed earlier on morphological grounds by Jeffrey (1977b). Nuclear DNA systematic studies in plants are few (e.g., Tanksley 1987), but most have focused on rDNA with equally promising results to that of cpDNA (cf. Doyle, Beachy, and Lewis 1984; Doyle, Soltis, and Soltis 1985; Hori, Lin, and Osawa 1985), and useful at lower levels in the hierarchy. Ribosomal RNA sequences have also been obtained for taxonomic comparisons (e.g., in the Gramineae; Hamby and Zimmer 1988).

Physiological Data

In addition to the above micromolecular and macromolecular data, differences that have taxonomic value exist in basic photosynthetic mechanisms. Although these data deal with CO_2 and small carbohydrates, they really are physiological data of sorts involving biosynthetic pathways of basic metabolites or *processes* rather than structures. Three types of photosynthesis occur in plants: C_3, C_4, and CAM. In the more common C_3 photosynthesis, the CO_2 from the air is fixed to a C_5 receptor (ribulose 1,5-diphosphate) yielding two C_3 molecules (phosphoglyceric acid). In C_4 photosynthesis, the CO_2 is fixed to phospho-enol-pyruvate to form malic and aspartic acids, both C_4 acids. Further, the fixed CO_2 is transferred to the bundle sheath cells of the leaves rather than being stored in the mesophyll (as in C_3). The suite of anatomical modifications associated with C_4 photosyn-

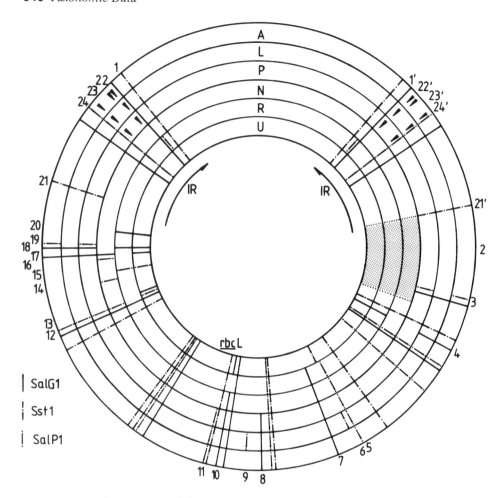

FIGURE 21.12 Comparison of the cpDNA restriction maps for eight species of *Linum* (Linaceae). IR=inverted repeat; rbcl=gene coding for large subunit of ribulose, 1,5-bisphosphate carboxylase. A=*L. alpinum;* L=*L. lewisii;* P=*L. perenne* and *L. boreale;* N=*L. narbonense;* R=*L. grandiflorum* var. *rubrum;* U=*L. usitatissimum* and *L. bienne.* Triangles indicate sites of short insertion/deletions relative to *L. usitatissimum* cpDNA. Shaded portion refers to a large deletion in the non-*perenne* group cpDNA as compared to the *perenne* group. (From Coates and Cullis 1987:263)

thesis, including agranal chloroplasts of the bundle sheath cells (Johnson and Brown 1973), is called the Kranz syndrome. This is usually applied to the anatomical modifications only, but Brown (1975) has pointed out that the term is sometimes applied to the entire physiological and anatomical condition. In CAM (Crassulacean Acid Metabolism) photosynthesis, malic and aspartic acids (C_4) are also the form of fixed CO_2, just as with the C_4 pathway, but they are produced at night rather than during the day. The photosynthetic reduction cycle of both, however, occurs in the day. CAM plants are generally succulent, but they do not display the consistency of anatomical adaptations found in C_4

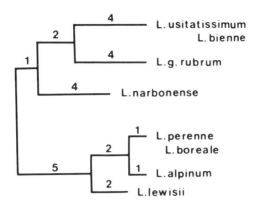

FIGURE 21.13 Cladogram showing relationships among taxa of *Linum* (Linaceae) based on cpDNA data. The numbers relate to the number of mutations in the branch that the cpDNAs have in common. See figure 21.12. (From Coates and Cullis 1987:266)

plants. Taxonomically, the combination of C_4 data with associated Kranz syndrome has provided helpful insights especially in the Gramineae (e.g., Ohsugi and Murata 1986). In this family, improved tribal delimitations were suggested based on this information (Brown 1977). Of lesser scale, but of equal interest, are similar surveys in *Euphorbia* (Euphorbiaceae) by Webster, Brown, and Smith (1975). Here it was suggested that CAM and C_4 photosynthesis have arisen independently within the genus. Surveys in the Compositae have also been done (Smith and Turner 1975), but with few positive correlations with existing tribal and subtribal limits. Overall, these photosynthetic data are only valuable within certain limits, and some taxa have great diversity including species with intermediate C_3-C_4 photosynthesis (Prendergast and Hattersley 1985).

Many examples of physiological ecology exist, and it is here that additional systematic ties to physiology occur. The work by Al-Aish and Brown (1958) on the germination response of grasses (Gramineae) to application of isopropyl-N-phenyl carbamate is one example. Here they found that representatives of the subfamily Festucoideae were inhibited from germinating whereas those of the Panicoideae were not. Rates of dark CO_2 uptake and acidification in 30 genera of the Bromeliaceae, Euphorbiaceae and Orchidaceae were examined by Mc-Williams (1970), who showed that this syndrome occurs in the advanced taxa of all three families, presumably reflecting evolutionary parallelism. The paper by Robichaux and Canfield (1985) deals with elastic properties of tissues in the Hawaiian *Dubautia* (Compositae), but mainly from the standpoint of their adaptations to different habitats. The species adapted to drier conditions showed lower tissue elastic properties than those adapted to mesic environments.

INVESTMENTS FOR GATHERING CHEMICAL DATA

The equipment, cost, time and expertise needed for chemical work are perhaps the most demanding of all the data-gathering approaches used in plant taxonomy. For flavonoids, monoterpenoids, serology, protein electrophoresis, and mustard oils, a list of equipment needed (although now out-of-date) with costs and time constraints is provided by Stuessy (1972b). Basically, flavonoids require chromatocabinets, flash evaporators and UV spectrophotometers at a general cost of $10,000 to $20,000. Terpenoid data require glassware for steam distillation, gas chromatographs and infrared analyzers at costs of $20,000 to $50,000. Serology requires animals, animal housing, a refrigerated centrifuge, and miscellaneous pieces of glassware and special trays at costs over $15,000. Electrophoresis of proteins will necessitate balances, refrigerators, incubators, and power supplies, pH meters, etc. to a cost of at least $10,000. Mustard oils have similar requirements to those of terpenoids. DNA techniques can be much more expensive, with ultracentrifuges bringing the total set-up lab costs to as much as $100,000. It takes considerable chemical knowledge and time commitment to do DNA work, but at the other extreme, flavonoid data can be generated much more rapidly by people trained as taxonomists with little or no chemical background. The continuing lab costs for supplies for macromolecular work are also greater than for work with micromolecules. The bottom line, however, is that chemical data do involve a considerable commitment of time and other resources to be effective, which is one reason why collaboration among workers is to be encouraged.

EFFICACY OF CHEMICAL DATA IN THE TAXONOMIC HIERARCHY

Chemical data are useful at all levels of the taxonomic hierarchy, the exact level depending upon the particular compounds employed. As a general rule the micromolecular data are most useful at the lower levels (such as specific and infraspecific) and the macromolecules are more efficaceous at the higher levels (generic and above). Flavonoids and terpenoids are clearly useful at the lower levels as demonstrated by numerous studies (e.g., Young 1979; Doyle 1983; Adams, Zanoni, and Hogge 1984; Bain and Denford 1985; Park 1987). Because of the relative ease of quantification of monoterpenoid data, these are especially useful at the infraspecific level whereby many populations over broad geographic areas need to be sampled. More recent use of HPLC (high performance liquid chromatography) with flavonoids offers hope that quantitative data with these compounds may be forthcoming in the near future. At the generic and familial levels the secondary compounds become generally less useful with some exceptions, e.g., flavonoids at the generic level in the Ulmaceae (Giannasi 1978) and flavonoids in determining familial affinities of the Idiospermaceae (Sterner and Young 1980). Betalains and glucosinolates, however, are the two micromolecular classes of compounds that are most useful at the ordinal level; in fact, these serve as the best examples of chemical data for ordinal circumscription! The real potential for revealing higher level relationships, however, must lie with macromolecular data, particularly DNA. Once the base pair sequences are obtained, we will need more ontogenetic and developmental data to understand how the sequences in one part of the DNA molecule interact with others to bring about the chemical and structural results seen in each cell, tissue, organ, and whole organism. Despite technical difficulties and expense, this holds much promise for the future.

SPECIAL CONCERNS WITH CHEMICAL DATA

One interesting focus with micromolecular data that should be mentioned is their ecological and adaptational role in the organism (e.g., Levin 1971b). It is generally believed that most secondary metabolites represent the plant's chemical arsenal to defend against environmental vagaries, especially predators (e.g., Cronquist 1977). That terpenoids inhibit feeding of some plants by insects and other herbivores has now been shown clearly (e.g., monoterpenes in *Hymenaea*, Leguminosae, Langenheim, Foster, and McGinley 1980; sesquiterpene lactones in *Vernonia*, Compositae, Burnett et al. 1974), and the same can be said for alkaloids (e.g., Barbosa and Krischik 1987). Plant pigments also obviously figure in pollination syndromes. Also we know that some terpenoids attract pollinators, as in the orchids and euglossine bees (e.g., Williams and Dodson 1972). Primary metabolic interactions of some secondary products have also been suggested (Seigler and Price 1976, Jacobs and Rubery 1988), although there is much less known about this aspect. This general area of function of secondary metabolites is significant and will no doubt receive deserved and increased attention in the years ahead. A recent example of this direction is seen in the symposium led off by Greger (1985).

A few comments on the use of cladistic techniques and DNA fragment analysis seem appropriate. DNA data can be treated in two ways: with base pair sequence data, phenetic algorithms are best; with cpDNA fragment map data, cladistic algorithms are more appropriate. The overall base pair sequences can be assessed by overall similarity or distance (or even maximum likelihood methods; Ritland and Clegg 1987) in much the same way as done with amino acid sequences in cytochrome *c* (e.g., Fitch 1984). With fragment maps, however, the comparisons to be made are based on unique lengths of DNA (caused by cleavage at unique short base pair codes) and hence may represent derived character states in comparison to an outgroup fragment map. If two taxa have the same fragment in common, and this is not found in the outgroup, it can be assumed that this is a macromolecular synapomorphy and may be treated by parsimony (or other tree-generating) techniques to obtain a cladogram. It is of historical interest, therefore, that many of the people earlier dedicated to isozyme studies at the populational level, who had little interest in broader questions of phylogeny or cladistic techniques, are now switching to cpDNA work in which the main questions are phylogenetic ones and the appropriate modes of analysis are cladistic! The cladists, on the other hand, earlier stressed the idea that morphology was the appropriate and totally adequate means of phylogenetic reconstruction via cladistic algorithms. Now the cladists are on the defensive from the cpDNA workers who are using cladistic techniques and claiming that these data are more diagnostic for reconstructing phylogeny than morphology!

Phylogenetic techniques have also been applied to the analysis of micromolecular data. The use of micromolecular data in the 3-D reconstruction of phylogeny has been advocated by Stuessy and Crawford (1983). This study stresses that flavonoids (and other similar data) can be treated phenetically, cladistically and patristically, as with morphology, in the reconstruction of a 3-D phylogram (see chapter 9 for example with morphology). That such data can be treated cladistically in a more limited context has also been shown on several occasions (e.g., Humphries and Richardson 1980; Seaman and Funk 1983). These data, in fact, have the additional power of known biosynthetic pathways. This is really ontogenetic information and is very helpful for suggesting ideas on the primitive versus advanced nature of particular compounds (e.g., Gottlieb 1980). Caution is required here, however, because the loss of just one enzyme can lead to markedly shortened pathways and spurious comparisons.

A final topic deals with paleochemotaxonomy. This book does not deal with details of fossil analysis, but it is of interest that some compounds (e.g., lipids and flavonoids) can persist (under unusual conditions of preservation) virtually unchanged in leaves and stems of fossil plants for many millions of years! One of the most amazing studies is that of Niklas and Giannasi in the Ulmaceae. The leaves of *Zelkova* from the Succor Creek flora of Oregon, ca. 20 million years old, are actually *green* when taken from the matrix! Most significantly, these studies have shown that the flavonoids still persist in these fossil leaves and can be compared with modern relatives. In this case a strong chemical connection has been shown between fossil taxa of Asia and North America and their modern relatives, confirming earlier suggestions of phytogeographers (Giannasi and Niklas 1977; Niklas and Giannasi 1977).

In conclusion, it should be obvious that chemical data in plant taxonomy are here to stay as a source of comparative information. Gone are the days of wrangling over the efficacy of chemical vs. morphological data popular in the late 1950s and early 1960s. Also largely gone are the extreme views by which taxa are circumscribed solely by chemical means (for an amazing early example see Fujita 1965, in *Mosla*, Labiatae). As Heslop-Harrison said two decades ago: "if anything is certain it is that they [chemical data] will in the future provide an increasing part of the evidence available to the working taxonomist in all groups. As such they have many potential advantages, including for some classes of compounds the particular merits of consistency, ease of assay and unambiguity" (1968:284).

CHAPTER 22

Reproductive Biology

Plants do not live in isolation; they constantly interact with biotic and abiotic elements of their environment. Furthermore, they possess differences in life history strategies which allow them to adapt to their environment and survive. The reproductive syndrome of plants is an important facet of their adaptational response. Many aspects can be included here, but some types of data are more strictly ecological or evolutionary and are not useful for comparative taxonomic purposes. For example, the allocation of resources to vegetative vs. reproductive organs is of interest ecologically and evolutionarily (e.g., Abrahamson and Hershey 1977; Abrahamson 1979; Abrahamson and Caswell 1982; Willson 1983), but few studies exist which compare taxa in this regard for purposes of classification. Hence, they are of limited value for taxonomic purposes, although they can lead to biological insights and stimulate future comparative studies. Many reproductive data *have* been more directly helpful, however, and the principal ones are phenology, UV patterns, pollinators, floral nectars, breeding systems, and dispersal agents. These are the ones to be discussed here.

HISTORY OF REPRODUCTIVE BIOLOGY IN PLANT TAXONOMY

The history of reproductive biology as it relates to plant taxonomy could be traced to Theophrastus or even to the ancient pictures of date palm workers in Egypt or Mesopotamia, but for our purposes it begins with Rudolph Jakob Camerarius (1665–1721), who first experimentally demonstrated sex in plants in 1694. This was followed by observations on the role of insects in flower pollination by Kölreuter (1761–1766). To Sprengel, however, must go the credit for more elaborate studies of the relationships of insects and flowers. His book *Das entdeckte Geheimniss der Natur im Bau und in der Befruchtung* in 1793 detailed the floral adaptations of 500 or more species. Darwin (1809–1882) was also much interested in pollination and breeding systems from an evolutionary perspective and his books *On the Various Contrivances by which Orchids are Fertilised by Insects* in 1862 and *The Effects of Cross and Self Fertilisation in the Vegetable Kingdom* in 1876 furthered this understanding. Hermann Müller (1829–1883) provided a powerful influence with his three publications (1879, 1881, 1883) on the details of insects and their flowers from central Europe. This was followed by Paul Knuth's three-volume book (1906–1909) on floral biology which summarized the available information and provided many original observations especially from Germany. Other vectors serving as pollinators, such as birds and bats, were recognized here too. Other important studies followed, such as those of Frisch (1914) on responses of bees, and Clements and Long (1923) on experimental pollination. For a more detailed history of pollination studies, see Baker (1983) and Ducker and Knox (1985).

GENERAL REPRODUCTIVE BIOLOGICAL TEXTS AND REFERENCES

From these beginnings has come a wide variety of works providing comparative data or perspectives on reproductive biology for taxonomists. General books on pollination biology include Percival (1965), Kugler (1970), Proctor and Yeo (1973), Richards (1978), Faegri and van der Pijl (1979), Meeuse and Morris (1984), and Barth (1985). Real (1983) has brought together a good collection of individual contributions, and Armstrong, Powell, and Richards (1982) have edited a symposium presented at the XIII International Botanical Congress (Sydney). Jones and Little (1983) edited a good series on experimental pollination, and more practical aspects of crop pollination can be examined in the work by Free (1970). For comprehensive lists of insect visitors to particular flowers, none is as impressive as the observations of Robertson (1928) who documented 15,172 visits to 453 different plant species! For visitors to more than 400 crop plants, one can consult Crane and Walker (1984). The pollination syndromes of particular plant families have been detailed by Grant and Grant (1965) for the Polemoniaceae, and by van der Pijl and Dodson (1966) and Dressler (1981) for the Orchidaceae. In-depth studies on pollinators abound, and these include a view of the hummingbirds and their flowers (Grant and Grant 1968), and intensive studies on bees, their senses, communications, pollination, and economics (Frisch 1950, 1953, 1967; Heinrich 1979). Breeding systems have been well covered recently by Richards (1986), and dispersal mechanisms have been discussed in depth by van der Pijl (1972) as well as in the symposium volumes by Kubitzki (1983), Estrada and Fleming (1986), and Murray (1987). Overall perspectives of reproductive biology can be obtained from the symposium edited by Godley (1979) as well as in the recent text on plant reproductive biology by Willson (1983). Aspects of coevolution that deal with reproductive biology can be found in Van Emden (1973), Davidse (1974), Gilbert and Raven (1975), Kraus (1978), and Abrahamson (1988).

TYPES OF REPRODUCTIVE DATA

Phenology

Phenological data from plants are useful for taxonomic purposes. If two closely related taxa do not flower at the same time (fig. 22.1), whether they are sympatric or allopatric, this suggests they will not interbreed and are likely to be good species (applying the biological species concept). Often there is an overlap period

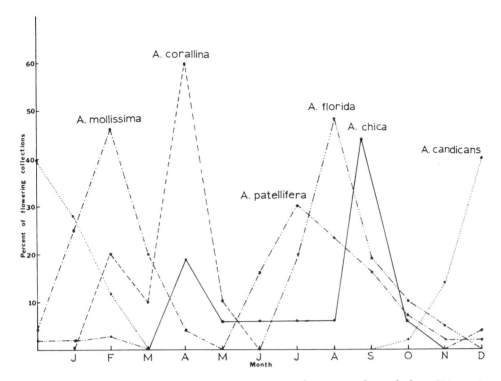

FIGURE 22.1 Summary of flowering durations of species of *Arrabidaea* (Bignoniaceae) in Costa Rica and Panama. (From Gentry 1974:65)

which must be evaluated in reproductive terms, such as the possible production of hybrids, for taxonomic import. Another important aspect is the comparative flowering "behavior" of certain flowers, especially those with complex morphologies. Such is the case in *Anthurium* of the Araceae (Croat 1980), with the characteristic spadix of unisexual flowers enclosed by the spathe (or bract). Protogyny (the earlier development of the female flowers), which promotes outcrossing, can occur here, and different patterns of stamen emergence (and retraction!) relate to differences in pollen presentation. How plants flower, therefore, can provide useful comparative data in certain groups.

Ultraviolet Patterns

Patterns of ultraviolet (UV) reflectance and absorbance from reproductive structures, especially petals, are important in pollination and useful as comparative data (Kevan 1983; fig. 22.2). These patterns derive from differences in cellular structure of the epidermis (e.g., see p. 228, chapter 15) and from different compounds such as carotenoids and flavonoids (e.g., Harborne and Smith 1978). Some workers have gone so far as to treat these as comparative data across broad groups (e.g., King and Krantz 1975), but this is not appropriate. These data are best used among very closely related taxa in the context of understanding their overall reproductive biology.

FIGURE 22.2 Ultraviolet reflectance patterns among three species of *Potentilla* (Rosaceae). A, *P. erecta;* B, *P. reptans;* C, *P. verna.* (From Daumer 1958:86)

Pollination Vectors

Many different vectors affect pollination, and the understanding of these can be very helpful taxonomically. Usually structural modifications accompany such variations, such as differences in corolla color and shape, anther size and shape, position of anthers relative to the stigma, and so forth. The major factors are both abiotic and biotic. Of the former we must count wind (anemophily) and water (hydrophily) as the agents to be considered. Wind pollination is very common in the angiosperms and many entire plant families are pollinated this way (e.g., Betulaceae, Ulmaceae); even most of the entire subclass Hamamelidae have flowers in catkins pollinated by wind. These inflorescences contain small unisexual flowers with no or only inconspicuous perianth parts. Water is a less common vector, and flowers here are similar to those of wind pollinated taxa in having reduced perianths and unisexuality. Sometimes pollination is along the surface of the water as in *Vallisneria* (Hydrocharitaceae; Proctor and Yeo 1973) or even through the water as in *Amphibolis antarctica* (Cymodoceaceae; Ducker and Knox 1976).

Of the biotic pollination factors we have insect (entomophily), bird (ornithophily), and mammal pollination (bats, marsupials and rodents). Numerous insects effect pollination in search of pollen and/or nectar as food sources. Bees (melittophily), wasps, butterflies (psychophily), moths (phalaenophily), flies (myophily), beetles (cantharophily), and ants (myrmecophily) are all important. Thrips are also pollinators (Norton, 1984). Flowers show adaptations to specific vectors, with bee flowers generally being blue or yellow, moth-pollinated flowers pale-colored or white with tubular corollas and fragrances (see Grant 1983b, 1985a, for recent data), fly flowers dull-colored (sometimes purple) and sometimes foul-smelling, beetle flowers with strong fragrances and food tissues, and ant flowers close to the ground with abundant nectar. The orientation of flowers (fig. 22.3) is related functionally to vector behavior. It must also be remembered that several species of visitors may be attracted to a particular flower (table 22.1), and this should be carefully investigated by long-term observations over several days or weeks. Bird pollinated flowers often have red or bright orange corollas with long tubes. Hummingbirds are important in the New World as pollinators (Grant and Grant 1968), as are honeycreepers in Hawaii (Lammers and Freeman 1986), sunbirds in Africa, and honeyeaters in Australia (e.g., Paton 1982). Bat pollination (chiropterophily) has been known in tropical regions for

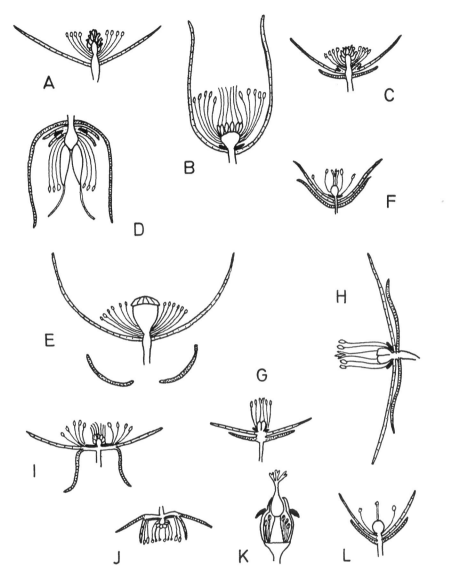

FIGURE 22.3 Sectional diagrams of flowers, showing more or less their natural orientations. Thick lines or black areas refer to nectaries; darkly stippled parts are sepals; lightly stippled are petals. A, *Anemone nemorosa* (Ranunculaceae); B, *A. pulsatilla;* C, *Ranunculus acris* (Ranunculaceae); D, *Helleborus foetidus* (Ranunculaceae); E, *Papaver rhoeas* (Papaveraceae); F, *Stellaria holostea* (Caryophyllaceae); G, *Geranium pyrenaicum* (Geraniaceae); H, *G. pratense;* I, *Rubus fruticosus* (Rosaceae); J, *R. idaeus;* K, *Euphorbia paralias* (Euphorbiaceae); L, *Lysimachia vulgaris* (Primulaceae). (From Procter and Yeo 1973:52)

TABLE 22.1 Oil Collecting *Anthophorinae* Bees Observed (X) During Several Flowering Periods on Several Species of Malpighiaceae at a Site Near Botucatu, Saõ Paulo, Brazil. (From Gottsberger 1986: 33)

Oil collecting Anthophorinae Plant species	Centridini							Epichoris bicolor	E. cockerelli
	Centris discolor	C. dorsata	C. mocsaryi	C. nitens	C. pectoralis	C. scopipes	C. sp. 1		
Banisteriopsis latifolia		X					X		
Byrsonima coccolobifolia	X		X	X			X	X	X
Byrsonima intermedia				X				X	X
Byrsonima vaccinifolia									
Byrsonima verbascifolia							X		
Tetrapterys ramiflora									
Malpighiaceae sp. 1									
Malpighiaceae sp. 2		X			X	X			
Malpighiaceae sp. 3		X							

some time. These flowers have strong scents, open at night, are often white or greenish, with stamens and styles often exserted, and they are strongly constructed so that the animals have a place to light or rest while feeding. More recently the role of other mammals in pollination has been acknowledged. Although more studies need to be done, it seems clear that some small marsupials in Australia do effect pollination (Wiens, Renfree, and Wooller 1979; Turner 1982; Paton and Turner 1985), as do small nonflying mammals (rodents) in South Africa (Wiens et al. 1983). Monkeys have also been implicated as pollinating agents (Prance 1980).

Nectar

The occurrence of nectaries in flowers and the composition of nectars are also useful comparative data for assessing relationships. Some decades ago it was realized that qualitative and quantitative differences in nectar sugar vary from family to family throughout the angiosperms (Percival 1961). There are also definite correlations with high levels of sucrose vs. hexose sugars with general classes of pollinators. For example, even without direct observations it was determined that species of *Clermontia* (Lobelioideae) of the Hawaiian Islands were most probably bird pollinated based upon the kinds of sugars present in the nectar (Lammers and Freeman 1986). This was an especially welcome suggestion because many of these birds are now extinct and no direct observations are possible. More recently amino acids have been found in nectar and these do have taxonomic significance (e.g., Baker and Baker 1986). Oil is also a food reward in a few selected systems (Vogel 1974), such as in the Krameriaceae (Simpson, Neff, and Seigler 1977). Sugar composition of nectar has also been shown to be subject to convergence (Freeman, Reid, Becvar, and Scogin 1984), as with many other features of angiosperms.

Breeding Systems

Breeding systems of plants should be documented during a taxonomic study if at all possible. Effective pollen transfer to the stigma and control of its germination and growth are most important for reproductive success of each species. Vectors are responsible for the transfer of pollen, but breeding systems determine if the pollen grain can germinate on a receptive stigma, penetrate the style, and effect fertilization. The different types of effective pollen transfer are shown in figure 22.4. Ramets are parts of an asexually reproducing population, whereas genets refer to genetically different individuals. Relating to these transfers are the arrangements of sexual parts within flowers and within single plants (table 22.2). An understanding of this phenetic complexity may allow us to interpret patterns of comparative data among taxa. Inbreeding and apomixis especially need documentation, because these often lead to pockets of variation in separate populations. A recent general survey of breeding systems for part of an entire flora is provided by McMullen (1987) for the Galapagos Islands. A technique to estimate some aspects of breeding systems is pollen/ovule ratio (Cruden 1977; Cruden and Miller-Ward 1981). The idea is that inbreeders tend to produce fewer

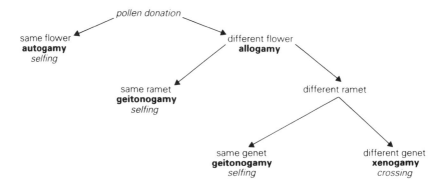

FIGURE 22.4 Patterns of pollen transfer within and between flowers and plants. (From Richards 1986:3)

pollen grains per ovule than outcrossers, and a rough estimate, therefore, can be obtained of the breeding systems even from herbarium material. A good example is the study of Philbrick and Anderson (1987) in which pollen/ovule ratios were surveyed in aquatic angiosperms. The results of the survey were contrasted with those of actual breeding system tests with consistent results, except that somewhat higher ratios were obtained in some groups due to variations in different modes of pollination. Competitive ability of pollen grains to germinate and grow through receptive stylar tissue had been used for assessing relationships within some species of *Haplopappus* (Compositae) by Smith (1968, 1970). Structures of the stigmas and styles themselves can also be used as comparative data and many useful differences occur (fig. 22.5). Good surveys include those of Heslop-

TABLE 22.2 Common Types of Sex Distribution Within and Between Flowers and Genets of Angiosperms. (From Richards 1986: 4)

Name	Distribution of sex organs		Breeding system	Angiosperm species (%)
	within a flower	within a plant		
dioecy	♂ or ♀	♂ or ♀	xenogamous (outcrossing)	4
gynodioecy	⚥, ♂ or ♀	⚥ or ♀	xenogamous, geitonogamous, autogamous	7
monoecy	♂ or ♀	⚥	allogamous, some selfing, some crossing	5
gynomonoecy	⚥ or ♀	⚥	allogamous and autogamous	3
hermaphrodity	⚥	⚥	allogamous and autogamous	72
(other)				9
				100

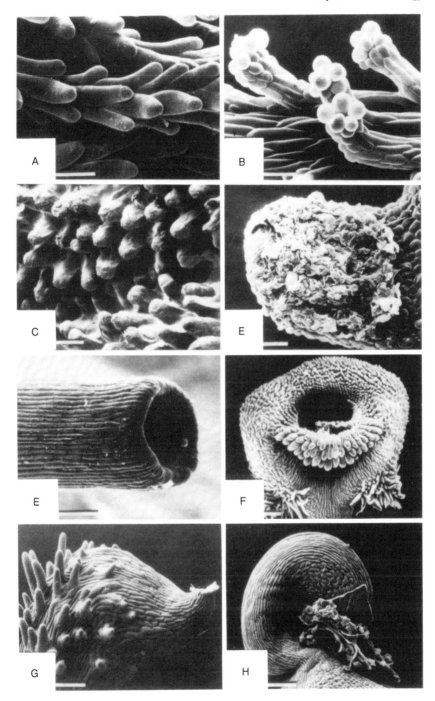

FIGURE 22.5 Examples of stigmatic surfaces in selected angiosperms. A, *Avena sativa* (Gramineae); B, *Ceratostigma plumbaginoides* (Plumbaginaceae); C, *Paris quadrifolia* (Liliaceae); D, *Corydalis bulbosa* (Fumariaceae); E, *Erica carnea* (Ericaceae); F, *Viola lutea* (Violaceae); G, H, *V. odorata* (chasmogamous and cleistogamous forms, respectively). (From Schill, Baumm, and Wolter 1985:189)

Harrison (1981), Small and Brookes (1983), Brown and Gilmartin (1984), Owens et al. (1984), and Schill, Baumm, and Wolter (1985). Cruden and Lyon (1985) point out, however, that such features which are so intimately connected with the reproductive system in some cases may be modified more by function than by phylogeny, although the latter is strongly implicated in most cases.

Dispersal Systems

Analogous to pollination syndromes, dispersal systems form a very important part of the reproductive process. Here, different vectors move the dispersal units away from the parent plant; structural modifications correlate with these different agents. Some seeds are distributed mechanically by means of the plant itself (autochory), such as the explosive fruits in *Impatiens* (Balsaminaceae). Others are distributed by animals (zoochory) such as birds (ornithochory), mammals (mammalichory), reptiles (saurochory), fish (ichthyochory), and ants (myrmecochory). Different terms are also used to indicate how and where the propagules (=disapores) are carried on the animals. If accidentally transported on the animal it is called epizoochory; if inside the animal, endozoochory; and if deliberately carried (usually in the mouth), synzoochory (van der Pijl 1972). Ants exemplify the latter when they carry off seeds for their food value. A food body or elaiosome is often involved (Beattie 1983, 1985; Slingsby and Bond 1984). In general, those fruits or seeds which are brightly colored, fleshy, or with much

---

TABLE 22.3 Definitions of Terms Describing Flower-Visitor Relationships. (From Faegri and van der Pijl 1979: 48)

Blossom Relationship	
Dealing principally with morphological adaptation of blossom	*Dealing principally with character of visits received*
No morphological adaptations for guiding visitors; can be utilized by unadapted, short-tongued visitors: *allophilic*	
Imperfectly adapted to being utilized by animals of intermediate degree of specialization: *hemiphilic*	Pollinated by many different taxa of visitors: *polyphilic*
	Pollinated by some related taxa of visitors: *oligophilic*
Strongly adapted to being utilized by specialized visitors: *euphilic*	Pollinated by one single or some closely related species only: *monophilic*

sugar are attractive to birds. Diaspores with hairs, parachutes, wings, and the like are usually wind dispersed, such as the samara fruits of *Acer* (Aceraceae) or the coma on the seeds of *Populus* (Salicaceae). In the case of floras of geologically youthful oceanic islands, summaries have been made of the importance of each of the dispersal mechanisms to help explain the modes of origin of the flora (e.g., in the Galapagos Islands; Porter 1984).

INVESTMENTS FOR GATHERING REPRODUCTIVE DATA

The financial investments needed to gather data from reproductive biology are much less costly than those discussed in other chapters, especially chemistry. The greatest investment here is *time:* time for observations of the plant materials in their natural surroundings and of the structural aspects of the plants themselves. Most useful is a greenhouse in which to grow the plants and make regular observations. Flowering behavior can be observed here most conveniently in combination with a careful morphological assessment of the floral structures. Ultraviolet patterns can be seen with special photographic filters or modified television cameras. Pollinators can be determined in the field by painstaking 24-hour watches over several different days. Nets, killing jars, movie cameras, and even infrared cameras can be useful equipment, the latter especially helpful for observing night-flying bats. Proper storage of insects involves routine entomo-

Visitor-Species Relationship	
Dealing with adaptation for blossom visits	*Dealing with character of visit activity*
Unadapted or counter-adapted; visits show no relation to the organization of the blossom, frequently destructive, but may cause pollination: *dystropic*	
Poorly adapted for utilization of blossoms; the food obtained from blossoms forms part of a mixed diet: *allotropic*	
Intermediate degree of specialization: *hemitropic (hemilectic)*	Visiting many different taxa of plants: *polytropic (polylectic)*
	Visiting some related taxa of plants only: *oligotropic (oligolectic)*
Fully adapted blossom visitors, taking their main food from blossoms: *eutropic (eulectic)*	Visiting one single or some closely related plant species only: *monotropic (monolectic)*

logical techniques of pinning, labeling, etc. Identification is likely to require the cooperation of a specialist in the insect group. Nectars can be drawn up into capillary tubes and analyzed for sugar composition by thin-layer or high-performance liquid chromatography, and for total sugar content by refractometry. Breeding system studies involve field and garden work with bagging of flowers, emasculations, keeping plants in insect-free cages, as well as considerable time at the microscope observing pollen germination, growth, and fertilization. Observations on dispersal structures may involve anatomical and/or SEM studies to relate structure to dispersal function. Field work with observations of transported diaspores is essential here, as are laboratory studies to determine seed survival and germination requirements. Once again, the costs of the techniques themselves are not great, but the time and travel commitment may be substantial.

EFFICACY OF REPRODUCTIVE DATA IN THE TAXONOMIC HIERARCHY

There is no question that reproductive data bear most strongly on taxonomic problems at the specific and infraspecific levels. If correlations exist at higher levels, such as with wind pollination in the subclass Hamamelidae, they only further substantiate other characters already used to delimit the taxa. Ordinarily, reproductive data are so closely allied with the overall fitness of an individual organism or population, or with closely related small groups of taxa, that the correlations are most useful at lower levels of the hierarchy. Knowledge that one species is an outcrosser and a close relative an inbreeder, for example, may not only give clues to other data by which to differentiate the two (e.g., more conspicuous petals in the outcrosser), but also to help explain narrower limits of character state variation in the inbreeder, and *suggest* evolutionary directionality from the outcrosser to the inbreeder (this is a common general trend in angiosperms; Wyatt 1983).

SPECIAL CONCERNS WITH REPRODUCTIVE DATA

Some aspects of reproductive biology fall clearly into the area of coevolution (Stone and Hawksworth 1986). Pollination and dispersal relate to this, and particularly pollination (Macior 1971, 1974). Here the flowers and pollinators have undergone a close modification leading to increased mutual fitness. The interrelationships are complex and some of the terms describing these interactions are given in table 22.3. A broad-scale view of coevolution in the Compositae has been presented by Leppik (1977). He emphasized that the evolution of head morphology in the family has paralleled the same kinds of trends of evolution in all the angiosperms from more regular flowers and generally more primitive pollinators (e.g., beetles) to irregular flowers with more advanced and specific pollinators (e.g., birds). Coevolution of flowers with insects in the explosive adaptive radiation of the angiosperms has been shown clearly by many workers, including recently Crepet (1983, 1984). Phytochemical aspects of coevolution could also be profitably discussed here, but this will be covered briefly in chapter 23.

In conclusion, reproductive data in taxonomy have not been used extensively over the years as a source of comparative information for making decisions on grouping and ranking, but they have been and still are invaluable for elucidating the biology of the characters and states of the features possessed by these plants. They give a better understanding of the function of structural characters used by taxonomists, they can help reveal proper homologies of characters states, and they can help explain patterns of partitioning of character state variation within and between populations. As Richards has well remarked: "The taxonomic and classificatory systems that we impose on plant variation are influenced by the pattern of variation that we perceive. No two plant populations have exactly similar breeding systems and exactly similar patterns of variation. Our taxonomic philosophy should take account of this" (1986:457).

CHAPTER 23

Ecology

Ecological data are different from other types of comparative data in taxonomy. These deal not with features of the plants themselves, which are the macro- and microstructures discussed in the previous eight chapters, but rather with the plant-environment interactions (Izco 1980). These interactions are the net effect of all the features of the plant with all the aspects of the environment, and they are of two basic types: (1) with the abiotic part of the environment, such as soils, temperature, moisture, etc.; and (2) with the biotic part of the environment, including aspects such as herbivores and competitors. Pollinators can also be treated as part of the biotic category, and this is sometimes called pollination ecology (e.g., Macior 1986a, b), but in this book I have treated it as part of reproductive biology (in the previous chapter). Most of the data of direct taxonomic utility have come from abiotic factors. The distributions of the plants themselves, which reflect the overall survival response of the plants to all the ecological factors, often have been used taxonomically because they not only can reveal spatial discontinuities, but they also relate directly to concepts of gene exchange and the biological species concept discussed earlier. Distributional patterns in the context of taxonomic affinities also relate to biogeography. (This is an interesting area to be mentioned briefly later, but a detailed treatment falls outside the scope of this book.) Finally, ecological studies allow for a manipulation of plants into different environments to determine the environmental vs. genetic components of observed patterns of variation which can be enormously helpful in arriving at taxonomic decisions.

HISTORY OF ECOLOGY IN PLANT TAXONOMY

The origins of ecology are almost as difficult to document clearly as those of taxonomy, perhaps more so. The term was coined by E. Haeckel in 1866, but it wasn't until the 1920s that it was regarded as a clearly identifiable and vigorous scientific field (McIntosh 1985). However, since the time of Theophrastus (300 B.C.), people have noticed and commented upon different distributions of plants and their environmental tolerances for temperatures, soils, rainfall, as well as on plant-animal interactions such as oak apple galls on oak leaves. The herbalists (1470–1670), the early classifiers such as J. Ray and A. Caesalpino, and even Linnaeus in the eighteenth century all noticed differences in plant distributions and environmental features. Candolle (1855) clearly set out views on "botanical geography" from the taxonomic perspective. Thus, in taxonomic works it was expected to find mention of geography and habitat at the very least, and so it has been to the present time.

In the development of plant ecology in this century, definite schools developed with the taxonomic plant geographers seeking more refined statements to explain historical aspects of plant distributions and the ecological plant geographers worrying more about the physiological bases for distributional patterns (Hagen 1986). F. E. Clements (1905, 1916) and H. C. Cowles (1909) were among the more physiologically oriented plant ecologists and helped establish ecology in the United States. In Europe the phytosociological (or sigmatist; McIntosh 1985) school of C. Schröter (1908) and finally J. Braun-Blanquet (1932) dominated in early years with the goal of classification of vegetation analogous to that in taxonomy. From these beginnings came input from population biology which yielded population ecology, influx of sophisticated mathematical approaches which yielded quantitative ecology, and so on.

GENERAL ECOLOGICAL TEXTS AND REFERENCES

Today there are many texts on different aspects of ecology of value to the taxonomist. Once again, a complete review is impossible, but several works can be mentioned. Books of direct taxonomic pertinence are the symposium volumes edited by Allen and James (1972), Valentine (1972), and Heywood (1973b). Good review chapters include Valentine (1978) and Moore (1984). The general ecology texts of Ricklefs (1979), Barbour, Burk, and Pitts (1987), and Ehrlich and Roughgarden (1987) are recommended for overviews of the entire field. Physiological plant ecology is covered by Chabot and Mooney (1985), quantitative plant ecology by Krebs (1978) and Kershaw and Looney (1985), and evolutionary ecology by Pianka (1988) and Crawley (1986). For plant-animal interactions see Abrahamson (1988). Plant population ecology is reviewed by the symposium volumes of Solbrig (1980) and Dirzo and Sarukhan (1984). General historical plant geography is covered well by the classic works of Cain (1944), Good (1953), Dansereau (1957), and Gleason and Cronquist (1964). More recent coverage is offered by Stott (1981) and the symposium volume of Sims, Price, and Whalley (1983). There are even texts on biophysical ecology (Gates 1980) or "ecophysics" (Wesley 1974)!

TYPES OF ECOLOGICAL DATA

Distribution

One of the most commonly used types of ecological data in taxonomy is distribution. Here the occurrence of taxa is usually shown on maps of political areas such as countries, states, and so on (fig. 23.1). Sometimes politically independent localities are also given, such as longitude, latitude, and smaller divisions (min-

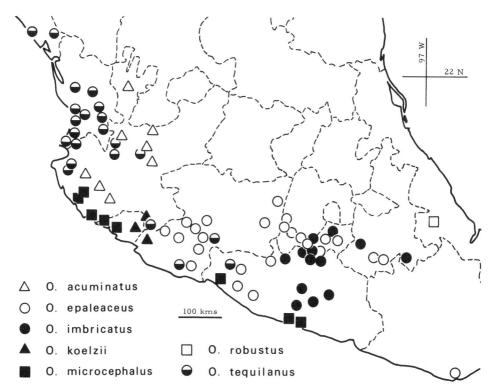

FIGURE 23.1 Distributions of species of *Otopappus* (Compositae) in Central Mexico. (From Hartman and Stuessy 1983:195)

utes, seconds, etc.). Dot maps themselves usually do not reveal much about the details of ecological differences among taxa (unless one is personally familiar with the area covered by the map); rather they give a summary of ecological data that show the differences (or similarities) of habitat tolerance. A difficulty with such data display is that dots may appear sympatric whereas the plants may not be! The taxa may be partitioned by elevation, soil, or other barriers not obvious on the large scale of the map. These data are valuable, however, for showing possible sympatry of species and infraspecific taxa and their overall ecological requirements. They are also valuable in the context of reproductive isolation.

Vegetation Zones

Correlation of dot maps with vegetative zone data enhances the understanding of the ecological similarities and differences of taxa. In taxonomic monographs it is customary to comment not only on the full range of a taxon but also on its habitat types and associated species. Vegetation zones on a broad scale have been mapped for the continents of the world and various parts of it (fig. 23.2; Riley and Young 1966; Walter 1985; Takhtajan 1986), based on latitude, elevation, and precipitation (e.g., Holdridge life zones, fig. 23.3). Again, on a broad

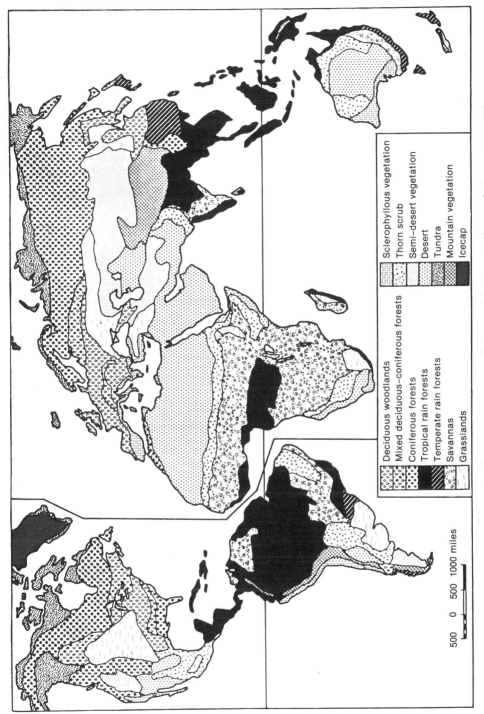

FIGURE 23.2 Generalized vegetation zones of the world. (From Riley and Young 1966:95)

Sclerophyllous vegetation
Thorn scrub
Semi-desert vegetation
Desert
Tundra
Mountain vegetation
Icecap

Deciduous woodlands
Mixed deciduous–coniferous forests
Coniferous forests
Tropical rain forests
Temperate rain forests
Savannas
Grasslands

500 0 500 1000 miles

367

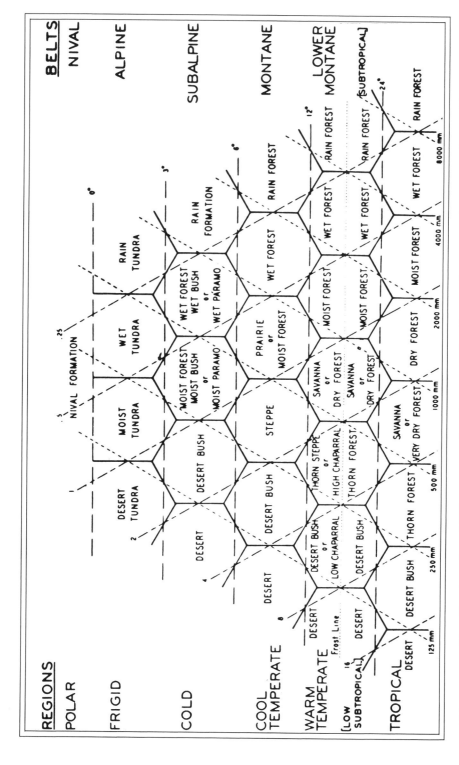

FIGURE 23.3 Classification of world life zones or plant formations by climate. (From Beard 1978:45; after Holdridge 1947)

scale this is only moderately helpful, because many closely related taxa occur in the same broad vegetational zones but are still clearly ecologically separated on a smaller scale. The dominant species in the particular local environment, therefore, need to be documented also. This often requires a clear view of classification of plant communities in a particular area, and this is not always available.

Local Environmental Factors

Soils. In the local environment both abiotic and biotic factors need to be considered. One of the most important of the abiotic factors is soil (e.g., Jeffrey 1987). Numerous studies have shown the effect on distributions of plants based on soil differences. Perhaps the most striking are those of the California serpentine soils (e.g., Kruckeberg 1984), which contain high levels of magnesium and chromium over calcium. Sandy soils, limestone-derived soils, and numerous other variations often can be factors in producing different distributional patterns. Good references to help classify soils are Eyre (1968) and Batten and Gibson (1977), and the USDA Soil Conservation Service classification system (e.g., Smith 1960). Major agricultural universities generally have soil testing laboratories which provide organic and elemental analyses at cost. Soils are so important that it is clear that they have played an important role as a stimulus for speciation (e.g., Wild and Bradshaw 1977; see review by Kruckeberg 1986). In some cases plants have apparently evolved tolerances to high levels of inorganic ions in soils (and hence also in their tissues) and maintained this high level of ionic concentration even when secondarily adapted to slightly different soil regimes (Rossner and Popp 1986).

Geology. Bedrock geology can also affect plant distributions, and clearly this relates to the development of soils discussed above. In some cases, however, no soil differences separate taxa, but patterns of underlying bedrocks do. For example, the distribution of naturalized *Carduus nutans* (Compositae) in northwest Ohio correlates positively with limestone bedrock below the soil surface (fig. 23.4). Warnock (1987) found that two closely related species of *Delphinium* (Ranunculaceae), *D. treleasei* and *D. alabamicum*, were restricted to dolomite and limestone outcrops, respectively, and not only represented edaphic seclusion from the rest of the genus, but also a case of parallel evolution into similar, but distinguishable, bedrock areas.

Other Abiotic Factors. Other abiotic factors are important as ecological data in taxonomy. These include rainfall, humidity, soil moisture, seasonal and diurnal temperatures, available light, and elevation. All of these have different components; for example, elevation relates to oxygen and UV levels as well as temperature (the temperature gets lower as one goes up in elevation). All these features can be regarded as part of the climate in which a taxon occurs, and many references exist to gain ideas of these parameters without new observations being necessary. (Such original observations are highly desirable, of course, especially at the microclimatic level, but resources sometimes might not allow for this.) A useful classical reference for the world is Kendrew (1922) which gives

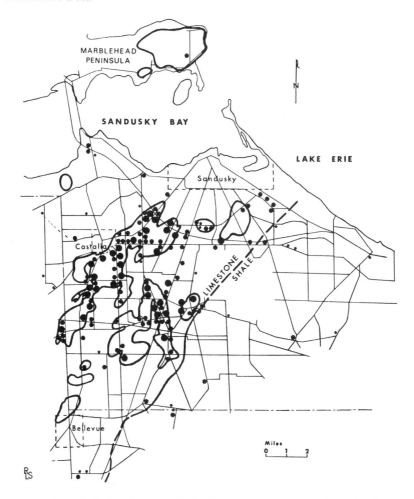

FIGURE 23.4 Detailed distribution of *Carduus nutans* (Compositae) in the Castalia-Bellevue-Sandusky area of northwestern Ohio. Large dots=numerous plants; medium-sized dots=several plants; small dots=one or two plants. The heavy solid black lines delimit limestone bedrock at six feet or less below the soil surface. (From Stuckey and Forsyth 1971:10)

all sorts of general data for continents on earth (figs. 23.5 and 23.6). Obviously, more local data are necessary for seeking meaningful correlations with distributions of specific and infraspecific taxa. The microclimate is really the most important factor here (see Franklin 1955). All species are affected by these climatic factors, and in some cases dramatic correlations exist which, although not yet tested experimentally, suggest strongly that particular dimensions may be the critical limiting factor in determining distributions of particular species (fig. 23.7).

Characteristics of past environments, especially past climates (= palaeoclimate), can also be useful data for the taxonomist (e.g., Nairn 1961; Davis 1968). These data are used more frequently for answering phytogeographical questions,

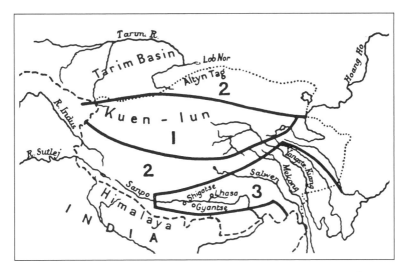

FIGURE 23.5 Map of the principal climatic regions of Tibet. (From Kendrew 1922:163)

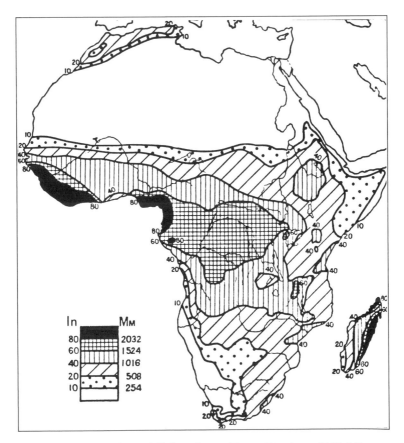

FIGURE 23.6 Mean annual rainfall for Africa. (From Kendrew 1922:20)

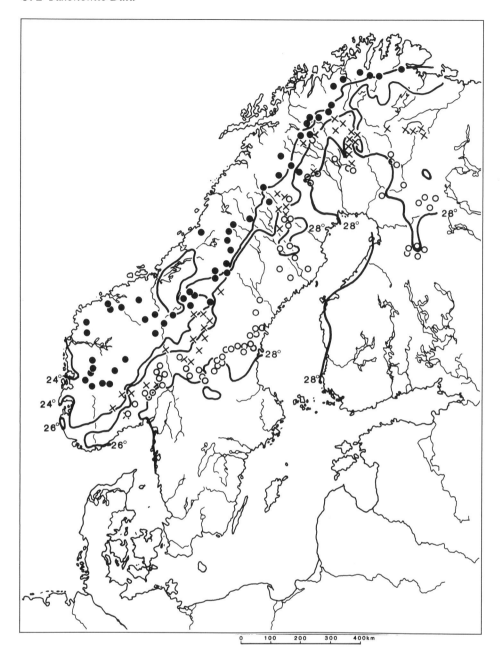

FIGURE 23.7 Distributions of *Gnaphalium norvegicum* (Compositae; circles), *G. su-pinum* (Xs), and *Ranunculus glacialis* (Ranunculaceae; dots) in relation to the 28°, 26°, and 24° C maximum summer isotherms in Scandinavia. (Redrawn from Dahl 1951:37)

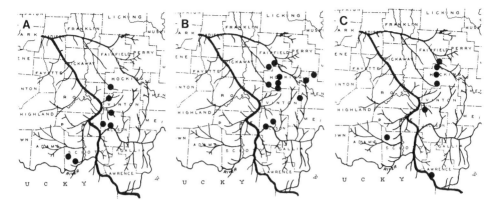

FIGURE 23.8 Distribution maps of south-central Ohio showing correlations of selected species of angiosperms with the old Teays River system. A, *Magnolia tripetala* (Magnoliaceae); B, *Phlox stolonifera* (Polemoniaceae); C, *Rhododendron maximum* (Ericaceae). (From R. L. Stuckey, with permission)

but they can be useful for understanding the overall ecology of a group and its relatives and to interpret present-day distributional patterns. Many plants in the northern hemisphere have distributions that correlate with past glacial advances. Another interesting correlation by way of specific example, but less well known, is with the glacial-age Teays River system in Ohio and neighboring states. Some species have present distributions that correlate well with the extent of the old river bed and its tributaries, suggesting that migration may have occurred along this river system, leaving the distributions now seen (fig. 23.8). Pollen diagrams (fig. 23.9) from bog cores and other static sedimentary deposits, as well as geological information, have also yielded many valuable data on palaeoclimate (e.g., Kapp 1970; Birks and Gordon 1985), leading to the reconstruction of environments during the past 10,000–20,000 years (fig. 23.10).

Biotic Factors

The biotic factors of the environment are equally important in determining plant distributions. The associated plants in the local community are very valuable as comparative data. These usually are given as general plant communities described by dominant taxa in temperate zones, such as beech-maple forest, pine-oak forest, etc., or as more general zones in tropical regions, e.g., tropical deciduous forest, rain forest, etc. The more complete the data on the associated species, the better the understanding of the environment. Although usually unavailable for most taxonomic work, the insect, bird, and mammal predators would be valuable to know too (for good recent volumes on herbivores, see Crawley 1983, and Abrahamson 1988). Pollinators are a part of this biotic environment, but we have already dealt with this under reproductive biology. Dispersal agents are also a source of comparative data and these are allied to the structural data of the fruits, seeds, and associated plant parts. Mutualistic associations also play a special role here and one can cite the dramatic tropical

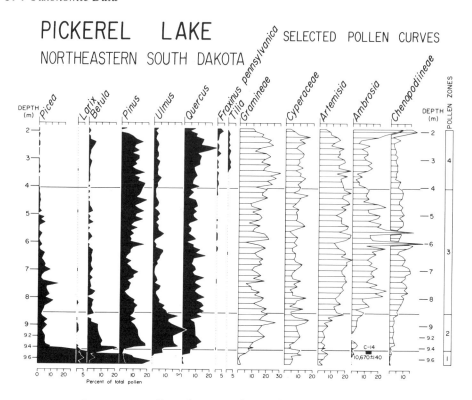

FIGURE 23.9 Summary pollen diagram for Pickerel Lake, northeastern South Dakota. (After Watts and Bright 1968; in Wright 1970:168)

ant-plant relationships as an example. Here the host plants have structures, genetically regulated and usually of modified stem tissue, which serve as homes for aggressive ants which protect their home territory and in the process keep the host plant free from other insect predators. The bull's horn *Acacia* (Leguminosae) of Central America is a good example of this phenomenon (Janzen 1967).

INVESTMENTS FOR GATHERING ECOLOGICAL DATA

The investments for gathering ecological data for taxonomic purposes are generally less than for most other types of data. Much of the data can be obtained by synthesizing available information on distributions and present and past environmental correlations (abiotic and biotic factors). It is fair to state that most taxonomists fail to look for such correlations even when much data are already available! This applies not only for more accurately characterizing the environments of each species, but also for biogeographic insights. Much more needs to be done here by taxonomists on a routine basis. Original data-gathering may be time-consuming, but usually not expensive. Bedrock geology, soil element and organic content analyses, moisture capacities, temperature, light, elevation, and so forth, all can be determined at low costs. The biotic aspects of

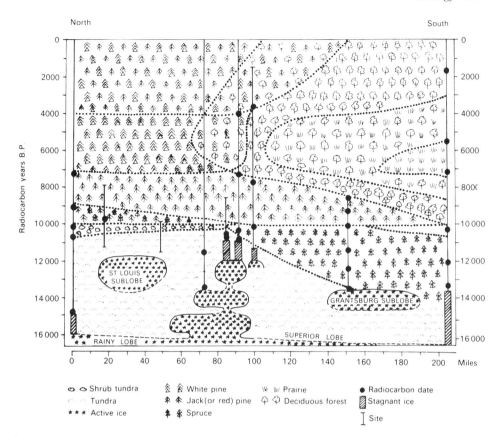

North   South

FIGURE 23.10 Reconstructed vegetation patterns in eastern Minnesota during the past 16,000 years. (From Birks and Birks 1980:170)

the environment can also be determined via much field work to document herbivores, pollinators, and mutualisms. The reconstruction of the palaeoclimate is more intensive of resources and requires bog cores, modern pollen reference collections, and computer analyses. Generally data of this type would ordinarily fall outside the scope of a normal taxonomic investigation, but data in the literature should be sought aggressively. Transplant data are neither difficult nor expensive to obtain, but field plot, greenhouse and/or growth chamber space is necessary. In summary, ecological data are not costly to synthesize for taxonomic purposes, and it is surprising that more has not been done in routine taxonomic efforts. Perhaps it is because these data are not of structures nor processes of the plants themselves and therefore are more difficult to interpret in consort with more traditional data.

EFFICACY OF ECOLOGICAL DATA IN THE TAXONOMIC HIERARCHY

Ecological data can be useful at different levels of the taxonomic hierarchy, but they have most impact at the specific and infraspecific levels. With some exceptions, most genera and families of flowering plants contain diverse species which have evolved in parallel in different environmental conditions. *Melampodium* (Compositae), for example, probably originated in the pine-oak habitat zone of Mexico, but different independent lines of the genus have evolved in tropical and desert zones (fig. 8.10). It would not be surprising to find taxa of related genera growing in the same environments, perhaps even right next to one another. Widely divergent species within the same genus, in fact, may often become sympatric without hybridizing. Distributional data, therefore, can show patterns of spatial isolating mechanisms, and they can help indicate the ecological bases for this isolation (for use of this approach see Sundberg and Stuessy, in press).

SPECIAL CONCERNS WITH ECOLOGICAL DATA

Experimental ecological studies which involve reciprocal transplants into different environments are worth special mention. Turesson (1922a, b, 1925) must be credited with much of the work which revealed that plant populations, even of broadly ranging species, are often adapted to local environmental conditions. These *ecotypes* sometimes have taxonomic value as varieties or subspecies if they form geographical units, but frequently they are regarded as physiological races worthy of comment but not of formal taxonomic recognition. A whole series of informal descriptors has been developed, however, such as demes, ecovarieties, etc. (see chapter 12 for discussion). Transplant studies were carried out in California by Clausen, Keck, and Hiesey (1939, 1940, 1941a) which gave further evidence of the genetic vs. environmental control of morphological features. They grew cloned individuals in different habitats at different elevations, and they also grew plants normally from different conditions together in the same environments. These types of data are important to help show which morphological characters and states are consistent and therefore useful taxonomically. Features which are under environmental regulation only are said to be *plastic* (rather than *variable*, which are due to genetic differences). Recent studies on plasticity have emphasized the adaptive and evolutionary role of this common phenomenon in plants (e.g., Schlichting and Levin 1984, 1986; Schlichting 1986; Sultan 1987), although many problems exist in measuring satisfactorily the degree to which a plant organ responds to different environments. Plants certainly are plastic, and the limits of this plasticity are important to understand before making final taxonomic decisions based on selected characters and states.

Sometimes distributional patterns not only reflect natural ecological differences but also historical introductions during very recent times (fig. 23.11). This is most important to understand before relying solely on these data in a comparative way. For example, Stuckey and Salamon (1987) have recently shown that *Typha angustifolia* (Typhaceae) is most certainly an introduction from Europe during the past 150 years. Hybridization widespread between this species and *T. latifolia* in North America, therefore, cannot be interpreted as a natural phenomenon, especially because the historical documentation of known hybrids parallels that of the migration of the introduced parent! Other studies of a similar nature include Stuckey (1968, 1970, 1974, 1985), Stuckey and Phillips (1970), and Les and Stuckey (1985).

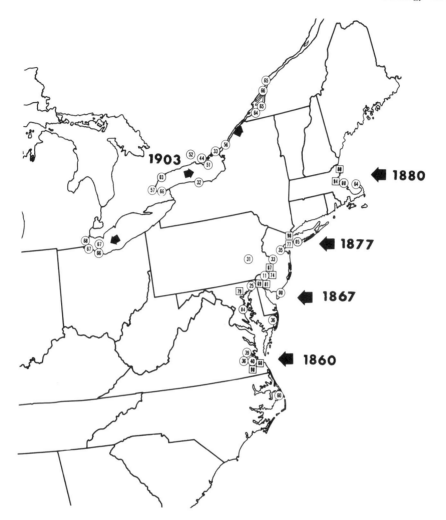

FIGURE 23.11 Known distribution of *Lycopus europaeus* (Labiatae) in eastern North America, based on herbarium specimens and other records. Numbers in the squares indicate the year in which the plant was collected before 1900, and in circles since 1900. Four major areas of apparent entry and establishment for the species are marked by large black arrows and dates. Smaller black arrows indicate possible migrations in Lakes Erie and Ontario and the St. Lawrence River. (From Stuckey and Phillips 1970:352)

A blend of chemical data and ecology yields the area of phytochemical ecology. Several symposium volumes give more than adequate introduction to these types of ecological data (Chambers 1970; Harborne 1972, 1977, 1978; Rosenthal and Janzen 1979). The comparative data are largely secondary plant products such as terpenoids, flavonoids, alkaloids, etc., and can serve as still additional characters and states if their role in the interaction is understood. The same compounds used for defense in one taxon and not in another provide additional data to be used for taxonomic purposes. It might also suggest that the compounds are not homologous, perhaps having come from different biosynthetic pathways.

A few comments on ecological data in cladistic and phyletic analyses seem pertinent. One of the most difficult problems in any evolutionarily based classification is the determination of primitive vs. derived character states (=polarity). Ecological data can sometimes provide such directionality by the second-level criterion of correlation (see chapter 8). Because of the great environmental changes on the earth during and after Pleistocene glaciation, studies of palaeoclimate have revealed that some habitats are more recently derived than others. Deserts of the western United States and glaciated boreal areas are examples. Taxa in these newly derived regions are also likely derived in relation to their allies in older and more stable habitats. A good example is the work of Kyhos (1965) on three closely related species of *Chaenactis* (Compositae) in California and the western United States. Cytogenetic data showed clearly the network of relationships among the species, but directionality was suggested largely by ecology, the ancestral *C. glabriuscula* being found in mesic areas and *C. fremontii* and *C. steviodes* in more recently derived xeric regions.

The decision to combine ecological data with structual information together in a basic data matrix for phenetic, cladistic or phyletic classification, must be based on the particular type of data available. Distributional data, in my opinion, should be kept separate from other data and compared with the overall results from structural information. Data on chemical interactions, predators, or abiotic factors, may well be included with other data, but they also may be analysed separately, with a final comparison between the results being done at the very end. If the ecological data are substantial, this seems a better alternative. In this fashion, data on relationships based on structures are compared with results based on ecological interactions. If the different evaluations of relationships coincide, the overall hypothesis is strengthened. I regard chromosomal information in largely the same way. These are such powerful reflections of the genome that to treat them as a single character with states seems to obscure much of their value. Both ecology and cytology have much to do with isolating mechanisms at the specific and infraspecific levels, and hence they deserve special emphasis.

Ecological data also relate directly to phytogeography, and therefore to the two principal schools of dispersal and vicariance. Although much has been said about these two approaches, it seems clear that plants have moved about the earth by dispersal as well as by the effects of major earth events such as continental drift, mountain building, etc. The point is not whether one or the other is the all-inclusive general explanation, but rather what combination of both of them helps explain the patterns of distribution in a particular group of taxa. A clear view of distributions and their ecological parameters, therefore, is essential for biogeographical analyses of any type (see chapter 8 for more discussion of vicariance biogeography).

In conclusion, ecological data are extremely valuable in taxonomy, but more use might be made of them in practical work. This lack of frequent use may have prompted Valentine to state: "It is probably true to say that ecological criteria are of comparatively little direct importance in taxonomy. The prime criterion, in classical taxonomy at least, has been and will doubtless remain morphological, though ecological criteria at the infraspecific level cannot be neglected" (1978:1). Ecological data are different from others in that they deal with the organism-environment interactions instead of with just the organism itself. As such they can be very helpful data for the assessment of affinities, hypotheses on evolutionary processes, and biogeographical considerations. Heywood has reminded us that: "Both taxonomy and ecology are regarded as sciences of synthesis in that they draw upon various subdisciplines and techniques for their information. Since these latter are largely the same, there is a major overlap between taxonomy and ecology. This, added to the fact that they are both to a large extent dependent upon each other, makes the relationships between them highly complex and multi-dimensional" (1973b:vii). It also means that one can have a positive influence on the other; more attention should be given to ecological data in routine taxonomic work.

Handling of Data

THE PREVIOUS nine chapters hopefully have shown clearly the wealth of comparative data that can be used for solving taxonomic problems. The amount and different kinds of data that potentially can be used in any particular case are almost overwhelming, and it is a challenge for the successful taxonomist to choose the data carefully and gather them in sufficient quantity to reveal clearly the relationships among the taxa. When data are gathered, they must be obtained in such a way as to maximize their utility for the problem at hand; this involves intelligent sampling. How to sample is an important consideration in solving any scientific problem, and taxonomic problems are no exception. How to measure the gathered data is also a concern, and this again relates to the nature of the taxonomic questions being posed. And after the data have been gathered and measured, they must be synthesized in some fashion numerically and/or graphically to allow relationships to be determined. To make matters even more challenging, the complexities of the evolutionary process sometimes obscure the recognition of discontinuities. But in some ways, it is this series of challenges that makes taxonomy so intriguing. As Turner has put it: "If all taxonomic problems were obvious and merely resolved themselves to cataloguing always discrete, easily placed entities, than many of us would long since have lost interest in the tabulation and turned to other fields" (1985:105).

CHAPTER 24

Gathering of Data

From a taxonomist's point of view, it is depressing to reflect upon the mass of published data that exists from plants dealing with cytology, chemistry, anatomy, and so forth, but which is of marginal use for gaining systematic insights because of the small sample size or lack of adequate documentation provided. In taxonomic studies, therefore, it is imperative to pay close attention to several considerations when gathering comparative data (table 24.1).

COLLECTION OF DATA

Sampling

The first consideration in gathering comparative data is sampling. This aspect of data-gathering is usually done in the field, and for this reason, the investigator must have a clear idea of what he or she is attempting to accomplish in the project before beginning field work. Three basic questions must be answerable before effective sampling can be done: (1) What parts of the plant should be sampled? (nature of sample); (2) What size should the sample be?; and (3) What techniques should be used to collect the desired plant parts? The ability to answer satisfactorily these three questions relates directly to the ability of the worker to know where he or she is going geographically, how long the trip needs to be, what will be collected once there, and what kind of equipment will be needed for making the desired types of collections. To understand these ideas, the investigator must know clearly what type of systematic problem is being addressed, and what kinds of data will be needed to solve the problem. Some types of studies, such as those oriented primarily toward classification in a single genus, will require probably fewer populational samples and fewer individuals per population than if intensive studies are being conducted on a closely-knit complex of taxa in which hybridization is suspected.

In general, a sample is adequate if it documents well the variability in character states at the next lower level in the hierarchy (whether formally recognized or not). For example, if relationships among genera are being investigated, the character variation among all the species within each genus should be known. If not all the species can be studied, then this does not mean that no comparisons among the genera can be made, but rather that the final interpretation of rela-

---

TABLE 24.1   Outline of Considerations When Gathering Comparative Data.

I. *Sampling* (Field)
    A. Nature of Sample
        1. Parts of the plant to sample
        2. Additional features of the plant worth noting that will not be sampled
    B. Size of Sample
        1. Number of parts per plant to sample
        2. Number of plants to sample
        3. Number of populations to sample
    C. Techniques in Sampling: methods for collecting desired plant parts
II. *Measurement* (Laboratory)
    A. Nature of Measurement: parts of plant to measure
    B. Size of Sample to Measure
        1. Number of parts per plant to measure
        2. Number of plants to measure
        3. Number of populations to measure
    C. Techniques in Measurement: methods for obtaining data

NOTE: Keep in mind that four factors will realistically control the nature and extent of any data-gathering attempt: (1) purpose of study; (2) time; (3) space; and (4) money.

tionships will be less convincing. If closely related species are being examined, the infraspecific variation (at the subspecific, varietal, or population levels) in each should be documented well before comparisons are made. In problems at any level, as new data are collected, the amount of additional variation that is added by the new samples needs to be examined. If variation of large deviations from the inferred (or calculated) mean continues to be obtained, more data probably should be collected for a better understanding of the distribution of character states before the final relationships are assessed.

Measurement

The second consideration in gathering comparative data is measurement. Although this aspect usually is completed in the laboratory, which gives the investigator more freedom from time restrictions, measurement is dependent upon plant parts collected in the field, and therefore, if the field sampling is inadequate for some reason, the measurements also will be incomplete. Before the taxonomist heads for the field, he or she must not only be able to answer the questions relating to sampling but must also ask and be able to answer these same questions relating to measurement: (1) what parts of the plant to measure (nature of measurement); (2) how many parts should be measured and from how many individuals in how many populations (size of sample to be measured); and (3) what techniques will be used to obtain the desired data?

    It is most important that the measurements be obtained in a consistent and acceptable manner. If the measurements are not properly made, the classifications based upon these data will be inaccurate or misleading. For measurements to be obtained consistently, the exact same type of equipment must be used in all cases to measure the exact same structures from precisely the same parts of the plants. For example, data on pubescence (hairs per square area) must be gathered using the same rule under the same magnification and from the same organs of each plant. For measurements to be gathered in an acceptable fashion, the worker must be thoroughly familiar with the type of data he or she is using. It must be certain that the techniques for obtaining the data will not cause alterations of the generated information. For example, some chemical data, such as essential oils (monoterpenoids), require delicate handling. If the compounds are not treated with great care (sealed vials, low temperatures, etc.), structural rearrangements can take place which, when the data are recorded, might lead to a distorted view of the relationships.

In the final analysis, four factors realistically control the nature and extent of any data-gathering attempt: purpose of the study; time; space; and money. As is true for the majority of our activities whether scientific or otherwise, inevitably the ultimate limiting factor is money. Time and space relate directly to money, and even the purpose of a particular study is usually delimited indirectly by financial considerations. However, these realistic considerations do not necessitate that a taxonomist must keep problems so narrow that no difficulty is ever encountered in obtaining the needed data. Rather, one determines first what the problem of interest and significance is, then what the realistic requirements are, and finally what time, space, and money are needed to gather the data to solve the problem.

EVALUATION OF DATA

In the process of collecting data, and even *after* data have been collected, it is important to evaluate them for their potential for helping solve the taxonomic problems at hand. Sometimes this evaluation can be done by simple inspection. Often, however, more sophisticated methods for evaluating these data are used, and these can be called techniques of mathematical analysis and summarization (Crovello 1970). These can help find pattern and structure in the data even though such patterns may be difficult to see by simple visual examination. Many statistical approaches exist, such as correlations, regressions, analysis of variance and covariance, and basic statistical measures of means, ranges, and so on. Various similarity coefficients will be used in phenetics, cladistics, or with explicit phyletics, and many different algorithms can be utilized before the final approaches are selected for presentation in the published report. Information theory can also be used to help determine the robust quality of classifications (e.g., Duncan and Estabrook 1976). The complex patterns of multidimensional variation can be reduced into fewer dimensions by ordination of different types, multidimensional scaling, and cluster analysis. Discriminant function analysis is also helpful, especially in situations involving hybridization, and other types of geometric and/or calculus evaluations also exist. In short, there is no lack of available sophisticated methods for data analysis to help make taxonomic decisions. The paper by Crovello (1970) is most helpful, and the text of Sokal and Rohlf (1981a) is also recommended as a starter in this area.

Clearly any particular set of comparative data should be examined rapidly at the earliest possible stage of a study for potential utility. If *no* variation exists in the groups with regard to a particular type of data, obviously this will be of no value whatsoever in helping to resolve relationships *within* that group. They may be extremely helpful, however, in delimiting that group from other groups at the next higher level in the hierarchy. The point is that sometimes it is difficult, if not impossible, to find taxonomically useful discontinuities within a particular set of data. This sometimes occurs even with the best of efforts by the taxonomist. The simple fact is that due to the dynamics of the evolutionary process, sometimes conditions are such that sharp discontinuities among taxa do not exist.

The evolutionary factors of speciation and divergence, which are responsible for the production of most of the diversity, at the same time may cause temporary (in an evolutionary sense) intergradations or continua to occur that obscure the usually observed discontinuities. A good knowledge of the evolutionary process, therefore, enables a taxonomist to approach more effectively and interpret more successfully the existing patterns of relationships in these more challenging situations. It is certainly true that a person does not have to know anything about evolution to be a good taxonomist (consider all the excellent pre-Darwinian workers, such as Jussieu and Candolle), but it is also true that the acquisition of such additional knowledge will help make him or her a much better worker.

It is not within the scope of this book to elaborate in detail all aspects of the processes of plant evolution that may contribute to the obscuring of discontinuities. Rather, it is the objective to outline briefly these various evolutionary aspects and refer the reader to references that will clarify this understanding. Table 24.2 lists the major aspects of the processes of plant evolution with which the plant systematist should be familiar to be well equipped to handle taxonomic problems. An excellent introduction to these processes is found in Stebbins (1977). For more advanced discussions, see Grant (1963, 1977, 1981, 1985b), Stebbins (1971b, 1974), Dobzhansky et al. (1977), White (1978), Templeton (1981), and Jeanmonod (1984).

As an aid to understanding how these processes might affect the comparison of data, outlined in table 24.3 are the levels in the taxonomic hierarchy at which the different evolutionary dynamics are likely to obscure discontinuities. In general, the greater the genomic alteration involved, the higher the level in the hierarchy that will be affected. For instance, the phenotypic effects of most mutations are minor, and therefore, will not usually obscure discontinuities above the population level. However, at the other end of the spectrum, hybridization and introgression involve large genomic alterations and these can be a common source of difficulty in resolving taxonomic problems at the specific level.

RELATIVE EFFICACY OF DIFFERENT KINDS OF DATA

Many claims have been made regarding the power of different types of data for solving taxonomic problems. Whenever new information appears on the scene, a brief "bandwagon" effect ensues, and the community passes through a vocifer-

---

TABLE 24.2 Outline of the Major Aspects of the Processes of Plant Evolution. (After Stebbins 1977)

I. Phenotypic plasticity
II. Genotypic variation
    A. Mutation (point mutations)
    B. Recombination
    C. Reproductive systems
        1. Asexual reproduction
           a. Vegetative reproduction
           b. Apomixis
        2. Sexual reproduction
    D. Chromosomal aberrations
        1. Alteration of linkage groups
           a. Translocation
               i. Reciprocal
               ii. Non-reciprocal
           b. Deletions
           c. Duplications
        2. Multiplication of genome
           a. Autopolyploidy
           b. Allopolyploidy
    E. Ecotypic differentiation
    F. Hybridization and introgression
        1. Hybridization
           a. Diploid level
           b. Polyploid level: allopolyploidy
        2. Introgression

---

ous period of advocacy only to settle down a decade or so later to a new type of integration of approaches. Such was the case with chemosystematics in the early 1960s, principally with secondary plant products, and we now see it with regard to chloroplast and nuclear DNA. People sometimes point to the complex metabolic and developmental interactions that exist from DNA to the final expression of morphological traits (fig. 24.1), and suggest that the further back one goes toward the absolute genetic material, the DNA, the closer one comes to having the "best" data for classification. From what is known of DNA at the present time, this simply will not provide a panacea any more than will any other single source of data. One of the difficulties with nuclear DNA is that numerous sites are inactive and many feedback mechanisms exist. Hence, until we know more (a lot more!) about the developmental interactions of the sequences, we will be unable to understand fully their evolutionary meaning. We certainly must push aggressively forward to obtain these data and understand their relevance, but we must also be ready to admit their limitations (as we also must always do with all other types of comparative data). A reasonable and balanced perspective on the positive value of both molecular and morphological data is given by Hillis (1987).

It is my opinion, therefore, that in a general sense no single type of data holds supremacy in determining relationships for purposes of classification. As A. J.

TABLE 24.3 Relationship of Sources of Variation in Characters to the Level in the Taxonomic Hierarchy at Which These Changes are Most Likely to Obscure Discontinuities.

Sources of variation in characters	Principal taxonomic levels at which discontinuities of characters are likely to be obscured by the different sources of variation.						
	Within populations	*Between populations (of the same taxon)*	*Between population systems (of the same species)*	*Between species*	*Between genera*	*Between families*	*Between higher taxa*
Phenotypic plasticity (not genomic)	X						
Recombination	X						
Mutation	X						
Reproductive systems asexual —apomixis		X					
—vegetative reproduction		X					
sexual —inbreeding (autogamy) —outbreeding (chasmogamy)							
Ecotypic differentiation		X	X				
Chromosomal aberrations			X	X			
Hybridization and introgression			X	X	⤬		

→ *INCREASING LEVELS IN THE TAXONOMIC HIERARCHY*

→ *INCREASING GENOMIC ALTERATION usually causing more marked phenotypic change*

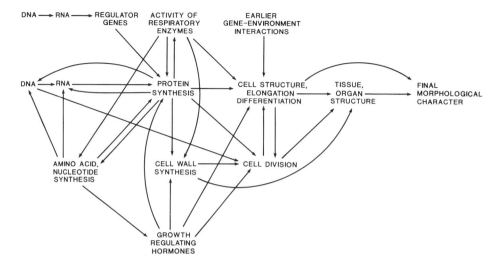

FIGURE 24.1 Diagram showing some of the metabolic processes which intervene between primary gene action and final expression of characters. (From Stebbins 1977:19)

Sharp pointed out years ago: "Should any botanist think he has the final technique or the final answer, may I remind him that science has taught us nothing more clearly in this century than that there are no absolutes and that everything is relative and can be predicted only within certain statistical limits" (1964:747). One might argue that if a person *had* to choose *only one* type of data for classification for use at all levels in the hierarchy, then probably morphology would be the wisest choice. The reason for this selection would be that in all probability the morphological attributes taken collectively would give the best indication of the evolutionarily significant features of the organism, and therefore, would be most useful for purposes of classification. However, the point to be stressed is that one rarely has to make such a decision. It is far better to remember that all types of data tell something about the genotype and adaptational and evolutionary history of the plants under study, and therefore, that all types of data should be used whenever possible. Obviously, every taxonomist has special training and interests, but whether by collaboration or by broadening one's perspective, an attempt should be made to bring as many different types of data to bear on a problem as possible. Only by this combined approach can the most useful and predictive classification of plants result. As Wagner has quipped, to deal with all these different types of data, the skilled taxonomist really needs to be a "chemo-cytohistomorphotaxonometrician" to be effective (1968:97).

A final point is that given the same comparative data, individual workers have different abilities to derive useful insights from them. Metcalfe, has summarized this perspective well: "In our search for taxonomic understanding let us by all means be aided by the hand lens, the microscope and indeed by any instruments of precision that bring the eye closer to the organisms that we are surveying. But in these operations let us preserve a proper sense of proportion by remembering that these appliances are there only to aid us in our search and that the organisms themselves are the most important item in our taxonomic exercises. True taxonomic insight is a comparatively rare faculty, and the faculty has to be cultivated if we are to make full use of it. Even in this mechanized age there is no really adequate substitute for this all important gift" (1967:131).

CHAPTER 25

Presentation of Data

After data have been collected and measured, some means must be devised for reducing them so that relationships among taxa are demonstrated in an understandable fashion. Sometimes all that is needed is a simple listing of the raw data, but more frequently some conversion is helpful such as means, ranges, or standard deviations, or even more complex statistics (see Sokal and Rohlf 1981a; and Tukey 1977). For the majority of systematists, however, a numerical treatment of data often is insufficient to illustrate relationships clearly. To comment on this point requires a momentary digression (also mentioned in chapter 3). In my experience, biologists tend to divide into two groups based upon their innate abilities and/or early training: (1) those who have ability to handle abstract concepts, such as mathematical or philosophical relationships that may or may not have anything to do with the world as we sense it; and (2) those who have ability to handle pattern data, particularly of the type that relates directly to objects and shapes of the world of our experience and that can be stored as visual images. This second type of mentality is of the kind usually possessed by plant taxonomists. A young taxonomist often is drawn into the field in the first place because of a strong ability to relate size, shape, and color to one another in the interpretation of relationships. To return to the original point, as a result of this orientation toward pattern data, it is useful not only to tabulate the data in numerical form, but also to display the data graphically so as to have the maximum impact on the reader (usually other taxonomists). When approaching the problem of data display, therefore, the most important consideration is which method (or methods) will have the strongest visual impact. The stronger the visual impression, the more effective will be the communication for the particular interpretation of relationships. The important point to be emphasized is *communication*. Published graphics are not ordinarily a means of allowing an investigator to gain fresh insights on relationships—they are to *communicate* the most important results of the work to the reader in a clear, concise, and attractive fashion. Preliminary types of data tabulation and display are done by the systematist to help generate insights, but these are not published directly. A selection is made, with refinement, and the most important points are illustrated graphically for maximum and lasting effect.

HISTORY OF GRAPHICS IN PLANT TAXONOMY

The history of use of graphics for representing plant relationships follows the natural development of the history of classification described earlier in this volume. During the age of the herbalists (1470–1670), new and often extremely realistic drawings of plants accompanied the texts in Latin, German, or other languages. Tabular material was used by John Ray, the English naturalist, and others toward the end of the 1600s. Linnaeus continued with descriptions and plates showing morphological features, but without graphics that attempted to show relationships among taxa *per se;* that is, the visual presentation of data was simply the direct representation of diversity, and it was not being used to make specific points about affinities. In fact, it wasn't until the end of the 1700s that more complex graphics came into general use (e.g., fig. 25.1). In part this was surely due to the many skills required, i.e., artistic, visual, empirical-statistic, mathematical, for the effective use of this mode of data presentation (Tufte 1983). William Playfair (1759–1823) was an important innovator in graphic design and his book, *The Statistical Breviary* (1801), set a new and relatively modern standard. The phylogenetic tree of Haeckel (1866; fig. 25.2) reveals a much more complex approach to data presentation.

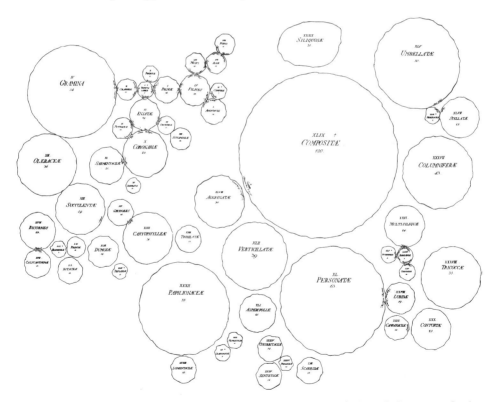

FIGURE 25.1   Early more complex graphic showing Giseke's (1792) diagram of relationships of natural "orders" of Linnaeus. (From Nelson and Platnick 1981:96)

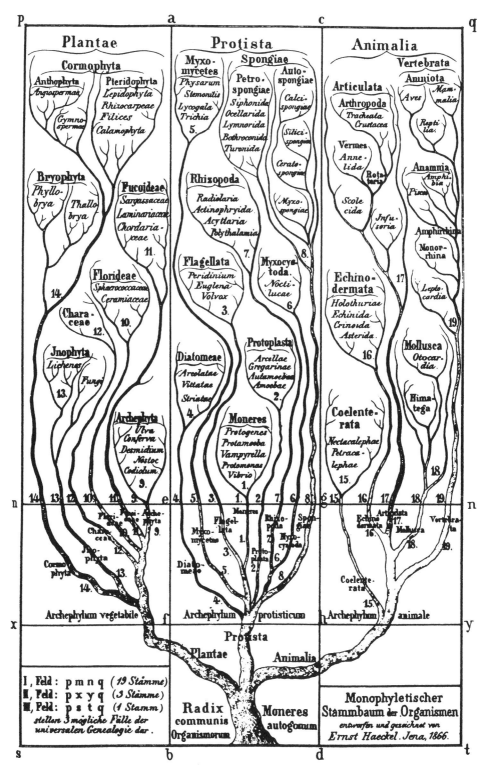

FIGURE 25.2 Early graphic of phylogenetic tree of organisms. (Haeckel 1866; from Mayr 1969c:63)

GENERAL GRAPHICS REFERENCES

From these beginnings have come many types of data presentation. For useful general references, see Crovello (1970), Tukey (1977), Chambers et al. (1983), Tufte (1983), Cleveland (1985), and Holmgren and Angell (1986). What will be done here is to sketch some of the principal approaches to both qualitative and quantitative data presentation to help stimulate an attitude about the importance of *communicating* the results of a taxonomic investigation to others.

TYPES OF GRAPHICS

Qualitative Data Comparisons

Qualitative data comparisons are frequently used by the plant taxonomist. Descriptions are obviously required to compare and contrast data, and for new entities the *International Code of Botanical Nomenclature* (Greuter et al. 1988) legislates that these must be published in Latin. It is worth asking if a tabular approach might not be a better way to present these descriptive data, particularly to ensure truly comparative data for all taxa in the treatment, but for the moment, descriptions are required. Drawings are also used for qualitative data display and these can be of all parts of taxa, such as in a new species presentation, or a selected series of morphological features (fig. 25.3). Tables are another common mode of qualitative data display. Often morphological features are highlighted, but sometimes distributional data are also given. Other tables are such that they combine some elements of graphs (table 25.1).

Quantitative Data Comparisons

Quantitative data comparisons offer many more possibilities for effective communication. Because of their nature, relationships can be drawn among them and compared and contrasted in many different ways; hence, the challenge is to select the proper vehicle for a particular case. Again, straightforward tables can be used with quantitative data, but generally it is more effective to offer the reader an improved synthesis to make the relationships easier to grasp. Pure tables place a strong burden on the reader to see the relationships with little visual help. Sometimes this is adequate; oftentimes it is not.

Graphs of quantitative data are called "summarization graphics" (Crovello 1970). As Tufte has put it: "Data graphics visually display measured quantities by means of the combined use of points, lines, a coordinate system, numbers, symbols, words, shading, and color" (1983:9). Such variation in method offers an enormous smorgasbord of approaches for any particular set of data. What method to use will depend to some extent on the personal preference of the author, and the artistic and/or drafting expertise available. An extensive list of different techniques is given in Crovello (1970; also summarized in Radford et al. 1974) and only some of these will be discussed here to help the reader begin

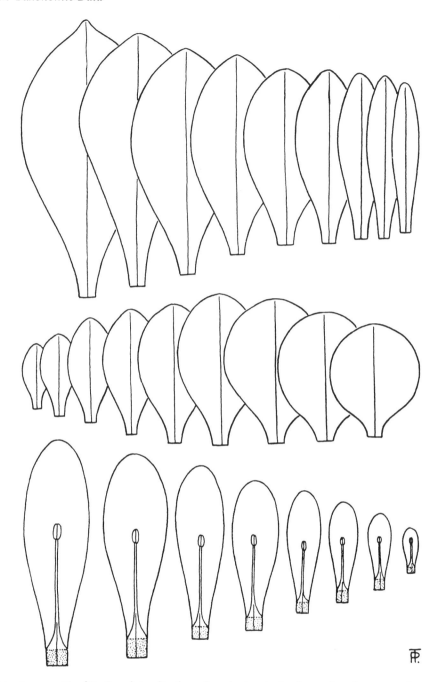

FIGURE 25.3 Qualitative data display of variation in leaf, petal and stamen features in *Anagallis serpens* subsp. *meyeri-johannis* (Primulaceae). (From Taylor 1955:333)

TABLE 25.1 Example of a Table Combining Elements of a Graph. Outline of Changes in the Subtribal Classification of the Heliantheae (Compositae). Arrows Indicate Continuity of Concepts from One System to the Other; Solid Arrows Show Transfer of Subtribes Directly; Dashed Arrows Indicate Remodeling of a Portion of a Subtribe. (From Stuessy 1977: 629)

Hoffmann (1890)		*Revised System*
HELIANTHEAE		
Ambrosiinae	⟶	Ambrosiinae
Fitchiinae	⟶	Fitchiinae
Madiinae	⟶	Madiinae
Zinniinae	⟶	Zinniinae
Milleriinae	⟶	Milleriinae
		Engelmanniinae
Melampodiinae	⟶	Melampodiinae
Coreopsidinae	⟶	Coreopsidinae
		Ecliptinae
Galinsoginae	⟶	Galinsoginae
Helianthinae (as "Verbesininae")	⟶	Helianthinae
		Verbesininae
Petrobiinae		Neurolaeninae
Lagasceinae		Bahiinae
		Gaillardiinae
HELENIEAE		
Riddelliinae		
Heleniinae	⟶	(Many genera to tribe Senecioneae)
Jaumeinae	⟶	(Other genera to tribes Astereae or Vernonieae)
Pectidinae (as "Tagetininae")	⟶	(All genera as subtribe or tribe in or near Senecioneae)

thinking about possibilities in his or her own work. Classificatory diagrams, i.e., phenograms, cladograms, phylograms, two- and three-dimensional ordinations, etc., also known as "directed graphs" (Crovello 1970) have already been presented earlier in this book (see chapters 7 to 9) and will not be repeated here.

Correlation of Variables. The most commonly used graphics of quantitative data deal with correlations of variables, and these usually are features of different taxa. *Scatter diagrams* relate points in space, often taxa, populations, or individ-

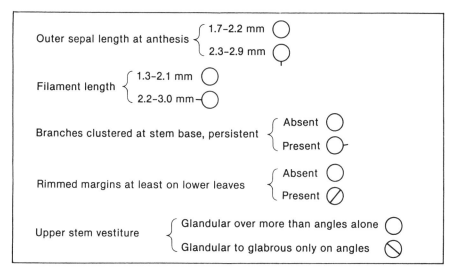

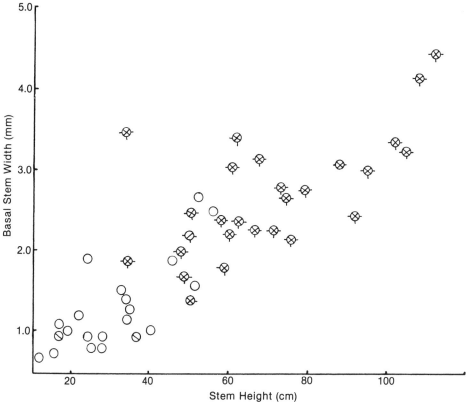

FIGURE 25.4 Scatter diagram comparing features of *Polygonella fimbriata* (hollow circles) with *P. robusta* (hatched circles; Polygonaceae). (From Nesom and Bates 1984:41)

uals, and usually in two dimensions, to features they possess which are scaled on the main axes of the graph. The use of this technique for the study of hybridization and introgression, pioneered by Edgar Anderson (1949), is now routine (fig. 25.4). These can be visually enhanced by "smoothing," in which a computer is used to draw a smooth curve among the data points, using the technique of "lowness" (=robust locally weighted regression; Cleveland and McGill 1985). This would not be particularly helpful in analysis of hybridization, but for other data plots it could prove useful. While on the topic of analyzing hybridization, another more recent graphic technique is *Wells' distance diagram* in which the position of taxa is by triangulated distance from two fixed parental end points at opposite ends of a hemicircle (Wells 1980).

Other quantitative graphics can also be used and a few examples will be given here. *Histograms* and *bar graphs* (figs. 25.5, 25.6) are well known to taxonomists and have been used for decades. They conveniently show the quantitative variation of one variable. *Polygonal graphs* were devised by Hutchinson (1936, 1940) to show the simultaneous variation of several features. This was originally used in ecology, but applied early to plant taxonomy (e.g., Davidson 1947; Voigt 1952; Löve and Nadeau 1961). Ellison, Alston, and Turner (1962) used these for showing relationships of biochemical data in *Bahia* (Compositae, fig. 25.7; see also Anuradha, Radhakrishnaiah, and Narayana 1987). These are visually effective, but they consume much space on the costly printed page (ca. $50 each). *Crossing*

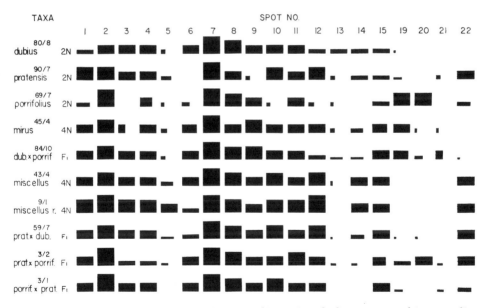

FIGURE 25.5 Histograms representing two-dimensional chromatographic spot distribution and intensity in natural populations of *Tragopogon* (Compositae). The vertical dimension gives average spot intensity as estimated in UV light. The horizontal dimension indicates the percentage of plants examined which had the spot. The numerator of the fraction above each taxon gives the number of individuals analyzed, denominator the number of populations. (From Brehm and Ownbey 1965:814)

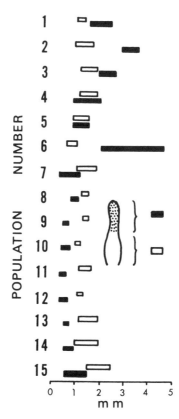

FIGURE 25.6 Bar graph showing variation in lengths of herbaceous tips (dark bar) and base (light bar) of lowermost phyllaries of representative collections of *Otopappus australis* (populations 1–6), *Zexmenia columbiana* (7–11), and *Z. mikanioides* (12–15; all Compositae). (From Anderson, Hartman, and Stuessy 1979:54)

diagrams for presenting the results of cytogenetic studies have already been discussed (see fig. 20.3). *Dot graphs* show points interconnected by various lines; these are extremely common with physiological data. More complex graphics include *"data-built data measures"* (Tufte 1983:141) in which the data themselves form the structure for the graph (fig. 25.8). Other graphics are designed to show relationships of major ideas in a study (fig. 25.9) more effectively than by words alone.

Geographical Plots. Geographical plots in taxonomy are extremely important to show not only simple presence of a taxon in an area, but also patterns of morphological (or other) variation across geographical areas. These data are extremely valuable in consideration of specific and infraspecific boundaries. The dot maps in revisions are familiar to all of us (fig. 23.1). Sometimes the symbols are elaborated upon to communicate even more information on date of collection, sight records or voucher specimens, and/or character variation (fig. 25.10; see also fig. 23.11). *Pie diagrams* can be used also to show intra-populational

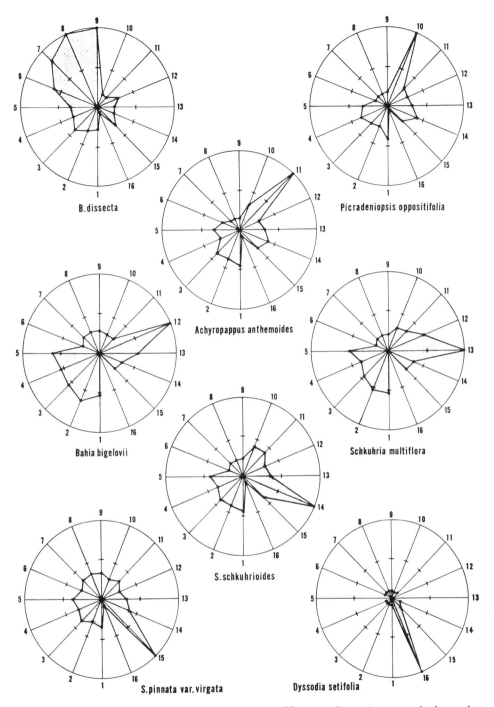

FIGURE 25.7 Polygonal graphs of the paired affinity indices (expressed along the radii 0–100 percent beginning from the center) of the phenolic compounds in species of *Bahia* (Compositae) and generic relatives with all 16 species of *Bahia* (numbered positions along outside of circles). (From Ellison, Alston, and Turner 1962:601)

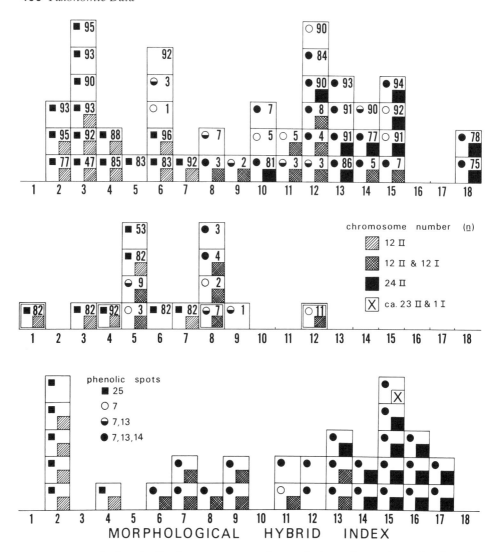

FIGURE 25.8 A data-built data measure of morphological, cytological, phenolic chemical, and pollen viability data in hybrid populations of *Picradeniopsis* (Compositae). The horizontal axis for all graphs is the morphological hybrid index. Each large square represents an individual plant. Numbers in squares are pollen data (percent stainable, and presumably viable, grains). (From Stuessy, Irving, and Ellison 1973:49)

variation within each population sampled across broad geographic areas (fig. 25.11). *Contour diagrams* are now popular for showing computer assisted analyses of trends of morphological (or other) variation graphically. Complex patterns of terpenoids can often be best presented in this fashion (see fig. 21.3). Finally, evolutionary trends can be superimposed on a map directly to show migrational directions within a phylogeny (fig. 25.12).

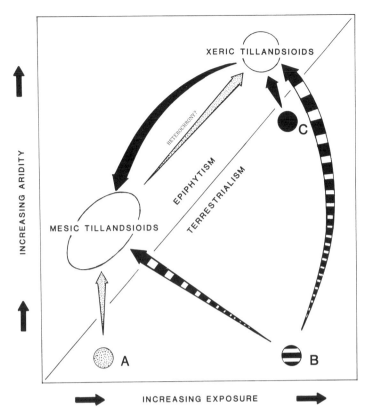

FIGURE 25.9 Diagram illustrating major evolutionary ideas, here showing three theories of tillandsioid (Bromeliaceae) evolution, each based on ancestors adapted to different degrees of exposure and humidity. (From Benzing, Givnish, and Bermudes 1985:87)

GRAPHIC DESIGN

Designing effective graphics is not easy, and it may necessitate seeking advice from a trained draftsman/artist for best results. The challenging nature of effective graphics leads to several pitfalls. First, sometimes they are unnecessary and should not be used at all. The text may be completely adequate to convey the points about the relationships. Second, care must be exercised so as not to distort the relationships. We all know that statistics can lie if used improperly, and so it is with graphics too. Third, charts should be "friendly" (Tufte 1983) so that the reader can comprehend what the points are without fighting laboriously through them. Too complex a graphic defeats the purpose of the visual presentation. Fourth, graphics should avoid "chartjunk" (Tufte 1983) which have so much visual stimuli in the way of shading and other artistic embellishments that they detract from the points being made. The whole topic of aesthetics in graphics is highly involved and goes beyond the scope of this book, but the reader should become at least somewhat sensitive to this issue if he or she is

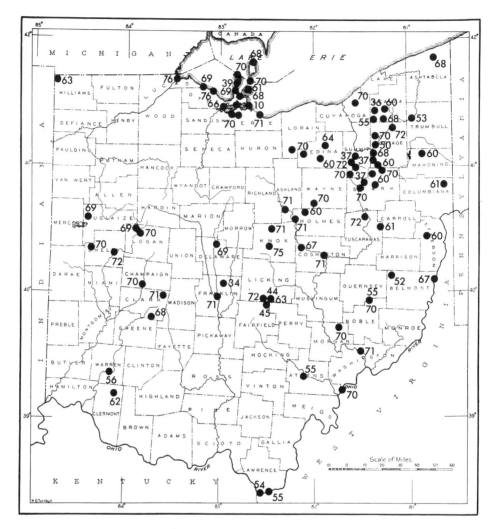

FIGURE 25.10 Dot map showing known distribution of *Potamogeton crispus* (Potamogetonaceae) in Ohio based on herbarium records. Numbers indicate the year (this century) in which the plants were collected. (From Stuckey 1979:33)

publishing actively. Let's face it; if we were so good at this we would be graphic artists and not taxonomists! As Tufte remarks: "Graphical competence demands three quite different skills: the substantive, statistical, and artistic" (1983:87). Few of us are talented in so many areas.

In conclusion, the presentation of data is enormously important for plant taxonomists to communicate the patterns of relationships they observe in their groups to other workers. This is most challenging and requires real thought, reflection, and imagination. Are graphics needed at all? What are the relationships to be stressed? How can these ideas be presented visually? "Design is

FIGURE 25.11 Pie-diagrams showing occurrence of uni- and bi-flowered heads within *Lagascea decipiens* (Compositae) in Mexico. Circles represent individual collections and 100 percent of the sample of heads analyzed; the shaded portion gives the percentage of uni-flowered heads, and the light portion the percentage of bi-flowered heads. (From Stuessy 1978:121)

choice," Tufte (1983:191) remarked correctly. Developing the ability to choose takes experience and a desire to learn some of the principles of design to be more effective. "What is to be sought in designs for the display of information is the clear portrayal of complexity. Not the complication of the simple" (Tufte 1983:191).

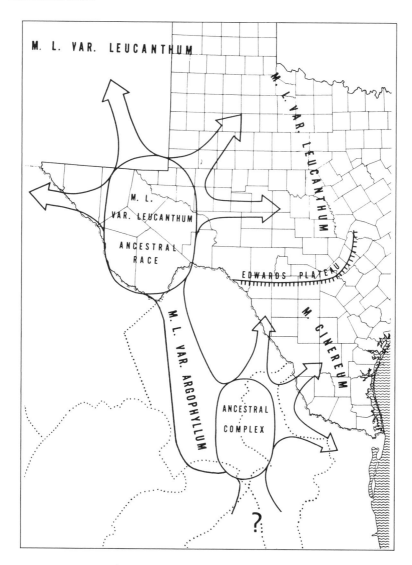

FIGURE 25.12 Diagram showing evolutionary relationships superimposed on a geographical area. Hypothetical phylogeny and migrations of taxa of the white-rayed complex of *Melampodium* (Compositae) in the southwestern United States and adjacent Mexico. (From Stuessy 1971b:188)

EPILOGUE

Taxonomy is one of the most important of the biological sciences. Without the predictive framework of classification developed over the past 2,000 years, human culture as we know it would have been very different and perhaps even still grossly underdeveloped. The need to understand the objects in our environment, both animate and inanimate, and to arrange these into a logical and predictive framework is essential to the human condition and doubtless fundamental to the development of language and of patterns of thinking in general.

There is also a powerful human curiosity to understand and explain the diversity of life on this planet. The natural interest in our own human origins ties closely to the development of a satisfactory world view integrating knowledge of origins of life, of patterns of change of life, and these integrated with religious beliefs. In fact, for one to feel comfortable with religion necessitates drawing conclusions about these biological issues at the same time. Taxonomy and, more broadly, systematics are fundamental for allowing us to develop these essential perspectives.

During the past thirty years, we have experienced a healthy reexamination and detailed dissection of classification, both philosophically and methodologically to learn what the process involves and how it might be done more effectively. The phenetic surge of the 1960s taught us that classification can be done quantitatively and more objectively. The cladistic push of the 1970s taught us that such quantitative and more precise approaches can be applied successfully to evolutionarily based classifications. Now it is clear that phyletics, or evolutionary classification, also can be done quantitatively, what I have labeled here as the New Phyletics, which is based on more evolutionary information than just branching patterns alone. This, in my opinion, will result in the most predictive and also the most generally useful classification for society as a whole.

Our use of taxonomic categories in the future will change little as these have served us well now for nearly 200 years. Special attention will be placed in botany, however, on achieving a consistent usage and attendant nomenclature for variety and subspecies. These are the only botanical categories which have had widely inconsistent usage over the history of the development of the taxonomic hierarchy. Despite recent attempts to view the problem more clearly (e.g., Stace 1986), no immediate solution is at hand. The near future must focus on this problem and provide workable alternatives. A good beginning would be to set a future start date (e.g., the year 2000) for using only one infraspecific unit in the taxonomic hierarchy (preferably the subspecies).

The amount and different kinds of data to be synthesized by taxonomists are now close to overwhelming. Chapters 15 to 23 have attempted to show the breadth of data available for reaching taxonomic conclusions in particular groups. For the modern taxonomist this task is enormous. Simply keeping up with the literature in nearly all fields of biology simultaneously is enough to fatigue even the most insomniacal taxonomic bibliophile! All data from organisms bear potentially on their evolutionary relationships and hence must be considered seriously. The recent interest in RNA and DNA and their exciting taxonomic applications, especially at the higher levels of the hierarchy, is yet another example of the need to keep abreast of all new developments in biology. If the challenge of keeping up with all the new data were not enough, it is necessary to assimilate all past literature, back at least to 1753 and in some cases even further back than that. The literature of the taxonomist never reduces or narrows—it continues to mushroom yearly.

We as taxonomists celebrate diversity. We celebrate the wildness of the planet. We celebrate the numerous human attempts to understand this wildness, and we mourn its loss through human miscalculation. We sense the aesthetic of life and much of our efforts are aimed at reflecting this composition. Above all we celebrate the challenges of being alive and dealing with the living world. There is no greater responsibility, privilege, nor satisfaction.

LITERATURE CITED

Abott, H. C. de S. 1886. Certain chemical constituents of plants considered in relation to their morphology and evolution. *Bot. Gaz.* 11:270–272.

Abbott, L. A., F. A. Bisby, and D. J. Rogers. 1985. *Taxonomic Analysis in Biology: Computers, Models, and Databases.* New York: Columbia Univ. Press.

Abbott, T. K. 1885. *Kant's Introduction to Logic, and His Essay on the Mistaken Subtilty of the Four Figures.* London: Longmans & Green.

Abrahamson, W. G. 1979. Patterns of resource allocation in wildflower populations of fields and woods. *Amer. J. Bot.* 66:71–79.

—— ed. 1988. *Plant-Animal Interactions.* New York: McGraw-Hill.

Abrahamson, W. G. and H. Caswell. 1982. On the comparative allocation of biomass, energy, and nutrients in plants. *Ecology* 63:982–991.

Abrahamson, W. G. and B. J. Hershey. 1977. Resource allocation and growth of *Impatiens capensis* (Balsaminaceae) in two habitats. *Bull. Torrey Bot. Club* 104:160–164.

Adams, E. N., III. 1972. Consensus techniques and the comparison of taxonomic trees. *Syst. Zool.* 21:390–397.

Adams, J., V. Dong, and N. Shelton, eds. 1980. *The World's Tropical Forests: A Policy, Strategy, and Program for the United States.* Publ. Dept. State (U.S.A.), 9117:1–53.

Adams, R. P. 1970. Contour mapping and differential systematics of geographic variation. *Syst. Zool.* 19:385–390.

—— 1972. Numerical analyses of some common errors in chemosystematics. *Brittonia* 24:9–21.

—— 1974a. On "numerical chemotaxonomy" revisited. *Taxon* 23:336–338.

—— 1974b. Computer graphic plotting and mapping of data in systematics. *Taxon* 23:53–70.

—— 1975a. Statistical character weighting and similarity stability. *Brittonia* 27:305–316.

—— 1975b. Numerical-chemosystematic studies of infraspecific variation in *Juniperus pinchotii*. *Biochem. Syst. Ecol.* 3:71–74.

—— 1977. Chemosystematics—analyses of populational differentiation and variability of ancestral and recent populations of *Juniperus ashei*. *Ann. Missouri Bot. Gard.* 64:184–209.

—— 1979. Diurnal variation in the terpenoids of *Juniperus scopulorum* (Cupressaceae)—summer versus winter. *Amer. J. Bot.* 66:986–988.

—— 1983. Infraspecific terpenoid variation in *Juniperus scopulorum:* Evidence for Pleistocene refugia and recolonization in western North America. *Taxon* 32:30–46.

Adams, R. P. and A. Hagerman. 1976. A comparison of the volatile oils of mature versus young leaves of *Juniperus scopulorum:* Chemosystematic significance. *Biochem. Syst. Ecol.* 4:75–79.

—— 1977. Diurnal variation in the volatile terpenoids of *Juniperus scopulorum* (Cupressaceae). *Amer. J. Bot.* 64:278–285.

Adams, R. P. and R. A. Powell. 1976. Seasonal variation of sexual differences in the volatile oil of *Juniperus scopulorum. Phytochemistry* 15:509–510.

Adams, R. P. and B. L. Turner. 1970. Chemosystematic and numerical studies of natural populations of *Juniperus ashei* Buch. *Taxon* 19:728–751.

Adams, R. P., T. A. Zanoni, and L. Hogge. 1984. Analyses of the volatile leaf oils of *Juniperus deppeana* and its infraspecific taxa: Chemosystematic implications. *Biochem. Syst. Ecol.* 12:23–27.

Adanson, M. 1763. *Familles des Plantes.* Paris.

Ahlstrand, L. 1978. Embryology of *Ursinia* (Compositae). *Bot. Not.* 131:487–496.

—— 1979a. Embryology of Arctoteae-Arctotinae (Compositae). *Bot. Not.* 132:109–116.

—— 1979b. Embryology of Arctotideae-Gundeliinae (Compositae). *Bot. Not.* 132:377–380.

—— 1979c. Embryology of Arctotideae-Gorteriinae (Compositae). *Bot. Not.* 132:371–376.

Aiken, S. G. 1978. Pollen morphology in the genus *Myriophyllum* (Haloragaceae). *Canad. J. Bot.* 56:976–982.

Airy Shaw, H. K., ed. 1966. *Willis's Dictionary of the Flowering Plants and Ferns.* 7th ed. Cambridge: Cambridge Univ. Press.

Al-Aish, M. and W. V. Brown. 1958. Grass germination responses to isopropyl-phenyl carbamate and classification. *Amer. J. Bot.* 45:16–23.

Alberch, P. 1985. Problems with the interpretation of developmental sequences. *Syst. Zool.* 34:46–58.

Albrecht, D. G., ed. 1982. *Recognition of Pattern and Form.* Berlin: Springer-Verlag.

Allen, R. T. and F. C. James, eds. 1972. *A Symposium on Ecosystematics.* Univ. Arkansas Museum Occas. Paper No. 4. Fayetteville, Ark.

Allen, T. F. H. and T. B. Starr. 1982. *Hierarchy: Perspectives for Ecological Complexity.* Chicago: Univ. Chicago Press.

Alston, R. E. 1965. Comparisons of the importance of basic metabolites, secondary compounds and macromolecules in systematic studies. *Lloydia* 28:300–312.

—— 1967. Biochemical systematics. *Evol. Biol.* 1:197–305.

Alston, R. E., T. J. Mabry, and B. L. Turner. 1963. Perspectives in chemotaxonomy. *Science* 142:545–552.

Amadon, D. 1949. The seventy-five per cent rule for subspecies. *Condor* 51:250–258.

—— 1966. Another suggestion for stabilizing nomenclature. *Syst. Zool.* 15:54–58.

Amadon, D. and L. L. Short. 1976. Treatment of subspecies approaching species status. *Syst. Zool.* 25:161–167.

Amici, J.-B. 1824. Observations microscopiques sur diverses espèces de plantes. *Ann. Sci. Nat.* (Paris) 2:41–70, 211–248.

—— 1830. Note sur le mode d'action du pollen sur le stigmate; extrait d'une lettre de M. Amici à M. Mirbel. *Ann. Sci. Nat.* (Paris) 21:329–332.

Amsden, T. W., ed. 1970. *The Genus: A Basic Concept in Paleontology.* Lawrence, Kansas: Allen Press.

Andersen, S. T. 1974a. Wind conditions and pollen deposition in a mixed deciduous forest. I. Wind conditions and pollen dispersal. *Grana* 14:57–63.

—— 1974b. Wind conditions and pollen deposition in a mixed deciduous forest. II. Seasonal and annual pollen deposition 1967–1972. *Grana* 14:64–77.

Anderson, E. 1940. The concept of the genus. II. A survey of modern opinion. *Bull. Torrey Bot. Club* 67:363–369.

—— 1949. *Introgressive Hybridization.* New York: Wiley.

—— 1956. Natural history, statistics, and applied mathematics. *Amer. J. Bot.* 43:882–889.

—— 1957. An experimental investigation of judgments concerning genera and species. *Evolution* 11:260–262.

—— 1969. Experimental studies of the species concept. *Ann. Missouri Bot. Gard.* 55:179–192.

Anderson, L. C. 1970. Embryology of *Chrysothamnus* (Astereae, Compositae). *Madroño* 20:337–342.

—— 1972. *Flaveria campestris* (Asteraceae): A case of polyhaploidy or relic ancestral diploidy? *Evolution* 26:671–673.

Anderson, L. C., R. L. Hartman, and T. F. Stuessy, 1979. Morphology, anatomy, and taxonomic relationships of *Otopappus australis* (Asteraceae). *Syst. Bot.* 4:44–56.

Anderson, L. C., D. W. Kyhos, T. Mosquin, A. M. Powell, and P. H. Raven. 1974. Chromosome numbers in Compositae. IX. *Haplopappus* and other Astereae. *Amer. J. Bot.* 61:665–671.

Anderson, L. C. and J. L. Reveal. 1966. *Chrysothamnus bolanderi*, an intergeneric hybrid. *Madroño* 18:225–233.

Anderson, S. 1974. Some suggested concepts for improving taxonomic dialogue. *Syst. Zool.* 23:58–70.

Anfinsen, C. B. 1959. *The Molecular Basis of Evolution.* New York: Wiley.

Anton, A. M. and A. E. Cocucci. 1984. The grass megagametophyte and its possible phylogenetic implications. *Pl. Syst. Evol.* 146:117–121.

Anuradha, S. M. J., M. Radhakrishnaiah, and L. L. Narayana. 1987. Numerical chemotaxonomy of some Mimosaceae. *Feddes Repert.* 98:247–252.

Arber, A. 1988. *Herbals: Their Origin and Evolution: A Chapter in the History of Botany 1470–1670.* 3d ed. Cambridge: Cambridge Univ. Press.

Archie, J. W. 1984. A new look at the predictive value of numerical classification. *Syst. Zool.* 33:30–51.

—— 1985. Methods for coding variable morphological features for numerical taxonomic analysis. *Syst. Zool.* 34:326–345.

Arends, J. C., J. D. Bastmeijer, and N. Jacobsen. 1982. Chromosome numbers and taxonomy in *Cryptocoryne* (Araceae). II. *Nord. J. Bot.* 2:453–463.

Armstrong, J. A., J. M. Powell, and A. J. Richards, eds. 1982. *Pollination and Evolution.* Sydney: Royal Botanic Gardens.

Armstrong, J. E. 1985. The delimitation of Bignoniaceae and Scrophulariaceae based on floral anatomy, and the placement of problem genera. *Amer. J. Bot.* 72:755–766.

Arnheim, N. and C. E. Taylor. 1969. Non-Darwinian evolution: Consequences for neutral allelic variation. *Nature* 223:900–902.

Arnold, E. N. 1981. Estimating phylogenies at low taxonomic levels. *Z. Zool. Syst. Evolut.-Forsch.* 19:1–35.

Arroyo, M. T. K. 1973. A taximetric study of infraspecific variation in autogamous *Limnanthes floccosa* (Limnanthaceae). *Brittonia* 25:177–191.

Ashlock, P. D. 1971. Monophyly and associated terms. *Syst. Zool.* 20:63–69.

—— 1979. An evolutionary systematist's view of classification. *Syst. Zool.* 28:441–450.

—— 1984. Monophyly: Its meaning and importance. In T. Duncan and T. F. Stuessy, eds., *Cladistics: Perspectives on the Reconstruction of Evolutionary History*, pp. 39–46. New York: Columbia Univ. Press.

Askew, R. R. 1970. Infraspecific categories in insects. *Biol. J. Linn. Soc.* 2:225–231.

Astolfi, P., A. Piazza, and K. K. Kidd. 1978. Testing of evolutionary independence in simulated phylogenetic trees. *Syst. Zool.* 27:391–400.

Atchley, W. R. and D. S. Woodruff, eds. 1981. *Evolution and Speciation: Essays in Honor of M. J. D. White.* Cambridge: Cambridge Univ. Press.

Atran, S. 1987. Origin of the species and genus concepts: An anthropological perspective. *J. Hist. Biol.* 20:195–279.

Auger, P. 1983. Hierarchically organized populations: Interactions between individual, population, and ecosystem levels. *Math. Biosci.* 65:269–289.

Averett, J. E. 1977. Chemosystematics: The Twenty-third Systematics Symposium. *Ann. Missouri Bot. Gard.* 64:145–146.

Avise, J. C. 1974. Systematic value of electrophoretic data. *Syst. Zool.* 23:465–481.

Avise, J. C. et al. 1987. Intraspecific phylogeography: The mitochondrial DNA bridge between population genetics and systematics. *Ann. Rev. Ecol. Syst.* 18:489–522.

Ayala, F. J. 1975. Genetic differentiation during the speciation proceess. *Evol. Biol.* 8:1–78.

—— ed. 1976. *Molecular Evolution.* Sunderland, Mass.: Sinauer Associates.

—— 1982. The genetic structure of species. In R. Milkman, ed., *Perspectives on Evolution*, pp. 60–82. Sunderland, Mass.: Sinauer Associates.

Ayensu, E. S. 1972. *Anatomy of the Monocotyledons.* Vol. 6: *Dioscoreales.* Oxford: Clarendon Press.

Baagøe, J. 1977a. Microcharacters in the ligules of the Compositae. In V. H. Heywood, J. B. Harborne, and B. L. Turner, eds., *The Biology and Chemistry of the Compositae*, pp. 119–139. London: Academic Press.

—— 1977b. Taxonomical application of ligule microcharacters in Compositae. I. Anthemideae, Heliantheae, and Tageteae. *Bot. Tidsskr.* 71:193–223.

—— 1980. SEM-studies in ligules of Lactuceae (Compositae). *Bot. Tidsskr.* 75:199–217.

Baas, P. 1976. Some functional and adaptive aspects of vessel member morphology. In P. Baas, A. J. Bolton, and D. M. Catling, eds., *Wood Structure in Biological and Technological Research*, pp. 157–181. Leiden: Leiden Univ. Press.

—— 1982a. Systematic, phylogenetic, and ecological wood anatomy—history and perspectives. In *New Perspectives in Wood Anatomy*, pp. 23–58. The Hague: Martinus Nijhoff/Junk.

—— ed. 1982b. *New Perspectives in Wood Anatomy.* The Hague: Martinus Nijhoff/Junk.

—— 1986. Ecological patterns of xylem anatomy. In T. J. Givnish, ed., *On the Economy of Plant Form and Function*, pp. 327–352. Cambridge: Cambridge Univ. Press.

Baba, M. L., L. L. Darga, M. Goodman, and J. Czelusniak. 1981. Evolution of cytochrome *c* investigated by the maximum parsimony method. *J. Mol. Evol.* 17:197–213.

Babcock, E. B. 1931. Cyto-genetics and the species-concept. *Amer. Naturalist* 65:5–18.

Babcock, E. B. and D. R. Cameron. 1934. Chromosomes and phylogeny in *Crepis.* II. The relationships of one hundred eight species. *Univ. California Publ. Agric. Sci.* 6:287–324.

Babcock, E. B. and J. A. Jenkins. 1943. Chromosomes and phylogeny in *Crepis.* III. The relationships of one hundred and thirteen species. *Univ. California Publ. Bot.* 18:241–291.

Babcock, E. B. and G. L. Stebbins, Jr. 1938. The American species of *Crepis:* Their interrelationships and distribution as affected by polyploidy and apomixis. *Publ. Carnegie Inst. Washington* 504:1–199.

Babcock, E. B., G. L. Stebbins, Jr., and J. A. Jenkins. 1937. Chromosomes and phylogeny in some genera of the Crepidinae. *Cytologia*, Fujii Jubilee vol.:188–210.

Bachmann, K., K. L. Chambers, H. J. Price, and A. Konig. 1983. Spatulate leaves: A marker gene for the evolution of *Microseris bigelovii* (Asteraceae–Lactuceae). *Beitr. Biol. Pflanzen* 57:167–179.

Bachmann, K., A. W. van Heusden, K. L. Chambers, and H. J. Price. 1987. A second gene determining spatulate leaf tips in *Microseris bigelovii* (Asteraceae–Lactuceae). *Beitr. Biol. Pflanzen* 62:97–106.

Bacon, J. D. 1978. Taxonomy of *Nerisyrenia* (Cruciferae). *Rhodora* 80:159–227.

Badr, A. and T. T. Elkington. 1978. Numerical taxonomy of species in *Allium* subgenus *Molium*. *New Phytol.* 81:401–417.

Bailey, I. W. 1953. The anatomical approach to the study of genera. *Chron. Bot.* 14:121–125.

Bain, J. F. and K. E. Denford. 1985. Flavonoid variation in the *Senecio streptanthifolius* complex. *Canad. J. Bot.* 63:1685–1690.

Baker, H. G. 1952. The ecospecies—prelude to discussion. *Evolution* 6:61–68.

—— 1983. An outline of the history of anthecology, or pollination biology. In L. Real, ed., *Pollination Biology*, pp. 7–28. Orlando: Academic Press.

Baker, H. G. and I. Baker. 1986. The occurrence and significance of amino acids in floral nectar. *Pl. Syst. Evol.* 151:175–186.

Baker, R. T. and H. G. Smith. 1920. *A Research on the Eucalypts Especially in Regard to Their Essential Oils.* 2d ed. Sydney: New South Wales Technological Museum.

Baldwin, J. T., Jr. 1938. *Kalanchoe:* The genus and its chromosomes. *Amer. J. Bot.* 25:572–579.

Banarescu, P. 1978. Some critical reflexions on Hennig's phyletical concepts. *Z. Zool. Syst. Evolut.-Forsch.* 16:91–101.

Banks, R. C., ed. 1979. *Museum Studies and Wildlife Management: Selected Papers.* Washington, D.C.: Smithsonian Institution Press.

Baranova, M. 1972. Systematic anatomy of the leaf epidermis in the Magnoliaceae and some related families. *Taxon* 21:447–469.

—— 1987. Historical development of the present classification of morphological types of stomates. *Bot. Rev.* 53:53–79.

Barber, H. N. 1955. Adaptive gene substitutions in Tasmanian eucalypts. I: Genes controlling the development of glaucousness. *Evolution* 9:1–14.

Barbosa, P. and V. A. Krischik. 1987. Influence of alkaloids on feeding preference of eastern deciduous forest trees by the gypsy moth *Lymantria dispar. Amer. Naturalist* 30:50–69.

Barbour, M. G., J. H. Burk, and W. D. Pitts. 1987. *Terrestrial Plant Ecology.* 2d ed. Menlo Park, Calif.: Benjamin/Cummings.

Barigozzi, C., ed. 1982. *Mechanisms of Speciation.* New York: Alan R. Liss.

Barlow, B. A. 1971. Cytogeography of the genus *Eremophila. Austral. J. Bot.* 19:295–310.

Bartcher, R. L. 1966. Fortran IV program for estimation of cladistic relationships using the IBM 7040. *Kansas Geol. Surv. Computer Contr.* 6:1–54.

Barth, F. G. 1985. *Insects and Flowers: The Biology of a Partnership.* Translated by M. A. Biederman-Thorson. Princeton, N.J.: Princeton Univ. Press.

Barth, O. M. 1965. Elektronen mikroskopische Beobachtungen am Sporoderm der Caryocaraceen. *Grana Palynol.* 6:7–25.

Barthlott, W. 1981. Epidermal and seed surface characters of plants: Systematic applicability and some evolutionary aspects. *Nord. J. Bot.* 1:345–355.

—— 1984. Microstructural features of seed surfaces. In V. H. Heywood and D. M. Moore, eds., *Current Concepts in Plant Taxonomy*, pp. 95–105. London: Academic Press.

Barthlott, W., N. Ehler, and R. Schill. 1976. Abtragung biologischer Oberflächen durch hochfrequenzaktivierten Saurstoff für die Raster-Elektronenmikroskopie. *Mikroskopie* 32:35–44.

Bartlett, H. H. 1940. The concept of the genus. I. History of the generic concept in botany. *Bull. Torrey Bot. Club* 67:349–362.

Bate-Smith, E. C. 1958. Plant phenolics as taxonomic guides. *Proc. Linn. Soc.* (London) 169:198–211.

Bate-Smith, E. C. and J. B. Harborne. 1971. Differences in flavonoid content between fresh and herbarium leaf tissue in *Dillenia. Phytochemistry* 10:1055–1058.

Bates, D. M. and O. J. Blanchard, Jr. 1970. Chromosome numbers in the Malvales. II. New or otherwise noteworthy counts relevant to classification in the Malvaceae, tribe Malveae. *Amer. J. Bot.* 57:927–934.

Batten, J. W. and J. S. Gibson. 1977. *Soils: Their Nature, Classes, Distribution, Uses, and Care.* Rev. ed. University: Univ. Alabama Press.

Bauhin, G. 1623. *Pinax Theatri Botanici.* Basel.

Baum, B. R. 1972. *Avena septentrionalis*, and the semispecies concept. *Canad. J. Bot.* 50:2063–2066.

—— 1973. The concept of relevance in taxonomy with special emphasis on automatic classification. *Taxon* 22:329–332.

—— 1975. Cladistic analysis of the diploid and hexaploid oats (*Avena*, Poaceae) using numerical techniques. *Canad. J. Bot.* 52:2115–2127.

—— 1976. Weighting character-states. *Taxon* 25:257–260.

—— 1977. Taxonomy of tribe Triticeae (Poaceae) using various numerical techniques. I. Historical perspectives, data accumulation, and character analysis. *Canad. J. Bot.* 55:1712–1740.

—— 1978a. Taxonomy of the tribe Triticeae (Poaceae) using various numerical techniques. II. Classification. *Canad. J. Bot.* 56:27–56.

—— 1978b. Taxonomy of the tribe Triticeae (Poaceae) using various numerical taxonomic techniques. III. Synoptic key to genera and synopses. *Canad. J. Bot.* 56:374–385.

—— 1978c. Generic relationships in Triticeae based on computations of Jardine and Sibson B$_k$ clusters. *Canad. J. Bot.* 56:2948–2954.

—— 1984. Application of compatibility and parsimony methods at the infraspecific, specific, and generic levels in Poaceae. In T. Duncan and T. F. Stuessy, eds., *Cladistics: Perspectives on the Reconstruction of Evolutionary History*, pp. 192–220. New York: Columbia Univ. Press.

Baum, B. R. and G. F. Estabrook. 1978. Application of compatibility analysis in numerical cladistics at the infraspecific level. *Canad. J. Bot.* 56:1130–1135.

Bayer, R. J. 1984. Chromosome numbers and taxonomic notes for North American species of *Antennaria* (Asteraceae: Inuleae). *Syst. Bot.* 9:74–83.

Bayer, R. J. and D. J. Crawford. 1986. Allozyme divergence among five diploid species of *Antennaria* (Asteraceae: Inuleae) and their allopolyploid derivations. *Amer. J. Bot.* 73:287–296.

Bayer, R. J. and G. L. Stebbins. 1982. A revised classification of *Antennaria* (Asteraceae: Inuleae) of the eastern United States. *Syst. Bot.* 7:300–313.

Beaman, J. H., D. C. D. De Jong, and W. P. Stoutamire. 1962. Chromosome studies in the alpine and subalpine floras of Mexico and Guatemala. *Amer. J. Bot.* 49:41–50.

Beard, J. S. 1978. The physiognomic approach. In R. H. Whittaker, ed., *Classification of Plant Communities*, pp. 33–64. 2d ed. The Hague: Junk.

Beattie, A. J. 1983. Distribution of ant-dispersed plants. In K. Kubitzki, ed., *Dispersal and Distribution: An International Symposium*, pp. 249–270. Hamburg: Verlag Paul Parey.

—— 1985. *The Ecology of Ant-Plant Mutualisms.* Cambridge: Cambridge Univ. Press.

Beatty, J. 1982. Classes and cladists. *Syst. Zool.* 31:25–34.

Beaudry, J. R. 1960. The species concept: Its evolution and present status. *Rev. Canad. Biol.* 19:219–240.

Beck, C. B. 1960. The identity of *Archaeopteris* and *Callixylon*. *Brittonia* 12:351–368.

—— 1962. Reconstructions of *Archaeopteris*, and further consideration of its phylogenetic position. *Amer. J. Bot.* 49:373–382.

—— 1970. Problems of generic delimitation in paleobotany. In T. W. Amsden, ed., *The Genus: A Basic Concept in Paleontology*, pp. 173–193. Lawrence, Kansas: Allen Press.

Becker, K. M. 1973. A comparison of angiosperm classification systems. *Taxon* 22:19–50.

Beckner, M. 1959. *The Biological Way of Thought.* New York: Columbia Univ. Press.

Behnke, H.-D. 1967. Über den Aufbau der Siebelement-Plastiden einiger Dioscoreaceen. *Z. Pflanzenphysiol.* 57:243–254.

—— 1968. Zum Feinbau der Siebröhren-Plastiden bei Monocotylen. *Naturwissenschaften* 3:140–141.

—— 1969. Die Siebröhren-Plastiden der Monocotyledonen: Vergleichende Untersuchungen über Feinbau und Verbreitung eines charakteristischen Plastidentyps. *Planta* 84:174–184.

—— 1971. Sieve-tube plastids of Magnoliidae and Ranunculidae in relation to systematics. *Taxon* 20:723–730.

—— 1972. Sieve-tube plastids in relation to angiosperm systematics—an attempt towards a classification by ultrastructural analysis. *Bot. Rev.* 38:155–197.

—— 1973. Sieve-tube plastids of Hamamelididae: Electron microscopic investigations with special reference to Urticales. *Taxon* 22:205–210.

—— 1974a. Sieve-element plastids of Gymnospermae: Their ultrastructure in relation to systematics. *Pl. Syst. Evol.* 123:1–12.

—— 1974b. Elektronenmikroskopische Untersuchungen an Siebröhren-Plastiden und ihre Aussage über die systematische Stellung von *Lophiocarpus*. *Bot. Jahrb. Syst.* 94:114–119.

—— 1974c. P- und S-Typ Siebelement-Plastiden bei Rhamnales. *Beitr. Biol. Pflanzen* 50:457–464.

—— 1975a. P-type sieve-element plastids: A correlative ultrastructural and ultrahistochemical study on the diversity and uniformity of a new reliable character in seed plant systematics. *Protoplasma* 83:91–101.

—— 1975b. Elektronenmikroskopische Untersuchungen zur Frage der verwandtschaftlichen Beziehungen zwischen *Theligonum* und *Rubiaceae:* Feinbau der Siebelement-Plastiden und Anmerkungen zur Struktur der Pollenexine. *Pl. Syst. Evol.* 123:317–326.

—— 1975c. *Hectorella caespitosa:* Ultrastructural evidence against its inclusion into Caryophyllaceae. *Pl. Syst. Evol.* 124:31–34.

—— 1975d. The bases of angiosperm phylogeny: Ultrastructure. *Ann. Missouri Bot. Gard.* 62:647–663.

—— 1976a. Ultrastructure of sieve-element plastids in Caryophyllales (Centrospermae), evidence for the delimitation and classification of the order. *Pl. Syst. Evol.* 126:31–54.

—— 1976b. Sieve-element plastids of *Fouquieria, Frankenia* (Tamaricales), and *Rhabdodendron* (Rutaceae), taxa sometimes allied with Centrospermae (Caryophyllales). *Taxon* 25:265–268.

—— 1976c. Die Siebelement-Plastiden der Caryophyllaceae, eine weitere spezifische Form der P-Typ Plastiden bei Centrospermen. *Bot. Jahrb. Syst.* 95:327–333.

—— 1977a. Transmission electron microscopy and systematics of flowering plants. In K. Kubitzki, ed., *Flowering Plants: Evolution and Classification of Higher Categories*, pp. 155–178. Berlin: Springer-Verlag.

—— 1977b. Phloem ultrastructure and systematic position of Gyrostemonaceae. *Bot. Not.* 130:255–260.

—— 1977c. Dilatierte ER-Zisternen, ein mikromorphologisches Merkmal der Capparales? *Ber. Deutsch. Bot. Ges.* 90:241–251.

—— 1978. Elektronenoptische Untersuchungen am Phloem sukkulenter Centrospermen (incl. Didiereaceen). *Bot. Jahrb. Syst.* 99:341–352.

—— 1981a. Siebelement-Plastiden, Phloem-Protein und Evolution der Blütenpflanzen: II. Monokotyledonen. *Ber. Deutsch. Bot. Ges.* 94:647–662.

—— 1981b. Sieve-element characters. *Nord. J. Bot.* 1:381–400.

—— ed. 1981c. Ultrastructure and systematics of seed plants. *Nord. J. Bot.* 1:341–460.

—— 1982a. Sieve-element plastids, exine sculpturing and the systematic affinities of the Buxaceae. *Pl. Syst. Evol.* 139:257–266.

—— 1982b. Sieve-element plastids of Cyrillaceae, Erythroxylaceae and Rhizophoraceae: Description and significance of subtype PV plastids. *Pl. Syst. Evol.* 141:31–39.

—— 1982c. *Geocarpon minimum:* Sieve-element plastids as additional evidence for its inclusion in the Caryophyllaceae. *Taxon* 31:45–47.

—— 1982d. Sieve-element plastids of Connaraceae and Oxalidaceae: A contribution to the knowledge of P-type plastids in dicotyledons and their significance. *Bot. Jahrb. Syst.* 103:1–8.

—— 1984. Ultrastructure of sieve-element plastids of Myrtales and allied groups. *Ann. Missouri Bot. Gard.* 71:824–831.

—— 1986a. Contributions to the knowledge of sieve-element plastids in Gunneraceae and allied families. *Pl. Syst. Evol.* 151:215–222.

—— 1986b. Sieve-element characters and the systematic position of *Austrobaileya*, Austrobaileyaceae—with comments to the distinction and definition of sieve cells and sieve-tube members. *Pl. Syst. Evol.* 152:101–121.

Behnke, H.-D. and W. Barthlott. 1983. New evidence for the ultrastructural and micromorphological fields in angiosperm classification. *Nord. J. Bot.* 3:43–66.

Behnke, H.-D., C. Chang, I. J. Eifert, and T. J. Mabry. 1974. Betalains and P-type sieve-tube plastids in *Petiveria* and *Agdestis* (Phytolaccaceae). *Taxon* 23:541–542.

Behnke, H.-D. and R. Dahlgren. 1976. The distribution of characters within an angiosperm system. 2. Sieve-element plastids. *Bot. Not.* 129:287–295.

Behnke, H.-D. and T. J. Mabry. 1977. S-type sieve-element plastids and anthocyanins in Vivianiaceae: Evidence against its inclusion into Centrospermae. *Pl. Syst. Evol.* 126:371–375.

Behnke, H.-D., L. Pop, and V. V. Sivarajan. 1983. Sieve-element plastids of Caryophyllales: Additional investigations with special reference to the Caryophyllaceae and Molluginaceae. *Pl. Syst. Evol.* 142:109–116.

Behnke, H.-D and B. L. Turner. 1971. On specific sieve-tube plastids in Caryophyllales. Further invesigations with special reference to the Bataceae. *Taxon* 20:731–737.

Bell, A. D. and P. B. Tomlinson. 1980. Adaptive architecture in rhizomatous plants. *Bot. J. Linn. Soc.* 80:125–160.

Bell, C. R. 1959. Mineral nutrition and flower to flower pollen size variation. *Amer. J. Bot.* 46:621–624.

—— 1967. *Plant Variation and Classification.* Belmont, Calif.: Wadsworth.

Bell, C. R. and L. Constance. 1957. Chromosome numbers in Umbelliferae. *Amer. J. Bot.* 44:565–572.

—— 1960. Chromosome numbers in Umbelliferae, II. *Amer. J. Bot.* 47:24–32.

—— 1966. Chromosome numbers in Umbelliferae, III. *Amer. J. Bot.* 53:512–520.

Bell, E. A. and B. V. Charlwood, eds. 1980. *Secondary Plant Products.* Berlin: Springer-Verlag.

Bendich, A. J. and R. S. Anderson. 1983. Repeated DNA sequences and species relatedness in the genus *Equisetum. Pl. Syst. Evol.* 143:47–52.

Bendich, A. J. and E. T. Bolton. 1967. Relatedness among plants as measured by the DNA-agar technique. *Pl. Physiol.* 42:959–967.

Bendz, G. and J. Santesson, eds. 1974. *Chemistry in Botanical Classification.* Nobel Foundation, Stockholm. New York: Academic Press.

Bennett, M. D. 1972. Nuclear DNA content and minimum generation time in herbaceous plants. *Proc. Roy. Soc.* (London), ser. B, 181:109–135.

—— 1984. The genome, the natural karyotype, and biosystematics. In W. F. Grant, ed., *Plant Biosystematics,* pp. 41–66. Toronto: Academic Press.

Benson, L. 1979. *Plant Classification.* 2d ed. Lexington, Mass.: D. C. Heath.

Bentham, G. and J. D. Hooker. 1862–1883. *Genera Plantarum.* 3 vols. London: Reeve.

Bentzer, B., R. Bothmer, L. Engstrand, M. Gustafsson, and S. Snogerup. 1971. Some sources of error in the determination of arm ratios of chromosomes. *Bot. Not.* 124:65–74.

Benzing, D. H. 1967a. Developmental patterns in stem primary xylem of woody Ranales. I. Species with unilacunar nodes. *Amer. J. Bot.* 54:805–813.

—— 1967b. Developmental patterns in stem primary xylem of woody Ranales. II. Species with trilacunar and multilacunar nodes. *Amer. J. Bot.* 54:813–820.

Benzing, D. H., T. J. Givnish, and D. Bermudes. 1985. Absorptive trichomes in *Briachinia reducta* (Bromeliaceae) and their evolutionary and systematic significance. *Syst. Bot.* 10:81–91.

Benzing, D. H., K. Henderson, B. Kessel, and J. Sulak. 1976. The absorptive capacities of bromeliad trichomes. *Amer. J. Bot.* 63:1009–1014.

Benzing, D. H. and A. M. Pridgeon. 1983. Foliar trichomes of Pleurothalliidinae (Orchidaceae): Functional significance. *Amer. J. Bot.* 70:173–180.

Benzing, D. H., J. Seemann, and A. Renfrow. 1978. The foliar epidermis in Tillandsioideae (Bromeliaceae) and its role in habitat selection. *Amer. J. Bot.* 65:359–365.

Berlin, B. 1973. Folk systematics in relation to biological classification and nomenclature. *Ann. Rev. Ecol. Syst.* 4:259–271.

Berlin, B., D. E. Breedlove, and P. H. Raven. 1966. Folk taxonomies and biological classification. *Science* 154:273–275.

—— 1974. *Principles of Tzeltal Plant Classification: An Introduction to the Botanical Ethnography of a Mayan-Speaking People* of *Highland Chiapas.* New York: Academic Press.

Berlyn, G. P. and J. P. Miksche. 1976. *Botanical Microtechnique and Cytochemistry.* Ames: Iowa State Univ. Press.

Bernier, R. 1984. The species as an individual: Facing essentialism. *Syst. Zool.* 33:460–469.

Bertalanffy, L. von. 1968. *General System Theory: Foundations, Development, Applications.* Rev. ed. New York: George Braziller.

Bessey, C. E. 1908. The taxonomic aspect of the species question. *Amer. Naturalist* 42:218–224.

—— 1915. The phylogenetic taxonomy of flowering plants. *Ann. Missouri Bot. Gard.* 2:109–164.

Beyer, W. A., M. L. Stein, T. F. Smith, and S. M. Ulam. 1974. A molecular sequence metric and evolutionary trees. *Math. Biosci.* 19:9–25.

Bhandari, N. N. 1971. Embryology of the Magnoliales and comments on their relationships. *J. Arnold Arbor.* 52:1–39, 285–304.

—— 1984. The microsporangium. In B. M. Johri, ed., *Embryology of Angiosperms,* pp. 53–121. Berlin: Springer-Verlag.

Bhojwani, S. S. and S. P. Bhatnagar. 1983. *The Embryology of Angiosperms.* 4th ed. New Delhi: Vikas.

Bierhorst, D. W. 1971. *Morphology of Vascular Plants.* New York: Macmillan.

Bierner, M. W., W. M. Dennis, and B. E. Wofford. 1977. Flavonoid chemistry, chromosome number and phylogenetic relationships of *Helenium chihuahuensis. Biochem. Syst. Ecol.* 5:23–28.

Biesboer, D. D. and P. G. Mahlberg. 1981. Laticifer starch grain morphology and laticifer evolution in *Euphorbia* (Euphorbiaceae). *Nord. J. Bot.* 1:447–457.

Bigelow, R. S. 1961. Higher categories and phylogeny. *Syst. Zool.* 10:86–91.

Birks, H. J. B. and H. H. Birks. 1980. *Quaternary Palaeoecology.* London: Arnold.

Birks, H. J. B. and A. D. Gordon. 1985. *Numerical Methods in Quaternary Pollen Analysis.* London: Academic Press.

Bisby, F. A. and R. M. Polhill. 1973. The role of taximetrics in angiosperm taxonomy II. Parallel taximetric and orthodox studies in *Crotalaria* L. *New Phytol.* 72:727–742.

Bisby, F. A., J. G. Vaughan, and C. A. Wright, eds. 1980. *Chemosystematics: Principles and Practice.* London: Academic Press.

Bisby, G. R. and G. C. Ainsworth. 1943. The numbers of fungi. *Trans. Brit. Mycol. Soc.* 26:16–19.

Blackith, R. E. and R. A. Reyment. 1971. *Multivariate Morphometrics.* London: Academic Press.

Blackmore, S. 1981. Palynology and intergeneric relationships in subtribe Hyoseridinae (Compositae: Lactuceae). *Bot. J. Linn. Soc.* 82:1–13.

—— 1982a. Palynology of subtribe Scorzonerinae (Compositae: Lactuceae) and its taxonomic significance. *Grana* 21:149–160.

—— 1982b. The apertures of Lactuceae (Compositae) pollen. *Pollen et Spores* 24:453–461.

—— 1982c. A functional interpretation of Lactuceae (Compositae) pollen. *Pl. Syst. Evol.* 141:153–168.

—— 1984. Pollen features and plant systematics. In V. H. Heywood and D. M. Moore, eds., *Current Concepts in Plant Taxonomy*, pp. 135–154. London: Academic Press.

Blackmore, S. and D. Claugher. 1984. Ion beam etching in palynology. *Grana* 23:85–89.

Blackmore, S. and H. G. Dickinson. 1981. A simple technique for sectioning pollen grains. *Pollen et Spores* 23:281–285.

Blackmore, S. and I. K. Ferguson, eds. 1986. *Pollen and Spores: Form and Function.* London: Academic Press.

Blackwelder, R. E. 1964. Phyletic and phenetic *versus* omnispective classification. In V. H. Heywood and J. McNeill, eds., *Phenetic and Phylogenetic Classification*, pp. 17–28. London: Systematics Association.

—— 1967a. *Taxonomy: A Text and Reference Book.* New York: Wiley.

—— 1967b. A critique of numerical taxonomy. *Syst. Zool.* 16:64–72.

Blackwelder, R. E. and A. Boyden. 1952. The nature of systematics. *Syst. Zool.* 1:26–33.

Blair, W. F. and B. L. Turner. 1972. The integrative approach to biological classification. In J. A. Behnke, ed., *Challenging Biological Problems: Directions Toward Their Solution*, pp. 193–217. New York: Oxford Univ. Press.

Blasdell, R. F. 1963. A monographic study of the fern genus *Cystopteris. Mem. Torrey Bot. Club* 21(4):[i–ii], 1–102.

Bloom, M. 1976. Evolution in the genus *Ruellia* (Acanthaceae): A discussion based on floral flavonoids. *Amer. J. Bot.* 63:399–405.

Blumenberg, B. and A. T. Lloyd. 1983. *Australopithecus* and the origin of the genus *Homo:* Aspects of biometry and systematics with accompanying catalog of tooth metric data. *Biosystems* 16:127–167.

Böcher, T. W. 1970. The present status of biosystematics. *Taxon* 19:3–5.

Bock, W. J. 1968. Phylogenetic systematics, cladistics and evolution. *Evolution* 22:646–648.

—— 1969. Discussion: The concept of homology. *Ann. New York Acad. Sci.* 167:71–73.

—— 1973. Philosophical foundations of classical evolutionary classification. *Syst. Zool.* 22:375–392.

—— 1977. Foundations and methods of evolutionary classification. In M. K. Hecht, P. C. Goody, and B. M. Hecht, eds., *Major Patterns in Vertebrate Evolution*, pp. 851–895. New York: Plenum Press.

—— 1979. The synthetic explanation of macroevolutionary change—a reductionistic approach. *Bull. Carnegie Mus. Nat. Hist.* 13:20–69.

—— 1986. Species concepts, speciation, and macroevolution. In K. Iwatsuki, P. H. Raven, and W. J. Bock, eds., *Modern Aspects of Species*, pp. 31–57. Tokyo: Univ. Tokyo Press.

Bocquet, G. 1959. The campylotropous ovule. *Phytomorphology* 9:222–227.

Bocquet, G. and J. D. Bersier. 1960. La valeur systématique de l'ovule: Développements teratologiques. *Arch. Sci.* 13:475–496.

Boesewinkel, F. D. and F. Bouman. 1984. The seed: Structure. In B. M. Johri, ed., *Embryology of Angiosperms*, pp. 567–610. Berlin: Springer-Verlag.

Bogert, C. M. 1954. The indication of infraspecific variation. *Syst. Zool.* 3:111–112.

Bogle, A. L. 1970. Floral morphology and vascular anatomy of the Hamamelidaceae: The apetalous genera of Hamamelidoideae. *J. Arnold Arbor.* 51:310–366.

Bohm, B. A. 1987. Intraspecific flavonoid variation. *Bot. Rev.* 53:197–279.

Bohm, B. A., S. W. Brim, R. J. Hebda, and P. F. Stevens. 1978. Generic limits in the tribe Cladothamneae (Ericaceae), and its position in the Rhododendroideae. *J. Arnold Arbor.* 59:311–341.

Bohm, D. 1980. *Wholeness and the Implicate Order.* London: Routledge & Kegan Paul.

Böhme, W. 1978. Das Kühnelt'sche Prinzip der regionalen Stenözie und seine Bedeutung für das Subspezies-Problem: Ein theoretischer Ansatz. *Z. Zool. Syst. Evolut.-Forsch.* 16:256–266.

—— 1979. Kühnelt's principle and the subspecies-problem: A reply to L. Botosaneanu. *Z. Zool. Syst. Evolut.-Forsch.* 17:243-246.

Boivin, B. 1950. The problem of generic segregates in the form-genus *Lycopodium. Amer. Fern J.* 40:32–41.

-—— 1962. Persoon and the subspecies. *Brittonia* 14:327–331.

Bold, H. C. 1967. *Morphology of Plants.* 2d ed. New York: Harper & Row.

Bold, H. C., C. J. Alexopoulos, and T. Delevoryas. 1987. *Morphology of Plants and Fungi.* 5th ed. New York Harper and Row.

Bold, H. C. and M. J. Wynne. 1985. *Introduction to the Algae: Structure and Reproduction.* 2d ed. Englewood Cliffs, N.J.: Prentice-Hall.

Bolick, M. R. 1978. Taxonomic, evolutionary, and functional considerations of Compositae pollen ultrastructure and sculpture. *Pl. Syst. Evol.* 130:209–218.

—— 1981. Mechanics as an aid to interpreting pollen structure and function. *Rev. Palaeobot. Palynol.* 35:61–79.

—— 1983. A cladistic analysis of the Ambrosiinae Less. and Engelmanniinae Stuessy. In N. I. Platnick and V. A. Funk, eds., *Advances in Cladistics,* 2:125–141. New York: Columbia Univ. Press.

Bolick, M. R. and J. J. Skvarla. 1976. A reappraisal of the pollen ultrastructure of *Parthenice mollis* Gray (Compositae). *Taxon* 25:261–264.

Booth, C. 1978. Do you believe in genera? *Trans. Brit. Mycol. Soc.* 71:1–9.

Borgen, L. 1970. Chromosome numbers of Macaronesian flowering plants. *Nytt Mag. Bot.* 17:145–161.

—— 1974. Chromosome numbers of Macaronesian flowering plants, II. *Norw. J. Bot.* 21:195–210.

—— 1975. Chromosome numbers of vascular plants from Macaronesia. *Norw. J. Bot.* 22:71–76.

—— 1977. *Check-List of Chromosome Numbers Counted in Macaronesian Vascular Plants.* Oslo: published by the author.

—— 1979. Karyology of the Canarian flora. In D. Bramwell, ed., *Plants and Islands,* pp. 329–346. London: Academic Press.

—— 1980. Chromosome numbers of Macaronesian flowering plants, III. *Bot. Macaronesica* 7:67–76.

Bossert, W. 1969. Computer techniques in systematics [with discussion]. In C. G. Sibley, chm. *Systematic Biology,* pp. 595–614. Washington, D.C.: National Academy of Sciences.

Bostock, C. J. and A. T. Sumner. 1978. *The Eukaryotic Chromosome.* Amsterdam: North-Holland.

Botosaneanu, L. 1979. Remarks about a paper devoted to the "subspecies problem." *Z. Zool. Syst. Evolut.-Forsch.* 17:242–243.

Boulter, D. 1972. The use of comparative animo acid sequence data in evolutionary studies of higher plants. In L. Reinhold and Y. Liwschitz, eds., *Progress in Phytochemistry,* 3:199–229. London: Interscience Publishers.

—— 1974. Amino acid sequences of cytochrome *c* and plastocyanins in phylogenetic studies of higher plants. *Syst. Zool.* 22:549–553.

—— 1980. The evaluation of present results and future possibilities of the use of amino acid sequence data in phylogenetic studies with specific reference to plant proteins. In F. A. Bisby, J. G. Vaughan, and C. A. Wright, eds., *Chemosystematics: Principles and Practice,* pp. 235–240. London: Academic Press.

Boulter, D., J. T. Gleaves, B. G. Haslett, and D. Peacock. 1978. The relationships of 8 tribes of the Compositae. *Phytochemistry* 17:1585–1589.

Boulter, D., D. Peacock, A. Guise, J. T. Gleaves, and G. Estabrook. 1979. Relationships between the partial amino acid sequences of plastocyanin from members of ten families of flowering plants. *Phytochemistry* 18:603–608.

Brady, R. H. 1985. On the independence of systematics. *Cladistics* 1:113–126.

Brandham, P. E. and M. D. Bennett, eds. 1983. *Kew Chromosome Conference II.* London: Allen & Unwin.

Brandham, P. E. and D. F. Cutler. 1978. Influence of chromosome variation on the organisation of the leaf epidermis in a hybrid *Aloe* (Liliaceae). *Bot. J. Linn. Soc.* 77:1–16.

—— 1981. Polyploidy, chromosome interchange and leaf surface anatomy as indicators of relationships within *Haworthia* section *Coarctatae* Baker (Liliaceae–Aloineae). *J. S. Afr. Bot.* 47:507–546.

Braun-Blanquet, J. 1932. *Plant Sociology: The Study of Plant Communities.* Translated and edited by G. D. Fuller and H. S. Conrad. New York: McGraw-Hill.

Brazier, J. D. 1968. The contribution of wood anatomy to taxonomy. *Proc. Linn. Soc.* (London) 179:271–274.

Brehm, B. G. and D. Krell. 1975. Flavonoid localization in epidermal papillae of flower petals: A specialized adaptation for ultraviolet absorption. *Science* 190:1221–1223.

Brehm, B. G. and M. Ownbey. 1965. Variation in chromatographic patterns in the *Tragopogon dubius-pratensis-porrifolius* complex (Compositae). *Amer. J. Bot.* 52:811–818.

Bremekamp, C. E. B. 1939. Phylogenetic interpretations and genetic concepts in taxonomy. *Chron. Bot.* 5:398–403.

Bremer, K. 1976. The genus *Relhania* (Compositae). *Opera Bot.* 40:1–85.

—— 1978. The genus *Leysera* (Compositae). *Bot. Not.* 131:369–383.

—— 1983a. Angiosperms and phylogenetic systematics—some problems and examples. *Abh. Verh. Natürwiss. Vereins* (Hamburg) 26:343–354.

—— 1983b. Vikarians-biogeografi. *Svensk Bot. Tidskr.* 77:33–40.

—— 1987. Tribal interrelationships of the Asteraceae. *Cladistics* 3:210–253.

Bremer, K., C. J. Humphries, B. D. Mishler, and S. P. Churchill. 1987. On cladistic relationships in green plants. *Taxon* 36:339–349.

Bremer, K. and H.-E. Wanntorp. 1978. Phylogenetic systematics in botany. *Taxon* 27:317–329.

—— 1979. Geographic populations or biological species in phylogeny reconstruction? *Syst. Zool.* 28:220–224.

—— 1981. The cladistic approach to plant classification. In V. A. Funk and D. R. Brooks, eds., *Advances in Cladistics*, pp. 87–94. New York: New York Botanical Garden.

—— 1982. Fylogenetisk systematik. *Svensk Bot. Tidskr.* 76:177–183.

Brewbaker, J. L. 1967. The distribution and phylogenetic significance of binucleate and trinucleate pollen grains in the angiosperms. *Amer. J. Bot.* 54:1069–1083.

Briggs, D. and M. Block. 1981. An investigation into the use of the "-deme" terminology. *New Phytol.* 89:729–735.

Brittan, N. H. 1970. A preliminary survey of the stem and leaf anatomy of *Thysanotus* R.Br. (Liliaceae). In N. K. B. Robson, D. F. Cutler, and M. Gregory, eds. *New Research in Plant Anatomy*, pp. 57–70. London: Academic Press.

Britton, N. L. 1908. The taxonomic aspect of the species question. *Amer. Naturalist* 42:225–242.

Brooks, D. R. 1981. Classifications as languages of empirical comparative biology. In V. A. Funk and D. R. Brooks, eds., *Advances in Cladistics*, pp. 61–70. New York: New York Botanical Garden.

Brooks, D. R., J. N. Caira, T. R. Platt, and M. R. Pritchard. 1984. *Principles and Methods of Phylogenetic Systematics: A Cladistics Workbook.* Univ. Kansas Mus. Nat. Hist. Spec. Publ. 12.

Brooks, J., P. R. Grant, M. Muir, P. van Gijzel, and G. Shaw, eds. 1971. *Sporopollenin.* London: Academic Press.

Brooks, J. and G. Shaw. 1978. Sporopollenin: A review of its chemistry, palaeochemistry and geochemistry. *Grana* 17:91–97.

Broome, C. R. 1978. Chromosome numbers and meiosis in North and Central American species of *Centaurium* (Gentianaceae). *Syst. Bot.* 3:299–312.

Brothers, D. J. 1983. Nomenclature at the ordinal and higher levels. *Syst. Zool.* 32:34–42.

Brown, D. F. M. 1964. A monographic study of the fern genus *Woodsia*. *Beih. Nova Hedwigia* 16:i–x, 1–154.

Brown, G. K. and A. J. Gilmartin. 1984. Stigma structure and variation in Bromeliaceae—neglected taxonomic characters. *Brittonia* 36:364–374.

Brown, H. 1956. *The Challenge of Man's Future: An Inquiry Concerning the Condition of Man During the Years that Lie Ahead.* New York: Viking Press.

Brown, R. 1811. On the Proteaceae of Jussieu. *Trans. Linn. Soc.* (London) 10:15–226.

Brown, W. L., Jr. and E. O. Wilson. 1954. The case against the trinomen. *Syst. Zool.* 3:174–176.

Brown, W. M., E. M. Prager, A. Wang, and A. C. Wilson. 1982. Mitochondrial DNA sequences of primates: Tempo and mode of evolution. *J. Mol. Evol.* 18:225–239.

Brown, W. V. 1972. *Textbook of Cytogenetics.* St. Louis: Mosby.

—— 1975. Variations in anatomy, associations, and origins of Kranz tissue. *Amer. J. Bot.* 62:395–402.

—— 1977. The Kranz syndrome and its subtypes in grass systematics. *Mem. Torrey Bot. Club* 23(3):1–97.

Brown, W. V. and C. Johnson. 1962. The fine structure of the grass guard cell. *Amer. J. Bot.* 49:110–115.

Brummitt, R. K., I. K. Ferguson, and M. M. Poole. 1980. A unique and extraordinary pollen type in the genus *Crossandra* (Acanthaceae). *Pollen et Spores* 22:11–16.

Brundin, L. 1976. A Neocomian chironomid and Podonominae-Aphroteniinae (Diptera) in the light of phylogenetics and biogeography. *Zoologica Scripta* 5:139–160.

Brunfels, O. 1530–1536. *Herbarum Vivae Eicones.* 3 vols. Stuttgart.

Brunken, J. N. 1979a. Cytotaxonomy and evolution in *Pennisetum* section *Brevivalvula* (Gramineae) in tropical Africa. *Bot. J. Linn. Soc.* 79:37–49.

—— 1979b. Morphometric variation and the classification of *Pennisetum* section *Brevivalvula* (Gramineae) in tropical Africa. *Bot. J. Linn. Soc.* 79:51–64.

Bryson, V. and H. J. Vogel, eds. 1965. *Evolving Genes and Proteins.* New York: Academic Press.

Buck, R. C. and D. L. Hull. 1966. The logical structure of the Linnaean hierarchy. *Syst. Zool.* 15:97–111.

Buck, W. R. 1980. A generic revision of the Entodontaceae. *J. Hattori Bot. Lab.* 48:71–159.

Bullini, L. and M. Coluzzi. 1972. Natural selection and genetic drift in protein polymorphism. *Nature* 239:160–161.

Burger, W. C. 1967. Families of Flowering Plants in Ethiopia. *Oklahoma Agric. Exp. Sta. Bull.* 45.

—— 1975. The species concept in *Quercus. Taxon* 24:45–50.

—— 1979. Cladistics: Useful tool or rigid dogma? *Taxon* 28:385–386.

Burgman, M. A. 1985. Cladistics, phenetics and biogeography of populations of *Boronia inornata* Turcz. (Rutaceae) and the *Eucalyptus diptera* Andrews (Myrtaceae) species complex in western Australia. *Austral. J. Bot.* 33:419–431.

Burma, B. H. 1954. Reality, existence, and classification: A discussion of the species problem. *Madroño* 12:193–209.

Burnett, W. C., Jr., S. B. Jones, Jr., T. J. Mabry, and W. G. Padolina. 1974. Sesquiterpene lactones—insect feeding deterrents in *Vernonia. Biochem. Syst. Ecol.* 2:25–29.

Burnham, C. R. 1962. *Discussions in Cytogenetics.* Minneapolis: Burgess.

Burns, J. M. 1968. A simple model illustrating problems of phylogeny and classification. *Syst. Zool.* 17:170–173.

Burns-Balogh, P. and V. A. Funk. 1986. A phylogenetic analysis of the Orchidaceae. *Smithsonian Contr. Bot.* 61:1–79.

Burt, W. H. 1954. The subspecies category in mammals. *Syst. Zool.* 3:99–104.

Burtt, B. L. 1964. Angiosperm taxonomy in practice. In V. H. Heywood and J. McNeill, eds., *Phenetic and Phylogenetic Classification,* pp. 5–16. London: Systematics Association.

—— 1965. Adanson and modern taxonomy. *Notes Roy. Bot. Gard. Edinburgh* 26:427–431.

—— 1970. Infraspecific categories in flowering plants. *Biol. J. Linn. Soc.* 2:233–238.

Bush, G. L. 1975. Modes of animal speciation. *Ann. Rev. Ecol. Syst.* 6:339–364.

Buss, P. A., Jr. and N. R. Lersten. 1975. Survey of tapetal nuclear number as a taxonomic character in Leguminosae. *Bot. Gaz.* 136:388–395.

Buth, D. G. 1984. The application of electrophoretic data in systematic studies. *Ann. Rev. Ecol. Syst.* 15:501–522.

Byatt, J. I., I. K. Ferguson, and B. G. Murray. 1977. Intergeneric hybrids between *Crataegus* L. and *Mespilus* L.: A fresh look at an old problem. *Bot. J. Linn. Soc.* 74:329–343.

Byerly, H. C. 1973. *A Primer of Logic.* New York: Harper & Row.

Caccone, A. and J. R. Powell. 1987. Molecular evolutionary divergence among North American cave crickets. II. DNA-DNA hybridization. *Evolution* 41:1215–1238.

Cain, A. J. 1953. Geography, ecology and coexistence in relation to the biological definition of the species. *Evolution* 7:76–83.

—— 1956. The genus in evolutionary taxonomy. *Syst. Zool.* 5:97–109.

—— 1958. Logic and memory in Linnaeus's system of taxonomy. *Proc. Linn. Soc.* (London) 169:144–163.

—— 1959a. The post-Linnaean development of taxonomy. *Proc. Linn. Soc.* (London) 170:234–244.

—— 1959b. Taxonomic concepts. *Ibis* 101:302–318.

—— 1959c. Deductive and inductive methods in post-Linnaean taxonomy. *Proc. Linn. Soc.* (London) 170:185–217.

—— 1962. The evolution of taxonomic principles. In G. C. Ainsworth and P. H. A. Sneath, eds., *Microbial Classification,* pp. 1–13. Cambridge: Cambridge Univ. Press.

—— 1976. The use of homology and analogy in evolutionary theory [with discussion]. In M. von Cranach, ed., *Methods of Inference from Animal to Human Behaviour,* pp. 25–38. The Hague: Mouton.

Cain, A. J. and G. A. Harrison. 1958. An analysis of the taxonomist's judgment of affinity. *Proc. Zool. Soc.* (London) 131:85–98.

—— 1960. Phyletic weighting. *Proc. Zool. Soc.* (London) 135:1–31.

Cain, S. A. 1943. Criteria for the indication of center of origin in plant geographical studies. *Torreya* 43:132–154.

—— 1944. *Foundations of Plant Geography*. New York: Harper & Row.

Calderon, C. E. and T. R. Soderstrom. 1973. Morphological and anatomical considerations of the grass subfamily Bambusoideae based on the new genus *Maclurolyra*. *Smithsonian Contr. Bot.* 11:1–55.

Calvin, M. 1969. *Chemical Evolution: Molecular Evolution Towards the Origin of Living Systems on the Earth and Elsewhere*. New York: Oxford Univ. Press.

Camerarius, R. J. 1694. *De Sexu Plantarum Epistola*. Tubingen.

Camin, J. H. and R. R. Sokal. 1965. A method for deducing branching sequences in phylogeny. *Evolution* 19:311–326.

Camp, W. H. 1940. The concept of the genus. V. Our changing generic concepts. *Bull. Torrey Bot. Club* 67:381–389.

—— 1951. Biosystematy. *Brittonia* 7:113–127.

Camp, W. H. and C. L. Gilly. 1943a. The structure and origin of species with a discussion of intraspecific variability and related nomenclatural problems. *Brittonia* 4:323–385.

—— 1943b. Polypetalous forms of *Vaccinium*. *Torreya* 42:168–173.

Campbell, A., S. Muncer, and D. Bibel. 1985. Taxonomies of aggressive behaviour—a preliminary report. *Aggressive Behavior* 11:217–222.

Campbell, C. S. 1986. Phylogenetic reconstructions and two new varieties in the *Andropogon virginicus* complex (Poaceae: Andropogoneae). *Syst. Bot.* 11:280–292.

Campbell, I. 1971. Numerical taxonomy of various genera of yeasts. *J. Gen. Microbiol.* 67:223–231.

—— 1972. Numerical analysis of the genera *Saccharomyces* and *Kluyveromyces*. *J. Gen. Microbiol.* 73:279–301.

Candolle, A. L. P. P. de, ed. 1844–1873. *Prodromus Systematis Naturalis Regni Vegetabilis*, vols. 8–17. Paris.

—— 1855. *Géographie Botanique Raisonnée*. 2 vols. Paris: Victor Masson.

—— 1867. Lois de la Nomenclature Botanique adoptées par le Congrès International de Botanique tenu à Paris en Aout 1867. 2d ed. Geneva: H. Georg.

Candolle, A. P. de. 1813. *Theorie Elementaire de la Botanique*. Paris.

—— ed. 1824–1838. *Prodromus Systematis Naturalis Regni Vegetabilis*, vols. 1–7. Paris.

Canne, J. M. 1979. A light and scanning electron microscope study of seed morphology in *Agalinis* (Scrophulariaceae) and its taxonomic significance. *Syst. Bot.* 4:281–296.

—— 1980. Seed surface features in *Aureolaria, Brachystigma, Tomanthera*, and certain South American *Agalinis* (Scrophulariaceae). *Syst. Bot.* 5:241–252.

Canright, J. E. 1963. Contributions of pollen morphology to the phylogeny of some Ranalean families. *Grana Palynol.* 4:64–72.

Cantino, P. D. 1982a. A monograph of the genus *Physostegia* (Labiatae). *Contr. Gray Herb.* 211:1–105.

—— 1982b. Affinities of the Lamiales: A cladistic analysis. *Syst. Bot.* 7:237–248.

—— 1985. Phylogenetic inference from nonuniversal derived character states. *Syst. Bot.* 10:119–122.

Carlock, J. R. and D. E. Fensholt. 1970. Problems in systematics considered as applications in set theory [abstract]. *Amer. J. Bot.* 57:767.

Carlquist, S. 1957. The genus *Fitchia* (Compositae). *Univ. California Publ. Bot.* 29:1–143.

—— 1958. Anatomy and systematic position of *Centaurodendron* and *Yunquea* (Compositae). *Brittonia* 10:78–93.

—— 1959. Studies on Madinae: Anatomy, cytology, and evolutionary relationships. *Aliso* 4:171–236.

—— 1961. *Comparative Plant Anatomy: A Guide to Taxonomic and Evolutionary Application of Anatomical Data in Angiosperms*. New York: Holt, Rinehart and Winston.

—— 1966. Wood anatomy of Compositae: A summary, with comments on factors controlling wood evolution. *Aliso* 6:25–44.

—— 1967. Anatomy and systematics of *Dendroseris* (sensu lato). *Brittonia* 19:99–121.

—— 1969a. Morphology and anatomy. In J. Ewan, ed., *A Short History of Botany in the United States*, pp. 49–57. New York: Hafner.

—— 1969b. Toward acceptable evolutionary interpretations of floral anatomy. *Phytomorphology* 19:332–362.

—— 1969c. Wood anatomy of Lobelioideae. *Biotropica* 1:47–72.

—— 1970. Wood anatomy of *Echium* (Boraginaceae). *Aliso* 7:183–199.

—— 1971. Wood anatomy of Macaronesian and other Brassicaceae. *Aliso* 7:365–384.

—— 1975. *Ecological Strategies of Xylem Evolution*. Berkeley: Univ. California Press.

—— 1977. Ecological factors in wood evolution: A floristic approach. *Amer. J. Bot.* 64:887–896.

—— 1978. Wood anatomy of Bruniaceae: Correlations with ecology, phylogeny, and organography. *Aliso* 9:323–364.

—— 1981a. Chance dispersal. *Amer. Scientist* 69:509–516.

—— 1981b. Wood anatomy of Chloanthaceae (Dicrastylidaceae). *Aliso* 10:19–34.

—— 1982a. Wood anatomy of Buxaceae: Correlations with ecology and phylogeny. *Flora* 172:463–491.

—— 1982b. Wood anatomy of Daphniphyllaceae: Ecological and phylogenetic considerations, review of Pittosporalean families. *Brittonia* 34:252–266.

—— 1982c. Wood anatomy of Dipsacaceae. *Taxon* 31:443–450.

—— 1982d. Wood anatomy of *Illicium* (Illiciaceae): Phylogenetic, ecological, and functional interpretations. *Amer. J. Bot.* 69:1587–1598.

—— 1982e. Wood anatomy of Onagraceae: Further species; root anatomy; significance of vestured pits and allied structures in dicotyledons. *Ann. Missouri Bot. Gard.* 69:755–769.

—— 1984a. Wood anatomy of some Gentianaceae: Systematic and ecological conclusions. *Aliso* 10:573–582.

—— 1984b. Wood and stem anatomy of Lardizabalaceae, with comments on the vining habit, ecology and systematics. *Bot. J. Linn. Soc.* 88:257–277.

—— 1985a. Wood anatomy of Begoniaceae, with comments on raylessness, paedomorphosis, relationships, vessel diameter, and ecology. *Bull. Torrey Bot. Club* 112:59–69.

—— 1985b. Wood anatomy of Coriariaceae: Phylogenetic and ecological implications. *Syst. Bot.* 10:174–183.

Carlquist, S. and D. R. Bissing. 1976. Leaf anatomy of Hawaiian geraniums in relation to ecology and taxonomy. *Biotropica* 8:248–259.

Carlquist, S. and L. Debuhr. 1977. Wood anatomy of Penaeaceae (Myrtales): Comparative, phylogenetic, and ecological implications. *Bot. J. Linn. Soc.* 75:211–227.

Carpenter, J. M. 1987. A report on the Society for the Study of Evolution workshop "Computer Programs for Inferring Phylogenies." *Cladistics* 3:52–55.

—— 1988. Choosing among multiple equally parsimonious cladograms. *Cladistics* 4:291–296.

Carr, G. D. 1978. Chromosome numbers of Hawaiian flowering plants and the significance of cytology in selected taxa. *Amer. J. Bot.* 65:236–242.

—— 1985. Additional chromosome numbers of Hawaiian flowering plants. *Pacific Sci.* 39:302–306.

Carr, S. G. M., L. Milkovits, and D. J. Carr. 1971. Eucalypt phytoglyphs: The microanatomical features of the epidermis in relation to taxonomy. *Austral. J. Bot.* 19:173–190.

Carroll, C. P. and M. Borrill. 1965. Tetraploid hybrids from crosses between diploid and tetraploid *Dactylis* and their significance. *Genetica* 36:65–82.

Cartmill, M. 1981. Hypothesis testing and phylogenetic reconstruction. *Z. Zool. Syst. Evolut.-Forsch.* 19:73–96.

Caspersson, T. and L. Zech, eds. 1973. *Chromosome Identification—Technique and Applications in Biology and Medicine.* Nobel Foundation, Stockholm. New York: Academic Press.

Castle, W. E. 1903. The laws of heredity of Galton and Mendel, and some laws governing race improvement by selection. *Proc. Amer. Acad. Arts* 39:223–242.

Cattell, R. B. and M. A. Coulter. 1966. Principles of behavioural taxonomy and the mathematical basis of the taxonome computer program. *Brit. J. Math. Stat. Psych.* 19:237–269.

Cave, M. S. 1948. Sporogenesis and embryo sac development of *Hesperocallis* and *Leucocrinum* in relation to their systematic position. *Amer. J. Bot.* 35:343–349.

—— 1953. Cytology and embryology in the delimitation of genera. *Chron. Bot.* 14:140–153.

—— 1955. Sporogenesis and the female gametophyte of *Phormium tenax. Phytomorphology* 5:247–253.

—— ed. 1958–1965. *Index to Plant Chromosome Numbers* [1956–1964 & Supplement]. Chapel Hill: Univ. North Carolina Press.

—— 1962. Embryological characters of taxonomic significance. *Lilloa* 31:171–181.

—— 1964. Cytological observations on some genera of the Agavaceae. *Madroño* 17:163–170.

—— 1968. The megagametophyte of *Androcymbium. Phytomorphology* 17:233–239.

—— 1974. Female gametophytes of *Chlorogalum* and *Schoenolirion (Hastingsia). Phytomorphology* 24:56–60.

—— 1975. Embryological studies in *Stypandra* (Liliaceae). *Phytomorphology* 25:95–99.

Cave, M. S. and L. Constance. 1942. Chromosome numbers in the Hydrophyllaceae. *Univ. California Publ. Bot.* 18:205–216.

—— 1944. Chromosome numbers in the Hydrophyllaceae: II. *Univ. California Publ. Bot.* 18:293–298.

—— 1947. Chromosome numbers in the Hydrophyllaceae: III. *Univ. California Publ. Bot.* 18:449–465.

—— 1950. Chromosome numbers in the Hydrophyllaceae: IV. *Univ. California Publ. Bot.* 23:363–381.

—— 1959. Chromosome numbers in the Hydrophyllaceae: V. *Univ. California Publ. Bot.* 30: 233–257.

Celebioglu, T., C. Favarger, and K.-L. Huynh. 1983. Contribution à la micromorphologie de la testa des graines du genre *Minuartia* (Caryophyllaceae). I. Sect. *Minuartia. Bull. Mus. Natl. Hist. Nat.* (Paris), Sect. B, *Adansonia*, ser. 4, 5:415–435.

Chabot, B. F. and H. A. Mooney. 1985. *Physiological Ecology of North American Plant Communities.* New York: Chapman and Hall.

Challice, J. S. and M. N. Westwood. 1973. Numerical taxonomic studies of the genus *Pyrus* using both chemical and botanical characters. *Bot. J. Linn. Soc.* 67:121–148.

Chaloner, W. G. and A. Sheerin. 1981. The evolution of reproductive strategies in early land plants. In G. G. E. Scudder and J. L. Reveal, eds., *Evolution Today*, pp. 93–100. Pittsburgh: Hunt Institute for Botanical Documentation.

Chambers, J. M., W. S. Cleveland, B. Kleiner, and P. A. Tukey. 1983. *Graphic Methods for Data Analysis.* Monterey, Calif.: Wadsworth.

Chambers, K. L., ed. 1970. *Biochemical Coevolution.* Corvallis: Oregon State Univ. Press.

Chance, G. D. and J. D. Bacon. 1984. Systematic implications of seed coat morphology in *Nama* (Hydrophyllaceae). *Amer. J. Bot.* 71:829–842.

Chang, C. P. and T. J. Mabry. 1973. The constitution of the order Centrospermae: rRNA-DNA hybridization studies among betalain- and anthocyanin-producing families. *Biochem. Syst.* 1:185–190.

Chapman, D. J. and R. K. Trench. 1982. Prochlorophyceae: Introduction and bibliography. In J. R. Rosowski and B. C. Parker, eds., *Selected Papers in Phycology II*, pp. 656–658. Lawrence, Kansas: Phycological Society of America.

Chapman, R. W., J. C. Avise, and M. A. Asmussen. 1979. Character space restrictions and boundary conditions in the evolution of quantitative multistate characters. *J. Theor. Biol.* 80:51–64.

Charig, A. J. 1982. Systematics in biology: A fundamental comparison of some major schools of thought. In K. A. Joysey and A. E. Friday, eds., *Problems of Phylogenetic Reconstruction*, pp. 363–440. London: Academic Press.

Charlesworth, B., R. Lande, and M. Slatkin. 1982. A neo-Darwinian commentary on macroevolution. *Evolution* 36:474–498.

Chater, A. O. and R. K. Brummitt. 1966a. Subspecies in the works of Friedrich Ehrhart. *Taxon* 15:95–106.

—— 1966b. Subspecies in the works of Christiaan Hendrik Persoon. *Taxon* 15:143–149.

Chatfield, C. and A. J. Collins. 1980. *Introduction to Multivariate Analysis.* London: Chapman and Hall.

Cheadle, V. I. 1943a. The origin and certain trends of specialization of the vessel in the Monocotyledoneae. *Amer. J. Bot.* 30:11–17.

—— 1943b. Vessel specialization in the late metaxylem of the various organs in the Monocotyledoneae. *Amer. J. Bot.* 30:484–490.

Chimphamba, B. B. 1973. Intergeneric hybridization between *Iris dichotoma*, Pall. and *Belamcanda chinensis*, Leman. *Cytologia* 38:539–547.

Chiribog, D. A. and S. Krystal. 1985. An empirical taxonomy of symptom types among divorcing persons. *J. Clinical Psych.* 41:601–613.

Chopra, R. N. and R. C. Sachar. 1963. Endosperm. In P. Maheshwari, ed., *Recent Advances in the Embryology of Angiosperms*, pp. 135–170. Delhi: International Society of Plant Morphologists, Univ. Delhi.

Christensen, J. E. and H. T. Horner, Jr. 1974. Pollen pore development and its spatial orientation during microsporogenesis in the grass *Sorghum bicolor. Amer. J. Bot.* 61:604–623.

Chuang, T.-I. and L. R. Heckard. 1972. Seed coat morphology in *Cordylanthus* (Scrophulariaceae) and its taxonomic significance. *Amer. J. Bot.* 59:258–265.

—— 1982. Chromosome numbers of *Orthocarpus* and related monotypic genera (Scrophulariaceae: subtribe Castillejinae). *Brittonia* 34:89–101.

—— 1983. Systematic significance of seed-surface features in *Orthocarpus* (Scrophulariaceae—subtribe Castillejinae). *Amer. J. Bot.* 70:877–890.

Churchill, S. P. 1981. A phylogenetic analysis, classification and synopsis of the genera of the Grimmiaceae (Musci). In V. A. Funk and D. R. Brooks, eds., *Advances in Cladistics*, pp. 127–144. New York: New York Botanical Garden.

Churchill, S. P. and E. O. Wiley. 1980. A comparison of Wagner's and Hennig's methods of phylogenetic analysis [abstract]. *Bot. Soc. Amer. Misc. Ser. Publ.* 158:23.

Churchill, S. P., E. O. Wiley, and L. A. Hauser. 1984. A critique of Wagner groundplan-divergence studies and a comparison with other methods of phylogenetic analysis. *Taxon* 33:212–232.

Chute, H. M. 1930. The morphology and anatomy of the achene. *Amer. J. Bot.* 17:703–723.

Clark, C. and D. J. Curran. 1986. Outgroup analysis, homoplasy, and global parsimony: A response to Maddison, Donoghue, and Maddison. *Syst. Zool.* 35:422–426.

Clark, P. J. 1952. An extension of the coefficient of divergence for use with multiple characters. *Copeia* 2:61–64.

Clark, R. B. 1956. Species and systematics. *Syst. Zool.* 5:1–10.

Clarke, A. E. and P. A. Gleeson. 1981. Molecular aspects of recognition and response in the pollen-stigma interaction. In F. A. Loewus and C. A. Ryan, eds., *Recent Advances in Phytochemistry*, Vol. 15: *The Phytochemistry of Cell Recognition and Cell Surface Interactions*, pp. 161–211. New York: Plenum Press.

Classen, D., C. Nozzolillo, and E. Small. 1982. A phenolic-taxometric study of *Medicago* (Leguminosae). *Canad. J. Bot.* 60:2477–2495.

Claugher, D. and J. J. Rowley. 1987. *Betula* pollen grain substructure revealed by fast atom etching. *Pollen et Spores* 29:5–20.

Clausen, C. P. 1942. The relation of taxonomy to biological control. *J. Econ. Entomol.* 35:744–748.

Clausen, J., D. D. Keck, and W. M. Hiesey. 1939. The concept of species based on experiment. *Amer. J. Bot.* 26:103–106.

—— 1940. Experimental studies on the nature of species I. Effect of varied environments on western North American plants. *Publ. Carnegie Inst.* (Washington) 520:i–viii, 1–452.

—— 1941a. Experimental taxonomy. *Carnegie Inst.* (Washington) *Year Book* 40:160–170.

—— 1941b. Regional differentiation in plant species. In J. Cattell, ed., *Biological Symposia*, 4:261–280. Lancaster, Pa.: Cattell Press.

—— 1945. Experimental studies on the nature of species II. Plant evolution through amphiploidy and autoploidy with examples from the Madiinae. *Publ. Carnegie Inst.* (Washington) 564:i–viii, 1–174.

—— 1948. Experimental studies on the nature of species III. Environmental responses of climatic races of *Achillea*. *Publ. Carnegie Inst.* (Washington) 581:i–iv, 1–129.

Clausen, R. T. 1941. On the use of the terms "subspecies" and "variety." *Rhodora* 43:157–167.

Clayton, W. D. 1970. Studies in the Gramineae: XXI. *Coelorhachis* and *Rhytachne:* A study in numerical taxonomy. *Kew Bull.* 24:309–314.

—— 1971. Studies in the Gramineae: XXVI. Numerical taxonomy of the Arundinelleae. *Kew Bull.* 26:111–123.

—— 1972. Some aspects of the genus concept. *Kew Bull.* 27:281–287.

—— 1974a. The logarithmic distribution of angiosperm families. *Kew Bull.* 29:271–279.

—— 1974b. A discriminant function for *Digitaria diagonalis*. Studies in the Gramineae: XXXVII. *Kew Bull.* 29:527–533.

—— 1975. Chorology of the genera of the Gramineae. *Kew Bull.* 30:111–132.

Cleland, R. E. 1923. Chromosome arrangements during meiosis in certain Oenotheras. *Amer. Naturalist* 57:562–566.

—— 1936. Some apsects of the cyto-genetics of *Oenothera*. *Bot. Rev.* 2:316–348.

—— 1944. The problem of species in *Oenothera*. *Amer. Naturalist* 78:5–28.

—— 1972. *Oenothera: Cytogenetics and Evolution*. London: Academic Press.

Clements, F. E. 1905. *Research Methods in Ecology*. Lincoln, Neb.: University Publishing.

—— 1916. Plant succession: An analysis of the development of vegetation. *Publ. Carnegie Inst.* (Washington) 242:i–xiv, 1–512.

Clements, F. E. and F. L. Long. 1923. Experimental pollination: An outline of the ecology of flowers and insects. *Publ. Carnegie Inst.* (Washington) 336:1–274.

Clench, W. J. 1954. The occurrence of clines in molluscan populations. *Syst. Zool.* 3:122–125.

Cleveland, W. S. 1985. *The Elements of Graphing Data*. Monterrey, Calif.: Wadsworth.

Cleveland, W. S. and R. McGill. 1985. Graphical perception and graphical methods for analyzing scientific data. *Science* 229:828–833.

Clifford, H. T. 1969. Attribute correlation in the Poaceae (grasses). *Bot. J. Linn. Soc.* 62:59–68.

—— 1970. Monocotyledon classification with special reference to the origin of the grasses (Poaceae). In N. K. B. Robson, D. F. Cutler, and M. Gregory, eds., *New Research in Plant Anatomy*, pp. 25–34. London: Academic Press.

—— 1977. Quantitative studies of inter-relationships amongst the Liliatae. In K. Kubitzki, ed.,

Flowering Plants: Evolution and Classification of Higher Categories, pp. 77–95. Berlin: Springer-Verlag.

Clifford, H. T. and D. W. Goodall. 1967. A numerical contribution to the classification of the Poaceae. *Austral. J. Bot.* 15:499–519.

Clifford, H. T. and W. Stephenson. 1975. *An Introduction to Numerical Classification.* New York: Academic Press.

Clowes, F. A. L. 1961. *Apical Meristems.* Oxford: Blackwell.

Coates, D. and C. A. Cullis. 1987. Chloroplast DNA variability among *Linum* species. *Amer. J. Bot.* 74:260–268.

Cocucci, A. E. 1983. New evidence from embryology in angiosperm classification. *Nord. J. Bot.* 3:67–73.

Cody, V., E. Middleton, Jr., and J. B. Harborne, eds. 1986. *Plant Flavonoids in Biology and Medicine: Biochemical, Pharmacological, and Structure-Activity Relationships.* New York: Alan R. Liss.

Cohen, D. M. and R. F. Cressey, eds. 1969. Natural history collections: Past, present, future. *Proc. Biol. Soc. Washington* 82:559–762.

Cohn, N. S. 1969. *Elements of Cytology.* 2d ed. New York: Harcourt, Brace & World.

Cole, A. J., ed. 1969. *Numerical Taxonomy.* London: Academic Press.

Cole, G. T. and H.-D. Behnke. 1975. Electron microscopy and plant systematics. *Taxon* 24:3–15.

Colless, D. H. 1967. The phylogenetic fallacy. *Syst. Zool.* 16:289–295.

—— 1969. The interpretation of Hennig's "Phylogenetic Systematics"—a reply to Dr. Schlee. *Syst. Zool.* 18:134–144.

—— 1970. The phenogram as an estimate of phylogeny. *Syst. Zool.* 19:352–362.

—— 1971. "Phenetic," "phylogenetic," and "weighting." *Syst. Zool.* 20:73–76.

—— 1977. A cornucopia of categories. *Syst. Zool.* 26:349–352.

—— 1981. Predictivity and stability in classifications: Some comments on recent studies. *Syst. Zool.* 30:325–331.

—— 1985. On the status of outgroups in phylogenetics. *Syst. Zool.* 34:364–366.

Comer, C. W., R. P. Adams, and D. F. van Haverbeke. 1982. Intra- and interspecific variation of *Juniperus virginiana* and *J. scopulorum* seedlings based on volatile oil composition. *Biochem. Syst. Ecol.* 10:297–306.

Constance, L. 1951. The versatile taxonomist. *Brittonia* 7:225–231.

—— 1957. Plant taxonomy in an age of experiment. *Amer. J. Bot.* 44:88–92.

—— 1963. Chromosome number and classification in Hydrophyllaceae. *Brittonia* 15:273–285.

—— 1964. Systematic botany—an unending synthesis. *Taxon* 13:257–273.

—— 1971. The uses of diversity. *Pl. Sci. Bull.* 17:22–23.

Constance, L. and T.-I. Chuang. 1982. Chromosome numbers of Umbelliferae (Apiaceae) from Africa south of the Sahara. *Bot. J. Linn. Soc.* 85:195–208.

Constance, L., T.-I Chuang, and C. R. Bell. 1971. Chromosome numbers in Umbelliferae. IV. *Amer. J. Bot.* 58:577–587.

—— 1976. Chromosome numbers in Umbelliferae. V. *Amer. J. Bot.* 63:608–625.

Constance, L., T.-I. Chuang, and R. A. Bye, Jr. 1976. Chromosome numbers in Chihuahuan Umbelliferae. *Bot. Mus. Leafl.* 24:241–247.

Cook, O. F. 1899. Four categories of species. *Amer. Naturalist* 33:287–297.

Cooper-Driver, G. A. and M. J. Balick. 1978. Effects of field preservation on the flavonoid content of *Jessenia bataua. Bot. Mus. Leafl.* 26:257–265.

Core, E. L. 1955. *Plant Taxonomy.* Englewood Cliffs, N.J.: Prentice-Hall.

Core, H. A., W. A. Côté, and A. C. Day. 1979. *Wood Structure and Identification.* 2d ed. Syracuse: Syracuse Univ. Press.

Correns, C. 1900. G. Mendel's Regel über das Verhalten der Nachkommenschaft der Rassenbastarde. *Ber. Deutsch. Bot. Ges.* 18:158–168.

Côté, W. A., Jr., ed. 1965. *Cellular Ultrastructure of Woody Plants.* Syracuse: Syracuse Univ. Press.

Coulter, J. M. and C. J. Chamberlain. 1903. *Morphology of Angiosperms.* New York: Appleton.

Cowan, S. T. 1962. The microbial species—a macromyth? *Symp. Soc. Gen. Microbiol.* 12:433–455.

Cowles, H. C. 1909. Present problems in plant ecology. The trend of ecological philosophy. *Amer. Naturalist* 43:356–368.

Cracraft, J. 1975. Historical biogeography and earth history: Perspectives for a future synthesis. *Ann. Missouri Bot. Gard.* 62:227–250.

—— 1982. Geographic differentiation, cladistics, and vicariance biogeography: Reconstructing the tempo and mode of evolution. *Amer. Zoologist* 22:411–424.

—— 1983. Cladistic analysis and vicariance biogeography. *Amer. Scientist* 71:273–281.

Cracraft, J. and N. Eldredge, eds. 1979. *Phylogenetic Analysis and Paleontology.* New York: Columbia Univ. Press.

Crane, E. and P. Walker. 1984. *Pollination Directory for World Crops.* London: International Bee Research Association.

Crane, P. R. 1985. Phylogenetic analysis of seed plants and the origin of angiosperms. *Ann. Missouri Bot. Gard.* 72:716–793.

Crang, R. E. and H. L. Dean. 1971. An intergeneric hybrid in the Sileneae (Caryophyllaceae). *Bull. Torrey Bot. Club* 98:214–217.

Craw, R. C. 1984. Leon Croizat's biogeographic works: A personal appreciation. *Tuatara* 27:8–13.

Crawford, D. J. 1979a. Allozyme studies in *Chenopodium incanum:* Intraspecific variation and comparison with *Chenopodium fremontii. Bull. Torrey Bot. Club* 106:257–261.

—— 1979b. Flavonoid chemistry and angiosperm evolution. *Bot. Rev.* 44:431–456.

—— 1983. Phylogenetic and systematic inferences from electrophoretic studies. In S. D. Tanksley and T. J. Orton, eds., *Isozymes in Plant Genetics and Breeding*, Part A, pp. 257–287. Amsterdam: Elsevier.

Crawford, D. J. and R. J. Bayer. 1981. Allozyme divergence in *Coreopsis cyclocarpa* (Compositae). *Syst. Bot.* 6:373–379.

Crawford, D. J. and K. A. Evans. 1978. The affinities of *Chenopodium flabellifolium* (Chenopodiaceae): Evidence from seed coat surface and flavonoid chemistry. *Brittonia* 30:313–318.

Crawford, D. J. and D. E. Giannasi. 1982. Plant chemosystematics. *BioScience* 32:114–124.

Crawford, D. J. and R. L. Hartman. 1972. Chromosome numbers and taxonomic notes for Rocky Mountain Umbelliferae. *Amer. J. Bot.* 59:344–392.

Crawford, D. J. and E. A. Julian. 1976. Seed protein profiles in the narrow-leaved species of *Chenopodium* of the western United States: Taxonomic value and comparison with distribution of flavonoid compounds. *Amer. J. Bot.* 63:302–308.

Crawford, D. J. and J. F. Reynolds. 1974. A numerical study of the common narrow-leaved taxa of *Chenopodium* occurring in the western United States. *Brittonia* 26:398–409.

Crawford, D. J. and E. B. Smith. 1982a. Allozyme variation in *Coreopsis nuecensoides* and *C. nuecensis* (Compositae), a progenitor-derivative species pair. *Evolution* 36:379–386.

—— 1982b. Allozyme divergence between *Coreopsis basalis* and *C. wrightii* (Compositae). *Syst. Bot.* 7:359–364.

—— 1984. Allozyme divergence and intraspecific variation in *Coreopsis grandiflora* (Compositae). *Syst. Bot.* 9:219–225.

Crawford, D. J., T. F. Stuessy, and M. Silva O. 1987. Allozyme divergence and the evolution of *Dendroseris* (Compositae: Lactuceae) on the Juan Fernandez Islands. *Syst. Bot.* 12:435–443.

Crawford, D. J. and H. D. Wilson. 1977. Allozyme variation in *Chenopodium fremontii. Syst. Bot.* 2:180–190.

—— 1979. Allozyme variation in several closely related diploid species of *Chenopodium* of the western United States. *Amer. J. Bot.* 66:237–244.

Crawley, M. J. 1983. *Herbivory: The Dynamics of Animal-Plant Interactions.* Oxford: Blackwell.

—— ed. 1986. *Plant Ecology.* Oxford: Blackwell.

Crepet, W. L. 1983. The role of insect pollination in the evolution of the angiosperms. In L. Real, ed., *Pollination Biology*, pp. 29–50. Orlando: Academic Press.

—— 1984. Advanced (constant) insect pollination mechanisms: Pattern of evolution and implications vis-a-vis angiosperm diversity. *Ann. Missouri Bot. Gard.* 71:607–630.

Cresti, M., F. Ciampolini, and G. Sarfatti. 1980. Ultrastructural investigations in *Lycopersicon peruvianum* pollen activation and pollen tube organization after self- and cross-pollination. *Planta* 150:211–217.

Cresti, M., E. Pacini, F. Ciampolini, and G. Sarfatti. 1977. Germination and early tube development in vitro of *Lycopersicon peruvianum* pollen: Ultrastructural features. *Planta* 136:239–247.

Crété, P. 1963. Embryo. In P. Maheshwari, ed., *Recent Advances in the Embryology of Angiosperms*, pp. 171–220. Delhi: International Society of Plant Morphologists, Univ. Delhi.

Crisci, J. V. 1974. A numerical-taxonomic study of the subtribe Nassauviinae (Compositae, Mutisieae). *J. Arnold Arbor.* 55:568–610.

—— 1980. Evolution in the subtribe Nassauviinae (Compositae, Mutisieae): A phylogenetic reconstruction. *Taxon* 29:213–224.

—— 1982. Parsimony in evolutionary theory: Law or methodological prescription? *J. Theor. Biol.* 97:35–41.

Crisci, J. V., J. H. Hunziker, R. A. Palacios, and C. A. Naranjo. 1979. A numerical-taxonomic study of the genus *Bulnesia* (Zygophyllaceae): Cluster analysis, ordination and simulation of evolutionary trees. *Amer. J. Bot.* 66:133–140.

Crisci, J. V. and M. F. López A. 1983. *Introducción a la Teoría y Práctica de la Taxonomía Numérica.* Washington, D.C.: Organization of American States.

Crisci, J. V. and T. F. Stuessy. 1980. Determining primitive character states for phylogenetic reconstruction. *Syst. Bot.* 5:112–135.

Croat, T. B. 1972. The role of overpopulation and agricultural methods in the destruction of tropical ecosystems. *BioScience* 22:465–467.

—— 1980. Flowering behavior of the neotropical genus *Anthurium* (Araceae). *Amer. J. Bot.* 67:888–904.

Croizat, L. 1962. *Space, Time, Form: The Biological Synthesis.* Caracas: published by the author.

—— 1978. Deduction, induction, and biogeography. *Syst. Zool.* 27:209–213.

—— 1982. Vicariance/vicariism, panbiogeography, "vicariance biogeography," etc.: A clarification. *Syst. Zool.* 31:291–304.

Croizat, L., G. Nelson, and D. E. Rosen. 1974. Centers of origin and related concepts. *Syst. Zool.* 23:265–287.

Cronquist, A. 1947. Revision of the North American species of *Erigeron*, north of Mexico. *Brittonia* 6:121–300.

—— 1957. Outline of a new system of families and orders of dicotyledons. *Bull. Jard. Bot. Bruxelles* 27:13–40.

—— 1963. The taxonomic significance of evolutionary parallelism. *Sida* 1:109–116.

—— 1964. The old systematics. In C. A. Leone, ed., *Taxonomic Biochemistry and Serology*, pp. 3–11. New York: Ronald Press.

—— 1968. *The Evolution and Classification of Flowering Plants.* Boston: Houghton Mifflin.

—— 1969. On the relationship between taxonomy and evolution. *Taxon* 18:177–187.

—— 1975. Some thoughts on angiosperm phylogeny and taxonomy. *Ann. Missouri Bot. Gard.* 62:517–520.

—— 1976. The taxonomic significance of the structure of plant proteins: A classical taxonomist's view. *Brittonia* 28:1–27.

—— 1977. On the taxonomic significance of secondary metabolites in angiosperms. In K. Kubitzki, ed., *Flowering Plants: Evolution and Classification of Higher Categories*, pp. 179–189. Berlin: Springer-Verlag.

—— 1978. Once again, what is a species? In L. V. Knutson, chm., *Biosystematics in Agriculture*, pp. 3–20. Montclair, N.J.: Allenheld Osmun.

—— 1980. *Vascular Flora of the Southeastern United States.* Vol. 1: *Asteraceae.* Chapel Hill: Univ. North Carolina Press.

—— 1981. *An Integrated System of Classification of Flowering Plants.* New York: Columbia Univ. Press.

—— 1983. Some realignments in the dicotyledons. *Nord. J. Bot.* 3:75–83.

—— 1987. A botanical critique of cladism. *Bot. Rev.* 53:1–52.

Cronquist, A., A. Takhtajan, and W. Zimmermann. 1966. On the higher taxa of Embryobionta. *Taxon* 15:129–134.

Crovello, T. J. 1968a. The effect of change of number of OTU's in a numerical taxonomic study. *Brittonia* 20:346–367.

—— 1968b. Key communality cluster analysis as a taxonomic tool. *Taxon* 17:241–258.

—— 1968c. A numerical taxonomic study of the genus *Salix*, section *Sitchenses. Univ. California Publ. Bot.* 44:1–61.

—— 1969. Effects of change of characters and of number of characters in numerical taxonomy. *Amer. Midl. Naturalist* 81:68–86.

—— 1970. Analysis of character variation in ecology and systematics. *Ann. Rev. Ecol. Syst.* 1:55–98.

—— 1974. Analysis of character variation in systematics. In A. E. Radford, W. C. Dickison, J. R. Massey, and C. R. Bell, *Vascular Plant Systematics*, pp. 451–484. New York: Harper & Row.

—— 1976. Numerical approaches to the species problem. *Pl. Syst. Evol.* 125:179–187.

Crovello, T. J. and W. W. Moss. 1971. A bibliography on classification in diverse disciplines. *Classification Soc. Bull.* 2(3):29–45.

Crowson, R. A. 1970. *Classification and Biology.* New York: Atherton Press.

—— 1972. A systematist looks at cytochrome *c. J. Mol. Evol.* 2:28–37.

Cruden, R. W. 1977. Pollen-ovule ratios: A conservative indicator of breeding systems in flowering plants. *Evolution* 31:32–46.

Cruden, R. W. and R. G. Jensen. 1979. Viscin threads, pollination efficiency and low pollen-ovule ratios. *Amer. J. Bot.* 66:875–879.

Cruden, R. W. and D. L. Lyon. 1985. Correlations among stigma depth, style length, and pollen grain size: Do they reflect function or phylogeny? *Bot. Gaz.* 146:143–149.

Cruden, R. W. and S. Miller-Ward. 1981. Pollen-ovule ratio, pollen size, and the ratio of stigmatic area to the pollen-bearing area of the pollinator: An hypothesis. *Evolution* 35:964–974.

Crum, H. 1985. Traditional make-do taxonomy. *Bryologist* 88:221–222.

Cullen, J. 1968. Botanical problems of numerical taxonomy. In V. H. Heywood, ed., *Modern Methods in Plant Taxonomy*, pp. 175–183. London: Academic Press.

Cullimore, D. R. 1969. The Adansonian classification using the heterotrophic spectra of *Chlorella vulgaris* by a simplified procedure involving a desk-top computer. *J. Appl. Bacteriol.* 32:439–447.

Cutler, D. F. 1969. *Anatomy of the Monocotyledons.* Vol. 4: *Juncales.* Oxford: Clarendon Press.

—— 1978a. *Applied Plant Anatomy.* London: Longman.

—— 1978b. The significance of variability in epidermal cell wall patterns of *Haworthia reinwardtii* var. *chalumnensis* (Liliaceae). *Revista Brasil. Bot.* 1:25–34.

—— 1979. Leaf surface studies in *Aloe* and *Haworthia* species (Liliaceae): Taxonomic implications. *Trop. Subtrop. Pflanzenw.* 28:449–471.

Cutler, D. F. and P. E. Brandham. 1977. Experimental evidence for the genetic control of leaf surface characters in hybrid Aloineae (Liliaceae). *Kew Bull.* 32:23–32.

Cutler, D. F., P. E. Brandham, S. Carter, and S. J. Harris. 1980. Morphological, anatomical, cytological and biochemical aspects of evolution in East African shrubby species of *Aloe* L. (Liliaceae). *Bot. J. Linn. Soc.* 80:293–317.

Cutter, E. G., ed. 1966. *Trends in Plant Morphogenesis.* New York: Wiley.

—— 1971. *Plant Anatomy: Experiment and Interpretation*, Part II: *Organs.* London: Arnold.

—— 1978. *Plant Anatomy: Experiment and Interpretation*, Part I: *Cells and Tissues.* 2d ed. London: Arnold.

Cuvier, G. 1835. *Le Règne Animal Distribué après son Organization.* 2d ed. Paris: Crochard.

Czaja, A. T. 1978. Structure of starch grains and the classification of vascular plant families. *Taxon* 27:463–470.

Czapik, R. 1974. Embryology of five species of the *Arabis hirsuta* complex. *Acta Biol. Cracov.* 17:13–25.

Dabinett, P. E. and A. M. Wellman. 1973. Numerical taxonomy of the genus *Rhizopus. Canad. J. Bot.* 51:2053–2064.

—— 1978. Numerical taxonomy of certain genera of Fungi Imperfecti and Ascomycotina. *Canad. J. Bot.* 56:2031–2049.

Dahl, E. 1951. On the relation between summer temperature and the distribution of alpine vascular plants in the lowlands of Fennoscandia. *Oikos* 3:22–52.

Dahlgren, R. 1975. A system of classification of the angiosperms to be used to demonstrate the distribution of characters. *Bot. Not.* 128:119–147.

—— 1980. A revised system of classification of the angiosperms. *Bot. J. Linn. Soc.* 80:91–124.

—— 1983. General aspects of angiosperm evolution and macrosystematics. *Nord. J. Bot.* 3:119–149.

Dahlgren, R. and H. T. Clifford. 1981. Some conclusions from a comparative study of the monocotyledons and related dicotyledonous orders. *Ber. Deutsch. Bot. Ges.* 94:203–227.

—— 1982. *The Monocotyledons: A Comparative Study.* London: Academic Press.

Dahlgren, R., H. T. Clifford, and P. F. Yeo. 1985. *The Families of the Monocotyledons: Structure, Evolution, and Taxonomy.* Berlin: Springer-Verlag.

Dahlgren, R. and F. N. Rasmussen. 1983. Monocotyledon evolution: Characters and phylogenetic estimation. *Evol. Biol.* 16:255–395.

Dahlgren, R., S. Rosendal-Jensen, and B. J. Nielson. 1981. A revised classification of the angiosperms with comments on correlation between chemical and other characters. In D. A. Young and D. S. Seigler, eds., *Phytochemistry and Angiosperm Phylogeny*, pp. 149–204. New York: Praeger.

Dale, J. E. and F. L. Milthorpe. 1983. *The Growth and Functioning of Leaves.* Cambridge: Cambridge Univ. Press.

Dale, M. B. 1968. On property structure, numerical taxonomy and data handling. In V. H. Heywood, ed., *Modern Methods in Plant Taxonomy*, pp. 185–197. London: Academic Press.

Dale, M. B., R. H. Groves, V. J. Hull, and J. F. O'Callaghan. 1971. A new method for describing leaf shape. *New Phytol.* 70:437–442.

Dallwitz, M. J. 1974. A flexible computer program for generating identification keys. *Syst. Zool.* 23:50–57.

D'Amato, F. 1984. Role of polyploidy in reproductive organs and tissues. In B. M. Johri, ed., *Embryology of Angiosperms*, pp. 519–566. Berlin: Springer-Verlag.

Danser, B. H. 1929. Ueber die Begriffe Komparium, Kommiskuum und Konvivium und ueber die Entstehungsweise der Konvivien. *Genetica* 11:399–450.

—— 1950. A theory of systematics. *Biblioth. Biotheor.* ser. D, 4:117–180.

Dansereau, P. 1957. *Biogeography: An Ecological Perspective.* New York: Ronald Press.

D'Arcy, W. G., ed. 1986. *Solanaceae: Biology and Systematics.* New York: Columbia Univ. Press.

D'Arcy, W. G. and R. C. Keating. 1979. Anatomical support for the taxonomy of *Calophyllum* (Guttiferae) in Panama. *Ann. Missouri Bot. Gard.* 66:557–571.

Darlington, C. D. 1973. *Chromosome Botany and the Origins of Cultivated Plants.* 3d ed. London: George Allen & Unwin.

Darlington, C. D. and L. La Cour. 1938. Differential reactivity of the chromosomes. *Ann. Bot.* n.s. 2:615–625.

—— 1975. *The Handling of Chromosomes.* 6th ed. New York: Halstead Press.

Darlington, C. D. and A. P. Wylie. 1955. *Chromosome Atlas of Flowering Plants.* 2d ed. London: George Allen & Unwin.

Darlington, P. J., Jr. 1970. A practical criticism of Hennig-Brundin "Phylogenetic Systematics" and Antarctic biogeography. *Syst. Zool.* 19:1–18.

—— 1971. Modern taxonomy, reality, and usefulness. *Syst. Zool.* 20:341–365.

—— 1972. What is cladism? *Syst. Zool.* 21:128–129.

Darwin, C. 1859. *On the Origin of Species by Means of Natural Selection.* London: Murray.

—— 1862. *On the Various Contrivances by which British and Foreign Orchids are Fertilised by Insects.* London: Murray.

—— 1876. *The Effects of Cross and Self Fertilisation in the Vegetable Kingdom.* London: Murray.

Daumer, K. 1958. Blumenfarben, wie sie die Bienen sehen. *Z. vergl. Physiol.* 41:49–110.

Davey, J. C. and W. D. Clayton. 1978. Some multiple discriminant function studies on *Oplismenus* (Gramineae). *Kew Bull.* 33:147–157.

Davidse, G. 1974. Plant-animal coevolution: The twentieth systematics symposium. *Ann Missouri Bot. Gard.* 61:674.

Davidson, J. F. 1947. The polygonal graph for simultaneous portrayal of several variables in population analysis. *Madroño* 9:105–110.

—— 1954. A dephlogisticated species concept. *Madroño* 12:246–251.

Davies, D. R. and R. A. Hopwood, eds. 1980 *The Plant Genome.* Norwich: John Innes Charity.

Davis, G. L. 1962a. Embryological studies in the Compositae I. Sporogenesis, gametogenesis, and embryogeny in *Cotula australis* (Less.) Hook. f. *Austral. J. Bot.* 10:1–12.

—— 1962b. Embryological studies in the Compositae II. Sporogenesis, gametogenesis, and embryogeny in *Ammobium elatum* R. Br. *Austral. J. Bot.* 10:65–75.

—— 1966. *Systematic Embryology of the Angiosperms.* New York: Wiley.

—— 1967. Apomixis in the Compositae. *Phytomorphology* 17:270–277.

Davis, G. M. 1982. Historical and ecological factors in the evolution, adaptive radiation, and biogeography of freshwater mollusks. *Amer. Zoologist* 22:375–395.

Davis, M. B. 1968. Climatic changes in southern Connecticut recorded by pollen deposition at Rogers Lake. *Ecology* 50:409–422.

Davis, P. H. 1978. The moving staircase: A discussion on taxonomic rank and affinity. *Notes Roy. Bot. Gard.* (Edinburgh) 36:325–340.

Davis, P. H. and V. H. Heywood. 1963. *Principles of Angiosperm Taxonomy.* Princeton, N.J.: Van Nostrand.

Day, W. H. E. 1983. The role of complexity in comparing classifications. *Math. Biosci.* 66:97–114.

—— 1986. Analysis of quartet dissimilarity measures between undirected phylogenetic trees. *Syst. Zool.* 35:325–333.

Day, W. H. E. and F. R. McMorris. 1985. A formalization of consensus index methods. *Bull. Math. Biol.* 47:215–229.

Day, W. H. E. and D. Sankoff. 1987. Computational complexity of inferring phylogenies from chromosome inversion data. *J. Theor. Biol.* 124:213–218.

DeBry, R. W. and N. A. Slade. 1985. Cladistic analysis of restriction endonuclease cleavage maps within a maximum-likelihood framework. *Syst. Zool.* 34:21–34.

DeLage, I. 1978. *Am I a Bunny?* Champaign, Ill.: Garrard.

Delevoryas, T. 1969. Paleobotany, phylogeny, and a natural system of classification. *Taxon* 18:204–212.

Denton, M. F. and J. L. Kerwin. 1980. Survey of vegetative flavonoids of *Sedum* section *Gormania* (Crassulaceae). *Canad. J. Bot.* 58:902–905.

Denton, M. F. and R. del Moral. 1976. Comparison of multivariate analyses using taxonomic data of *Oxalis. Canad. J. Bot.* 54:1637–1646.

De Robertis, E. D. P. and E. M. F. De Robertis, Jr. 1987. *Cell and Molecular Biology.* 8th ed. Philadelphia: Lea & Febiger.

Desch, H. E. 1973. *Timber: Its Structure and Properties.* 5th ed. New York: St. Martin's Press.

Deumling, B. and J. Greilhuber. 1982. Characterization of heterochromatin in different species of the *Scilla siberica* group (Liliaceae) by *in situ* hybridization of satellite DNAs and fluorochrome banding. *Chromosoma* (Berlin) 84:535–555.

De Vries, H. 1900a. Das Spaltungsgesetz der Bastarde. *Ber. Deutsch. Bot. Ges.* 18:83–90. Summarized as "Sur la loi de disjonction des hybrides." *Compt. Rend. Hebd. Seances Acad. Sci.* 130:845–847. 1900.

—— 1900b. Sur les unités des caractères spécifiques et leur application a l'étude des hybrides. *Rev. Gén. Bot.* 12:257–271.

De Wet, J. M. J. 1965. Diploid races of tetraploid *Dichanthium* species. *Amer. Naturalist* 99:167–171.

—— 1968. Diploid-tetraploid-haploid cycles and the origin of variability in *Dichanthium* agamospecies. *Evolution* 22:394–397.

—— 1971. Reversible tetraploidy as an evolutionary mechanism. *Evolution* 25:545–548.

De Wet, J. M. J. and J. R. Harlan. 1972. Chromosome pairing and phylogenetic affinities. *Taxon* 21:67–70.

Dewey, D. R. 1967a. Synthetic hybrids of *Elymus canadensis* x *Sitanion hystrix*. *Bot. Gaz.* 128:11–16.

—— 1967b. Genome relations between *Agropyron scribneri* and *Sitanion hystrix*. *Bull. Torrey Bot. Club.* 94:395–404.

—— 1967c. Synthetic hybrids of *Agropyron scribneri* x *Elymus junceus*. *Bull. Torrey Bot. Club* 94:388–395.

—— 1970. Genome relations among *Elymus canadensis*, *Elymus triticoides*, *Elymus dasystachys*, and *Agropyron smithii*. *Amer. J. Bot.* 57:861–866.

—— 1983. Historical and current taxonomic perspectives of *Agropyron*, *Elymus*, and related genera. *Crop Sci.* 23:637–642.

—— 1984. The genomic system of classification as a guide to intergeneric hybridization within the perennial Triticeae. In J. P. Gustafson, ed., *Gene Manipulation in Plant Improvement*, pp. 209–279. New York: Plenum.

Dewey, D. R. and A. H. Holmgren. 1962. Natural hybrids of *Elymus cinereus* x *Sitanion hystrix*. *Bull. Torrey Bot. Club.* 89:217–228.

Dickinson, H. G. 1982. The development of pollen. *Rev. Cytol. Biol. Veg. Bot.* 5:5–19.

Dickinson, T. A., W. H. Parker, and R. E. Strauss. 1987. Another approach to leaf shape comparisons. *Taxon* 36:1–20.

Dickison, W. C. 1969. Comparative morphological studies in Dilleniaceae,. IV. Anatomy of the node and vascularization of the leaf. *J. Arnold Arbor.* 50:384–400.

—— 1975. The bases of angiosperm phylogeny: Vegetative anatomy. *Ann. Missouri Bot. Gard.* 62:590–620.

Dickison, W. C., P. M. Rury, and G. L. Stebbins. 1978. Xylem anatomy of *Hibbertia* (Dilleniaceae) in relation to ecology and evolution. *J. Arnold Arbor.* 59:32–49.

Diels, L. 1924. Die Methoden der Phytographie und der Systematik der Pflanzen. In E. Abderhalden, ed., *Handbuch der biologischen Arbeitsmethoden.* Abt. 11: *Methoden zur Erforschung der Leistungen des Pflanzenorganismus*, 1:67–190. Berlin: Urban & Schwarzenburg.

Dilcher, D. L. 1974. Approaches to the identification of angiosperm leaf remains. *Bot. Rev.* 40:1–157.

Dilcher, D. L. and W. Crepet. 1984. Historical perspectives of angiosperm evolution. *Ann. Missouri Bot. Gard.* 71:348–350.

Dirzo, R. and J. Sarukhan, eds. 1984. *Perspectives on Plant Population Ecology.* Sunderland, Mass.: Sinauer.

Dixon, W. J., ed. 1981. *BMDP Statistical Software.* Los Angeles: Univ. California Press.

Dobzhansky, T. 1935. A critique of the species concept in biology. *Philos. Sci.* 2:344–355.

—— 1937. What is a species? *Scientia* 61:280–286.

—— 1972. Species of *Drosophila*. *Science* 177:664–669.

Dobzhansky, T., F. J. Ayala, G. L. Stebbins, and J. W. Valentine. 1977. *Evolution.* San Francisco: Freeman.

Doebley, J. F. 1983. The maize and teosinte male inflorescence: A numerical taxonomic study. *Ann. Missouri Bot. Gard.* 70:32–70.

Doebley, J. F., M. M. Goodman, and C. W. Stuber. 1984. Isoenzymatic variation in *Zea* (Gramineae). *Syst. Bot.* 9:203–218.

Donoghue, M. J. 1981. Growth patterns in woody plants with examples from the genus *Viburnum. Arnoldia* 41:2–23.

—— 1983a. The phylogenetic relationships of *Viburnum.* In N. I. Platnick and V. A. Funk, eds., *Advances in Cladistics*, 2:144–166. New York: Columbia Univ. Press.

—— 1983b. A preliminary analysis of phylogenetic relationships in *Viburnum* (Caprifoliaceae s. l.). *Syst. Bot.* 8:45–58.

—— 1985a. Pollen diversity and exine evolution in *Viburnum* and the Caprifoliaceae sensu lato. *J. Arnold Arbor.* 66:421–469.

—— 1985b. A critique of the biological species concept and recommendations for a phylogenetic alternative. *Bryologist* 88:172–181.

—— 1987. Experiments and hypotheses in systematics. *Taxon* 36:584–587.

Donoghue, M. J. and P. D. Cantino. 1984. The logic and limitations of the outgroup substitution approach to cladistic analysis. *Syst. Bot.* 9:192–202.

—— 1988. Paraphyly, ancestors, and the goals of taxonomy: A botanical defense of cladism. *Bot. Rev.* 54:107–128.

Dormer, C. 1972. *Shoot Organization in Vascular Plants.* London: Chapman and Hall.

Dormer, K. J. 1962. The fibrous layer in the anthers of Compositae. *New Phytol.* 61:150–153.

Douglas, M. E. and J. C. Avise. 1982. Speciation rates and morphological divergence in fishes: Tests of gradual versus rectangular modes of evolutionary change. *Evolution* 36:224–232.

Dover, G. A. and R. B. Flavell, eds. 1982. *Genome Evolution.* London: Academic Press.

Doyen, J. T. and C. N. Slobodchikoff. 1974. An operational approach to species classification. *Syst. Zool.* 23:239–247.

Doyle, J. A. 1978. Origin of angiosperms. *Ann. Rev. Ecol. Syst.* 9:365–392.

—— 1987. The origin of angiosperms: A cladistic approach. In E. M. Friis, W. G. Chaloner, and P. R. Crane, eds., *The Origin of Angiosperms and Biological Consequences*, pp. 17–49. Cambridge: Cambridge Univ. Press.

Doyle, J. A. and M. J. Donoghue. 1986a. Relationships of angiosperms and Gnetales: A numerical cladistic analysis. In R. A. Spicer and B. A. Thomas, eds., *Systematic and Taxonomic Approaches in Palaeobotany*, pp. 177–198. Oxford: Clarendon Press.

—— 1986b. Seed plant phylogeny and the origin of angiosperms: An experimental cladistic approach. *Bot. Rev.* 52:321–431.

—— 1987. The importance of fossils in elucidating seed plant phylogeny and macroevolution. *Rev. Paleobot. Palynol.* 50:63–95.

Doyle, J. J. 1983. Flavonoid races of *Claytonia virginica* (Portulacaceae). *Amer. J. Bot.* 70:1085–1091.

Doyle, J. J., R. N. Beachy, and W. H. Lewis. 1984. Evolution of rDNA in *Claytonia* polyploid complexes. In W. F. Grant, ed., *Plant Biosystematics*, pp. 321–341. Toronto: Academic Press.

Doyle, J. J. and E. E. Dickson. 1987. Preservation of plant samples for DNA restriction endonuclease analysis. *Taxon* 36:715–722.

Doyle, J. J., D. E. Soltis, and P. S. Soltis. 1985. An intergeneric hybrid in the Saxifragaceae: Evidence from ribosomal RNA genes. *Amer. J. Bot.* 72:1388–1391.

Dressler, R. L. 1981. *The Orchids: Natural History and Classification.* Cambridge, Mass.: Harvard Univ. Press.

Ducker, S. C. and R. B. Knox. 1976. Submarine pollination in seagrasses. *Nature* 263:705–706.

—— 1985. Pollen and pollination: A historical review. *Taxon* 34:401–419.

Ducker, S. C., W. T. Williams, and G. N. Lance. 1965. Numerical classification of the Pacific forms of *Chlorodesmis* (Chlorophyta). *Austral. J. Bot.* 13:489–499.

Duek, J. J., S. P. Sinha, and L. Muxica. 1979. Comparisons of similarity criteria in a numerical classification of the fern genus *Lygodium* in America. *Feddes Repert.* 90:11–18.

Duke, J. A. and E. E. Terrell. 1974. Crop diversification matrix: Introduction. *Taxon* 23:759–799.

Dunbar, M. J. 1980. The blunting of Occam's Razor, or to hell with parsimony. *Canad. J. Zool.* 58:123–128.

Duncan, T. 1980a. Cladistics for the practicing taxonomist—an eclectic view. *Syst. Bot.* 5:136–148.

—— 1980b. A cladistic analysis of the *Ranunculus hispidus* complex. *Taxon* 29:441–454.

—— 1980c. A taxonomic study of the *Ranunculus hispidus* Michaux complex in the western hemisphere. *Univ. California Publ. Bot.* 77:1–125.

—— 1984. Willi Hennig, character compatibility, Wagner parsimony, and the "Dendrogrammaceae" revisited. *Taxon* 33:698–704.

Duncan, T. and B. R. Baum. 1981. Numerical phenetics: Its uses in botanical systematics. *Ann. Rev. Ecol. Syst.* 12:387–404.

Duncan, T. and G. F. Estabrook. 1976. An operational method for evolutionary classifications. *Syst. Bot.* 1:373–382.

Duncan, T., R. B. Phillips, and W. H. Wagner, Jr. 1980. A comparison of branching diagrams derived by various phenetic and cladistic methods. *Syst. Bot.* 5:264–293.

Duncan, T. and T. F. Stuessy, eds. 1984. *Cladistics: Perspectives on the Reconstruction of Evolutionary History.* New York: Columbia Univ. Press.

—— 1985. *Cladistic Theory and Methodology.* New York: Van Nostrand Reinhold.

Dunford, M. P. 1984. Cytotype distribution of *Atriplex canescens* (Chenopodiaceae) of southern New Mexico and adjacent Texas. *Southw. Naturalist* 29:223–228.

Dunn, G. and B. S. Everitt. 1982. *An Introduction to Mathematical Taxonomy.* Cambridge: Cambridge Univ. Press.

Dunn, R. A. and R. A. Davidson. 1968. Pattern recognition in biologic classification. *Pattern Recognition* 1:75–93.

DuPraw, E. J. 1964. Non-Linnean taxonomy. *Nature* 202:849–852.

—— 1965. Non-Linnean taxonomy and the systematics of honeybees. *Syst. Zool.* 14:1–24.

Dupuis, C. 1978. Permanence et actualité de la systématique: La "Systématique Phylogénétique" de W. Hennig (historique, discussion, choix de références). *Cah. Naturalistes* n.s. 34:1–69.

—— 1984. Willi Hennig's impact on taxonomic thought. *Ann. Rev. Ecol. Syst.* 15:1–24.

Du Rietz, G. E. 1930. The fundamental units of biological taxonomy. *Svensk Bot. Tidskr.* 24:333–428.

Durrant, S. D. 1955. In defense of the subspecies. *Syst. Zool.* 4:186–190.

Dyer, A. F. 1979. *Investigating Chromosomes.* New York: Wiley.

Eames, A. J. 1953. Floral anatomy as an aid in generic limitation. *Chron. Bot.* 14:126–132.

Eames, A. J. and L. H. MacDaniels. 1947. *An Introduction to Plant Anatomy.* 2d ed. New York: McGraw-Hill.

East, E. M. 1928. The genetics of the genus *Nicotiana. Bibliogr. Genet.* 4:243–320.

Echlin, P. 1968. The use of the scanning reflection electron microscope in the study of plant and microbial material. *J. Roy. Microscop. Soc.* (London) 88:407–418.

Eck, R. V. and M. O. Dayhoff. 1966. *Atlas of Protein Sequence and Structure 1966.* National Biomedical Research Foundation, Silver Spring, Md.

Eckenwalder, J. E. 1977. North American cottonwoods (*Populus*, Salicaceae) sections *Abaso* and *Aigeiros. J. Arnold Arbor.* 58:193–207.

Edmonds, J. M. 1978. Numerical taxonomic studies on *Solanum* L. section *Solanum (Maurella). Bot. J. Linn. Soc.* 76:27–51.

Edwards, A. W. F. and L. L. Cavalli-Sforza. 1964. Reconstruction of evolutionary trees. In V. H. Heywood and J. McNeill, eds., *Phenetic and Phylogenetic Classification,* pp. 67–76. London: Systematics Association.

Edwards, J. G. 1954. A new approach to infraspecific categories. *Syst. Zool.* 3:1–20.

—— 1955. Clarification of certain aspects of infraspecific systematics. *Syst. Zool.* 5:92–94.

Eggers Ware, D. M. 1983. Genetic fruit polymorphism in North American *Valerianella* (Valerianaceae) and its taxonomic implications. *Syst. Bot.* 8:33–44.

Eggli, U. 1984. Stomatal types of Cactaceae. *Pl. Syst. Evol.* 146:197–214.

Eglinton, G. and R. J. Hamilton. 1967. Leaf epicuticular waxes. *Science* 156:1322–1335.

Ehler, N. 1976. Mikromorphologie der Samenoberflachen der Gattung *Euphorbia. Pl. Syst. Evol.* 126:189–207.

Ehrendorfer, F. 1976. Evolutionary significance of chromosomal differentiation patterns in gymnosperms and primitive angiosperms. In C. B. Beck, ed., *Origin and Early Evolution of Angiosperms,* pp. 220–240. New York: Columbia Univ. Press.

—— 1980. Polyploidy and distribution. In W. H. Lewis, ed., *Polyploidy: Biological Relevance,* pp. 45–60. New York: Plenum Press.

—— 1984. Artbegriff und Artbildung in botanischer Sicht. *Z. Zool. Syst. Evolut.-Forsch.* 22:234–263.

Ehrendorfer, F., F. Krendl, E. Habeler, and W. Sauer. 1968. Chromosome numbers and evolution in primitive angiosperms. *Taxon* 17:337–353.

Ehrendorfer, F., D. Schweizer, H. Greger, and C. Humphries. 1977. Chromosome banding and synthetic systematics in *Anacyclus* (Asteraceae-Anthemideae). *Taxon* 26:387–394.

Ehrlich, P. R. 1961. Has the biological species concept outlived its usefulness? *Syst. Zool.* 10:167–176.

—— 1964. Some axioms of taxonomy. *Syst. Zool.* 13:109–123.

—— 1968. *The Population Bomb.* New York: Ballantine Books.

Ehrlich, P. R. and P. H. Raven. 1969. Differentiation of populations. *Science* 165:1228–1232.

Ehrlich, P. R. and J. Roughgarden. 1987. *The Science of Ecology.* New York: Macmillan.

Ehrlich, P. R. and R. R. White. 1980. Colorado checkerspot butterflies: Isolation, neutrality, and the biospecies. *Amer. Naturalist* 115:328–341.

Eichler, A. W. 1883. *Syllabus der Vorlesungen über Phanerogamenkunde.* 3d ed. Berlin: Gebrüder Borntraeger.

Eigen, M. and R. Winkler-Oswatitsch. 1983. The origin and evolution of life at the molecular level. In C. Helene, ed., *Structure, Dynamics, Interactions and Evolution of Biological Macromolecules*, pp. 353–370. Dordrecht: Reidel.

Eldredge, N. 1985. *Time Frames: The Rethinking of Darwinian Evolution and the Theory of Punctuated Equilibria.* New York: Simon and Schuster.

Eldredge, N. and J. Cracraft. 1980. *Phylogenetic Patterns and the Evolutionary Process: Method and Theory in Comparative Biology.* New York: Columbia Univ. Press.

Eldredge, N. and S. J. Gould. 1972. Punctuated equilibria: An alternative to phyletic gradualism. In T. J. M. Schopf, ed., *Models in Paleobiology*, pp. 82–115. San Francisco: Freeman, Cooper.

Eleftheriou, E. P. 1984. Sieve-element plastids of *Triticum* and *Aegilops* (Poaceae). *Pl. Syst. Evol.* 145:119–133.

El-Gadi, A. and T. T. Elkington. 1977. Numerical taxonomic studies on species in *Allium* subgenus *Rhizirideum. New Phytol.* 79:183–201.

El-Ghazaly, G. 1980. Palynology of Hypochoeridinae and Scolyminae (Compositae). *Opera Bot.* 58:1–48.

—— 1982. Ontogeny of pollen wall of *Leontodon autumnalis* (Hypochoeridinae, Compositae). *Grana* 21:103–113.

Eliasson, U. 1972. Studies in Galapagos plants XI. Embryology of *Macraea laricifolia* Hook. f. (Compositae). *Svensk Bot. Tidskr.* 66:43–47.

Elisens, W. J. 1985. The systematic significance of seed coat anatomy among New World species of tribe Antirrhineae (Scrophulariaceae). *Syst. Bot.* 10:282–299.

Ellison, W. L., R. E. Alston, and B. L. Turner. 1962. Methods of presentation of crude biochemical data for systematic purposes, with particular reference to the genus *Bahia* (Compositae). *Amer. J. Bot.* 49:599–604.

Elsal, J. A. 1985. Illustrations of the use of higher plant taxa in biogeography. *J. Biogeogr.* 12:433–444.

Emboden, W. A., Jr. 1969. Detection of palynological introgression in *Salvia* (Labiatae). *Los Angeles County Mus. Contr. Sci.* 156:1–10.

Emig, C. C. 1985. A new method for representing trees. *Syst. Zool.* 34:234–238.

Endler, J. A. 1973. Gene flow and population differentiation. *Science* 179:243–250.

—— 1982. Problems in distinguishing historical from ecological factors in biogeography. *Amer. Zoologist* 22:441–452.

Endress, P. K. 1980a. Floral structure and relationships of *Hortonia* (Monimiaceae). *Pl. Syst. Evol.* 133:199–221.

—— 1980b. Ontogeny, function and evolution of extreme floral construction in Monimiaceae. *Pl. Syst. Evol.* 134:79–120.

Endress, P. K. and R. Honegger. 1980. The pollen of the Austrobaileyaceae and its phylogenetic significance. *Grana* 19:177–182.

Endress, P. K. and D. H. Lorence. 1983. Diversity and evolutionary trends in the floral structure of *Tambourissa* (Monimiaceae). *Pl. Syst. Evol.* 143:53–81.

Engel, H. 1953. The species concept of Linnaeus. *Arch. Int. Hist. Sci.* 6:249–259.

Engler, A. 1886. *Führer durch den königlich botanischen Garten der Universität zu Breslau.* Breslau: J. U. Kern's Verlag.

Engler, A. and K. Prantl, eds. 1887–1915. *Die natürlichen Pflanzenfamilien.* Leipzig: Wilhelm Engelmann.

Erdtman, G. 1952. *Pollen Morphology and Plant Taxonomy: Angiosperms* (An Introduction to Palynology. I.) Stockholm: Almqvist & Wiksell; revised ed. New York: Hafner, 1971.

—— 1960. The acetolysis method: A revised description. *Svensk Bot. Tidskr.* 54:561–564.

—— 1966. Sporoderm morphology and morphogenesis: A collocation of data and suppositions. *Grana Palynol.* 6:317–323.

—— 1969. *Handbook of Palynology: Morphology-Taxonomy-Ecology: An Introduction to the Study of Pollen Grains and Spores.* New York: Hafner.

Erdtman, G., ed. 1970. *World Pollen Flora.* New York: Hafner.

Erdtman, H. G. H. 1968. Chemical principles in chemosystematics. In T. J. Mabry, R. E. Alston, and V. C. Runeckles, eds., *Recent Advances in Phytochemistry*, 1:13–56. New York: Appleton-Century-Crofts.

Esau, K. 1953. *Plant Anatomy.* New York: Wiley. 2d ed. 1965.

—— 1960. *Anatomy of Seed Plants.* New York: Wiley. 2d ed. 1977.

—— 1965. *Vascular Differentiation in Plants.* New York: Holt, Rinehart and Winston.

—— 1969. *The Phloem.* Berlin: Gebrüder Borntraeger.

Estabrook, G. F. 1967. An information theory model for character analysis. *Taxon* 16:86–97.

—— 1968. A general solution in partial orders for the Camin-Sokal model in phylogeny. *J. Theor. Biol.* 21:421–438.

—— 1971. Some information theoretic optimality criteria for general classification. *Math. Geol.* 3:203–207.

—— 1972. Cladistic methodology: A discussion of the theoretical basis for the induction of evolutionary history. *Ann. Rev. Ecol. Syst.* 3:427–456.

Estabrook, G. F., ed. 1975. *Proceedings of the Eighth International Conference on Numerical Taxonomy.* San Francisco: Freeman.

—— 1978. Some concepts for the estimation of evolutionary relationships in systematic botany. *Syst. Bot.* 3:146–158.

—— 1980. The compatibility of occurrence patterns of chemicals in plants. In F. A. Bisby, J. G. Vaughan, and C. A. Wright, eds., *Chemosystematics: Principles and Practice,* pp. 379–397. London: Academic Press.

—— 1986. Evolutionary classification using convex phenetics. *Syst. Zool.* 35:560–570.

Estabrook, G. F. and W. R. Anderson. 1978. An estimate of phylogenetic relationships within the genus *Crusea* (Rubiaceae) using character compatibility analysis. *Syst. Bot.* 3:179–196.

Estabrook, G. F., C. S. Johnson, Jr., and F. R. McMorris. 1975. An idealized concept of the true cladistic character. *Math. Biosci.* 23:263–272.

—— 1976a. An algebraic analysis of cladistic characters. *Discrete Math.* 16:141–147.

—— 1976b. A mathematical foundation for the analysis of cladistic character compatibility. *Math Biosci.* 29:181–187.

Estabrook, G. F. and L. Landrum. 1975. A simple test for the possible simultaneous evolutionary divergence of two amino acid positions. *Taxon* 24:609–613.

Estabrook, G. F., F. R. McMorris, and C. A. Meacham. 1985. Comparison of undirected phylogenetic trees based on subtrees of four evolutionary units. *Syst. Zool.* 34:193–200.

Estabrook, G. F. and C. A. Meacham. 1979. How to determine the compatibility of undirected character state trees. *Math. Biosci.* 46:251–256.

Estabrook, G. F., J. G. Strauch, Jr., and K. L. Fiala. 1977. An application of compatibility analysis to the Blackiths' data on orthopteroid insects. *Syst. Zool.* 26:269–276.

Estrada, A. and T. H. Fleming, eds. 1986. *Frugivores and Seed Dispersal.* Dordrecht: Junk.

Ettlinger, M. G. and A. Kjaer. 1968. Sulfur compounds in plants. In T. J. Mabry, R. E. Alston, and V. C. Runeckles, eds., *Recent Advances in Phytochemistry,* 1:59–144. New York: Appleton-Century-Crofts.

Evans, A. M. 1968. Interspecific relationships in the *Polypodium pectinatum-plumula* complex. *Ann. Missouri Bot. Gard.* 55:193–293.

Evert, R. F. 1984. Comparative structure of phloem. In R. A. White and W. C. Dickison, eds. *Contemporary Problems in Plant Anatomy,* pp. 145–234. Orlando: Academic Press.

Ewan, J., ed. 1969. *A Short History of Botany in the United States.* New York: Hafner.

Eyde, R. H. 1967. The peculiar gynoecial vasculature of Cornaceae and its systematic significance. *Phytomorphology* 17:172–182.

—— 1971. Evolutionary morphology: Distinguishing ancestral structure from derived structure in flowering plants. *Taxon* 20:63–73.

—— 1975a. The bases of angiosperm phylogeny: Floral anatomy. *Ann. Missouri Bot. Gard.* 62:521–537.

—— 1975b. The foliar theory of the flower. *Amer. Scientist* 63:430–437.

Eyde, R. H. and C. C. Tseng. 1969. Flower of *Tetraplasandra gymnocarpa:* Hypogyny with epigynous ancestry. *Science* 166:506–508.

Eyre, S. R. 1968. *Vegetation and Soils: A World Picture.* 2d ed. London: Edward Arnold.

Faden, R. B. and Y. Suda. 1980. Cytotaxonomy of Commelinaceae: Chromosome numbers of some African and Asiatic species. *Biol. J. Linn. Soc.* 81:301–325.

Faegri, K. and J. Iversen. 1950. *Text-Book of Modern Pollen Analysis.* Copenhagen: Munksgaard.

—— 1964. *Textbook of Pollen Analysis.* 2d ed. New York: Hafner.

Faegri, K. and L. van der Pijl. 1979. *The Principles of Pollination Ecology.* 3d ed. Oxford: Pergamon Press.

Fahn, A. 1982. *Plant Anatomy.* 3d ed. Oxford: Pergamon Press.

—— 1986. Structural and functional properties of xeromorphic leaves. *Ann. Bot.* 57:631–637.

Fair, F. 1977. On interpreting a philosophy of science: A response to Gareth Nelson. *Syst. Zool.* 26:89–91.

Fairbrothers, D. E. 1983. Evidence from nucleic acid and protein chemistry, in particular serology, in angiosperm classification. *Nord. J. Bot.* 3:35–41.

Fairbrothers, D. E., T. J. Mabry, R. L. Scogin, and B. L. Turner. 1975. The bases of angiosperm phylogeny: Chemotaxonomy. *Ann. Missouri Bot. Gard.* 62:765–800.

Fairbrothers, D. E. and F. P. Petersen. 1983. Serological investigation of the Annoniflorae (Magnoliiflorae, Magnoliidae). In U. Jensen and D. E. Fairbrothers, eds., *Proteins and Nucleic Acids in Plant Systematics*, pp. 301–310. Berlin: Springer-Verlag.

Fairbrothers, D. E. and J. A. Quinn. 1970. Habitat ecology and chromosome numbers of natural populations of the *Danthonia sericea* complex. *Amer. Midl. Naturalist* 85:531–536.

Falconer, D. S. 1981. *Introduction to Quantitative Genetics.* 2d ed. New York: Wiley.

Farber, P. L. 1976. The type-concept in zoology during the first half of the nineteenth century. *J. Hist. Biol.* 9:93–119.

Farkas, L., M. Gabor, and F. Kallay, eds. 1986. *Flavonoids and Bioflavonoids, 1985.* Amsterdam: Elsevier.

Farris, J. S. 1966. Estimation of conservatism of characters by constancy within biological populations. *Evolution* 20:587–591.

—— 1967. Definitions of taxa. *Syst. Zool.* 16:174–175.

—— 1968. Categorical ranks and evolutionary taxa in numerical taxonomy. *Syst. Zool.* 17:151–159.

—— 1969. On the cophenetic correlation coefficient. *Syst. Zool.* 18:279–285.

—— 1970. Methods for computing Wagner trees. *Syst. Zool.* 19:83–92.

—— 1971. The hypothesis of nonspecificity and taxonomic congruence. *Ann. Rev. Ecol. Syst.* 2:277–302.

—— 1972. Estimating phylogenetic trees from distance matrices. *Amer. Naturalist* 106:645–668.

—— 1976. Phylogenetic classification of fossils with recent species. *Syst. Zool.* 25:271–282.

—— 1977a. On the phenetic approach to vertebrate classification. In M. K. Hecht, P. C. Goody, and B. M. Hecht, eds., *Major Patterns in Vertebrate Evolution*, pp. 823–850. New York: Plenum Press.

—— 1977b. Phylogenetic analysis under Dollo's Law. *Syst. Zool.* 26:77–88.

—— 1978. Inferring phylogenetic trees from chromosome inversion data. *Syst. Zool.* 27:275–284.

—— 1979a. On the naturalness of phylogenetic classification. *Syst. Zool.* 28:200–214.

—— 1979b. The information content of the phylogenetic system. *Syst. Zool.* 28:483–519.

—— 1982. Outgroups and parsimony. *Syst. Zool.* 31:328–334.

—— 1985. The pattern of cladistics. *Cladistics* 1:190–201.

Farris, J. S., A. G. Kluge, and M. J. Eckardt. 1970a. A numerical approach to phylogenetic systematics. *Syst. Zool.* 19:172–191.

—— 1970b. On predictivity and efficiency. *Syst. Zool.* 19:363–372.

Farris, J. S., A. G. Kluge, and M. F. Mickevich. 1979. Paraphyly of the *Rana boylii* species group. *Syst. Zool.* 28:627–634.

Farwell, O. A. 1927. Botanical gleanings in Michigan—IV. *Amer. Midl. Naturalist* 10:199–219.

Favarger, C. 1978. Philosophie des comptages de chromosomes. *Taxon* 27:441–448.

Favre-Duchartre, M. 1984. Homologies and phylogeny. In B. M. Johri, ed., *Embryology of Angiosperms*, pp. 697–734. Berlin: Springer-Verlag.

Featherly, H. I. 1954. *Taxonomic Terminology of the Higher Plants.* Ames: Iowa State College Press.

Fedorov, A. A., ed. 1969. *Khromosomnye Chisla Tsvetkovykh Rastenii* [Chromosome Numbers of Flowering Plants]. Leningrad: Acad. Sci. U.S.S.R.

Felsenstein, J. 1973. Maximum likelihood and minimum-steps methods for estimating evolutionary trees from data on discrete characters. *Syst. Zool.* 22:240–249.

—— 1978. Cases in which parsimony or compatibility methods will be positively misleading. *Syst. Zool.* 27:401–410.

—— 1979. Alternative methods of phylogenetic inference and their interrelationship. *Syst. Zool.* 28:49–62.

—— 1981. Evolutionary trees from DNA sequences: A maximum likelihood approach. *J. Mol. Evol.* 17:368–376.

—— 1982. Numerical methods for inferring evolutionary trees. *Quart. Rev. Biol.* 57:379–404.

—— 1983a. Parsimony in systematics: Biological and statistical issues. *Ann. Rev. Ecol. Syst.* 14:313–333.

—— ed. 1983b. *Numerical Taxonomy.* Berlin: Springer-Verlag.

—— 1984. The statistical approach to inferring evolutionary trees and what it tells us about parsimony and compatibility. In T. Duncan and T. F. Stuessy, eds., *Cladistics: Perspectives on the Reconstruction of Evolutionary History*, pp. 169–191. New York: Columbia Univ. Press.

—— 1985. Confidence limits on phylogenies: An approach using the bootstrap. *Evolution* 35:783–791.

Fennah, R. G. 1955. Subspecific nomenclature: The proposed method of Wilson and Brown. *Syst. Zool.* 4:140.

Ferguson, A. 1980. Biochemical Systematics and Evolution. New York: Halstead Press.

Ferguson, I. K. 1978. Technique utilisant le méthylate de sodium comme solvant de la résine époxy des blocs d'inclusion "type MET" pour les observations de l'exine des grains de pollen. *Ann. Mines Belg.* 2:153–157.

Ferguson, I. K. and J. Muller, eds. 1976. *The Evolutionary Significance of the Exine*. London: Academic Press.

Ferguson, I. K. and J. J. Skvarla. 1982. Pollen morphology in relation to pollinators in Papilionoideae (Leguminosae). *Bot. J. Linn. Soc.* 84:183–193.

Fernald, M. L. 1918. An intergeneric hybrid in the Cyperaceae. *Rhodora* 20:189–191.

—— 1936. Minor forms and transfers. *Rhodora* 38:233–239.

Fernandes, A. and F. Franca. 1972. Contribution a la connaissance cytotaxinomique des Spermatophyta du Portugal VI. Plantaginaceae. *Bol. Soc. Brot.* ser. 2, 46:465–501.

Fernandes, A. and M. T. Leitão. 1971. Contribution a la connaissance cytotaxinomique des Spermatophyta du Portugal III. Caryophyllaceae. *Bol. Soc. Brot.* ser. 2, 45:143–176.

—— 1972. Contribution a la connaissance cytotaxinomique des Spermatophyta du Portugal V. Boraginaceae. *Bol. Soc. Brot.* ser. 2, 46:389–405.

Fernandes, A. and M. Queirós. 1971. Sur la caryologie de quelques plantes récoltées pendant la III^{ene} reunion de botanique péninsulaire. *Mem. Soc. Brot.* 21:343–385.

Fernandes, A. and M. de F. Santos. 1971. Contribution a la connaissance cytotaxinomique des Spermatophyta du Portugal IV. Leguminosae. *Bol. Soc. Brot.* ser. 2, 45:177–225.

Feulgen, R. and H. Rossenbeck. 1924. Mikroskopisch-chemischer Nachweis einer Nuclein-säure von Typus der Thymonucleinsäure und die darauf beruhende elektive Farbung von Zellkernen in mikroskopischer Präparater. *Hoppe Seyler's Z. Physiol. Chem.* 135:203–248.

Fiala, K. L. and R. R. Sokal. 1985. Factors determining the accuracy of cladogram estimation: Evaluation using computer simulation. *Evolution* 39:609–622.

Findlay, G. W. D. and J. F. Levy. 1970. Wood anatomy in three dimensions. In N. K. B. Robson, D. F. Cutler, and M. Gregory, eds., *New Research in Plant Anatomy*, pp. 71–74. London: Academic Press.

Fink, W. L. 1982. The conceptual relationship between ontogeny and phylogeny. *Paleobiology* 8:254–264.

—— 1986. Microcomputers and phylogenetic analysis. *Science* 234:1135–1139.

Fioroni, P. 1980. Ontogenie-Phylogenie: Eine Stellungnahme zu einigen neuen entwicklungsgeschichtlichen Theorien. *Z. Zool. Syst. Evolut.-Forsch.* 18:90–103.

Firmage, D. H. 1981. Environmental influences on the monoterpene variaton in *Hedeoma drummondii. Biochem. Syst. Ecol.* 9:53–58.

Firmage, D. H. and R. Irving. 1979. Effect of development on monoterpene composition of *Hedeoma drummondii. Phytochemistry* 18:1827–1829.

Fischer, C. A. H. 1890. *Beiträge zur vergleichenden Morphologie der Pollen-Körener*. Berslau: Kern.

Fischer, N. H., E. J. Olivier, and H. D. Fischer. 1979. The biogenesis and chemistry of sesquiterpene lactones. *Fortschr. Chem. Org. Naturstoffe* 38:47–390.

Fish, R. K. 1970. Megagametogenesis in *Clematis* and its taxonomic and phylogenetic implications. *Phytomorphology* 20:317–327.

Fisher, D. R. and F. J. Rohlf. 1969. Robustness of numerical taxonomic methods and errors in homology. *Syst. Zool.* 18:33–36.

Fisher, F. J. F. 1960. A discussion of leaf morphogenesis in *Ranunculus hirtus. New Zealand J. Sci.* 3:685–693.

Fisher, J. B. 1974. Axillary and dichotomous branching in the palm *Chamaedorea. Amer. J. Bot.* 61:1046–1056.

—— 1984. Tree architecture: Relationships between structure and function. In R. A. White and W. C. Dickison, eds., *Contemporary Problems in Plant Anatomy*, pp. 541–589. Orlando: Academic Press.

Fisher, R. A. 1930. *The Genetical Theory of Natural Selection*. Oxford: Oxford Univ. Press.

Fitch, W. M. 1971. Toward defining the course of evolution: Minimum change for a specific tree topology. *Syst. Zool.* 20:406–416.

—— 1977. The phyletic interpretation of macromolecular sequence information: Simple methods. In M. K. Hecht, P. C. Goody, and B. M. Hecht, eds., *Major Patterns in Vertebrate Evolution*, pp. 169–204. New York: Plenum Press.

—— 1980. Estimating the total number of nucleotide substitutions since the common ancestor of a pair of homologous genes: Comparison of several methods and three beta hemoglobin messenger RNA's. *J. Mol. Evol.* 16:153–209.

—— 1984. Cladistic and other methods: Problems, pitfalls, and potentials. In T. Duncan and T. F. Stuessy, eds., *Cladistics: Perspectives on the Reconstruction of Evolutionary History*, pp. 221–252. New York: Columbia Univ. Press.

Fitch, W. M. and E. Margoliash. 1967. Construction of phylogenetic trees. *Science* 155:279–284.

Flake, R. H., E. von Rudloff, and B. L. Turner. 1969. Quantitative study of clinal variation in *Juniperus virginiana* using terpenoid data. *Proc. Natl. Acad. U.S.A.* 64:487–494.

Flavell, R. B., M. O'Dell, and W. F. Thompson. 1983. Cytosine methylation of ribosomal RNA genes and nucleolus organiser activity in wheat. In P. E. Brandham and M. D. Bennett, eds., *Kew Chromosome Conference II*, pp. 11–17. London: George Allen & Unwin.

Flemming, W. 1882. *Zellsubstanz, Kern und Zellteilung*. Leipzig: Vogel.

Fosberg, F. R. 1942. Subspecies and variety. *Rhodora* 44:153–157.

—— 1972. The value of systematics in the environmental crisis. *Taxon* 21:631–634.

—— 1986. Biodiversity. *Environmental Awareness* 9:125–129.

Foster, A. S. 1942. *Practical Plant Anatomy*. New York: Van Nostrand. 2d ed., 1949.

Foster, A. S. and E. M. Gifford, Jr. 1974. *Comparative Morphology of Vascular Plants*. 2d ed. San Francisco: Freeman.

Fox, G. E., L. J. Magrum, W. E. Balch, R. S. Wolfe, and C. R. Woese. 1977. Classification of methanogenic bacteria by 16S ribosomal RNA characterization. *Proc. Natl. Acad. U.S.A.* 74:4537–4541.

Fox, R. M. 1955. On subspecies. *Syst. Zool.* 4:93–95.

Franklin, T. B. 1955. *Climates in Miniature: A Study of Micro-Climate and Environment*. London: Faber and Faber.

Free, J. B. 1970. *Insect Pollination of Crops*. London: Academic Press.

Freeman, C. E., W. H. Reid, J. E. Becvar, and R. Scogin. 1984. Similarity and apparent convergence in the nectar-sugar composition of some hummingbird-pollinated flowers. *Bot. Gaz.* 145:132–135.

French, J. C. 1985a. Patterns of endothecial wall thickenings in Araceae: Subfamilies Pothoideae and Monsteroideae. *Amer J. Bot.* 72:472–486.

—— 1985b. Patterns of endothecial wall thickening in Araceae: Subfamilies Calloideae, Lasioideae, and Philodendroideae. *Bot. Gaz.* 146:521–533.

Friedmann, H. 1966. The significance of the unimportant in studies of nature and of art. *Proc. Amer. Philos. Soc.* 110:256–260.

Friis, I. 1978. A reconsideration of the genera *Monotheca* and *Spiniluma* (Sapotaceae). *Kew Bull.* 33:91–98.

Frisch, K. von. 1914. Der Farbensinn und Formensinn der Biene. *Zool. Jahrb. Abt. Allg. Physiol. Tiere* 35:1–182.

—— 1950. *Bees: Their Vision, Chemical Senses, and Language*. Ithaca: Cornell Univ. Press.

—— 1953. *The Dancing Bees: An Account of the Life and Senses of the Honey Bee*. 5th ed. Translated by Dora Ilse. Berlin: Springer Verlag.

—— 1967. *The Dance Language and Orientation of Bees*. Translated by L. E. Chadwick. Cambridge, Mass.: Harvard Univ. Press.

Fritzsche, C. J. 1837. Ueber den Pollen. *Mém. Sav. Êtrang. Acad. St. Petersburg* 3:649–672.

Frohlich, M. W. 1987. Common-is-primitive: A partial validation by tree counting. *Syst. Bot.* 12:217–237.

Frohne, D. and U. Jensen, eds. 1973. *Systematik des Pflanzenreichs*. Stuttgart: Fischer.

Frost, F. H. 1930a. Specialization in secondary xylem of dicotyledons. I: Origin of vessel. *Bot. Gaz.* 89:67–94.

—— 1930b. Specialization in secondary xylem of dicotyledons II. Evolution of end wall of vessel segment. *Bot. Gaz.* 90:198–212.

—— 1931. Specialization in secondary xylem of dicotyledons III. Specialization of lateral wall of vessel segment. *Bot. Gaz.* 91:88–96.

Frye, N. 1981. The bridge of language. *Science* 212:127–132.

Fryns-Claessens, E. and W. van Cotthem. 1973. A new classification of the ontogenetic types of stomata. *Bot. Rev.* 39:71–138.

Fryxell, P. A. 1971. Phenetic analysis and the phylogeny of the diploid species of *Gossypium* L. (Malvaceae). *Evolution* 25:554–562.

Fu, K. S., ed. 1982. *Applications of Pattern Recognition*. Boca Raton, Fla.: CRC Press.

Fuchs, H. P. 1938. Historische Bemerkungen zum Begriff der Subspezies. *Taxon* 7:44–52.

Fujita, Y. 1965. Classification and phylogeny of the genus *Mosla* (=*Orthodon*) (Lamiaceae) based on the constituents of essential oil I. *Bot. Mag.* (Tokyo) 78:212–219.

Fukuda, I. 1984. Chromosome banding and biosystematics. In W. F. Grant, ed., *Plant Biosystematics*, pp. 97–116. Toronto: Academic Press.

Fukuda, I. and W. F. Grant. 1980. Chromosome variation and evolution in *Trillium grandiflorum*. *Canad. J. Genet. Cytol.* 22:81–91.

Funk, V. A. 1981. Special concerns in estimating plant phylogenies. In V. A. Funk and D. R. Brooks, eds., *Advances in Cladistics*, pp. 73–86. New York: New York Botanical Garden.

—— 1982. The systematics of *Montanoa* (Asteraceae, Heliantheae). *Mem. New York Bot. Gard.* 36:1–133.

—— 1985a. Phylogenetic patterns and hybridization. *Ann. Missouri Bot. Gard.* 72:681–715.

—— 1985b. Cladistics and generic concepts in the Compositae. *Taxon* 34:72–80.

Funk, V. A. and D. R. Brooks. 1981. Foreword. In *Advances in Cladistics*, pp. v–vi. New York: New York Botanical Garden.

Funk, V. A. and P. H. Raven. 1980. Polyploidy in *Montanoa* Cerv. (Compositae, Heliantheae). *Taxon* 29:417–419.

Funk, V. A. and T. F. Stuessy. 1978. Cladistics for the practicing plant taxonomist. *Syst. Bot.* 3:159–178.

Funk, V. A. and W. H. Wagner, Jr. 1982. A bibliography of botanical cladistics: I. 1981. *Brittonia* 34:118–124.

Funk, V. A. and Q. D. Wheeler. 1986. Symposium: Character weighting, cladistics, and classification. *Syst. Zool.* 35:100–101.

Funkhouser, J. W. 1969. Factors that affect sample reliability. In R. H. Tschudy and R. A. Scott, eds., *Aspects of Palynology*, pp. 97–102. New York: Wiley.

Furlow, J. J. 1987. The *Carpinus caroliniana* complex in North America. I. A multivariate analysis of geographical variation. *Syst. Bot.* 12:21–40.

Gabriel, B. L. 1982a. *Biological Electron Microscopy*. New York: Van Nostrand Reinhold.

—— 1982b. *Biological Scanning Electron Microscopy*. New York: Van Nostrand Reinhold.

Gabriel, K. R. and R. R. Sokal. 1969. A new statistical approach to geographic variation analysis. *Syst. Zool.* 18:259–278.

Gaffney, E. S. 1979. An introduction to the logic of phylogeny reconstruction. In J. Cracraft and N. Eldredge, eds., *Phylogenetic Analysis and Paleontology*, pp. 79–111. New York: Columbia Univ. Press.

Gangadhara, M. and J. A. Inamdar. 1977. Trichomes and stomata, and their taxonomic significance in the Urticales. *Pl. Syst. Evol.* 127:121–137.

Garay, L. A. and H. R. Sweet. 1966. Natural and artificial hybrid generic names of orchids 1887–1965. *Bot. Mus. Leafl.* 21:141–212.

—— 1969. Natural and artificial hybrid generic names of orchids. Supplement I: 1966–1969. *Bot. Mus. Leafl.* 22:273–296.

Gardner, E. J. and D. P. Snustad. 1984. *Principles of Genetics*. 7th ed. New York: Wiley.

Gardner, R. C. 1976. Evolution and adaptive radiation in *Lipochaeta* (Compositae) of the Hawaiian Islands. *Syst. Bot.* 1:383–391.

Gardner, R. C. and J. C. La Duke. 1978. Phyletic and cladistic relationships in *Lipochaeta* (Compositae). *Syst. Bot.* 3:197–207.

Gardner, R. O. 1975. A survey of the distribution of binucleate and trinucleate pollen in the New Zealand flora. *New Zealand J. Bot.* 13:361–366.

Gasc, J. P. 1978. Relations entre la phylogénie et la classification: Évocation des débats actuel entre phénéticiens et cladistes. *Bull Soc. Zool.* (France) 103:167–178.

Gasson, P. 1979. The identification of eight woody genera of the Caprifoliaceae by selected features of their root anatomy. *Bot. J. Linn. Soc.* 78:267–284.

Gastony, G. J. 1986. Electrophoretic evidence for the origin of fern species by unreduced spores. *Amer. J. Bot.* 73:1563–1569.

Gates, D. M. 1980. *Biophysical Ecology*. New York: Springer-Verlag.

Gates, R. R. 1938. The species concept in the light of cytology and genetics. *Amer. Naturalist* 72:340–349.

Gatlin, L. L. 1972. *Information Theory and the Living System*. New York: Columbia Univ. Press.

Gauld, I. D. and L. A. Mound. 1982. Homoplasy and the delineation of holophyletic genera in some insect groups. *Syst. Entomol.* 7:73–86.

Geesink, R. 1984. Scala Millettiearum: A survey of the genera of the Millettieae (Legum. - Pap.) with methodological considerations. *Leiden Bot. Ser.* 8(8):1–131.

Geissman, T. A., ed. 1962. *The Chemistry of Flavonoid Compounds*. New York: Macmillan.

Geissman, T. A. and D. H. G. Crout. 1969. *Organic Chemistry of Secondary Plant Metabolism.* San Francisco: Freeman, Cooper.

Genermont, M. 1980. Trois conceptions modernes en taxinomie: Taxinomie cladistique, taxinomie évolutive, taxinomie phénétique. *Ann. Biol.* 19:19–40.

Gentry, A. H. 1974. Flowering phenology and diversity in tropical Bignoniaceae. *Biotropica.* 6:64–68.

Gentry, A. H. and A. S. Tomb. 1979. Taxonomic implications of Bignoniaceae palynology. *Ann. Missouri Bot. Gard.* 66:756–777.

George, T. N. 1956. Biospecies, chronospecies and morphospecies. In P. C. Sylvester-Bradley, ed., *The Species Concept in Palaeontology*, pp. 123–137. London: Systematics Association.

Gerassimova-Navashina, H. 1960. A contribution to the cytology of fertilization in flowering plants. *Nucleus* (Calcutta) 3:111–120.

Germeraad, J. H. and J. Muller. 1970. A computer-based numerical coding system for the description of pollen grains and spores. *Rev. Palaeobot. Palynol.* 10:175–202.

Gershenzon, J., D. E. Lincoln, and J. H. Langenheim. 1978. The effect of moisture stress on monoterpenoid yield and composition in *Satureja douglasii*. *Biochem. Syst. Ecol.* 6:33–43.

Gershenzon, J. and T. J. Mabry. 1983. Secondary metabolites and the higher classification of angiosperms. *Nord. J. Bot.* 3:5–34.

Geslot, A. and J. Medus. 1971. Morphologie pollinique et nombre chromosomique dans la sous-section *Heterophylla* du genre *Campanula*. *Canad. J. Genet. Cytol.* 13:888–894.

Ghiselin, M. T. 1966a. An application of the theory of definitions to systematic principles. *Syst. Zool.* 15:127–130.

—— 1966b. On psychologism in the logic of taxonomic controversies. *Syst. Zool.* 15:207–215.

—— 1974. A radical solution to the species problem. *Syst. Zool.* 23:536–544.

—— 1977. On paradigms and the hypermodern species concept. *Syst. Zool.* 26:437–438.

—— 1984. "Definition," "character," and other equivocal terms. *Syst. Zool.* 33:104–110.

Giannasi, D. E. 1975. The flavonoid systematics of the genus *Dahlia* (Compositae). *Mem. New York Bot. Gard.* 26(2):1–125.

—— 1978. Generic relationships in the Ulmaceae based on flavonoid chemistry. *Taxon* 27:331–334.

—— 1979. Systematic aspects of flavonoid biosynthesis and evolution. *Bot. Rev.* 44:399–429.

Giannasi, D. E. and D. J. Crawford. 1986. Biochemical systematics II. A reprise. *Evol. Biol.* 20:25–248.

Giannasi, D. E. and K. J. Niklas. 1977. Flavonoid and other chemical constituents of fossil Miocene *Celtis* and *Ulmus* (Succor Creek Flora). *Science* 197:765–767.

Gibbs, R. D. 1963. History of chemical taxonomy. In T. Swain, ed., *Chemical Plant Taxonomy*, pp. 41–88. London: Academic Press.

—— 1974. *Chemotaxonomy of Flowering Plants.* 4 vols. Montreal: McGill-Queen's Univ. Press.

Gilbert, L. E. and P. H. Raven, eds. 1975. *Coevolution of Animals and Plants.* Austin: Univ. Texas Press.

Gill, L. S. and C. C. Chinnappa. 1977. Chromosome numbers from herbarium sheets in some Tanzanian *Impatiens* L. (Balsaminaceae). *Caryologia* 30:375–379.

Gillham, N. W. 1956. Geographic variation and the subspecies concept in butterflies. *Syst. Zool.* 5:110–120.

Gillis, W. T. 1971. The systematics and ecology of poison-ivy and the poison-oaks (*Toxicodendron*, Anacardiaceae). *Rhodora* 73:72–237, 370–540.

Gilmartin, A. J. 1967. Numerical taxonomy—an eclectic viewpoint. *Taxon* 16:8–12.

—— 1974. Variation within populations and classification. *Taxon* 23:523–536.

—— 1976. Effect of changes in character-sets upon within-group phenetic distance. *Syst. Zool.* 25:129–136.

—— 1986. Experimental systematics today. *Taxon* 35:118–119.

Gilmour, J. S. L. 1940. Taxonomy and philosophy. In J. Huxley, ed., *The New Systematics*, pp. 461–474. Oxford: Oxford Univ. Press.

—— 1951. The development of taxonomic theory since 1851. *Nature* 168:400–402.

Gilmour, J. S. L. and J. W. Gregor. 1939. Demes: A suggested new terminology. *Nature* 144:333.

Gilmour, J. S. L. and J. Heslop-Harrison. 1954. The deme terminology and the units of micro-evolutionary change. *Genetica* 27:147–161.

Gingeras, T. R. and R. J. Roberts. 1980. Steps toward computer analysis of nucleotide sequences. *Science* 209:1322–1328.

Gingerich, P. D. 1979a. The stratophenetic approach to phylogeny reconstruction in vertebrate paleontology. In J. Cracraft and N. Eldredge, eds., *Phylogenetic Analysis and Paleontology*, pp. 41–77. New York: Columbia Univ. Press.

—— 1979b. Paleontology, phylogeny, and classification: An example from the mammalian fossil record. *Syst. Zool.* 28:451–464.

Giseke, P. D. 1792. *Praelectiones in Ordines Naturales Plantarum.* Hamburg.

Gisin, H. 1966. Signification des modalités de l'évolution pour la théorie de la systématique. *Z. Zool. Syst. Evolut.-Forsch.* 4:1–12.

Glassman, S. F. 1972. Systematic studies in the leaf anatomy of palm genus *Syagrus. Amer. J. Bot.* 59:775–788.

Gleason, H. A. and A. Cronquist. 1964. *The Natural Geography of Plants.* New York: Columbia Univ. Press.

Goddijn, W. A. 1934. On the species conception in relation to taxonomy and genetics. *Blumea* 1:75–89.

Godley, E. J., ed. 1979. Reproduction in flowering plants. *New Zealand J. Bot.* 17:425–671.

Goldblatt, P. 1980. Polyploidy in angiosperms: Monocotyledons. In W. H. Lewis, ed., *Polyploidy: Biological Relevance,* pp. 219–239. New York: Plenum Press.

Goldblatt, P., ed. 1981. *Index to Plant Chromosome Numbers 1975–1978.* St. Louis: Missouri Botanical Garden.

—— ed. 1984. *Index to Plant Chromosome Numbers 1979–1981.* St. Louis: Missouri Botanical Garden.

—— ed. 1985. *Index to Plant Chromosome Numbers 1982–1983.* St. Louis: Missouri Botanical Garden.

—— ed. 1988. *Index to Plant Chromosome Numbers 1984–1985.* St. Louis: Missouri Botanical Garden.

Goldblatt, P. and D. E. Johnson. 1988. Frequency of descending versus ascending aneuploidy and its phylogenetic implications [abstract]. *Amer. J. Bot.* 75(6), pt. 2:175–176.

Goldman, N. 1988. Methods for discrete coding of morphological characters for numerical analysis. *Cladistics* 4:59–71.

Goldstein, I. J. and M. E. Etzler, eds. 1983. *Chemical Taxonomy, Molecular Biology, and Function of Plant Lectins.* New York: Liss.

Gómez-Pompa, A., C. Vázquez-Yanes, and S. Guevara. 1972. The tropical rain forest: A nonrenewable resource. *Science* 177:762–765.

Gómez-Pompa, A., R. Villalobos-Pietrini, and A. Chimal. 1971. Studies in the Agavaceae. I. Chromosome morphology and number of seven species. *Madroño* 21:208–221.

Good, R. 1953. *The Geography of the Flowering Plants.* 2d ed. London: Longmans, Green.

—— 1956. *Features of Evolution in the Flowering Plants.* London: Longmans, Green.

Goodall, D. W. 1964. A probabilistic similarity index. *Nature* 203:1098.

—— 1966. A new similarity index based on probability. *Biometrics* 22:882–907.

Goodman, M., ed. 1982. *Macromolecular Sequences in Systematics and Evolutionary Biology.* New York: Plenum Press.

Goodman, M. and J. Pechere. 1977. The evolution of muscular parvalbumins investigated by the maximum parsimony method. *J. Mol. Evol.* 9:131–158.

Goodman, M. M. and R. M. Bird. 1977. The races of maize IV: Tentative grouping of 219 Latin American races. *Econ. Bot.* 31:204–221.

Goodspeed, T. H. 1934. *Nicotiana* phylesis in the light of chromosome number, morphology, and behavior. *Univ. California Publ. Bot.* 17:369–398.

Goodwin, T. W., ed. 1965. *Chemistry and Biochemistry of Plant Pigments.* London: Academic Press.

—— ed. 1971. *Aspects of Terpenoid Chemistry and Biochemistry.* London: Academic Press.

Goodwin, T. W. and E. I. Mercer. 1983. *Introduction to Plant Biochemistry.* 2d ed. Oxford: Pergamon Press.

Gordon, A. D. 1981. *Classification: Methods for the Exploratory Analysis of Multivariate Data.* London: Chapman and Hall.

Gordon, A. D. and H. J. B. Birks. 1972. Numerical methods in Quaternary palaeoecology. I. Zonation of pollen diagrams. *New Phytol.* 71:961–979.

Goronzy, F. 1969. A numerical taxonomy on business enterprises. In A. J. Cole, ed., *Numerical Taxonomy,* pp. 42–52. London: Academic Press.

Gosline, W. A. 1954. Further thoughts on subspecies and trinomials. *Syst. Zool.* 3:92–94.

Gottlieb, L. D. 1971. Gel electrophoresis: New approach to the study of evolution. *BioScience* 21:939–944.

—— 1972. Levels of confidence in the analysis of hybridization in plants. *Ann. Missouri Bot. Gard.* 59:435–446.

—— 1973. Genetic differentiation, sympatric speciation, and the origin of a diploid species of *Stephanomeria. Amer. J. Bot.* 60:545–553.

—— 1977a. Electrophoretic evidence and plant systematics. *Ann. Missouri Bot. Gard.* 64:161–180.

—— 1977b. Phenotypic variation in *Stephanomeria exigua* ssp. *coronaria* (Compositae) and its recent derivative species *"malheurensis." Amer. J. Bot.* 64:873–880.

—— 1978. *Stephanomeria malheurensis* (Compositae), a new species from Oregon. *Madroño* 25:44–46.

—— 1981a. Electrophoretic evidence and plant populations. *Progr. Phytochem.* 7:1–46.

—— 1981b. Gene number in species of Astereae that have different chromosome numbers. *Proc. Natl. Acad. (U.S.A.)* 78:3726–3729.

—— 1984. Genetics and morphological evolution in plants. *Amer. Naturalist* 123:681–709.

Gottlieb, O. R. 1980. Micromolecular systematics: Principles and practice. In F. A. Bisby, J. G. Vaughan, and C. A. Wright, eds., *Chemosystematics: Principles and Practice*, pp. 329–352. London: Academic Press.

Gottsberger, G. 1986. Some pollination strategies in neotropical savannas and forests. *Pl. Syst. Evol.* 152:29–45.

Gould, F. W. 1963. Cytotaxonomy of *Digitaria sanguinalis* and *D. adscendens. Brittonia* 15:241–244.

Gould, S. J. 1977. *Ontogeny and Phylogeny.* Cambridge, Mass.: Harvard Univ. Press.

—— 1980. Is a new and general theory of evolution emerging? *Paleobiology* 6:119–130.

Gould, S. J. and N. Eldredge. 1977. Punctuated equilibria: The tempo and mode of evolution reconsidered. *Paleobiology* 3:115–151.

Gower, J. C. 1971. A general coefficient of similarity and some of its properties. *Biometrics* 27:857–871.

—— 1974. Maximal predictive classification. *Biometrics* 30:643–654.

Graham, A. and G. Barker. 1981. Palynology and tribal classification in the Caesalpinioideae. In R. M. Polhill and P. H. Raven, eds., *Advances in Legume Systematics*, pp. 801–834. Kew: Royal Botanic Gardens.

Graham, A., S. A. Graham, and D. Geer. 1968. Palynology and systematics of *Cuphea* (Lythraceae). I. Morphology and ultrastructure of the pollen wall. *Amer. J. Bot.* 55:1080–1088.

Graham, A. and A. S. Tomb. 1974. Palynology of *Erythrina* (Leguminosae: Papilionoideae): Preliminary survey of the subgenera. *Lloydia* 37:465–481.

Graham, S. A. and A. Graham. 1971. Palynology and systematics of *Cuphaea* (Lythraceae). II. Pollen morphology and infrageneric classification. *Amer. J. Bot.* 58:844–857.

Grande, L. 1985. The use of paleontology in systematics and biogeography, and a time control refinement for historical biogeography. *Paleobiology* 11:234–243.

Grant, K. A. and V. Grant. 1968. *Hummingbirds and Their Flowers.* New York: Columbia Univ. Press.

Grant, V. 1950. The protection of the ovules in flowering plants. *Evolution* 4:179–201.

—— 1957. The plant species in theory and practice. In E. Mayr, ed., *The Species Problem*, pp. 39–80. Washington, D.C.: American Association for the Advancement of Science.

—— 1959. *Natural History of the Phlox Family.* Vol. 1: *Systematic Botany.* The Hague: Martinus Nijhoff.

—— 1963. *The Origin of Adaptations.* New York: Columbia Univ. Press.

—— 1966a. Selection for vigor and fertility in the progeny of a highly sterile species hybrid in *Gilia. Genetics* 53:757–776.

—— 1966b. The origin of a new species of *Gilia* in a hybridization experiment. *Genetics* 54:1189–1199.

—— 1975. *Genetics of Flowering Plants.* New York: Columbia Univ. Press.

—— 1977. *Organismic Evolution.* San Francisco: Freeman.

—— 1980. Gene flow and the homogeneity of species populations. *Biol. Zentralbl.* 99:157–169.

—— 1981. *Plant Speciation.* 2d ed. New York: Columbia Univ. Press.

—— 1982a. Periodicities in the chromosome numbers of the angiosperms. *Bot. Gaz.* 143:379–389.

—— 1982b. Chromosome number patterns in primitive angiosperms. *Bot. Gaz.* 143:390–394.

—— 1983a. The synthetic theory strikes back. *Biol. Zentralbl.* 102:149–158.

—— 1983b. The systematic and geographical distribution of hawkmoth flowers in the temperate North American flora. *Bot. Gaz.* 144:439–449.

—— 1985a. Additional observations on temperate North American hawkmoth flowers. *Bot. Gaz.* 146:517–520.

—— 1985b. *The Evolutionary Process: A Critical Review of Evolutionary Theory.* New York: Columbia Univ. Press.

Grant, V. and K. A. Grant. 1965. *Flower Pollination in the Phlox Family.* New York: Columbia Univ. Press.

Grant, W. F. 1960. The categories of classical and experimental taxonomy and the species concept. *Rev. Canad. Biol.* 19:241–262.

—— 1961. Speciation and basic chromosome number in the genus *Celosia. Canad. J. Bot.* 39:45–50.

—— 1965. A chromosome atlas and interspecific hybridization index for the genus *Lotus* (Leguminosae). *Canad. J. Genet. Cytol.* 7:457–471.

—— 1969. Decreased DNA content of birch *(Betula)* chromosomes at high ploidy as determined by cytophotometry. *Chromosoma* 26:326–336.

—— ed. 1984. *Plant Biosystematics.* Toronto: Academic Press.

Grant, W. F. and B. S. Sidhu. 1967. Basic chromosome number, cyanogenic glucoside variation, and geographic distribution of *Lotus* species. *Canad. J. Bot.* 45:639–647.

Grashoff, J. L. 1975. *Metastevia* (Compositae: Eupatorieae): A new genus from Mexico. *Brittonia* 27:69–73.

Grashoff, J. L. and B. L. Turner. 1970. "The new synantherology"—A case in point for points of view. *Taxon* 19:914–917.

Gray, A. 1836. *Elements of Botany.* New York: Carvill.

—— 1856. *Manual of the Plants of the Northeastern United States and Adjacent Canada.* 2d ed. New York: Putnam.

—— 1868. Laws of botanical nomenclature adopted by the International Botanical Congress held at Paris in August, 1867; together with an historical introduction and commentary. By Alphonse DeCandolle. Translated from the French [by Weddell]. *Amer. J. Sci.* 46:63–77.

Gray, J. and W. Smith. 1962. Fossil pollen and archaeology. *Archaeology* 15:16–26.

Gray, J. R., J. A. Quinn, and D. E. Fairbrothers. 1969. Leaf epidermis morphology in populations of the *Danthonia sericea* complex. *Bull. Torrey Bot. Club* 96:525–530.

Grayum, M. H. 1986. Phylogenetic implications of pollen nuclear number in the Araceae. *Pl. Syst. Evol.* 151:145–161.

Greene, E. L. 1909. *Landmarks of Botanical History.* Smithsonian Misc. Collect. 54(1):1–329.

—— 1983. *Landmarks of Botanical History.* Edited by F. N. Egerton, with contributions by R. P. McIntosh and R. McVaugh. 2 vols. Stanford, Calif.: Stanford Univ. Press.

Greenman, J. M. 1940. The concept of the genus. III. Genera from the standpoint of morphology. *Bull. Torrey Bot. Club* 67:371–374.

Greger, H. 1985. Vergleichende Phytochemie als biologische Disziplin. *Pl. Syst. Evol.* 150:1–13.

Gregg, J. R. 1950. Taxonomy, language and reality. *Amer. Naturalist* 84:419–435.

—— 1954. *The Language of Taxonomy: An Application of Symbolic Logic to the Study of Classificatory Systems.* New York: Columbia Univ. Press.

—— 1967. Finite Linnaean structures. *Bull. Math. Biophysics* 29:191–206.

Gregor, J. W. 1930. Experiments on the genetics of wild populations. I. *Plantago maritima. J. Genet.* 22:15–25.

—— 1938. Experimental taxonomy. II Initial population differentiation in *Plantago maritima* L. of Britain. *New Phytol.* 37:1–49.

Gregor, J. W., V. M. Davey, and J. M. S. Lang. 1936. Experimental taxonomy. I Experimental garden techniques in relation to the recognition of the small taxonomic units. *New Phytol.* 35:323–350.

Greig-Smith, P. 1983. *Quantitative Plant Ecology.* 3d ed. Oxford: Blackwell.

Greilhuber, J. 1977. Nuclear DNA and heterochromatin contents in the *Scilla hohenackeri* group, *S. persica*, and *Puschkinia scilloides* (Liliaceae). *Pl. Syst. Evol.* 128:243–257.

—— 1984. Chromosomal evidence in taxonomy. In V. H. Heywood and D. M. Moore, eds., *Current Concepts in Plant Taxonomy*, pp. 157–180. London: Academic Press.

Greilhuber, J. and F. Ehrendorfer. 1975. Chromosome numbers and evolution in *Ophrys* (Orchidaceae). *Pl. Syst. Evol.* 124:125–138.

Gresson, R. A. R. 1948. *Essentials of General Cytology.* Edinburgh: Edinburgh University Press.

Greuter, W. 1979. Mediterranean conservation as viewed by a plant taxonomist. *Webbia* 34:87–99.

Greuter, W. et al., eds. 1988. International Code of Botanical Nomenclature as Adopted by the Fourteenth International Botanical Congress, Berlin, July–August 1987. *Regnum Veg.* 118:i–xiv, 1–328.

Grew, N. 1682. *The Anatomy of Plants.* London.

Grierson, D. and S. N. Covey. 1984. *Plant Molecular Biology.* Glasgow: Blackie.

Griffith, J. G. 1968. A taxonomic study of the manuscript tradition of Juvenal. *Mus. Helveticum* 25:101–138.

—— 1969. Numerical taxonomy and some primary manuscripts of the Gospels. *J. Theol. Stud.* n.s. 20:389–406.

Griffiths, G. C. D. 1974a. On the foundations of biological systematics. *Acta Biotheor.* 23:85–131.

—— 1974b. Some fundamental problems in biological classification. *Syst. Zool.* 22:338–343.

—— 1976. The future of Linnaean nomenclature. *Syst. Zool.* 25:168–173.

Guédès, M. 1979. *Morphology of Seed-Plants*. Vaduz: Cramer.

—— 1982. Nothing new with cladistics. *Taxon* 31:95–96.

Guignard, L. 1899. Sur les anthérozoides et la double copulation sexuelle chez les végétaux angiospermes. *Rev. Gén. Bot.* 11:129–135.

Gunn, C. R. 1972. Seed collecting and identification. In T. T. Kozlowski, ed., *Seed Biology*. Vol. 3: *Insects, and Seed Collection, Storage, Testing, and Certification*, pp. 55–143. New York: Academic Press.

Gunning, B. E. S. and M. W. Steer. 1975. *Ultrastructure and the Biology of Plant Cells*. London: Arnold.

Guralnik, D. B. and J. H. Friend, eds. 1953. *Webster's New World Dictionary of the American Language*. College ed. Cleveland: World.

Gustafsson, M. 1973. Evolutionary trends in the *Atriplex triangularis* group of Scandinavia I. Hybrid sterility and chromosomal differentiation. *Bot. Not.* 126:345–392.

Haapala, O. 1983. A prologue to the study of metaphase chromosome structure. In P. E. Brandham and M. D. Bennett, eds., *Kew Chromosome Conference II*, pp. 19–25. London: Allen & Unwin.

Haeckel, E. 1866. *Generelle Morphologie der Organismen*. Vol. 2: *Allgemeine Entwickelungsgeschichte der Organismen*. Berlin: Reiner.

Hagen, J. B. 1983. The development of experimental methods in plant taxonomy. *Taxon* 32:406–416.

—— 1984. Experimentalists and naturalists in twentieth-century biology: Experimental taxonomy, 1920–1950. *J. Hist. Biol.* 17:249–270.

—— 1986. Ecologists and taxonomists: Divergent traditions in twentieth-century plant geography. *J. Hist. Biol.* 19:197–214.

Hagmeier, E. M. 1958. Inapplicability of the subspecies concept to North American marten. *Syst. Zool.* 7:1–7.

Haldane, J. B. S. 1924a. A mathematical theory of natural and artificial selection. Part I. *Trans. Cambridge Philos. Soc.* 23:19–41.

—— 1924b. A mathematical theory of natural and artificial selection. Part II. *Biol. Proc. Cambridge Philos. Soc.* 1:158–163.

—— 1925–1932. A mathematical theory of natural and artificial selection. Parts III–IX. *Proc. Cambridge Philos. Soc.* 23:158–163, 363–372, 607–615, 838–844; 26:220–230; 27:131–142; 28:244–248.

—— 1932. *The Causes of Evolution*. New York: Harper.

Hale, W. G. 1970. Infraspecific categories in birds. *Biol. J. Linn. Soc.* 2:239–255.

Hall, A. V. 1988. A joint phenetic and cladistic approach for systematics. *Biol. J. Linn. Soc.* 33:367–382.

Hall, H. M. 1928. The genus *Haplopappus*: A phylogenetic study in the Compositae. *Publ. Carnegie Inst.* (Washington) 389:i–viii, 1–391.

—— 1929a. The taxonomic treatment of units smaller than species. *Proc. Internatl. Congr. Pl. Sci.* 2:1461–1468.

—— 1929b. Significance of taxonomic units and their natural basis. *Proc. Internatl. Congr. Pl. Sci.* 2:1571–1574.

—— 1932. Heredity and environment—as illustrated by transplant studies. *Sci. Monthly* 35:289–302.

Hall, H. M. and F. E. Clements. 1923. The phylogenetic method in taxonomy: The North American species of *Artemisia, Chrysothamnus,* and *Atriplex. Publ. Carnegie Inst. Washington* 326:i–iv, 1–355.

Hall, H. M., D. D. Keck, and W. M. Heusi [Hiesey]. 1931. Experimental taxonomy. *Carnegie Inst. Washington Year Book* 30:250–256.

Hall, I. V. and L. E. Aalders. 1961. Cytotaxonomy of lowbush blueberries in eastern Canada. *Amer. J. Bot.* 48:199–201.

Hall, J. B., A. J. Morton, and S. S. Hooper. 1976. Application of principal components analyses with constant character number in a study of the *Bulbostylis/Fimbristylis* (Cyperaceae) complex in Nigeria. *Bot. J. Linn. Soc.* 73:333–354.

Hallam, N. D. and T. C. Chambers. 1970. The leaf waxes of the genus *Eucalyptus* L'Heritier. *Austral. J. Bot.* 18:335–386.

Halle, F., R. A. A. Oldeman, and P. B. Tomlinson. 1978. *Tropical Trees and Forests: An Architectural Analysis*. Berlin: Springer-Verlag.

Halle, M., J. Bresnan, and G. A. Miller, eds. 1978. *Linguistic Theory and Psychological Reality*. Cambridge, Mass.: MIT Press.

Halmos, P. R. 1960. *Naive Set Theory*. Princeton, N.J.: Van Nostrand.

Hamby, R. K. and E. A. Zimmer. 1988. Ribosomal RNA sequences for inferring phylogeny within the grass family (Poaceae). *Pl. Syst. Evol.* 160:29–39.

Hamerton, J. L. 1973. Chromosome band nomenclature—the Paris Conference, 1971. In T. Caspersson and L. Zech, eds., *Chromosome Identification—Technique and Applications in Biology and Medicine*, pp. 90–96. Nobel Foundation, Stockholm. New York: Academic Press.

Hansen, B. and K. Rahn. 1969. Determination of angiosperm families by means of a punched-card system. *Dansk Botanisk Arkiv*. 26.

Hanson, E. D. 1977. *The Origin and Early Evolution of Animals*. Middletown, Conn.: Wesleyan Univ. Press.

Harborne, J. B. 1967. *Comparative Biochemistry of the Flavonoids*. London: Academic Press.

——, ed. 1970. *Phytochemical Phylogeny*. London: Academic Press.

——, ed. 1972. *Phytochemical Ecology*. London: Academic Press.

—— 1977. *Introduction to Ecological Biochemistry*. London: Academic Press.

——, ed. 1978. *Biochemical Aspects of Plant and Animal Coevolution*. Academic Press, London.

—— 1984a. Chemical data in practical taxonomy. In V. H. Heywood and D. M. Moore, eds., *Current Concepts in Plant Taxonomy*, pp. 237–261. London: Academic Press.

—— 1984b. *Phytochemical Methods: A Guide to Modern Techniques of Plant Analysis*. 2d ed. London: Chapman and Hall.

Harborne, J. B., D. Boulter, and B. L. Turner, eds. 1971. *Chemotaxonomy of the Leguminosae*. London: Academic Press.

Harborne, J. B. and T. J. Mabry, eds. 1982. *The Flavonoids: Advances in Research*. London: Chapman and Hall.

Harborne, J. B., T. J. Mabry, and H. Mabry, eds. 1975. *The Flavonoids*. London: Chapman and Hall.

Harborne, J. B. and D. M. Smith. 1978. Anthochlors and other flavonoids as honey guides in the Compositae. *Biochem. Syst. Ecol.* 6:287–291.

Harborne, J. B. and B. L. Turner. 1984. *Plant Chemosystematics*. London: Academic Press.

Hardin, J. W. 1957. A revision of the American Hippocastanaceae. *Brittonia* 9:145–171, 173–195.

—— 1976. Terminology and classification of *Quercus* trichomes. *J. Elisha Mitchell Sci. Soc.* 92:157–161.

Hardy, G. H. 1908. Mendelian proportions in a mixed population. *Science* n.s. 28:49–50.

Harlan, J. R. and J. M. J. De Wet. 1963. The compilospecies concept. *Evolution* 17:497–501.

Harling, G. 1960. Further embryological and taxonomical studies in *Anthemis* L. and some related genera. *Svensk Bot. Tidskr.* 54:571–590.

Harms, L. J. 1968. Cytotaxonomic studies in *Eleocharis* subser. *Palustres:* Central United States taxa. *Amer. J. Bot.* 55:966–974.

Harper, R. A. 1923. The species concept from the point of view of a morphologist. *Amer. J. Bot.* 10:229–233.

Harrington, H. D. 1957. *How to Identify Plants*. Chicago: Swallow Press.

Hartigan, J. A. 1975. *Clustering Algorithms*. New York: Wiley.

Hartl, D. 1987. *Principles of Population Genetics*. 2d ed. Sunderland, Mass.: Sinauer.

Hartman, H., J. G. Lawless, and P. Morrison, eds. 1987. *Search for the Universal Ancestors: The Origins of Life*. Palo Alto, Calif.: Blackwell Scientific Publications.

Hartman, R. L. and T. F. Stuessy. 1983. Revision of *Otopappus* (Compositae, Heliantheae). *Syst. Bot.* 8:185–210.

Hasegawa, M., H. Kishino, and T. Yano. 1985. Dating the human-ape splitting by a molecular clock of mitochondrial DNA. *J. Mol. Evol.* 22:160–174.

Haskell, G. and A. B. Wills. 1968. *Primer of Chromosome Practice: Plant and Animal Chromosomes Under the Microscope*. Edinburgh: Oliver and Boyd.

Hassall, D. C. 1976. Numerical and cytotaxonomic evidence for generic delimitation in Australian Euphorbieae. *Austral. J. Bot.* 24:633–640.

Hatch, M. H. 1941. The logical basis of the species concept. In J. Cattell, ed., *Biological Symposia*, 4:223–242. Lancaster, Pa.: Jacques Cattell Press.

Hauke, R. 1963. A taxonomic monograph of the genus *Equisetum* subgenus *Hippochaete*. *Beih. Nova Hedwigia* 8:1–123.

Hauser, E. J. P. and J. H. Morrison. 1964. The cytochemical reduction of nitro blue tetrazolium as an index of pollen viability. *Amer. J. Bot.* 51:748–752.

Hawkes, J. G., ed. 1968. *Chemotaxonomy and Serotaxonomy*. London: Academic Press.

Hawkes, J. G. 1978. The taxonomist's role in the conservation of genetic diversity. In H. E. Street, ed., *Essays in Plant Taxonomy*, pp. 125–142. New York: Academic Press.

Hayat, M. A. 1970. *Principles and Techniques of Electron Microscopy. Biological Applications,* vol. 1. New York: Van Nostrand Reinhold.

—— 1978. *Introduction to Biological Scanning Electron Microscopy.* Baltimore: University Park Press.

—— 1981. *Fixation for Electron Microscopy.* New York: Academic Press.

Hayward, H. E. 1938. *The Structure of Economic Plants.* New York: Macmillan.

Hecht, M. K. and J. L. Edwards. 1977. The methodology of phylogenetic inferences above the species level. In M. K. Hecht, P. C. Goody, and B. M. Hecht, eds., *Major Patterns in Vertebrate Evolution,* pp. 3–51. New York: Plenum Press.

Hedberg, I. and O. Hedberg. 1972. Ecology, taxonomy and rational land use in Africa. *Bot. Not.* 125:483–486.

—— 1977. Chromosome numbers of afroalpine and afromontane angiosperms. *Bot. Not.* 130:1–24.

Hedberg, O. 1957. Afroalpine vascular plants: A taxonomic revision. *Symb. Bot. Upsal.* 15(1):1–411.

—— 1978. Preface. In I. Hedberg, ed., *Systematic Botany, Plant Utilization, and Biosphere Conservation,* p. 7. Stockholm: Almqvist & Wiksell.

Hegnauer, R. 1954. Gedanken über die theoretische Bedeutung der chemisch-ontogenetischen und chemisch-systematischen Betrachtung von Arzneipflanzen. *Pharm. Acta Helv.* 29:203–220.

—— 1958. Chemotaxonomische Betrachtungen. VI. Phytochemie und Systematik: Eine Ruck- und Vorausschau auf die Entwicklung einer Chemotaxonomie. *Pharm. Acta Helv.* 33:287–305.

—— 1962–1986. *Chemotaxonomie der Pflanzen,* vols. 1–7. Basel: Birkhauser.

—— 1975. Biologische und systematische Bedeutung von chemischen Rassen. *Pl. Med.* 28:230–243.

Heimburger, M. 1959. Cytotaxonomic studies in the genus *Anemone. Canad. J. Bot.* 37:588–612.

Heinrich, B. 1979. *Bumblebee Economics.* Cambridge Mass.: Harvard Univ. Press.

Heintzelman, C. E., Jr. and R. A. Howard. 1948. The comparative morphology of the Icacinaceae. V. The pubescence and the crystals. *Amer. J. Bot.* 35:42–52.

Heiser, C. B., Jr. 1949. Natural hybridization with particular reference to introgression. *Bot. Rev.* 15:645–687.

—— 1963a. Modern species concepts: Vascular plants. *Bryologist* 66:120–124.

—— 1963b. Artificial intergeneric hybrids of *Helianthus* and *Viguiera. Madroño* 17:118–127.

—— 1973. Introgression re-examined. *Bot. Rev.* 39:347–366.

Heneen, W. K. and H. Runemark. 1972. Chromosomal polymorphism in isolated populations of *Elymus (Agropyron)* in the Aegean. *Bot. Not.* 125:419–429.

Hengeveld, P. 1988. Mayr's ecological species criterion. *Syst. Zool.* 37:47–55.

Hennig, W. 1950. *Grundzuge einer Theorie der phylogenetischen Systematik.* Berlin: Deutscher Zentralverlag.

—— 1965. Phylogenetic systematics. *Ann. Rev. Entomol.* 10:97–116.

—— 1966. *Phylogenetic Systematics.* Translated by D. D. Davis and R. Zangerl. Urbana: Univ. Illinois Press.

—— 1969. *Die Stammesgeschichte der Insekten.* Frankfurt: Waldemar Kramer.

—— 1975. "Cladistic analysis or cladistic classification?" A reply to Ernst Mayr. *Syst. Zool.* 24:244–256.

—— 1981. *Insect Phylogeny.* Edited and translated by A. C. Pont; revisionary notes by D. Schlee. New York: Wiley.

Herr, J. M., Jr. 1971. A new clearing-squash technique for the study of ovule development in angiosperms. *Amer. J. Bot.* 58:785–790.

—— 1982. An analysis of methods for permanently mounting ovules cleared in four-and-a-half type clearing fluids. *Stain Technol.* 57:161–169.

—— 1984. Embryology and taxonomy. In B. M. Johri, ed., *Embryology of Angiosperms,* pp. 647–696. Berlin: Springer-Verlag.

Herrnstadt, I. and C. C. Heyn. 1975. A study of *Cachrys* populations in Israel and its application to generic delimitation. *Bot. Not.* 128:227–234.

Heslop-Harrison, J. 1952. A reconsideration of plant teratology. Phyton (Horn) 4:19–34.

—— 1962. Purposes and procedures in the taxonomic treatment of higher organisms. In G. C. Ainsworth and P. H. A. Sneath, eds., *Microbial Classification,* pp. 14–36. Cambridge: Cambridge Univ. Press.

—— 1968. Chairman's summing up. In J. G. Hawkes, ed., *Chemotaxonomy and Serotaxonomy,* pp. 279–284. London: Academic Press.

—— ed. 1971. *Pollen: Development and Physiology.* London: Butterworths.

—— 1976. The adaptive significance of the exine. In I. K. Ferguson and J. Muller, eds., *The Evolutionary Significance of the Exine,* pp. 27–37. London: Academic Press.

—— 1979. Pollen walls as adaptive systems. *Ann. Missouri Bot. Gard.* 66:813–829.

Heslop-Harrison, Y. 1970. Scanning electron microscopy of fresh leaves of *Pinguicula. Science* 167:172–174.

—— 1981. Stigma characteristics and angiosperm taxonomy. *Nord. J. Bot.* 1:401–420.

Heusser, C. J. 1971. *Pollen and Spores of Chile: Modern Types of the Pteridophyta, Gymnospermae, and Angiospermae.* Tucson: Univ. Arizona Press.

Heyn, C. C. and I. Herrnstadt. 1977. Seed coat structure of Old World *Lupinus* species. *Bot. Not.* 130:427–435.

Heywood, V. H. 1958a. The interpretation of binary nomenclature for subdivisions of species. *Taxon* 7:89–93.

Heywood, V. H., comp. 1958b. *The Presentation of Taxonomic Information: A Short Guide for Contributors to Flora Europaea.* Leicester: Leicester Univ. Press.

——, ed. 1960. Problems of taxonomy and distribution in the European flora; discussion on genera and generic criteria. *Feddes Rep. Spec. Nov. Regni Veg.* 63:206–211.

—— 1967. Variation in species concepts. *Bull. Jard. Bot. Etat* 37:31–36.

—— 1968. Scanning electron microscopy and micro-characters in the fruits of the Umbelliferae-Caucalideae. *Proc. Linn. Soc. London:* 179:287–289.

—— 1969. Scanning electron microscopy in the study of plant materials. *Micron* 1:1–14.

——, ed. 1971a. *The Biology and Chemistry of the Umbelliferae.* London: Academic Press.

——, ed. 1971b. *Scanning Electron Microscopy: Systematic and Evolutionary Applications.* London: Academic Press.

—— 1973a. Taxonomy in crisis? or taxonomy is the digestive system of biology. *Acta Bot. Acad. Sci. Hung.* 19:139–146.

——, ed. 1973b. *Taxonomy and Ecology.* London: Academic Press.

—— 1974. Systematics—The stone of Sisyphus. *Biol. J. Linn. Soc.* 6:169–178.

—— 1983. The mythology of taxonomy. *Trans. Bot. Soc. Edinburgh* 44:79–94.

—— 1985. Linnaeus—the conflict between science and scholasticism. In J. Weinstock, ed., *Contemporary Perspectives on Linnaeus,* pp. 1–15. Lanham: Univ. Press of America.

Heywood, V. H. and K. M. M. Dakshini. 1971. Fruit structure in the Umbelliferae-Caucalideae. In V. H. Heywood, ed., *The Biology and Chemistry of the Umbelliferae,* pp. 215–232. London: Academic Press.

Heywood, V. H., J. B. Harborne, and B. L. Turner, eds. 1977. *The Biology and Chemistry of the Compositae.* London: Academic Press.

Heywood, V. H. and J. McNeill. 1964a. Preface. In V. H. Heywood and J. McNeill, eds., *Phenetic and Phylogenetic Classification,* pp. iii–vi. London: Systematics Association.

Heywood, V. H. and J. McNeill, eds. 1964b. *Phenetic and Phylogenetic Classification.* London: Systematics Association.

Hickey, L. J. 1973. Classification of the architecture of dicotyledonous leaves. *Amer. J. Bot.* 60:17–33.

—— 1979. A revised classification of the architecture of dicotyledonous leaves. In C. R. Metcalfe and L. Chalk, eds., *Anatomy of the Dicotyledons.* Vol. 1: *Systematic Anatomy of Leaf and Stem, with a Brief History of the Subject,* pp. 25–39. 2d ed. Oxford: Clarendon Press.

Hickey, L. J. and J. A. Doyle. 1977. Early Cretaceous fossil evidence for angiosperm evolution. *Bot. Rev.* 43:2–104.

Hickey, L. J. and J.A. Wolfe. 1975. The bases of angiosperm phylogeny: Vegetative morphology. *Ann. Missouri Bot. Gard.* 62:538–589.

Hickman, J. C. and M. P. Johnson. 1969. An analysis of geographical variation in western North American *Menziesia* (Ericaceae). *Madroño* 20:1–11.

Hideux, M. 1972. Techniques d'étude du pollen au MEB: Effets comparés des différents traitements physico-chimiques. *Micron* 3:1–31.

—— 1979. *Le Pollen données nouvelles de la microscopie électronique et de l'informatique: Structure du sporoderme des Rosidae-Saxifragales, étude comparative et dynamique.* Paris: Agence de Coopération Culturelle et Technique.

Hideux, M. and I. K. Ferguson. 1975. Stéréostructure de l'exine des Saxifragales: Proposition d'une description logique et schématique. *Soc. Bot. Fr. Coll. Palynologie* 122:57–67.

—— 1976. The stereostructure of the exine and its evolutionary significance in Saxifragaceae sensu lato. In I. K. Ferguson and J. Muller, eds., *The Evolutionary Significance of the Exine,* pp. 327–377. New York: Academic Press.

Hideux, M. and J. Mahe. 1977. Traitement par la taxinomie numérique de données palynologiques: Saxifragacées ligneuses australes. *Rev. Gen. Bot.* 84:21–59.

Hideux, M. and L. Marceau. 1972. Techniques d'étude du pollen au MEB: Méthode simple de coupes. *Adansonia* n.s. 12:609–618.

Higgins, C. A. and F. R. Safayeni. 1984. A critical-appraisal of task taxonomies as a tool for studying office activities. *ACM Trans. Office Inform. Systems* 2:331–339.

Hill, C. R. and P. R. Crane. 1982. Evolutionary cladistics and the origin of angiosperms. In K. A. Joysey and A. E. Friday, eds., *Problems of Phylogenetic Reconstruction*, pp. 269–361. London: Academic Press.

Hill, J. 1770. *The Construction of Timber*. London.

Hill, M. O. and A. J. E. Smith. 1976. Principal component analysis of taxonomic data with multi-state discrete characters. *Taxon* 25:249–255.

Hill, R. J. 1976. Taxonomic and phylogenetic significance of seed coat microsculpturing in *Mentzelia* (Loasaceae) in Wyoming and adjacent western states. *Brittonia* 28:86–112.

Hill, R. S. 1980. A numerical taxonomic approach to the study of angiosperm leaves. *Bot. Gaz.* 141:213–229.

Hillis, D. 1987. Molecular versus morphological approaches to systematics. *Ann. Rev. Ecol. Syst.* 18:23–42.

Hillson, C. J. 1963. Hybridization and floral vascularization. *Amer. J. Bot.* 50:971–978.

Hilu, K. W. 1983. The role of single-gene mutations in the evolution of flowering plants. *Evol. Biol.* 16:97–128.

Hilu, K. W. and K. Wright. 1982. Systematics of Gramineae: A cluster analysis study. *Taxon* 31:9–36.

Hitchcock, A. S. 1916. The scope and relations of taxonomic botany. *Science* 43:331–342.

Hochachka, P. W. and G. N. Somero. 1973. *Strategies of Biochemical Adaptation*. Philadelphia: Saunders.

Hoefert, L. L. 1975. Tubules in dilated cisternae of endoplasmic reticulum of *Thlaspi arvense* (Cruciferae). *Amer. J. Bot.* 62:756–760.

Hoenigswald, H. M. 1960. *Language Change and Linguistic Reconstruction*. Chicago: Univ. Chicago Press.

Hoenigswald, H. M. and L. F. Wiener, eds. 1987. *Biological Metaphor and Cladistic Classification: An Interdisciplinary Perspective*. Philadelphia: Univ. Pennsylvania Press.

Hofmeister, W. 1847. Untersuchungen des Vorgangs bei der Befruchtung der Oenothereen. *Bot. Zeitung* (Berlin) 5:785–792.

—— 1848. Ueber die Entwicklung des Pollens. *Bot. Zeitung* (Berlin) 6:425–434, 649–658, 670–674.

—— 1849. *Die Entstehung des Embryo der Phanerogamen: Eine Reihe mikroskopischer Untersuchungen*. Leipzig: published by the author.

—— 1859. Neue Beiträge zur Kenntnis der Embryobildung der Phanerogamen. I. Dikotyledonen mit ursprünglich einzelligem, nur durch Zellentheilung wachsendem Endosperm. *Abh. Königl. Sachs. Ges. Wiss.* 6:533–672.

—— 1861. Neue Beiträge zur Kenntniss der Embryobildung der Phanerogamen. II. Monokotyledonen. *Abh. Königl. Sachs. Ges. Wiss.* 7:629–760.

Holdridge, L. R. 1947. Determination of world plant formations from simple climatic data. *Science* 105:367–368.

Hollingshead, L. and E. B. Babcock. 1930. Chromosomes and phylogeny in *Crepis*. *Univ. California Publ. Agric. Sci.* 6:1–53.

Holmgren, N. H. and B. Angell. 1986. *Botanical Illustration: Preparation for Publication*. New York: New York Botanical Garden.

Honda, H. and J. B. Fisher. 1978. Tree branch angle: Maximizing effective leaf area. *Science* 199:888–889.

Honda, H., P. B. Tomlinson, and J. B. Fisher. 1981. Computer simulation of branch interaction and regulation by unequal flow rates in botanical trees. *Amer. J. Bot.* 68:569–585.

—— 1982. Two geometrical models of branching of botanical trees. *Ann. Bot.* n.s. 49:1–11.

Hooke, R. 1665. *Micrographia*. London.

Hopfinger, J. A., J. Kumamoto, and R. W. Scora. 1979. Diurnal variation in the essential oils of Valencia orange leaves. *Amer. J. Bot.* 66:111–115.

Hopwood, A. T. 1959. The development of pre-Linnaean taxonomy. *Proc. Linn. Soc.* (London) 170:230–234.

Hori, H., B.-L. Lin, and S. Osawa. 1985. Evolution of green plants as deduced from 5S rRNA sequences. *Proc. Natl. Acad. U.S.A.* 82:820–823.

Horn, H. S. 1971. *The Adaptive Geometry of Trees*. Princeton: Princeton University Press.

Horner, H. T., Jr. and C. Pearson. 1978. Pollen wall and aperture development in *Helianthus annus* (Compositae: Heliantheae). *Amer. J. Bot.* 65:293–309.

Hoshizaki, B. J. 1972. Morphology and phylogeny of *Platycerium* species. *Biotropica* 4:93–117.

Howard, R. A. 1970. Some observations on the nodes of woody plants with special reference to the problem of the "split-lateral" versus the "common gap." In N. K. B. Robson, D. F. Cutler, and M. Gregory, eds., *New Research in Plant Anatomy*, pp. 195–214. London: Academic Press.

Howe, T. D. 1975. The female gametophyte of three species of *Grindelia* and of *Prionopsis ciliata* (Compositae). *Amer. J. Bot.* 62:273–279.

Hsiao, J.-Y. 1973. A numerical taxonomic study of the genus *Platanus* based on morphological and phenolic characters. *Amer. J. Bot.* 60:678–684.

Hsu, C.-C. 1967. Preliminary chromosome studies on the vascular plants of Taiwan. *Taiwania* 13:117–130.

—— 1968. Preliminary chromosome studies on the vascular plants of Taiwan (II). *Taiwania* 14:11–27.

Hu, C. C., T. J. Crovello, and R. R. Sokal. 1985. The numerical taxonomy of some species of *Populus* based only on vegetative characters. *Taxon* 34:197–206.

Hubbell, T. H. 1954. The naming of geographically variant populations or what is all the shooting about? *Syst. Zool.* 3:113–121.

Hubbs, C. L. 1934. Racial and individual variation in animals, especially fishes. *Amer. Naturalist* 68:115–128.

Hudson, G. E., R. A. Parker, J. Vanden Berge, and P. J. Lanzillotti. 1966. A numerical analysis of the modifications of the appendicular muscles in various genera of gallinaceous birds. *Amer. Midl. Naturalist* 76:1–73.

Hudson, H. J. 1970. Infraspecific categories in fungi. *Biol. J. Linn. Soc.* 2:211–219.

Hughes, N. F. 1972. Suggestions for better handling of the genus in palaeo-palynology. *Grana Palynol.* 9:137–146.

——, ed. 1973. *Organisms and Continents through Time: Methods of Assessing Relationships between Past and Present Biologic Distributions and the Positions of Continents.* London: Palaeontological Association.

—— 1976. *Palaeobiology of Angiosperm Origins: Problems of Mesozoic Seed-Plant Evolution.* Cambridge: Cambridge Univ. Press.

Hull, D. L. 1965. The effect of essentialism on taxonomy—two thousand years of stasis. *Brit. J. Philos. Sci.* 15:314–326; 16:1–18.

—— 1966. Phylogenetic numericlature. *Syst. Zool.* 15:14–17.

—— 1967. Certainty and circularity in evolutionary taxonomy. *Evolution* 21:174–189.

—— 1970a. Contemporary systematic philosophies. *Ann. Rev. Ecol. Syst.* 1:19–54.

—— 1970b. Morphospecies and biospecies: A reply to Ruse. *Brit. J. Philos. Sci.* 21:280–282.

—— 1976. Are species really individuals? *Syst. Zool.* 25:174–191.

—— 1978. A matter of individuality. *Philos. Sci.* 45:335–360.

—— 1979. The limits of cladism. *Syst. Zool.* 28:416–440.

—— 1984. Cladistic theory: Hypotheses that grow and blur. In T. Duncan and T. F. Stuessy, eds., *Cladistics: Perspectives on the Reconstruction of Evolutionary History*, pp. 5–23. New York: Columbia Univ. Press.

Humphrey, H. B. 1961. *Makers of North American Botany.* New York: Ronald Press.

Humphries, C. J. 1979. A revision of the genus *Anacyclus* L. (Compositae: Anthemideae). *Bull. Brit. Mus. (Nat. Hist.), Bot.* 7:83–142.

—— 1980. Cytogenetic and cladistic studies in *Anacyclus* (Compositae: Athemideae). *Nord. J. Bot.* 1:93–96.

—— 1981. Biogeographical methods and the southern beeches (Fagaceae: *Nothofagus*). In V. A. Funk and D. R. Brooks, eds., *Advances in Cladistics*, pp. 177–207. New York: New York Botanical Garden.

Humphries, C. J. and J. A. Chappill. 1988. Systematics as science: A response to Cronquist. *Bot. Rev.* 54:129–144.

Humphries, C. J. and V. A. Funk. 1984. Cladistic methodology. In V. H. Heywood and D. M. Moore, eds., *Current Concepts in Plant Taxonomy*, pp. 323–362. London: Academic Press.

Humphries, C. J. and L. R. Parenti. 1986. *Cladistic Biogeography.* Oxford: Clarendon Press.

Humphries, C. J. and P. M. Richardson. 1980. Hennig's methods and phytochemistry. In F. A. Bisby, J. G. Vaughan, and C. A. Wright, eds., *Chemosystematics: Principles and Practice*, pp. 353–378. London: Academic Press.

Hunt, E. 1983. On the nature of intelligence. *Science* 219:141–146.

Hunter, I. J. 1964. Parology, a concept complementary to homology and analogy. *Nature* 204:604.

Huntley, B. and H. J. B. Birks. 1983. *An Atlas of Past and Present Pollen Maps for Europe: 0-13000 Years Ago.* Cambridge: Cambridge Univ. Press.

Hunziker, J. H. 1968. Protein electrophoresis as an aid in genome analysis. *The Nucleus* (Calcutta) 12 (suppl.):226–236.

Hunziker, J. H., H.-D. Behnke, I. J. Eifert, and T. J. Mabry. 1974. *Halophytum ameghinoi:* A betalain-containing and P-type sieve-tube plastid species. *Taxon* 23:537–539.

Hutchinson, A. H. 1936. The polygonal presentation of polyphase phenomena. *Trans. Roy. Soc. Canada,* ser. 3, 30 (sect. 5):19–26.

—— 1940. Polygonal graphing of ecological data. *Ecology* 21:475–487.

Hutchinson, J. 1983. *In situ* hybridisation mapping of plant chromosomes. In P. E. Brandham and M. D. Bennett, eds., *Kew Chromosome Conference II,* pp. 27–34. London: Allen & Unwin.

Hutchinson, John. 1926. *The Families of Flowering Plants* I. Dicotyledons. Arranged According to a New System Based on Their Probable Phylogeny. London: Macmillan.

—— 1934. *The Families of Flowering Plants* II. Monocotyledons. *Arranged According to a New System Based on Their Probable Phylogeny.* London: Macmillan.

—— 1969. *Evolution and Phylogeny of Flowering Plants. Dicotyledons: Facts and Theory.* London: Academic Press.

Huxley, J., ed. 1940. *The New Systematics.* Oxford: Oxford Univ. Press.

—— 1957. The three types of evolutionary process. *Nature* 180:454–455.

Huynh, K.-L. 1972. The original position of the generative nucleus in the pollen tetrads of *Agropyron, Itea, Limnanthes,* and *Onosma,* and its phylogenetic significance in the angiosperms. *Grana* 12:105–112.

Hyde, H. A. and D. A. Williams. 1945. Pollen of lime (*Tilia* spp.). *Nature* 155:457.

Hyland, F., B. F. Graham, Jr., F. H. Steinmetz, and M. A. Vickers. 1953. *Maine Air-Borne Pollen and Fungous Spore Survey.* Orono: Univ. Maine.

Iltis, H. H. 1982. Discovery of No. 832: An essay in defense of the National Science Foundation. *Desert Plants* 2:175–192.

Inamdar, J. A., J. S. S. Mohan, and R. Bagavathi Subramanian. 1986. Stomatal classification—a review. *Feddes Repert.* 97:147–160.

Inglis, W. G. 1966. The observational basis of homology. *Syst. Zool.* 15: 219–228.

—— 1986. Stratigramy: Biological classifications through spontaneous self-assembly. *Austral. J. Bot.* 34:411–437.

—— 1988. Cladogenesis and anagenesis: A confusion of synapomorphies. *Z. Zool. Syst. Evolut.-Forsch.* 26:1–11.

Isely, D. 1947. Investigations in seed classification by family characteristics. *Iowa Agric. Exp. Sta. Res. Bull.* 351:317–380.

—— 1972. The disappearance. *Taxon* 21:3–12.

Iversen, T.-H. 1970a. Cytochemical localization of myrosinase (β-thioglucosidase) in root tips of *Sinapis alba. Protoplasma:* 71:451–466.

—— 1970b. The morphology, occurrence, and distribution of dilated cisternae of the endoplasmic reticulum in tissues of plants of the Cruciferae. *Protoplasma* 71:467–477.

Izco, J. 1980. The role of phytosociological data in floras and taxonomy. *Bot. J. Linn. Soc.* 80:179–190.

Jaccard, P. 1908. Nouvelles recherches sur la distribution florale. *Bull. Soc. Vaud. Sci. Nat.* 44:223–270.

Jackson, R. C. 1957. New low chromosome number for plants. *Science* 126:1115–1116.

—— 1962. Interspecific hybridization in *Haplopappus* and its bearing on chromosome evolution in the *Blepharodon* section. *Amer. J. Bot.* 49:119–132.

—— 1971. The karyotype in systematics. *Ann. Rev. Ecol. Syst.* 2:327–368.

—— 1976. Evolution and systematic significance of polyploidy. *Ann. Rev. Ecol. Syst.* 7:209–234.

—— 1982. Polyploidy and diploidy: New perspectives on chromosome pairing and its evolutionary implications. *Amer. J. Bot.* 69:1512–1523.

—— 1984a. Chromosome pairing in species and hybrids. In W. F. Grant, ed., *Plant Biosystematics,* pp. 67–86. Toronto: Academic Press.

—— 1984b. Chromosome pairing, hybrid sterility, and polyploidy: Comments on G. L. Stebbins's reply. *Syst. Bot.* 9:121–123.

Jackson, R. C. and J. Casey. 1980. Cytogenetics of polyploids. In W. H. Lewis, ed., *Polyploidy: Biological Relevance,* pp. 17–44. New York: Plenum Press.

—— 1982. Cytogenetic analyses of autopolyploids: Models and methods for triploids to octoploids. *Amer. J. Bot.* 69:487–501.

Jackson, R. C. and T. J. Crovello. 1971. A comparison of numerical and biosystematic studies in *Haplopappus. Brittonia* 23:54–70.

Jackson, R. C. and C. T. Dimas. 1981. Experimental evidence for systematic placement of the *Haplopappus phyllocephalus* complex (Compositae). *Syst. Bot.* 6:8–14.

Jackson, R. C. and D. P. Hauber. 1982. Autotriploid and autotetraploid cytogenetic analyses: Correction coefficients for proposed binomial models. *Amer. J. Bot.* 69:644–646.

—— eds. 1983. *Polyploidy*. Stroudsburg, Pa.: Hutchinson Ross.

Jacobs, M. 1966. Adanson—the first neo-Adansonian? *Taxon* 15:51–55.

—— 1969. Large families—not alone! *Taxon* 18:253–262.

Jacobs, M. and P. H. Rubery. 1988. Naturally occurring auxin transport regulators. *Science*. 241:346–349.

Jacobsen, N. 1977. Chromosome numbers and taxonomy in *Cryptocoryne* (Araceae). *Bot. Not.* 130:71–87.

James, M. T. 1953. An objective aid in determining generic limits. *Syst. Zool.* 2:136–137.

James, S. H. 1979. Chromosome numbers and genetic systems in the trigger plants of western Australia (*Stylidium;* Stylidiaceae). *Austral. J. Bot.* 27:17–25.

Jancey, R. C. 1977. A hyperspatial model for complex group structure. *Taxon* 26:409–411.

Jane, F. W. 1970. *The Structure of Wood.* 2d ed. Revised by K. Wilson and D. J. B. White. London: Adam & Charles Black.

Jansen, R. K. 1981. Systematics of *Spilanthes* (Compositae: Heliantheae). *Syst. Bot.* 6:231–257.

—— 1985. The systematics of *Acmella* (Asteraceae-Heliantheae). *Syst. Bot. Monographs* 8:1–115.

Jansen, R. K. and J. D. Palmer. 1987. A chloroplast DNA inversion marks an ancient evolutionary split in the sunflower family (Asteraceae). *Proc. Natl. Acad. U.S.A.* 84:5818–5822.

Jansen, R. K. and T. F. Stuessy. 1980. Chromosome counts of Compositae from Latin America. *Amer. J. Bot.* 67:585–594.

Jansen, R. K., T. F. Stuessy, S. Díaz-Piedrahita, and V. A. Funk. 1984. Recuentos cromosómicos en Compositae de Colombia. *Caldasia* 14:7–20.

Janssen, C. R. 1970. Problems in the recognition of plant communities in pollen diagrams. *Vegetatio* 20:187–198.

Janvier, P. 1984. Cladistics: Theory, purpose, and evolutionary implications. In J. W. Pollard, ed., *Evolutionary Theory: Paths into the Future*, pp. 39–75. Chichester: Wiley.

Janzen, D. H. 1967. Interaction of the bull's-horn acacia (*Acacia cornigera* L.) with an ant inhabitant (*Pseudomyrmex ferruginea* F. Smith) in eastern Mexico. *Univ. Kansas Sci. Bull.* 47:315–558.

Jardine, C. J., N. Jardine, and R. Sibson. 1967. The structure and construction of taxonomic hierarchies. *Math. Biosci.* 1:173–179.

Jardine, N. 1967. The concept of homology in biology. *Brit. J. Philos. Sci.* 18:125–139.

—— 1969a. A logical basis for biological classification. *Syst. Zool.* 18:37–52.

—— 1969b. The observational and theoretical components of homology: A study based on the morphology of the dermal skull-roofs of rhipidistian fishes. *Biol. J. Linn. Soc.* 1:327–361.

Jardine, N. and J. M. Edmonds. 1974. The use of numerical methods to describe population differentiation. *New Phytol.* 73:1259–1277.

Jardine, N. and R. Sibson. 1971. *Mathematical Taxonomy*. London: Wiley.

Jeanmonod, D. 1984. La speciation: Aspects divers et modeles recents. *Candollea* 39:151–194.

Jeffrey, C. 1977a. *Biological Nomenclature.* 2d ed. London: Arnold.

—— 1977b. Corolla forms in Compositae—some evolutionary and taxonomic speculations. In V. H. Heywood, J. B. Harborne, and B. L. Turner, eds., *The Biology and Chemistry of the Compositae*, pp. 111–118. London: Academic Press.

Jeffrey, D. W. 1987. *Soil-Plant Relationships: An Ecological Approach.* London: Croom Helm.

Jeffrey, E. C. 1917. *The Anatomy of Woody Plants.* Chicago: Univ. Chicago Press.

Jenik, J. 1978. Roots and root systems in tropical trees: Morphologic and ecologic aspects. In P. B. Tomlinson and M. H. Zimmermann, eds., *Tropical Trees as Living Systems*, pp. 323–349. Cambridge: Cambridge Univ. Press.

Jensen, R. J. 1977. A preliminary numerical analysis of the red oak complex in Michigan and Wisconsin. *Taxon* 26:399–407.

—— 1981. Wagner networks and Wagner trees: A presentation of methods for estimating most parsimonious solutions. *Taxon* 30:576–590.

Jensen, R. J. and W. H. Eshbaugh. 1976. Numerical taxonomic studies of hybridization in *Quercus.* II. Populations with wide areal distributions and high taxonomic diversity. *Syst. Bot.* 1:11–19.

Jensen, R. J., M. J. McLeod, W. H. Eshbaugh, and S. I. Guttman. 1979. Numerical taxonomic analyses of allozymic variation in *Capsicum* (Solanaceae). *Taxon* 28:315–327.

Jensen, U. and D. E. Fairbrothers, eds. 1983. *Proteins and Nucleic Acids in Plant Systematics.* Berlin: Springer-Verlag.

Jensen, U. and B. Greven. 1984. Serological aspects and phylogenetic relationships of the Magnoliidae. *Taxon* 33:563–577.

Jensen, W. A. 1962. *Botanical Histochemistry: Principles and Practice*. San Francisco: Freeman.

—— 1965a. The composition and ultrastructure of the nucellus in cotton. *J. Ultrastruct. Res.* 13:112–128.

—— 1965b. The ultrastructure and histochemistry of the synergids of cotton. *Amer. J. Bot.* 52:238–256.

—— 1965c. The ultrastructure and composition of the egg and central cell of cotton. *Amer. J. Bot.* 52:781–797.

—— 1968a. Cotton embryogenesis: The tube-containing endoplasmic reticulum. *J. Ultrastruct. Res.* 22:296–302.

—— 1968b. Cotton embryogenesis: The zygote. *Planta* 79:346–366.

—— 1968c. Cotton embryogenesis: Polysome formation in the zygote. *J. Cell Biol.* 36:403–406.

Jensen, W. A., M. Ashton, and L. R. Heckard. 1974. Ultrastructural studies of the pollen of subtribe Castilleiinae, family Scrophulariaceae. *Bot. Gaz.* 135:210–218.

Jensen, W. A. and D. B. Fisher. 1968. Cotton embryogenesis: The entrance and discharge of the pollen tube in the embryo sac. *Planta* 78:158–183.

Johansen, D. A. 1940. *Plant Microtechnique*. New York: McGraw-Hill.

—— 1950. *Plant Embryology: Embryogeny of the Spermatophyta*. Waltham, Mass.: Chronica Botanica.

John, B. 1976. *Population Cytogenetics*. London: Arnold.

Johnson, A. M. 1931. *Taxonomy of the Flowering Plants*. New York: Century.

Johnson, C. and W. V. Brown. 1973. Grass leaf ultrastructural variations. *Amer. J. Bot.* 60:727–735.

Johnson, G. B. 1973. Enzyme polymorphism and biosystematics: The hypothesis of selective neutrality. *Ann. Rev. Ecol. Syst.* 4:93–116.

Johnson, H. B. 1975. Plant pubescence: An ecological perspective. *Bot. Rev.* 41:233–258.

Johnson, L. A. S. 1968. Rainbow's end: The quest for an optimal taxonomy. *Proc. Linn. Soc. New South Wales* 93:8–45. Expanded and reprinted, 1970. *Syst. Zool.* 19:203–239.

—— 1970. Biosystematics alive—a discussion. *Taxon* 19:152–153.

—— 1973. [Discussion of symposium papers on contemporary systematic philosophies.] *Syst. Zool.* 22:399.

—— 1976. Problems of species and genera in *Eucalyptus* (Myrtaceae). *Pl. Syst. Evol.* 125:155–167.

Johnson, L. A. S. and B. G. Briggs. 1975. On the Proteaceae—the evolution and classification of a southern family. *Bot. J. Linn. Soc.* 70:83–182.

—— 1984. Myrtales and Myrtaceae—a phylogenetic analysis. *Ann. Missouri Bot. Gard.* 71:700–756.

Johnson, M. P. and R. W. Holm. 1968. Numerical taxonomic studies in the genus *Sarcostemma* R. Br. (Asclepiadaceae). In V. H. Heywood, ed., *Modern Methods in Plant Taxonomy*, pp. 199–217. London: Academic Press.

Johnson, R. 1982. Parsimony principles in phylogenetic systematics: A critical re-appraisal. *Evol. Theory* 6:79–90.

Johnson, R. W. 1982. Effect of weighting and the size of the attribute set in numerical classification. *Austral. J. Bot.* 30:161–174.

Johri, B. M. 1963. Embryology and taxonomy. In P. Maheshwari, ed., *Recent Advances in the Embryology of Angiosperms*, pp. 395–444. International Society of Plant Morphologists. Delhi: Univ. Delhi.

——, ed. 1982. *Experimental Embryology of Vascular Plants*. Berlin: Springer-Verlag.

——, ed. 1984. *Embryology of Angiosperms*. Berlin: Springer-Verlag.

Johri, B. M. and K. B. Ambegaokar. 1984. Embryology: Then and now. In B. M. Johri, ed., *Embryology of Angiosperms*, pp. 1–52. Berlin: Springer-Verlag.

Johri, B. M. and P. S. Rao. 1984. Experimental embryology. In B. M. Johri, ed., *Embryology of Angiosperms*, pp. 735–802. Berlin: Springer-Verlag.

Jones, A. 1985. Chromosomal features as generic criteria in the Astereae. *Taxon* 34:44–54.

Jones, C. E. and R. J. Little, eds. 1983. *Handbook of Experimental Pollination Biology*. New York: Scientific and Academic Editions.

Jones, K. 1974. Chromosome evolution by Robertsonian translocation in *Gibasis* (Commelinaceae). *Chromosoma* (Berlin) 45:353–368.

—— 1978. Aspects of chromosome evolution in higher plants. *Advances Bot. Res.* 6:119–194.

—— 1984. Cytology and biosystematics: 1983. In W. F. Grant, ed., *Plant Biosystematics*, pp. 25–39. Toronto: Academic Press.

Jones, K. and P. E. Brandham, eds. 1976. *Current Chromosome Research*. Amsterdam: North-Holland.

Jones, K. and C. Jopling. 1972. Chromosomes and the classification of the Commelinaceae. *Bot. J. Linn. Soc.* 65:129–162.

Jones, K., A. Kenton, and D. R. Hunt. 1981. Contributions to the cytotaxonomy of the Commelinaceae. Chromosome evolution in *Tradescantia* section *Cymbispatha. Bot. J. Linn. Soc.* 83:157–188.

Jones, R. N. and H. Rees. 1982. *B Chromosomes.* London: Academic Press.

Jones, S. B., Jr. 1979. Synopsis and pollen morphology of *Vernonia* (Compositae: Vernonieae) in the New World. *Rhodora* 81:425–447.

Jones, S. B., Jr., and A. E. Luchsinger. 1986. *Plant Systematics.* 2d ed. New York: McGraw-Hill.

Jong, R. de. 1980. Some tools for evolutionary and phylogenetic studies. *Z. Zool. Syst. Evolut.-Forsch.* 18:1–23.

Jonsell, B. 1978. Linnaeus's views on plant classification and evolution. *Bot. Not.* 131:523–530.

Jordan, D. E. and P. E. Swartz. 1976. Word frequency as a determinant of clustering when taxonomic frequency is held constant. *Canad. J. Psychol.* 30:187–192.

Jordan, K. 1905. Der Gegensatz zwischen geographischer und nichtgeographischer Variation. *Z. Wiss. Zool.* 83:151–210.

Joysey, K. A. and A. E. Friday, eds. 1982. *Problems of Phylogenetic Reconstruction.* London: Academic Press.

Judd, W. S. 1979. Generic relationships in the Andromedeae (Ericaceae). *J. Arnold Arbor.* 60:477–503.

—— 1982. A taxonomic revision of *Pieris* (Ericaceae). *J. Arnold Arbor.* 63:103–144.

Jukes, T. H. 1966. *Molecules and Evolution.* New York: Columbia Univ. Press.

Juniper, B. E. and D. E. Bradley. 1958. The carbon replica technique in the study of the ultrastructure of leaf surfaces. *J. Ultrastruct. Res.* 2: 16–27.

Juniper, B. E. and C. E. Jeffree. 1983. *Plant Surfaces.* London: Arnold.

Jussieu, A. L. de. 1789. *Genera Plantarum.* Paris.

Just, T. 1946. The relative value of taxonomic characters. *Amer. Midl. Naturalist* 36:291–297.

—— 1953. Generic synopses and modern taxonomy. *Chron. Bot.* 14:103–114.

Kam, Y. K. and J. Maze. 1974. Studies on the relationships and evolution of supraspecific taxa utilizing developmental data. II. Relationships and evolution of *Oryzopsis hymenoides, O. virescens, O. kingii, O. micrantha,* and *O. asperifolia. Bot. Gaz.* 135:227–247.

Kamemoto, H. and K. Shindo. 1962. Genome relationships in interspecific and intergeneric hybrids of *Renanthera. Amer. J. Bot.* 49:737–748.

Kanal, L. N. and A. Rosenfeld, eds. 1981. *Progress in Pattern Recognition,* vol. 1. Amsterdam: North-Holland.

Kaneko, K.-I. and N. Hashimoto. 1982. Five biovars of *Yersinia enterocolitica* delineated by numerical taxonomy. *Internatl. J. Syst. Bacteriol.* 32:275–287.

Kapadia, Z. J. 1963. Varietas and subspecies: A suggestion towards greater uniformity. *Taxon* 12:257–259.

Kapadia, Z. J. and F. W. Gould. 1964. Biosystematic studies in the *Bouteloua curtipendula* complex. III. Pollen size as related to chromosome numbers. *Amer. J. Bot.* 51:166–172.

Kapil, R. N. and R. S. Vani. 1966. *Nyctanthes arbor-tristis* Linn.: Embryology and relationships. *Phytomorphology* 16:553–563.

Kaplan, D. R. 1971. On the value of comparative development in phylogenetic studies—a rejoinder. *Phytomorphology* 21:134–140.

—— 1984. The concept of homology and its central role in the elucidation of plant systematic relationships, pp. 51–70. In T. Duncan and T. F. Stuessy, eds., *Cladistics: Perspectives on the Reconstruction of Evolutionary History,* pp. 51–70. New York: Columbia Univ. Press.

Kapp, R. O. 1969. *How to Know Pollen and Spores.* Dubuque, Iowa: Brown.

—— 1970. Pollen analysis of pre-Wisconsin sediments from the Great Plains. In W. Dort, Jr. and J. K. Jones, Jr., eds., *Pleistocene and Recent Environments of the Central Great Plains,* pp. 143–155. Lawrence: Univ. Kansas Press.

Kasha, K. J., ed. 1974. *Haploids in Higher Plants: Advances and Potentials.* Guelph, Ontario: Univ. Guelph.

Kaul, R. B. 1969. Morphology and development of the flowers of *Boottia cordata, Ottelia alismoides,* and their synthetic hybrid (Hydrocharitaceae). *Amer. J. Bot.* 56: 951–959.

Kavanaugh, D. H. 1972. Hennig's principles and methods of phylogenetic systematics. *Biologist* 54:115–127.

—— 1978. Hennigian phylogenetics in contemporary systematics: Principles, methods, and uses. In L. Knutson, chm., *Biosystematics in Agriculture,* pp. 139–150. Montclair, N.J.: Allenheld, Osmun.

Keating, R. C. 1970. Comparative morphology of the Cochlospermaceae. II. Anatomy of the young vegetative shoot. *Amer. J. Bot.* 57:889–898.

—— 1979. Palynology and systematics: The Twenty-fifth Systematics Symposium. *Ann. Missouri Bot. Gard.* 66:591–592.

—— 1984. Leaf histology and its contribution to relationships in the Myrtales. *Ann. Missouri Bot. Gard.* 71:801–823.

Kee, D. W. and L. Helfend. 1977. Assessment of taxonomic encoding categories in different populations. *J. Educ. Psychol.* 69:344–348.

Keefe, J. M. and M. F. Moseley, Jr. 1978. Wood anatomy and phylogeny of *Paeonia* section *Moutan. J. Arnold Arbor.* 59:274–297.

Keener, C. S. 1967. A biosystematic study of *Clematis* subsection *Integrifoliae* (Ranunculaceae). *J. Elisha Mitchell Sci. Soc.* 83:1–41.

Keil, D. J. and T. F. Stuessy. 1975. Chromosome counts of Compositae from the United States, Mexico, and Guatemala. *Rhodora* 77:171–195.

—— 1977. Chromosome counts of Compositae from Mexico and the United States. *Amer. J. Bot.* 64:791–798.

—— 1981. Systematics of *Isocarpha* (Compositae: Eupatorieae). *Syst. Bot.* 6:258–287.

Kemp, T. S. 1985. Models of diversity and phylogenetic reconstruction. *Oxford Surv. Evol. Biol.* 2:135–158.

Kendrew, W. G. 1922. *The Climates of the Continents.* Oxford: Clarendon Press.

Kendrick, W. B. 1964. Quantitative characters in computer taxonomy. In V. H. Heywood and J. McNeill, eds., *Phenetic and Phylogenetic Classification*, pp. 105–114. London: Systematics Association.

—— 1965. Complexity and dependence in computer taxonomy. *Taxon* 14:141–154.

—— 1974. The generic iceberg. *Taxon* 23:747–753.

Kendrick, W. B. and L. K. Weresub. 1966. Attempting neo-Adansonian computer taxonomy at the ordinal level in the Basidiomycetes. *Syst. Zool.* 15:307–329.

Kershaw, A. P. 1970. Pollen morphological variation within the Casuarinaceae. *Pollen et Spores* 12:145–161.

Kershaw, K. A. and J. H. H. Looney. 1985. *Quantitative and Dynamic Plant Ecology.* 3d ed. Baltimore: Arnold.

Kethley, J. B. 1977. A review of the higher categories of *Trigynaspida* (Acari: Parasitiformes). *Internatl. J. Acar.* 3:129–149.

Kevan, P. G. 1983. Floral colors through the insect eye: What they are and what they mean. In C. E. Jones and R. J. Little, eds., *Handbook of Experimental Pollination Biology*, pp. 3–30. New York: Scientific and Academic Editions.

Kimura, M. 1968. Evolutionary rate at the molecular level. *Nature* 217:624–626.

—— 1983. *The Neutral Theory of Molecular Evolution.* New York: Cambridge Univ. Press.

King, D. S. 1976. Systematics of *Conidiobolus* (Entomophthorales) using numerical taxonomy. I. Biology and cluster analysis. *Canad. J. Bot.* 54:45–65.

—— 1977. Systematics of *Conidiobolus* (Entomophthorales) using numerical taxonomy. III. Descriptions of recognized species. *Canad. J. Bot.* 55:718–729.

King, G. J. and M. J. Ingrouille. 1987. Genome heterogeneity and classification of the Poaceae. *New Phytol.* 107:633–644.

King, J. L. and T. H. Jukes. 1969. Non-Darwinian evolution. *Science* 164:788–798.

King, P. J. H. 1976. A taxonomy of computer science. *Computer Bull.* 2(8):28–30.

King, R. M. and V. E. Krantz. 1975. Ultraviolet reflectance patterns in the Asteraceae, I. Local and cultivated species. *Phytologia* 31:66–114.

King, R. M., D. W. Kyhos, A. M. Powell, P. H. Raven, and H. Robinson. 1976. Chromosome numbers in Compositae. XIII. Eupatorieae. *Ann. Missouri Bot. Gard.* 63:862–888.

King, R. M. and H. Robinson. 1970. The new synantherology. *Taxon* 19:6–11.

—— 1987. *The Genera of the Eupatorieae (Asteraceae).* St. Louis: Missouri Botanical Garden.

Kirkbride, M. C. G. 1982. A preliminary phylogeny for the neotropical *Rubiaceae. Pl. Syst. Evol.* 141:115–122.

Kirkpatrick, J. B. 1974. The numerical intraspecific taxonomy of *Eucalyptus globulus* Labill. (Myrtaceae). *Bot. J. Linn. Soc.* 69:89–104.

Kitts, D. B. 1977. Karl Popper, verifiability, and systematic zoology. *Syst. Zool.* 26:185–194.

—— 1983. Can baptism alone save a species? *Syst. Zool.* 32:27–33.

Kjellqvist, E. and A. Löve. 1963. Chromosome numbers of some *Carex* species from Spain. *Bot. Not.* 116:241–248.

Kliphuis, E. and J. H. Wieffering. 1972. Chromosome numbers of some angiosperms from the south of France. *Acta Bot. Neerl.* 21:598–604.

Kluge, A. G. 1967. Higher taxonomic categories of gekkonid lizards and their evolution. *Bull. Amer. Mus. Nat. Hist.* 135:1–59.

—— 1976. Phylogenetic relationships in the lizard family, Phygopodidae: An evaluation of theory, methods and data. *Univ. Michigan Mus. Zool. Misc. Publ.* 152:1–72.

—— 1982. The cladistic perspective [review of E. O. Wiley's *Phylogenetics: The Theory and Practice of Phylogenetic Systematics*]. *Science* 215:51–52.

—— 1984. The relevance of parsimony to phylogenetic inference. In T. Duncan and T. F. Stuessy, eds., *Cladistics: Perspectives on the Reconstruction of Evolutionary History*, pp. 24–38. New York: Columbia Univ. Press.

—— 1985. Ontogeny and phylogenetic systematics. *Cladistics* 1:13–27.

Kluge, A. G. and J. S. Farris. 1969. Quantitative phyletics and the evolution of anurans. *Syst. Zool.* 18:1–32.

Kluge, A. G. and W. C. Kerfoot. 1973. The predictability and regularity of character divergence. *Amer. Naturalist* 107:426–442.

Kluge, A. G. and R. E. Strauss. 1985. Ontogeny and systematics. *Ann. Rev. Ecol. Syst.* 16:247–268.

Knight, D. 1981. *Ordering the World: A History of Classifying Man*. London: Burnett Books.

Knobloch, I. W. 1968. *A Check List of Crosses in the Gramineae*. East Lansing, Mich.: published by the author.

—— 1972. Intergeneric hybridization in flowering plants. *Taxon* 21:97–103.

Knobloch, I. W., H. P. Rasmussen, and W. S. Johnson. 1975. Scanning electron microscopy of trichomes of *Cheilanthes* (Sinopteridaceae). *Brittonia* 27:245–250.

Knox, R. B. 1984. The pollen grain. In B. M. Johri, ed., *Embryology of Angiosperms*, pp. 197–271. Berlin: Springer-Verlag.

Knuth, P. 1906–1909. *Handbook of Flower Pollination: Based upon Hermann Muller's Work "The Fertilisation of Flowers by Insects."* Translated by J. R. Ainsworth Davis. 3 vols. Oxford: Clarendon Press.

Koch, M. F. 1930a. Studies in the anatomy and morphology of the Compositae flower I. The corolla. *Amer. J. Bot.* 17:938–952.

—— 1930b. Studies in the anatomy and morphology of the Compositae flower II. The corollas of the Heliantheae and Mutisieae. *Amer. J. Bot.* 17:995–1010.

Kölreuter, J. G. 1761–1766. *Vorläufige Nachricht von einigen das Geschlecht der Pflanzen betreffenden Versuchen und Beobachtungen*. 4 vols. Leipzig.

Koponen, T. 1968. Generic revision of Mniaceae Mitt. (Bryophyta). *Ann. Bot. Fenn.* 5:117–151.

—— 1973. *Rhizomnium* (Mniaceae) in North America. *Ann. Bot. Fenn.* 10:1–26.

—— 1980. A synopsis of Mniaceae (Bryophyta). II: Orthomnion. *Ann. Bot. Fenn.* 17:35–55.

Köstler, J. N., E. Brückner, and H. Bibelriether. 1968. *Die Wurzeln der Waldbäume: Untersuchungen zur Morphologie der Waldbäume in Mitteleuropa*. Hamburg: Parey.

Kowal, R. R., S. A. Mori, and J. A. Kallunki. 1977. Chromosome numbers of Panamanian Lecythidaceae and their use in subfamilial classification. *Brittonia* 29:399–410.

Kraus, F. 1988. An empirical evaluation of the use of the ontogeny polarization in phylogenetic inference. *Syst. Zool.* 37:106–141.

Kraus, O., ed. 1978. *Co-Evolution*. Hamburg: Parey.

Krebs, C. J. 1978. *Ecology: The Experimental Analysis of Distribution and Abundance*. 2d ed. New York: Harper & Row.

Kremp, G. O. W. 1965. *Morphologic Encyclopedia of Palynology*. Tucson: Univ. Arizona Press.

Kress, W. J. 1986. Exineless pollen structure and pollination systems of tropical *Heliconia* (Heliconiaceae). In S. Blackmore and I. K. Ferguson, eds., *Pollen and Spores: Form and Function*, pp. 329–345. London: Academic Press.

Kress, W. J., D. E. Stone and S. C. Sellars. 1978. Ultrastructure of exine-less pollen: *Heliconia* (Heliconiaceae). *Amer. J. Bot.* 65:1064–1076.

Kruckeberg, A. R. 1954. Chromosome numbers in *Silene* (Caryophyllaceae), I. *Madroño* 12:238–246.

—— 1962. Intergeneric hybrids in the Lychnideae (Caryophyllaceae). *Brittonia* 14:311–321.

—— 1969. Ecological aspects of the systematics of plants [with discussion]. In C. G. Sibley, chm., *Systematic Biology*, pp. 161–212. Washington, D.C.: National Academy of Sciences.

—— 1984. California serpentines: Flora, vegetation, geology, soils, and management problems. *Univ. California Publ. Bot.* 78:i–xiv, 1–180.

—— 1986. An essay: The stimulus of unusual geologies for plant speciation. *Syst. Bot.* 11:455–463.

Kruckeberg, A. R. and F. L. Hedglin. 1963. Natural and artificial hybrids of *Besseya* and *Synthyris* (Scrophulariaceae). *Madroño* 17:109–115.

Kruskal, J. B., Jr. 1956. On the shortest spanning subtree of a graph and the traveling salesman problem. *Proc. Amer. Math. Soc.* 7:48–50.

Kubitzki, K., ed. 1983. *Dispersal and Distribution: An International Symposium.* Hamburg: Parey.

Kubitzki, K. 1984. Phytochemistry in plant systematics and evolution. In V. H. Heywood and D. M. Moore, eds., *Current Concepts in Plant Taxonomy*, pp. 263–277. London: Academic Press.

Kubler, G. 1962. *The Shape of Time: Remarks on the History of Things.* New Haven: Yale Univ. Press.

Kugler, H. 1970. *Blütenokölogie.* Stuttgart: Gustav Fisher Verlag.

Kujat, R. and J. N. Rafinski. 1978. Seed coat structure of *Crocus vernus* agg. (Iridaceae). *Pl. Syst. Evol.* 129:255–260.

Kukkonen, I. 1967. Spikelet morphology and anatomy of *Uncinia* Pers. (Cyperaceae). *Kew Bull.* 21:93–97.

Kupchan, S. M., J. H. Zimmerman, and A. Afonso. 1961. The alkaloids and taxonomy of *Veratrum* and related genera. *Lloydia* 24:1–26.

Kuprianova, L. A. 1969. On the evolutionary levels in the morphology of pollen grains and spores. *Pollen et Spores* 11:333–351.

Kutschera, L. 1960. *Wurzelatlas mitteleuropäischer Ackerunkräuter und Kulturpflanzen.* Frankfurt: DLG-Verlags-GmbH.

Kyhos, D. W. 1965. The independent aneuploid origin of two species of *Chaenactis* (Compositae) from a common ancestor. *Evolution* 19:26–43.

—— 1967. Natural hybridization between *Encelia* and *Geraea* (Compositae) and some related experimental investigations. *Madroño* 19:33–43.

Lackey, J. A. 1978. Leaflet anatomy of Phaseoleae (Leguminosae: Papilionoideae) and its relation to taxonomy. *Bot. Gaz.* 139:436–446.

Ladiges, P. Y. 1984. A comparative study of trichomes in *Angophora* Cav. and *Eucalyptus* L'Herit.—a question of homology. *Austral. J. Bot.* 32:561–574.

Ladizinsky, G. and D. Zohary. 1968. Genetic relationships between diploids and tetraploids in series *Eubarbatae* of *Avena. Canad. J. Genet. Cytol.* 10:68–81.

La Duke, J. C. 1982. Revision of *Tithonia. Rhodora* 84:453–522.

—— 1987. The existence of hypotheses in plant systematics or biting the hand that feeds you. *Taxon* 36:60–64.

La Duke, J. C. and D. J. Crawford. 1979. Character compatibility and phyletic relationships in several closely related species of *Chenopodium* of the western United States. *Taxon* 28:307–314.

Lakshmanan, K. K. and K. B. Ambegaokar. 1984. Polyembryony. In B. M. Johri, ed., *Embryology of Angiosperms*, pp. 445–474. Berlin: Springer-Verlag.

Lam, H. J. 1936. Phylogenetic symbols, past and present (being an apology for genealogical trees). *Acta Biotheor.* 2:153–194.

Lambert, J. M., S. E. Meacock, J. Barrs, and P. F. M. Smartt. 1973. AXOR and MONIT: Two new polythetic-divisive strategies for hierarchical classification. *Taxon* 22:173–176.

Lambinon, J. 1959. Brèves considérations sur la nomenclature des taxa infraspécifiques. *Bull. Soc. Roy. Bot. Belgique* 91:213–215.

Lammers, T. G. and C. E. Freeman. 1986. Ornithophily among the Hawaiian Lobelioideae (Campanulaceae): Evidence from floral nectar sugar compositions. *Amer. J. Bot.* 73:1613–1619.

Lammers, T. G., T. F. Stuessy, and M. Silva O. 1986. Systematic relationships of Lactoridaceae, an endemic family of the Juan Fernandez Islands, Chile. *Pl. Syst. Evol.* 152:243–266.

Lance, G. N. and W. T. Williams. 1967. A general theory of classificatory sorting strategies 1. Hierarchical systems. *Computer J.* 9:373–380.

—— 1968. Mixed-data classificatory programs II. Divisive systems. *Austral. Computer J.* 1:82–85.

Landrum, L. R. 1981. The phylogeny and geography of *Myrceugenia* (Myrtaceae). *Brittonia* 33:105–129.

Lane, M. A. 1985. Features observed by electron microscopy as generic criteria. *Taxon* 34:38–43.

Lane, M. A. and B. L. Turner, eds. 1985. The generic concept in the Compositae: A symposium. *Taxon* 34:5.

Langenheim, J. H., C. E. Foster, and R. B. McGinley. 1980. Inhibitory effects of different quantitative compositions of *Hymenaea* leaf resins on a generalist herbivore *Spodoptera exigua. Biochem. Syst. Ecol.* 8:385–396.

Lanham, U. 1965. Uninominal nomenclature. *Syst. Zool.* 14:144.

Larson, J. L. 1971. *Reason and Experience: The Representation of Natural Order in the Work of Carl von Linné.* Berkeley: Univ. California Press.

Laszlo, E. 1972. *The Systems View of the World: The Natural Philosophy of the New Developments in the Sciences.* New York: Braziller.

Lavania, U. C. 1986. High bivalent frequencies in artificial autopolyploids of *Hyoscyamus muticus* L. *Canad. J. Genet. Cytol.* 28:7–11.

Lawrence, G. H. M. 1951. *Taxonomy of Vascular Plants.* New York: Macmillan.

—— 1953. Plant genera, their nature and definition: The need for an expanded outlook. *Chron. Bot.* 14:117–120.

Lawrence, M. E. 1985. *Senecio* L. (Asteraceae) in Australia: Nuclear DNA amounts. *Austral. J. Bot.* 33:221–232.

Leaver, C. J., ed. 1980. *Genome Organization and Expression in Plants.* New York: Plenum Press.

Legendre, P. 1971. Circumscribing the concept of the genus. *Taxon* 20:137–139.

—— 1972. The definition of systematic categories in biology. *Taxon* 21:381–406.

Legendre, P. and P. Vaillancourt. 1969. A mathematical model for the entities species and genus. *Taxon* 18:245–356.

Lehman, H. 1967. Are biological species real? *Philos. Sci.* 34:157–167.

Leins, P. 1971. Pollensystematische Studien an Inuleen I. Tarchonanthinae, Plucheinae, Inulinae, Buphthalminae. *Bot. Jahrb. Syst.* 91:91–146.

Lemen, C. A. and P. W. Freeman. 1984. The genus: A macroevolutionary problem. *Evolution* 38:1219–1237.

Lennox, J. G. 1980. Aristotle on genera, species, and "the more and the less." *J. Hist. Biol.* 13:321–346.

Leone, C. A., ed. 1964. *Taxonomic Biochemistry and Serology.* New York: Ronald Press.

Leppik, E. E. 1956. The form and function of numerical patterns in flowers. *Amer. J. Bot.* 43:445–455.

—— 1968a. Morphogenic classification of flower types. *Phytomorphology* 18:451–466.

—— 1968b. Directional trend of floral evolution. *Acta Biotheor.* 18:87–102.

—— 1972. Origin and evolution of bilateral symmetry in flowers. *Evol. Biol.* 5:49–85.

—— 1977. The evolution of capitulum types of the Compositae in the light of insect-flower interaction. In V. H. Heywood, J. B. Harborne, and B. L. Turner, eds., *The Biology and Chemistry of the Compositae,* pp. 61–90. London: Academic Press.

Le Quesne, W. J. 1969. A method of selection of characters in numerical taxonomy. *Syst. Zool.* 18:201–205.

—— 1972. Further studies based on the uniquely derived character concept. *Syst. Zool.* 21:281–288.

—— 1974. The uniquely evolved character concept and its cladistic application. *Syst. Zool.* 23:513–517.

Leroi-Gourhan, A. 1975. The flowers found with Shanidar IV, a Neanderthal burial in Iraq. *Science* 190:562–564.

Lersten, N. R. and J. D. Curtis. 1977. Anatomy and distribution of secretory glands and other emergences in *Tofieldia* (Liliaceae). *Ann. Bot.* n.s. 41:879–882.

Les, D. H. and R. L. Stuckey. 1985. The introduction and spread of *Veronica beccabunga* (Scrophulariaceae) in eastern North America. *Rhodora* 87:503–515.

Levin, D. A. 1968. The genome constitutions of eastern North American *Phlox* amphiploids. *Evolution* 22:612–632.

—— 1971a. The origin of reproductive isolating mechanisms in flowering plants. *Taxon* 20:91–113.

—— 1971b. Plant phenolics: An ecological perspective. *Amer. Naturalist* 105:157–181.

—— 1973. The role of trichomes in plant defense. *Quart. Rev. Biol.* 48:3–15.

—— 1975. Somatic cell hybridization: Application in plant systematics. *Taxon* 24:261–270.

—— 1978. The origin of isolating mechanisms in flowering plants. *Evol. Biol.* 11:185–317.

—— 1979a. The nature of plant species. *Science* 204:381–384.

——, ed. 1979b. *Hybridization: An Evolutionary Perspective.* Stroudsburg, Pa.: Dowden, Hutchinson & Ross.

Levin, D. A. and H. W. Kerster. 1974. Gene flow in seed plants. *Evol. Biol.* 7:139–220.

Levin, D. A. and B. A. Schall. 1970. Reticulate evolution in *Phlox* as seen through protein electrophoresis. *Amer. J. Bot.* 57:977–987.

Levin, D. A. and A. C. Wilson. 1976. Rates of evolution in seed plants: Net increase in diversity of chromosome numbers and species numbers through time. *Proc. Natl. Acad. U.S.A.* 73:2086–2090.

Levinton, J. 1988. *Genetics, Paleontology, and Macroevolution.* Cambridge: Cambridge Univ. Press.

Lewin, B. 1987. *Genes.* 3d ed. New York: Wiley.

Lewin, R. A. 1981. The Prochlorophytes. In M. P. Starr, H. Stolp, H. G. Truper, A. Balows, and H. G. Schlegel, eds., *The Prokaryotes: A Handbook on Habitats, Isolation, and Identification of Bacteria,* 1:257–266. Berlin: Springer-Verlag.

Lewis, H. 1955. Specific and infraspecific categories in plants. In *Biological Systematics,* pp. 13–20. 16th Ann. Biol. Colloq., Oregon State Univ., Corvallis.

—— 1957. Genetics and cytology in relation to taxonomy. *Taxon* 6:42–46.

—— 1959. The nature of plant species. *J. Arizona Acad. Sci.* 1:3–7.

—— 1967. The taxonomic significance of autopolyploidy. *Taxon* 16:265–271.

—— 1969. Comparative cytology in systematics [with discussion]. In C. G. Sibley, chm., *Systematic Biology,* pp. 523–535. Washington, D.C.: National Academy of Sciences.

Lewis, H. and M. R. Roberts. 1956. The origin of *Clarkia lingulata. Evolution* 10:126–138.

Lewis, W. H. 1962. Aneusomaty in aneuploid populations of *Claytonia virginica. Amer. J. Bot.* 49:918–928.

—— 1964. Meiotic chromosomes in African Commelinaceae. *Sida* 1:274–293.

—— 1967. Cytocatalytic evolution in plants. *Bot. Rev.* 33:105–115.

—— 1970a. Extreme instability of chromosome number in *Claytonia virginica. Taxon* 19:180–182.

—— 1970b. Chromosomal drift, a new phenomenon in plants. *Science* 168:1115–1116.

—— 1980a. Polyploidy in angiosperms: Dicotyledons. In W. H. Lewis, ed., *Polyploidy: Biological Relevance,* pp. 241–268. New York: Plenum Press.

——, ed. 1980b. *Polyploidy: Biological Relevance.* New York: Plenum Press.

Lewis, W. H. and S. A. Davis. 1962. Cytological observations of *Polygala* in eastern North America. *Rhodora* 64:102–113.

Lewis, W. H., R. L. Oliver, and Y. Suda. 1967. Cytogeography of *Claytonia virginica* and its allies. *Ann. Missouri Bot. Gard.* 54:153–171.

Lewis, W. H. and J. C. Semple. 1977. Geography of *Claytonia virginica* cytotypes. *Amer. J. Bot.* 64:1078–1082.

Lewis, W. H. and Y. Suda. 1968. Karyotypes in relation to classification and phylogeny in *Claytonia. Ann. Missouri Bot. Gard.* 55:64–67.

Lewis, W. H., Y. Suda, and B. MacBryde. 1967. Chromosome numbers of *Claytonia virginica* in the St. Louis, Missouri area. *Ann. Missouri Bot. Gard.* 54:147–152.

Lewis, W. H., P. Vinay, and V. E. Zenger. 1983. *Airborne and Allergenic Pollen of North America.* Baltimore: Johns Hopkins Univ. Press.

Li, H.-L. 1974. Plant taxonomy and the origin of cultivated plants. *Taxon* 23:715–724.

Li, N. and R. C. Jackson. 1961. Cytology of supernumerary chromosomes in *Haplopappus spinulosus* ssp. *cotula. Amer. J. Bot.* 48:419–426.

Li, Y.-H. C. and J. B. Phipps. 1973. Studies in the Arundinelleae (Gramineae). XV: Taximetrics of leaf anatomy. *Canad. J. Bot.* 51:657–680.

Lincoln, D. E. and J. H. Langenheim. 1978. Effect of light and temperature on monoterpenoid yield and composition in *Satureja douglasii. Biochem. Syst. Ecol.* 6:21–32.

Linde-Laurson, I. and R. von Bothmer. 1986. Comparison of the karyotypes of *Psathyrostachyi juncea* and *P. huashanica* (Poaceae) studied by banding techniques. *Pl. Syst. Evol.* 151:203–213.

Lindley, J. 1830a. *An Introduction to the Natural System of Botany.* London: Longman, Rees, Orme, Brown, and Green.

—— 1830b. *An Outline of the First Principles of Botany.* London: Longman.

—— 1830–1840. *The Genera and Species of Orchidaceous Plants.* London: Ridgways.

—— 1832. Book IV. Glossology; or, Of the terms used in botany. In *An Introduction to Botany,* pp. 369–431. London: Longman, Rees, Orme, Brown, Green, & Longman.

—— 1833. *Nixus Plantarum.* London: Ridgway.

Link, H. F. 1798. *Philosophiae Botanicae Novae seu Institutionum Phytographicarum Prodromus* Gottingen.

Linnaeus, C. 1729. *Praeludia Sponsalia Plantarum.* Uppsala. Reprinted, 1746, as *Sponsalia Plantarum.* Stockholm.

—— 1735. *Systema Naturae.* Leiden.

—— 1737a. *Critica Botanica.* Leiden. English ed., 1938. Translated by A. Hort. London: Ray Society.

—— 1737b. *Genera Plantarum.* Leiden.

—— 1751. *Philosophia Botanica.* Stockholm.

—— 1753. *Species Plantarum.* 2 vols. Stockholm.

—— 1754. *Genera Plantarum.* Stockholm.

—— 1787. *The Families of Plants, with Their Natural Characters, According to the Number, Figure,*

Situation, and Proportion of All the Parts of Fructification. Translated anonymously. Lichfield, England: Lichfield Botanical Society.

Lionni, L. 1977. *Parallel Botany.* Translated by Patrick Creagh. New York: Knopf.

Little, F. J., Jr. 1964. The need for a uniform system of biological numericlature. *Syst. Zool.* 13:191–194.

Littlejohn, M. J. 1981. Reproductive isolation: A critical review. In W. R. Atchley and D. S. Woodruff, eds., *Evolution and Speciation: Essays in Honor of M. J. D. White*, pp. 298–334. Cambridge: Cambridge Univ. Press.

Lockhardt, W. R. and J. Liston, eds. 1970. *Methods for Numerical Taxonomy.* Bethesda, Md.: American Society for Microbiology.

Lomax, A. and N. Berkowitz. 1972. The evolutionary taxonomy of culture. *Science* 177:228–239.

Long, R. W. 1973. A biosystematic approach to generic delimitation in *Ruellia* (Acanthaceae). *Taxon* 22:543–555.

Lotsy, J. P. 1925. Species or linneon? *Genetica* 7:487–506.

—— 1931. On the species of the taxonomist in its relation to evolution. *Genetica* 13:1–16.

Lott, J. N. A. 1981. Protein bodies in seeds. *Nord. J.Bot.* 1:421–432.

Louie, A. H. 1983a. Categorical system theory and the phenomenological calculus. *Bull. Math. Biol.* 45:1029–1045.

—— 1983b. Categorical system theory. *Bull. Math. Biol.* 45:1047–1072.

Löve, A. 1954. Cytotaxonomical evaluation of corresponding taxa (with 4 tables). *Vegetatio* 5–6: 212–224.

—— 1960. Biosystematics and classification of apomicts. *Feddes Repert.* 63:136–149.

—— 1962. The biosystematic species concept. *Preslia* 34:127–139.

—— 1963. Cytotaxonomy and generic delimitation. *Regnum Veget.* 27:45–51.

—— 1964. The biological species concept and its evolutionary structure. *Taxon* 13:33–45.

Löve, A. and B. M. Kapoor. 1967a. A chromosome atlas of the collective genus *Rumex. Cytologia* 32:328–342.

—— 1967b. The highest plant chromosome number in Europe. *Svensk Bot. Tidskr.* 61:29–32.

Löve, A. and D. Löve. 1949. The geobotanical significance of polyploidy. I.: Polyploidy and latitude. *Portug. Acta Biol.* ser. A, R. B. Goldschmidt vol.: 273–352.

—— 1974. *Cytotaxonomical Atlas of the Slovenian Flora.* Lehre: Cramer.

—— 1975a. *Cytotaxonomical Atlas of the Arctic Flora.* Vaduz: Cramer.

—— 1975b. *Plant Chromosomes.* Vaduz: Cramer.

Löve, A., D. Löve, and M. Raymond. 1957. Cytotaxonomy of *Carex* section *Capillares. Canad. J. Bot.* 35:715–761.

Löve, D. and L. Nadeau. 1961. The Hutchinson polygraph, a method for simultaneous expression of multiple and variable characters. *Canad. J. Genet. Cytol.* 3:289–294.

Lovtrup, S. 1978. On von Baerian and Haeckelian recapitulation. *Syst. Zool.* 27:348–352.

—— 1979. The evolutionary species: Fact or fiction? *Syst. Zool.* 28:386–392.

Lowrey, T. K. 1986. A biosystematic revision of Hawaiian *Tetramolopium* (Compositae: Astereae). *Allertonia* 4:203–265.

Lubke, R. A. and J. B. Phipps. 1973. Taximetrics of *Loudetia* (Gramineae) based on leaf anatomy. *Canad. J. Bot.* 51:2127–2146.

Lundberg, J. G. 1972. Wagner networks and ancestors. *Syst. Zool.* 21:398–413.

Luteyn, J. L. 1976. A revision of the Mexican-Central American species of *Cavendishia* (Vacciniaceae). *Mem. New York Bot. Gard.* 28:1–138.

Mabry, T. J. 1970. Infraspecific variation of sesquiterpene lactones in *Ambrosia* (Compositae): Applications to evolutionary problems at the populational level. In J. B. Harborne, ed., *Phytochemical Phylogeny*, pp. 269–300. London: Academic Press.

—— 1974. The chemistry of disjunct taxa. In G. Bendz and J. Santesson, eds., *Chemistry in Botanical Classification*, pp. 63–66. Nobel Foundation, Stockholm. New York: Academic Press.

—— 1977. The order Centrospermae. *Ann. Missouri Bot. Gard.* 64:210–220.

Mabry, T. J. and H.-D. Behnke. 1976. Betalains and P-type sieve-element plastids: The systematic position of *Dysphania* R. Br. (Centrospermae). *Taxon* 25:109–111.

Mabry, T. J., H.-D. Behnke, and I. J. Eifert. 1976. Betalains and P-type sieve-element plastids in *Gisekia* L. (Centrospermae). *Taxon* 25:112–114.

Mabry, T. J. and A. S. Dreiding. 1968. The betalains. In T. J. Mabry, R. G. Alston, and V. C. Runeckles, eds., *Recent Advances in Phytochemistry*, 1:145–160. New York: Appleton-Century-Crofts.

Mabry, T. J., K. R. Markham and M. B. Thomas. 1970. *The Systematic Identification of Flavonoids.* New York: Springer Verlag.

McComb, J. A. 1975. Is intergeneric hybridization in the Leguminosae possible? *Euphytica* 24:497–502.

MacDonald, N. 1983. *Trees and Networks in Biological Models.* Chichester: Wiley.

McDowall, R. M. 1978. Generalized tracks and dispersal in biogeography. *Syst. Zool.* 27:88–104.

McGuire, R. F. 1984. A numerical taxonomic study of *Nostoc* and *Anabaena. J. Phycol.* 20:454–460.

Machol, R. E. and R. Singer. 1971. Bayesian analysis of generic relations in Agaricales. *Nova Hedwigia* 21:753–787.

McIntosh, R. P. 1985. *The Background of Ecology: Concept and Theory.* Cambridge: Cambridge Univ. Press.

Macior, L. W. 1971. Co-evolution of plants and animals—systematic insights from plant-insect interactions. *Taxon* 20:17–28.

—— 1974. Behavioral aspects of coadaptations between flowers and insect pollinators. *Ann. Missouri Bot. Gard.* 61:760–769.

—— 1986a. Pollination ecology and endemic adaptation of *Pedicularis howellii* Gray (Scrophulariaceae). *Pl. Sp. Biol.* 1:163–172.

—— 1986b. Pollinator ecology and endemism of *Pedicularis pulchella* Pennell (Scrophulariaceae). *Pl. Sp. Biol.* 1:173–180.

McKelvey, B. 1982. *Organizational Systematics: Taxonomy, Evolution, Classification.* Berkeley: Univ. California Press.

McKusick, V. A. 1969. On lumpers and splitters, or the nosology of genetic disease. *Perspect. Biol. Med.* 12:298–312.

McMorris, F. R. 1975. Compatibility criteria for cladistic and qualitative taxonomic characters. In G. F. Estabrook, ed., *Proceedings of the Eighth International Conference on Numerical Taxonomy,* pp. 399–415. San Francisco: Freeman.

—— 1977. On the compatibility of binary qualitative taxonomic characters. *Bull. Math. Biol.* 39:133–138.

—— 1985. Axioms for consensus functions on undirected phylogenetic trees. *Math. Biosci.* 74:17–21.

McMorris, F. R., D. B. Meronk, and D. A. Neumann. 1983. A view of some consensus methods for trees. In J. Felsenstein, ed., *Numerical Taxonomy,* pp. 122–126. Berlin: Springer-Verlag.

McMullen, C. K. 1987. Breeding systems of selected Galapagos Islands angiosperms. *Amer. J. Bot.* 74:1706–1708.

McNair, J. B. 1934. The evolutionary status of plant families in relation to some chemical properties. *Amer. J. Bot.* 21:427–452.

—— 1935. Angiosperm phylogeny on a chemical basis. *Bull. Torrey Bot. Club* 62:515–532.

—— 1945. Some comparisons of chemical ontogeny with chemical phylogeny in vascular plants. *Lloydia* 8:145–169.

McNeill, J. 1972. The hierarchical ordering of characters as a solution to the dependent character problem in numerical taxonomy. *Taxon* 21:71–82.

—— 1974. The handling of character variation in numerical taxonomy. *Taxon* 23:699–705.

—— 1975. A generic revision of Portulacaceae tribe Montieae using techniques of numerical taxonomy. *Canad. J. Bot.* 53:789–809.

—— 1979a. Purposeful phenetics. *Syst. Zool.* 28:465–482.

—— 1979b. Structural value: A concept used in the construction of taxonomic classifications. *Taxon* 28:481–504.

McVaugh, R. 1945. The genus *Triodanis* Rafinesque, and its relationships to *Specularia* and *Campanula. Wrightia* 1:13–52.

—— 1972a. Botanical exploration in Nueva Galicia, Mexico from 1790 to the present time. *Contr. Univ. Michigan Herb.* 9:205–357.

—— 1972b. Compositarum Mexicanarum pugillus. *Contr. Univ. Michigan Herb.* 9:359–484.

—— 1982. The new synantherology vs. *Eupatorium* in Nueva Galicia. *Contr. Univ. Michigan Herb.* 15:181–190.

McWilliams, E. L. 1970. Comparative rates of CO_2 uptake and acidification in Bromeliaceae, Orchidaceae, and Euphorbiaceae. *Bot. Gaz.* 131:285–290.

Maddison, W. P., M. J. Donoghue, and D. R. Maddison. 1984. Outgroup analysis and parsimony. *Syst. Zool.* 33:83–103.

Maheshwari, P. 1948. The angiosperm embryo sac. *Bot. Rev.* 14:1–56.

—— 1950. *An Introduction to the Embryology of Angiosperms.* New York: McGraw-Hill.

——, ed. 1963. *Recent Advances in the Embryology of Angiosperms.* Delhi: International Society of Plant Morphologists.

Maheshwari, P., B. M. Johri, and S. N. Dixit. 1957. The floral morphology and embryology of the Loranthoideae (Loranthaceae). *J. Madras Univ.* ser. B, 27:121–136.

Maheswari Devi, H. 1975. Embryology of Jasminums and its bearing on the composition of Oleaceae. *Acta Bot. Indica* 3:52–61.

Maheswari Devi, H. and K. Lakshminarayana. 1977. Embryology of *Oxystelma esculentum*. *Phytomorphology* 27:59–67.

Maheswari Devi, H. and K. L. Narayana. 1975. Embryology of two cultivars of *Nerium indicum* Mill. *Curr. Sci.* 44:641–642.

Maheswari Devi, H. and T. Pullaiah. 1976. Embryological investigations in the Melampodiinae. I. *Melampodium divaricatum. Phytomorphology* 26:77–86.

Mahlberg, P. G. 1975. Evolution of the laticifer in *Euphorbia* as interpreted from starch grain morphology. *Amer. J. Bot.* 62:577–583.

Mahlberg, P. G. and J. Pleszczynska. 1983. Phylogeny of *Euphorbia* interpreted from sterol composition of latex. In J. Felsenstein, ed., *Numerical Taxonomy*, pp. 500–504. Berlin: Springer-Verlag.

Mahlberg, P.G., J. Pleszczynska, W. Rauh, and E. Schnepf. 1983. Evolution of succulent *Euphorbia* as interpreted from latex composition. *Bothalia* 14:857–863.

Maksymowych, R. 1973. *Analysis of Leaf Development.* Cambridge: Cambridge Univ. Press.

Malecka, J. 1971a. Processes of degeneration in the anthers' tapetum of two male-sterile species of *Taraxacum. Act. Biol. Cracov., Bot.* 14:1–10.

—— 1971b. Cyto-taxonomical and embryological investigations on a natural hybrid between *Taraxacum kok-saghyz* Rodin and *T. officinale* Web. and their putative parent species. *Acta Biol. Cracov., Bot.* 14:179–197.

—— 1973. Problems of the mode of reproduction in microspecies of *Taraxacum* section *Palustria* Dahlstedt. *Acta Biol. Cracov., Bot.* 16:37–84.

Malmgren, B. A., W. A. Berggren, and G. P. Lohmann. 1983. Evidence for punctuated gradualism in the Late Neogene *Globorotalia tumida* lineage of planktonic foraminifera. *Paleobiology* 9:377–389.

—— 1984. Species formation through punctuated gradualism in planktonic foraminifera. *Science* 225:317–318.

Malpighi, M. 1675–1679. *Anatome Plantarum.* London.

Manischewitz, J. R. 1973. Prediction and alternative procedures in numerical taxonomy. *Syst. Zool.* 22:176–184.

Manitz, H. 1975. Friedrich Ehrhart's botanische Publikationen. *Taxon* 24:469–474.

Manton, I. 1958. The concept of aggregate species. *Uppsala Univ. Arsskr. 1958*(6):104–112.

Marchi, E. and R. I. C. Hansell. 1973. A framework for systematic zoological studies with game theory. *Math. Biosci.* 16:31–58.

Markgraf, V. and H. L. D'Antoni. 1978. *Pollen Flora of Argentina: Modern Spore and Pollen Types of Pteridophyta, Gymnospermae, and Angiospermae.* Tucson: Univ. Arizona Press.

Markham, K. R. 1982. *Techniques of Flavonoid Identification.* London: Academic Press.

Markham, K. R. and L. J. Porter. 1969. Flavonoids in the green algae (Chlorophyta). *Phytochemistry* 8:1777–1781.

—— 1978. Chemical constituents of the bryophytes. *Progr. Phytochem.* 5:181–272.

Marks, G. E. 1966. The origin and significance of intraspecific polyploidy: Experimental evidence from *Solanum chacoense. Evolution* 20:552–557.

Marticorena, C. and O. Parra. 1974. Morfología de los granos de polen y posición sistemática de *Anisochaeta* DC., *Chionopappus* Benth., *Feddea* Urb. y *Gochnatia gomeriflora* Gray (Compositae). *Bol. Soc. Biol. Concepción* 47:187–197.

Martin, A. C. and W. D. Barkley. 1961. *Seed Identification Manual.* Berkeley: Univ. California Press.

Martin, J. T. and B. E. Juniper. 1970. *The Cuticles of Plants.* London: Arnold.

Martin, P. G., D. Boulter, and D. Penny. 1985. Angiosperm phylogeny studied using sequences of five macromolecules. *Taxon* 34:393–400.

Martin, P. G. and J. M. Dowd. 1986. A phylogenetic tree for some monocotyledons and gymnosperms derived from protein sequences. *Taxon* 35:469–475.

Martin, P. G. and A. C. Jennings. 1983. The study of plant phylogeny using amino acid sequences of ribulose-1,5-bisphosphate carboxylase. I. Biochemical methods and the patterns of variability. *Austral. J. Bot.* 31:395–409.

Martin, P. G. and S. J. L. Stone. 1983. The study of plant phylogeny using amino acid sequences of ribulose-1,5-bisphosphate carboxylase. II. The analysis of small subunit data to form phylogenetic trees. *Austral. J. Bot.* 31:411–418.

Martin, P. S. 1969. Pollen analysis and the scanning electron microscope. In *Scanning Electron Microscopy 1969*, pp. 89–103. Chicago: ITT Research Institute.

Martin, P. S. and C. M. Drew. 1969. Scanning electron photomicrographs of southwestern pollen grains. *J. Arizona Acad. Sci.* 5:147–176.

—— 1970. Additional scanning electron photomicrographs of southwestern pollen grains. *J. Arizona Acad. Sci.* 6:140–161.

Marx, H. and G. B. Rabb. 1972. Phyletic analysis of fifty characters of advanced snakes. *Fieldiana, Zool.* 63:i–viii, 1–321.

Mascherpa, J.-M. 1975. Taxonomie numerique: Chimie taxonomique et ordinateur au service de la systematique. *Saussurea* 6:171–185.

Mascherpa, J.-M. and G. Boquet. 1981. Deux programmes interactifs de détermination automatique. Un idée, un but. *Candollea* 36:463–483.

Maslin, T. P. 1952. Morphological criteria of phyletic relationships. *Syst. Zool.* 1:49–70.

Mason, H. L. 1950. Taxonomy, systematic botany and biosystematics. *Madroño* 10:193–208.

—— 1953. Plant geography in the delimitation of genera: The role of plant geography in taxonomy. *Chron. Bot.* 14:154–159.

Mathew, L. and G. L. Shah. 1984. Crystals and their taxonomic significance in some Verbenaceae. *Bot. J. Linn. Soc.* 88:279–289.

Mathew, P. M. and O. Philip. 1986. The distribution and systematic significance of pollen nuclear number in the Rubiaceae. *Cytologia* 51:117–124.

Matthews, J. F. and P. A. Levins. 1986. The systematic significance of seed morphology in *Portulaca* (Portulacaceae) under scanning electron microscopy. *Syst. Bot.* 11:302–308.

Mauseth, J. D. 1988. *Plant Anatomy*. Menlo Park, Calif.: Benjamin/Cummings.

Maynard Smith, J. 1976. Evolution and the theory of games. *Amer. Scientist* 64: 41–45.

—— 1982. *Evolution and the Theory of Games*. Cambridge: Cambridge Univ. Press.

Mayr, E. 1931. Birds collected during the Whitney South Sea Expedition. XII: Notes on *Halcyon chloris* and some of its subspecies. *Amer. Mus. Novitates* 469:1–10.

—— 1940. Speciation phenomena in birds. *Amer. Naturalist* 74:249–278.

—— 1942. *Systematics and the Origin of Species from the Viewpoint of a Zoologist*. New York: Columbia Univ. Press.

—— 1954. Notes on nomenclature and classification. *Syst. Zool.* 3:86–89.

—— 1955. Karl Jordan's contribution to current concepts in systematics and evolution. *Trans. Roy. Entomol. Soc. London* 107:45–66.

—— 1957a. Preface. In E. Mayr, ed., *The Species Problem*, pp. iii–v. Washington, D.C.: American Association for the Advancement of Science.

—— 1957b. Species concepts and definitions. In E. Mayr, ed., *The Species Problem*, pp. 1–22. Washington, D.C.: American Association for the Advancement of Science.

——, ed. 1957c. *The Species Problem*. Washington, D.C.: American Association for the Advancement of Science.

—— 1963. *Animal Species and Evolution*. Cambridge, Mass.: Harvard Univ. Press.

—— 1964. The new systematics. In C. A. Leone, ed., *Taxonomic Biochemistry and Serology*, pp. 13–32. New York: Ronald Press.

—— 1965. Numerical phenetics and taxonomic theory. *Syst. Zool.* 14:73–97.

—— 1966. The proper spelling of taxonomy. *Syst. Zool.* 15:88.

—— 1968a. The role of systematics in biology. *Science* 159:595–599.

—— 1968b. Illiger and the biological species concept. *J. Hist. Biol.* 1:163–178.

—— 1969a. The biological meaning of species. *Biol. J. Linn. Soc.* 1:311–320.

—— 1969b. Introduction: The role of systematics in biology. In C. G. Sibley, chm., *Systematic Biology*, pp. 4–15. Washington, D.C.: National Academy of Sciences.

—— 1969c. *Principles of Systematic Zoology*. New York: McGraw-Hill.

—— 1974a. The challenge of diversity. *Taxon* 23:3–9.

—— 1974b. Cladistic analysis or cladistic classification? *Z. Zool. Syst. Evolut.-Forsch.* 12:94–128.

—— 1976. Is the species a class or an individual? *Syst. Zool.* 25:192.

—— 1978. Origin and history of some terms in systematic and evolutionary biology. *Syst. Zool.* 27:83–88.

—— 1981. Biological classification: Toward a synthesis of opposing methodologies. *Science* 214:510–516.

—— 1982. *The Growth of Biological Thought: Diversity, Evolution, and Inheritance*. Cambridge, Mass.: Harvard Univ. Press.

—— 1985. Darwin and the definition of phylogeny. *Syst. Zool.* 34:97–98.

Maze, J., L. R. Bohm, and C. E. Beil. 1972. Studies on the relationships and evolution of supraspecific taxa utilizing developmental data. I: *Stipa lemmonii* (Gramineae). *Canad. J. Bot.* 50:2327–2352.

Maze, J., W. H. Parker, and G. E. Bradfield. 1981. Generation-dependent patterns of variation and population differentiation in *Abies amabilis* and *A. lasiocarpa* (Pinaceae) from north-coastal British Columbia. *Canad. J. Bot.* 59:275–282.

Meacham, C. A. 1980. Phylogeny of the Berberidaceae with an evaluation of classifications. *Syst. Bot.* 5:149–172.

—— 1981. A manual method for character compatibility analysis. *Taxon* 30:591–600.

—— 1984a. Evaluating characters by character compatibility analysis. In T. Duncan and T. F. Stuessy, eds., *Cladistics: Perspectives on the Reconstruction of Evolutionary History*, pp. 152–165. New York: Columbia Univ. Press.

—— 1984b. The role of hypothesized direction of characters in the estimation of evolutionary history. *Taxon* 33:26–38.

—— 1986. More about directed characters: A reply to Donoghue and Maddison. *Taxon* 35:538–540.

Meacham, C. A. and T. Duncan. 1987. The necessity of convex groups in biological classification. *Syst. Bot.* 12:78–90.

Meacham, C. A. and G. F. Estabrook. 1985. Compatibility methods in systematics. *Ann. Rev. Ecol. Syst.* 16:431–446.

Meeuse, A. D. J. 1964. A critique of numerical taxonomy. In V. H. Heywood and J. McNeill, eds., *Phenetic and Phylogenetic Classification*, pp. 115–121. London: Systematics Association.

—— 1966. The homology concept in phytomorphology—some moot points. *Acta Bot. Neerl.* 15:451–476.

—— 1982. Cladistics, wood anatomy and angiosperm phylogeny—a challenge. *Acta Bot. Neerl.* 31:345–354.

Meeuse, B. and S. Morris. 1984. *The Sex Life of Flowers*. New York: Facts on File Publications.

Meikle, R. D., comp. 1957. "What is the subspecies?" *Taxon* 6:102–105.

Meisel, W. S. 1972. *Computer-Oriented Approaches to Pattern Recognition*. New York: Academic Press.

Melchert, T. E. 1966. Chemo-demes of diploid and tetraploid *Thelesperma simplicifolium* (Heliantheae, Coreopsidineae). *Amer. J. Bot.* 53:1015–1020.

Melchior, H. 1964. *A. Engler's Syllabus der Pflanzenfamilien*. Vol. 2: *Angiospermae*. 12th ed. Berlin: Borntraeger.

Melkó, E. 1976. Numerical taxonomic studies on *Iris pumila* L. by cluster analysis. *Acta Bot. Acad. Sci. Hung.* 22:403–414.

Melville, R. 1976. The terminology of leaf architecture. *Taxon* 25:549–561.

—— 1981. Surface tension, diffusion and the evolution and morphogenesis of pollen aperture patterns. *Pollen et Spores* 23:179–203.

Mendel, G. 1866. Versuche über Pflanzenhybriden. *Verh. Naturf. Vereins Brunn* 4:3–47.

Mendelson, D. and D. Zohary. 1972. Behaviour and transmission of supernumerary chromosomes in *Aegilops speltoides*. *Heredity* 29:329–339.

Menzel, M. Y. 1962. Pachytene chromosomes of the intergeneric hybrid *Lycopersicon esculentum* x *Solanum lycopersicoides*. *Amer. J. Bot.* 49:605–615.

Merxmüller, H. 1970. Provocation of biosystematics. *Taxon* 19:140–145.

—— 1972. Systematic botany—an unachieved synthesis. *Biol. J. Linn. Soc.* 4:311–321.

Metcalfe, C. R. 1954. An anatomist's views on angiosperm classification. *Kew Bull.* 9:427–440.

—— 1960. *Anatomy of the Monocotyledons*. Vol. 1: *Gramineae*. Oxford: Clarendon Press.

—— 1967. Some current problems in systematic anatomy. *Phytomorphology* 17:128–132.

—— 1968. Current developments in systematic plant anatomy. In V. H. Heywood, ed., *Modern Methods in Plant Taxonomy*, pp. 45–57. London: Academic Press.

—— 1971. *Anatomy of the Monocotyledons*. Vol. 5: *Cyperaceae*. Oxford: Clarendon Press.

—— 1979. History of systematic anatomy, part I: General anatomy. In C. R. Metcalfe and L. Chalk, eds., *Anatomy of the Dicotyledons*. Vol. 1: *Systematic Anatomy of Leaf and Stem, with a Brief History of the Subject*, pp. 1–4. 2d ed. Oxford: Clarendon Press.

Metcalfe, C. R. and L. Chalk. 1950. *Anatomy of the Dicotyledons: Leaves, Stems, and Wood in Relation to Taxonomy with Notes on Economic Uses*. 2 vols. Oxford: Clarendon Press.

—— 1979. *Anatomy of the Dicotyledons*. Vol. 1: *Systematic Anatomy of Leaf and Stem, with a Brief History of the Subject*. 2d ed. Oxford: Clarendon Press.

—— 1983. *Anatomy of the Dicotyledons*. Vol. 2: *Wood Structure and Conclusion of the General Introduction*. 2d ed. Oxford: Clarendon Press.

Meyer, D. E. 1964. Zum Aussagewert des Chromosomenbildes für die Systematik. *Bot. Jahrb. Syst.* 83:107–114.

Meyer-Abich, A. 1926. *Logik der Morphologie*. Berlin: Springer.

Mez, C. 1922. Anleitung zu sero-diagnostischen Untersuchungen für Botaniker. *Bot. Arch.* 1:177–200.

Michener, C. D. 1963. Some future developments in taxonomy. *Syst. Zool.* 12:151–172.

—— 1964. The possible use of uninominal nomenclature to increase the stability of names in biology. *Syst. Zool.* 13:182–190.

—— 1970. Diverse approaches to systematics. *Evol. Biol.* 4:1–38.

—— 1978. Dr. Nelson on taxonomic methods. *Syst. Zool.* 27:112–118.

Michener, C.D. and R. R. Sokal. 1957. A quantitative approach to a problem in classification. *Evolution* 11:130–162.

—— 1966. Two tests of the hypothesis of nonspecificity in the *Hoplitis* complex (Hymenoptera: Megachilidae). *Ann. Entomol. Soc. Amer.* 59:1211–1217.

Mickel, J. T. 1962. A monographic study of the fern genus *Anemia*, subgenus *Coptophyllum. Iowa State J. Sci.* 36:349–482.

Mickevich, M. F. 1978. Taxonomic congruence. *Syst. Zool.* 27:143–158.

—— 1980. Taxonomic congruence: Rohlf and Sokal's misunderstanding. *Syst. Zool.* 29:162–176.

—— 1981. Quantitative phylogenetic biogeography. In V. A. Funk and D. R. Brooks, eds., *Advances in Cladistics*, pp. 209–222. New York: New York Botanical Garden.

Miescher, F. 1871. Ueber die chemische Zusammensetzung der Eiterzellen. *Hoppe-Seyler's Med.-Chem. Untersuch.* 4:441–460.

Miller, H. E., T. J. Mabry, B. L. Turner, and W. W. Payne. 1968. Infraspecific variation of sesquiterpene lactones in *Ambrosia psilostachya* (Compositae). *Amer. J. Bot.* 55:316–324.

Miller, J. M. 1976. Variation in populations of *Claytonia perfoliata* (Portulacaceae). *Syst. Bot.* 1:20–34.

Miller, J. M., K. L. Chambers, and C. E. Fellows. 1984. Cytogeographic patterns and relationships in the *Claytonia sibirica* complex (Portulacaceae). *Syst. Bot.* 9:266–271.

Milne, R. G. 1985. Alternatives to the species concept for virus taxonomy. *Intervirology* 24:94–98.

Mirov, N. T. 1948. The terpenes (in relation to the biology of genus *Pinus*). *Ann. Rev. Biochem.* 17:521–540.

Mishler, B. D. 1985a. The phylogenetic relationships of *Tortula*—an SEM survey and a preliminary cladistic analysis. *Bryologist* 88:388–403.

—— 1985b. The morphological, developmental, and phylogenetic bases of species concepts in bryophytes. *Bryologist* 88:207–214.

Mishler, B. D. and S. P. Churchill. 1984. A cladistic approach to the phylogeny of the "bryophytes." *Brittonia* 36:406–424.

—— 1985a. Cladistics and the land plants: A response to Robinson. *Brittonia* 37:212–285.

—— 1985b. Transition to a land flora: Phylogenetic relationshps of the green algae and bryophytes. *Cladistics* 1:305–328.

—— 1987. Transition to a land flora: A reply. *Cladistics* 3:65–71.

Mishler, B. D. and M. J. Donoghue. 1982. Species concepts: A case for pluralism. *Syst. Zool.* 31:491–503.

Miyamoto, M. M. 1985. Consensus cladograms and general classifications. *Cladistics* 1:186–189.

Mohan Ram, H. Y. and M. Wadhi. 1964. Endosperm in Acanthaceae. *Phytomorphology* 14:388–413.

Mohl, H. von. 1835. Sur la structure et les formes des grains de pollen. *Ann. Sci. Nat. Bot.* 3:148–180, 220–236, 304–346.

—— 1851. *Grundzüge der Anatomie und Physiologie der Vegetabilischen Zelle.* Braunschweig: Vieweg.

Molisch, H. and K. Hofler. 1961. *Anatomie der Pflanze.* Jena: Fischer.

Mooney, H. A. and W. A. Emboden, Jr. 1968. The relationship of terpene composition, morphology, and distribution of populations of *Bursera microphylla* (Burseraceae). *Brittonia* 20:44–51.

Moore, D. M. 1968. The karyotype in taxonomy. In V. H. Heywood, ed., *Modern Methods in Plant Taxonomy*, pp. 61–75. London: Academic Press.

—— 1978. The chromosomes and plant taxonomy. In H. E. Street, ed., *Essays in Plant Taxonomy*, pp. 39–56. London: Academic Press.

—— 1984. Taxonomy and geography. In V. H. Heywood and D. M. Moore, eds., *Current Concepts in Plant Taxonomy*, pp. 219–234. London: Academic Press.

Moore, G. W. and M. Goodman. 1977. Alignment statistic for identifying related protein sequences. *J. Mol. Evol.* 9:121–130.

Moore, I. M. 1954. Nomenclatorial treatment of specific and infraspecific caegories. *Syst. Zool.* 3:90–91.

Moore, P. D. and J. A. Webb. 1978. *An Illustrated Guide to Pollen Analysis.* London: Hodder and Stoughton.

Moore, R. J., ed. 1970–1977. Index to plant chromosome numbers [1968–1974]. *Regnum Veg.* 68:1–119; 77:1–116; 84:1–138; 91:1–108; 96:1–257.

Mooring, J. S. 1965. Chromosome studies in *Chaenactis* and *Chamaechaenactis* (Compositae; Helenieae). *Brittonia* 17:17–25.

Morgan, T. H. 1911. An attempt to analyze the constitution of the chromosomes on the basis of sex-limited inheritance in *Drosophila. J. Exper. Zool.* 11:365–413.

Morishima, H. 1969. Phenetic similarity and phylogenetic relationships among strains of *Oryza perennis*, estimated by methods of numerical taxonomy. *Evolution* 23:429–443.

Morse, L. E. 1971. Specimen identification and key construction with time-sharing computers. *Taxon* 20:269–282.

Morton, A. G. 1981. *History of Botanical Science: An Account of the Development of Botany from Ancient Times to the Present Day.* London: Academic Press.

Mosquin, T. 1964. Chromosomal repatterning in *Clarkia rhomboidea* as evidence for post-Pleistocene changes in distribution. *Evolution* 18:12–25.

—— 1966. Toward a more useful taxonomy for chromosomal races. *Brittonia* 18:203–214.

—— 1971. Systematics as an educational and political force. *BioScience* 21:1166–1170.

Mosquin, T. and D. E. Hayley. 1966. Chromosome numbers and taxonomy of some Canadian arctic plants. *Canad. J. Bot.* 44:1209–1218.

Moss, W. W. 1971. Taxonomic repeatability: An experimental approach. *Syst. Zool.* 20:309–330.

—— 1972. Some levels of phenetics. *Syst. Zool.* 21:236–239.

—— 1978. Numerical taxonomy and evolution. *Proc. Acad. Nat. Sci. Philadelphia* 129:87–98.

—— 1979. Phenetic approaches to classification. *Amer. Zoologist* 19:1217–1223.

Moss, W. W. and J. A. Henrickson, Jr. 1973. Numerical taxonomy. *Ann. Rev. Entomol.* 18:227–258.

Moss, W. W. and D. M. Power. 1975. Semi-automatic data recording. *Syst. Zool.* 24:199–208.

Mueller, L. E., P. H. Carr, and W. E. Loomis. 1954. The submicroscopic structure of plant surfaces. *Amer. J. Bot.* 41:593–600.

Muhammad, A. F. and R. Sattler. 1982. Vessel structure of *Gnetum* and the origin of angiosperms. *Amer. J. Bot.* 69:1004–1021.

Muir, J. W. 1968. The definition of taxa. *Syst. Zool.* 17:345.

Mulcahy, D. L. 1965. Interpretation of crossing diagrams. *Rhodora* 67:146–154.

—— ed. 1975. *Gamete Competition in Plants and Animals.* Amsterdam: North-Holland.

Mulcahy, D. L. and E. Ottaviano, eds. 1983. *Pollen: Biology and Implications for Plant Breeding.* New York: Elsevier.

Müller, H. L. H. 1879. Weitere Beobachtungen über Befruchtung der Blumen durch Insekten. II. *Verh. Naturhist. Vereines Preuss. Rheinl. Westphalens* 36:198–268.

—— 1881. *Alpenblumen, ihre Befruchtung durch Insekten und ihre Anpassungen an dieselben.* Leipzig: Engelmann.

—— 1883. *The Fertilisation of Flowers.* Translated and edited by D'Arcy Thompson. London: Macmillan.

Muller, J. 1969. A palynological study of the genus *Sonneratia* (Sonneratiaceae). *Pollen et Spores* 11:223–298.

—— 1973. Pollen morphology of *Barringtonia calyptrocalyx* K. Sch. (Lecythidaceae). *Grana* 13:29–44.

—— 1978. New observations on pollen morphology and fossil distribution of the genus *Sonneratia* (Sonneratiaceae). *Rev. Palaeobot. Palynol.* 26:277–300.

—— 1979. Form and function in angiosperm pollen. *Ann. Missouri Bot. Gard.* 66:593–632.

—— 1981. Fossil pollen records of extant angiosperms. *Bot. Rev.* 47:1–142.

—— 1984. Significance of fossil pollen for angiosperm history. *Ann. Missouri Bot. Gard.* 71:419–443.

Mulligan, G. A. 1971a. Cytotaxonomic studies of the closely allied *Draba cana, D. cinerea,* and *D. groenlandica* in Canada and Alaska. *Canad. J. Bot.* 49:89–93.

—— 1971b. Cytotaxonomic studies of *Draba* species of Canada and Alaska: *D. ventosa, D. ruaxes,* and *D. paysonii. Canad. J. Bot.* 49:1455–1460.

Muniyamma, M. and J. B. Phipps. 1979. Meiosis and polyploidy in Ontario species of *Crataegus* in relation to their systematics. *Canad. J. Genet. Cytol.* 21:231–241.

Müntzing, A. 1969. On the methods of experimental taxonomy. *Amer. J. Bot.* 56:791–798.

Murray, D. R., ed. 1987. *Seed Dispersal.* Sydney: Academic Press.

Musselman, L. J. and W. F. Mann, Jr. 1976. A survey of surface characteristics of seeds of Scrophulariaceae and Orobanchaceae using scanning electron microscopy. *Phytomorphology* 26:370–378.

Myers, G. S. 1952. The nature of systematic biology and of a species description. *Syst. Zool.* 1:106–111.

Myers, N. 1980. *Conversion of Tropical Moist Forests.* Washington, D.C.: National Academy of Sciences.

Myint, T. 1966. Revision of the genus *Stylisma* (Convolvulaceae). *Brittonia* 18:97–117.

Nagl, W., V. Hemleben, and F. Ehrendorfer, eds. 1979. *Genome and Chromatin: Organization, Evolution, Function.* London: Academic Press.

Nair, P. K. K. 1965. *Pollen Grains of Western Himalayan Plants.* New York: Asia Publishing House.

—— 1966. *Essentials of Palynology.* Bombay: Asia Publishing House.

—— 1970. *Pollen Morphology of Angiosperms: A Historical and Phylogenetic Study.* Lucknow: Scholar Publishing House; Delhi: Vikas.

Nairn, A. E. M., ed. 1961. *Descriptive Palaeoclimatology.* New York: Wiley.

Nanzetta, P. and G. E. Strecker. 1971. *Set Theory and Topology.* Tarrytown-on-Hudson, N.Y.: Bogden & Quigley.

Nastansky, L., S. M. Selkow, and N. F. Stewart. 1974. An improved solution to the generalized Camin-Sokal model for numerical cladistics. *J. Theor. Biol.* 48:413–424.

Natesh, S. and M. A. Rau. 1984. The embryo. In B. M. Johri, ed., *Embryology of Angiosperms*, pp. 377–443. Berlin: Springer-Verlag.

Nawaschin, S. 1898. Resultate einer Revision der Befruchtungsvorgange bei *Lilium martagon* und *Fritillaria tenella. Bull. Acad. Imp. Sci. Saint. Petersbourg* ser. 5, 9:377–382.

Naylor, B. G. 1982. Vestigial organs are evidence of evolution. *Evol. Theory* 6:91–96.

Neff, N. A. 1986. A rational basis for a priori character weighting. *Syst. Zool.* 35:110–123.

Neff, N. A. and L. F. Marcus. 1980. *A Survey of Multivariate Methods for Systematics.* New York: American Museum of Natural History.

Neff, N. A. and G. R. Smith. 1979. Multivariate analysis of hybrid fishes. *Syst. Zool.* 28:176–196.

Nei, M. 1972. Genetic distance between populations. *Amer. Naturalist* 106:283–292.

—— 1987. *Molecular Evolutionary Genetics.* New York: Columbia Univ. Press.

Nelson, C. H. and G. S. Van Horn. 1975. A new simplified method for constructing Wagner networks and the cladistics of *Pentachaeta* (Compositae, Astereae). *Brittonia* 27:362–372.

Nelson, G. J. 1970. Outline of a theory of comparative biology. *Syst. Zool.* 19:373–384.

—— 1971. "Cladism" as a philosophy of classification. *Syst. Zool.* 20:373–376.

—— 1972a. Phylogenetic relationship and classification. *Syst. Zool.* 21:227–231.

—— 1972b. Comments on Hennig's "Phylogenetic Systematics" and its influence on ichthyology. *Syst. Zool.* 21:364–374.

—— 1973a. The higher-level phylogeny of vertebrates. *Syst. Zool.* 22:87–91.

—— 1973b. Comments on Leon Croizat's biogeography. *Syst. Zool.* 22:312–320.

—— 1973c. Classification as an expression of phylogenetic relationships. *Syst. Zool.* 22:344–359.

—— 1978. Ontogeny, phylogeny, paleontology, and the biogenetic law. *Syst. Zool.* 27:324–345.

—— 1979. Cladistic analysis and synthesis: Principles and definitions, with a historical note on Adanson's *Familles des Plantes* (1763–1764). *Syst. Zool.* 28:1–21.

—— 1984. Cladistics and biogeography. In T. Duncan and T. F. Stuessy, eds., *Cladistics: Perspectives on the Reconstruction of Evolutionary History*, pp. 273–293. New York: Columbia Univ. Press.

—— 1985. Outgroups and ontogeny. *Cladistics* 1:29–45.

Nelson, G. J. and N. I. Platnick. 1978. The perils of plesiomorphy: Widespread taxa, dispersal, and phenetic biogeography. *Syst. Zool.* 27:474–477.

—— 1981. *Systematics and Biogeography: Cladistics and Vicariance.* New York: Columbia Univ. Press.

Nelson, G. J. and D. E. Rosen, eds. 1981. *Vicariance Biogeography: A Critique.* New York: Columbia Univ. Press.

Nesom, G. L. 1983. *Galax* (Diapensiaceae): Geographic variation in chromosome number. *Syst. Bot.* 8:1–14.

Nesom, G. L. and V. M. Bates. 1984. Reevaluations of infraspecific taxonomy in *Polygonella* (Polygonaceae). *Brittonia* 36:37–44.

Neticks, F. G. 1978. Derived characters I have known. *Syst. Zool.* 27:238–239.

Neumann, D. A. 1983. Faithful consensus methods for *n*-trees. *Math. Biosci.* 63:271–287.

Newcomb, W. 1973a. The development of the embryo sac of sunflower *Helianthus annuus* before fertilization. *Canad. J. Bot.* 51:863–878.

—— 1973b. The development of the embryo sac of sunflower *Helianthus annuus* after fertilization. *Canad. J. Bot.* 51:879–890.

Newell, C. A. and T. Hymowitz. 1978. Seed coat variation in *Glycine* Willd. subgenus *Glycine* (Leguminosae) by SEM. *Brittonia* 30:76–88.

Newell, I. M. 1970. Construction and use of tabular keys. *Pacific Insects* 12:25–37.

—— 1972. Tabular keys: Further notes on their construction and use. *Trans. Connecticut Acad. Arts* 44:257–267.

—— 1976. Construction and use of tabular keys: Addendum. *Syst. Zool.* 25:243–250.

Newman, A. A., ed. 1972. *Chemistry of Terpenes and Terpenoids.* London: Academic Press.

Newton, L. E. 1972. Taxonomic use of the cuticular surface features in the genus *Aloe* (Liliaceae). *Bot. J. Linn. Soc.* 65:335–339.

Niehaus, T. F. 1971. A biosystematic study of the genus *Brodiaea* (Amaryllidaceae). *Univ. California Publ. Bot.* 60:i–vi, 1–66.

Niklas, K. J. 1980. Commentary, in panel discussion on cladistics and plant systematics. *Syst. Bot.* 5:227–228.

Niklas, K. J. and P. G. Gensel. 1978. Chemotaxonomy of some Paleozoic vascular plants. Part III. Cluster configurations and their bearing on taxonomic relationships. *Brittonia* 30:216–232.

Niklas, K. J. and D. E. Giannasi. 1977. Flavonoids and other chemical constituents of fossil Miocene *Zelkova* (Ulmaceae). *Science* 196:877–878.

Nilsson, S., ed. 1973–continuing. *World Pollen and Spore Flora.* Stockholm: Almqvist & Wiksell.

Nilsson, S. and J. Muller. 1978. Recommended palynological terms and definitions. *Grana* 17:55–58.

Nilsson, S. and J. J. Skvarla. 1969. Pollen morphology of saprophytic taxa in the Gentianaceae. *Ann. Missouri Bot. Gard.* 56:420–438.

Nitecki, M. H., ed. 1984. *Extinctions.* Chicago: Univ. Chicago Press.

Nogler, G. A. 1984. Gametophytic apomixis. In B. M. Johri, ed., *Embryology of Angiosperms*, pp. 475–518. Berlin: Springer-Verlag.

Nordenstam, B. 1977. Senecioneae and Liabeae—systematic review. In V. H. Heywood, J. B. Harborne, and B. L. Turner, eds., *The Biology and Chemistry of the Compositae*, pp. 799–830. London: Academic Press.

—— 1978. Taxonomic studies in the tribe Senecioneae (Compositae). *Opera Bot.* 44:1–83.

Northington, D. K. 1976. Evidence bearing on the origin of infraspecific disjunction in *Sophora gypsophila* (Fabaceae). *Pl. Syst. Evol.* 125:233–244.

Norton, S. A. 1984. Thrips pollination in the lowland forest of New Zealand. *New Zealand J. Ecol.* 7:157–163.

Nowicke, J. W. 1970. Pollen morphology in the Nyctaginaceae. I. Nyctagineae (Mirabileae). *Grana* 10:79–88.

Nowicke, J. W., J. L. Bittner, and J. J. Skvarla. 1986. *Paeonia*, exine substructure and plasma ashing. In S. Blackmore and I. K. Ferguson, eds., *Pollen and Spores: Form and Function*, pp. 81–95. London: Academic Press.

Nowicke, J. W. and J. J. Skvarla. 1974. A palynological investigation of the genus *Tournefortia* (Boraginaceae). *Amer. J. Bot.* 61:1021–1036.

—— 1977. Pollen morphology and the relationship of the Plumbaginaceae, Polygonaceae, and Primulaceae to the order Centrospermae. *Smithsonian Contr. Bot.* 37:1–64.

—— 1979. Pollen morphology: The potential influence in higher order systematics. *Ann. Missouri Bot. Gard.* 66:633–700.

Nuttall, T. 1827. *An Introduction to Systematic and Physiological Botany.* Boston: Hilliard, Gray, Little, and Wilkins.

O'Brien, T. P. and M. E. McCully. 1969. *Plant Structure and Development: A Pictorial and Physiological Approach.* New York: Macmillan.

—— 1981. *The Study of Plant Structure: Principles and Selected Methods.* Melbourne: Termarcarphi.

O'Grady, R. T. and G. B. Deets. 1987. Coding multistate characters, with special reference to the use of parasites as characters of their hosts. *Syst. Zool.* 36:268–279.

Ohsugi, R. and T. Murata. 1986. Variations in the leaf anatomy among some C4 *Panicum* species. *Ann. Bot.* n.s. 58:443–453.

Okamura, J. K. and R. B. Goldberg. 1985. Tobacco single-copy DNA is highly homologous to sequences present in the genomes of its diploid progenitors. *Mol. Gen. Genet.* 198:290–298.

Olsen, J. S. 1979. Systematics of *Zaluzania* (Asteraceae: Heliantheae). *Rhodora* 81:449–501.

Olson, E. C. 1964. Morphological integration and the meaning of characters in classification systems. In V. H. Heywood and J. McNeill, eds., *Phenetic and Phylogenetic Classification*, pp. 123–156. London: Systematics Association.

Orchard, A. E. 1981. The generic limits of *Ixodia* R.Br. ex Ait. (Compositae—Inuleae). *Brunonia* 4:185–197.

Ornduff, R., ed. 1967–1969. Index to plant chromosome numbers [1965–1967]. *Regnum Veg.* 50:i–viii, 1–128; 55:1–126; 59:1–129.

Ornduff, R., T. Mosquin, D. W. Kyhos, and P. H. Raven. 1967. Chromosome numbers in Compositae. VI. Senecioneae, II. *Amer. J. Bot.* 54:205–213.

Ornduff, R., P. H. Raven, D. W. Kyhos, and A. R. Kruckeberg. 1963. Chromosome numbers in Compositae. III. Senecioneae. *Amer. J. Bot.* 50:131–139.

Ornduff, R., N. A. M. Saleh and B. A. Bohm. 1973. The flavonoids and affinities of *Blennosperma* and *Crocidium* (Compositae). Taxon 22:407–412.

Owens, S. J., S. McGrath, M. A. Fraser, and L. R. Fox. 1984. The anatomy, histochemistry and ultrastructure of stigmas and styles in Commelinaceae. *Ann. Bot.* 54:591–603.

Ownbey, M. 1950. Natural hybridization and amphiploidy in the genus *Tragopogon*. *Amer. J. Bot.* 37:487–499.

Oxford, G. S. and D. Rollinson, eds. 1983. *Protein Polymorphism: Adaptive and Taxonomic Significance*. London: Academic Press.

Pacheco, P., D. J. Crawford, T. F. Stuessy, and M. Silva O. 1985. Flavonoid evolution in *Robinsonia* (Compositae) of the Juan Fernandez Islands. *Amer. J. Bot.* 72:989–998.

Pacqué, M. 1980. The structure of apertures in some Passifloraceae and their possible functioning. *5th Internatl. Palynol. Confer. Abstr.*, p. 298.

Page, R. D. M. 1987. Graphs and generalized tracks: Quantifying Croizat's panbiogeography. *Syst. Zool.* 36:1–17.

Pahlson, C., F. Bergqvis and U. Forsum. 1985. Numerical taxonomy of motile anaerobic curved rods isolated from vaginal discharge. *Scand. J. Urol. & Nephrol.* S86:251–256.

Palma-Otal, M., W. S. Moore, R. P. Adams, and G. R. Joswiak. 1983. Morphological, chemical, and biogeographical analyses of a hybrid zone involving *Juniperus virginiana* and *J. horizontalis* in Wisconsin. *Canad. J. Bot.* 61:2733–2746.

Palmer, J. D. 1986. Chloroplast DNA and phylogenetic relationships. In S. K. Dutta, ed., *DNA Systematics*, Vol. 2. *Plants*, pp. 63–80. Boca Raton, Fl.: CRC Press.

—— 1987. Chloroplast DNA evolution and biosystematic uses of chloroplast DNA variation. *Amer. Naturalist* 130 (suppl.):s6–s29.

Palmer, J. D. and D. Zamir. 1982. Chloroplast DNA evolution and phylogenetic relationships in *Lycoperiscon*. *Proc. Natl. Acad. U.S.A.* 79:5006–5010.

Palmer, P. G., S. Gerbeth-Jones, and S. Hutchinson. 1985. A scanning electron microscope survey of the epidermis of East African grasses, III. *Smithsonian Contr. Bot.* 55:1–136.

Palser, B. F. 1975. The bases of angiosperm phylogeny: Embryology. *Ann. Missouri Bot. Gard.* 62:621–646.

Panchen, A. L. 1982. The use of parsimony in testing phylogenetic hypotheses. *Zool. J. Linn. Soc.* 74:305–328.

Pandey, A. K. and S. Chopra. 1979. Development of seed and fruit in *Gerbera jamisonii* Bolus. *Geophytology* 9:171–174.

Pandey, A. K., S. Chopra, and R. P. Singh. 1982. Anatomy of seeds and fruits in some Astereae (Compositae). *Geophytology* 12:105–110.

Pandy, A. K. and R. P. Singh. 1978. Development of seed and fruit in *Dimorphotheca sinuata* DC. *Flora* 167:57–64.

Pankhurst, R. J. 1974. Automated identification in systematics. *Taxon* 23:45–51.

——, ed. 1975. *Biological Identification with Computers*. London: Academic Press.

—— 1978. *Biological Identification: The Principles and Practice of Identification Methods in Biology*. London: Arnold.

Panshin, A. J. and C. de Zeeuw. 1980. *Textbook of Wood Technology: Structure, Identification, Properties, and Uses of the Commercial Woods of the United States and Canada*. 4th ed. New York: McGraw-Hill.

Parenti, L. R. 1980. A phylogenetic analysis of the land plants. *Biol. J. Linn. Soc.* 13:225–242.

Park, C.-W. 1987. Flavonoid chemistry of *Polygonum* sect. *Echinocaulon*: A systematic survey. *Syst. Bot.* 12:167–179.

Parker, W. H. 1976. Comparison of numerical taxonomic methods used to estimate flavonoid similarities in the Limnanthaceae. *Brittonia* 28:390–399.

Parker, W. H. and B. A. Bohm. 1979. Flavonoids and taxonomy of the Limnanthaceae. *Amer. J. Bot.* 66:191–197.

Parkes, K. C. 1955. Sympatry, allopatry, and the subspecies in birds. *Syst. Zool.* 4:35–40.

Parkhurst, D. F. 1976. Effects of *Verbascum thapsus* leaf hairs on heat and mass transfer: A reassessment. *New Phytol.* 76:453–457.

—— 1978. The adaptive significance of stomatal occurrence on one or both surfaces of leaves. *J. Ecol.* 66:367–383.

Parra, O. and C. Marticorena. 1972. Granos de polen de plantas Chilenas. II. Compositae-Mutisieae. *Gayana, Bot.* 21:3–107.

Pasteur, G. 1976. The proper spelling of taxonomy. *Syst. Zool.* 25:192–193.

Patel, V. C., J. J. Skvarla, and P. H. Raven. 1984. Pollen characters in relation to the delimitation of Myrtales. *Ann. Missouri Bot. Gard.* 71:858–969.

Patil, V. P. and G. B. Deodikar. 1972. Inter-specific variations in chiasma frequencies and terminalization in emmer wheats. *Cytologia* 37:225–234.

Paton, D. C. 1982. The influence of honeyeaters on flowering strategies of Australian plants. In J. A. Armstrong, J. M. Powell, and A. J. Richards, ed., *Pollination and Evolution*, pp. 95–108. Sydney: Royal Botanic Gardens.

Paton, D. C. and V. Turner. 1985. Pollination of *Banksia ericifolia* Smith: Birds, mammals and insects as pollen vectors. *Austral. J. Bot.* 33:271–286.

Pattee, H. H., ed. 1973. *Hierarchy Theory: The Challenge of Complex Systems.* New York: Braziller.

Patterson, C. 1980. Cladistics. *Biologist* 27:234–240.

—— 1981. Significance of fossils in determining evolutionary relationships. *Ann. Rev. Ecol. Syst.* 12:195–223.

—— 1982a. Morphological characters and homology. In K. A. Joysey and A. E. Friday, eds., *Problems of Phylogenetic Reconstruction*, pp. 21–74. New York: Academic Press.

——, ed. 1982b. Methods of phylogenetic reconstruction. *Zool. J. Linn. Soc.* 74:197–344.

Patterson, R. 1977. The generic status of perennial species of *Linanthus* (Polemoniaceae). *Taxon* 26:507–511.

Payne, W. W. 1964. A re-evaluation of the genus *Ambrosia* (Compositae). *J. Arnold Arbor.* 45:401–430.

—— 1970. Helicocytic and allelocytic stomata: Unrecognized patterns in the Dicotyledonae. *Amer. J. Bot.* 57:140–147.

—— 1978. A glossary of plant hair terminology. *Brittonia* 30:239–255.

—— 1979. Stomatal patterns in embryophytes: Their evolution, ontogeny and interpretation. *Taxon* 28:117–132.

—— 1981. Structure and function in angiosperm pollen wall evolution. *Rev. Palaeobot. Palynol.* 35:39–59.

Payne, W. W., P. H. Raven, and D. W. Kyhos. 1964. Chromosome numbers in Compositae. IV. Ambrosieae. *Amer. J. Bot.* 51:419–424.

Payne, W. W. and J. J. Skvarla. 1970. Electron microscope study of *Ambrosia* pollen (Compositae: Ambrosineae). *Grana* 10:87–100.

Pearson, K. 1904a. A Mendelian's view of the law of ancestral inheritance. *Biometrika* 3:109–112.

—— 1904b. Mathematical contributions to the theory of evolution.—XII. On a generalised theory of alternative inheritance, with special reference to Mendel's laws. *Philos. Trans. Roy. Soc.* (London) ser. A, 203:53–86.

Pennell, F. W. 1931. Genotypes of the Scrophulariaceae in the first edition of Linne's "Species Plantarum." *Proc. Acad. Nat. Sci.* (Philadelphia) 82:9–26.

—— 1949. Toward a simple and clear nomenclature. *Amer. J. Bot.* 36:19–22.

Penny, D. 1976. Criteria for optimising phylogenetic trees and the problem of determining the root of a tree. *J. Mol. Evol.* 8:95–116.

Penzias, A. A. 1979. The origin of the elements. *Science* 205:549–554.

Percival, M. S. 1961. Types of nectar in angiosperms. *New Phytol.* 60:235–281.

—— 1965. *Floral Biology.* Oxford: Pergamon Press.

Persoon, C. H. 1805. *Synopsis Plantarum, seu Enchiridium Botanicum*, vol. 1. Paris: Cramerum.

Peters, D. S. 1978. Phylogeny reconstruction and classificatory insufficiency. *Syst. Zool.* 27:225–227.

Petersen, F. P. and D. E. Fairbrothers. 1983. A serotaxonomic appraisal of *Amphipterygium* and *Leitneria*—two amentiferous taxa of Rutiflorae (Rosidae). *Syst. Bot.* 8:134–148.

—— 1985. A serotaxonomic appraisal of the "Amentiferae." *Bull. Torrey Bot. Club* 112:43–52.

Petersen, R. H. 1971. Interfamilial relationships in the clavarioid and cantharelloid fungi [and discussion]. In R. H. Petersen, ed., *Evolution in the Higher Basidiomycetes*, pp. 345–371. Knoxville: Univ. Tennessee Press.

Pettigrew, C. J. and L. Watson. 1975. On the classification of Australian Acacias. *Austral. J. Bot.* 23:833–847.

Philbrick, C. T. and G. J. Anderson. 1987. Implications of pollen/ovule ratios and pollen size for the reproductive biology of *Potamogeton* and autogamy in aquatic angiosperms. *Syst. Bot.* 12:98–105.

Philipson, M. N. 1970. Cotyledons and the taxonomy of *Rhododendron*. *Notes Roy. Bot. Gard.* (Edinburgh) 30:55–77.

Philipson, W. R. 1987. The treatment of isolated genera. *Bot. J. Linn. Soc.* 95:19–25.

Philipson, W. R., J. M. Ward, and B. G. Butterfield. 1971. *The Vascular Cambium: Its Development and Activity.* London: Chapman and Hall.

Phipps, J. B. 1984. Problems of hybridity in the cladistics of *Crataegus* (Rosaceae). In W. F. Grant, ed., *Plant Biosystematics*, pp. 417–438. Toronto: Academic Press.

Pianka, E. R. 1988. *Evolutionary Ecology.* 4th ed. New York: Harper & Row.

Pimentel, R. A. 1958. Taxonomic methods, their bearing on subspeciation. *Syst. Zool.* 7:139–156.

—— 1959. Mendelian infraspecific divergence levels and their analysis. *Syst. Zool.* 8:139–159.

—— 1979. *Morphometrics: The Multivariate Analysis of Biological Data.* Dubuque, Iowa: Kendall/Hunt.

Pinkava, D. J. and M. G. McLeod. 1971. Chromosome numbers in some cacti of western North America. *Brittonia* 23:171–176.

Pinkava, D. J., M. G. McLeod, L. A. McGill, and R. C. Brown. 1973. Chromosome numbers in some cacti of western North America—II. *Brittonia* 25:2–9.

Platnick, N. I. 1976. Are monotypic genera possible? *Syst. Zool.* 25:198–199.

—— 1977a. Review of C. N. Slobodchikoff's *Concepts of Species. Syst. Zool.* 26:96–98.

—— 1977b. Paraphyletic and polyphyletic groups. *Syst. Zool.* 26:195–200.

—— 1977c. Cladograms, phylogenetic trees, and hypothesis testing. *Syst. Zool.* 26:438–442.

—— 1978. Gaps and prediction in classification. *Syst. Zool.* 27:472–474.

—— 1979. Philosophy and the transformation of cladistics. *Syst. Zool.* 28:537–546.

—— 1981. Widespread taxa and biogeographic congruence. In V. A. Funk and D. R. Brooks, eds., *Advances in Cladistics,* pp. 223–227. New York: New York Botanical Garden.

—— 1985. Philosophy and the transformation of cladistics revisited. *Cladistics* 1:87–94.

—— 1987. An empirical comparison of microcomputer parsimony programs. *Cladistics* 3:121–144.

Platnick, N. I. and H. D. Cameron. 1977. Cladistic methods in textual, linguistic, and phylogenetic analysis. *Syst. Zool.* 26:380–385.

Platnick, N. I. and V. A. Funk, eds. 1983. *Advances in Cladistics,* vol. 2. New York: Columbia Univ. Press.

Platnick, N. I. and G. Nelson. 1978. A method of analysis for historical biogeography. *Syst. Zool.* 27:1–16.

Playfair, W. 1801. *The Statistical Breviary.* London: Wallis.

Plumier, C. 1703. *Nova Plantarum Americanarum Genera.* Paris: Boudot.

Pohl, R. W. 1966. X *Elyhordeum iowense,* a new intergeneric hybrid in the Triticeae. *Brittonia* 18:250–255.

Pojar, J. 1973. Levels of polyploidy in four vegetation types of southwestern British Columbia. *Canad. J. Bot.* 51:621–628.

Ponzi, R., P. Pizzolongo and G. Caputo. 1978. Ultrastructural particularities in ovular tissues of some Rhoeadales taxa and their probable taxonomic value. *J. Submicroscop. Cytol.* 10:81–88.

Popper, K. R. 1959. *The Logic of Scientific Discovery.* New York: Harper & Row.

—— 1962. *Conjectures and Refutations: The Growth of Scientific Knowledge.* New York: Basic Books.

Porter, C. L. 1967. *Taxonomy of Flowering Plants.* 2d ed. San Francisco: Freeman.

Porter, D. M. 1984. Relationships of the Galapagos flora. *Biol. J. Linn. Soc.* 21:243–251.

Posluszny, U. 1983. Re-evaluation of certain key relationships in the Alismatidae: Floral organogenesis of *Scheuchzeria palustris* (Scheuchzeriaceae). *Amer. J. Bot.* 70:925–933.

Potvin, C., Y. Bergeron, and J.-P. Simon. 1983. A numerical taxonomic study of selected *Citrus* species (Rutaceae) based on biochemical characters. *Syst. Bot.* 8:127–133.

Powell, A. M. 1985. Crossing data as generic criteria in the Asteraceae. *Taxon* 34:55–60.

Powell, A. M., D. W. Kyhos, and P. H. Raven. 1974. Chromosome numbers in Compositae. X. *Amer. J. Bot.* 61:909–913.

Powell, A. M. and S. A. Sloan. 1975. Polyploid percentages in gypsum and non-gypsum floras of the Chihuahuan Desert. *Sci. Biol. J.* 1:37–38.

Powell, A. M. and B. L. Turner. 1963. Chromosome numbers in the Compositae. VII. Additional species from the southwestern United States and Mexico. *Madroño* 17:128–140.

Prager, E. M. and A. C. Wilson. 1978. Construction of phylogenetic trees for proteins and nucleic acids: Empirical evaluation of alternative matrix methods. *J. Mol. Evol.* 11:129–142.

Prakash, N. 1967. Aizoaceae—a study of its embryology and systematics. *Bot. Not.* 120:305–323.

Prance, G. T. 1980. A note on the probable pollination of *Combretum* by *Cebus* monkeys. *Biotropica* 12:239.

Prance, G. T. and T. S. Elias, eds. 1977. *Extinction Is Forever.* New York: New York Botanical Garden.

Prance, G. T., D. J. Rogers, and F. White. 1969. A taximetric study of an angiosperm family: General delimitation in the Chrysobalanaceae. *New Phytol.* 68:1203–1234.

Pratt, V. 1972a. Biological classification. *Brit. J. Philos. Sci.* 23:305–327.

—— 1972b. Numerical taxonomy—a critique. *J. Theor. Biol.* 36:581–592.

—— 1974. Numerical taxonomy: On the incoherence of its rationale. *J. Theor. Biol.* 48:497–499.

Prendergast, H. D. V. and P. W. Hattersley. 1985. Distribution and cytology of Australian *Neurachne* and its allies (Poaceae), a group containing C_3, C_4 and C_3-C_4 intermediate species. *Austral. J. Bot.* 33:317–336.

Presch, W. 1979. Phenetic analysis of a single data set: Phylogenetic implications. *Syst. Zool.* 28:366–371.

Price, H. J., K. L. Chambers, K. Bachmann, and J. Riggs. 1986. Patterns of mean nuclear DNA content in *Microseris douglasii* (Asteraceae) populations. *Bot. Gaz.* 147:496–507.

Pridgeon, A. M. 1981. Absorbing trichomes in the Pleurothallidinae (Orchidaceae). *Amer. J. Bot.* 68:64–71.

Pridham, J. B., ed. 1967. *Terpenoids in Plants.* London: Academic Press.

Prim, R. C. 1957. Shortest connection networks and some generalizations. *Bell Syst. Tech. J.* 36:1389–1401.

Prior, M., D. Boulton, C. Gajzago, and D. Perry. 1975. The classification of childhood psychoses by numerical taxonomy. *J. Child Psychol. Psychiat.* 16:321–330.

Proctor, J. R. 1966. Some processes of numerical taxonomy in terms of distance. *Syst. Zool.* 15:131–140.

Proctor, M. and P. Yeo. 1973. *The Pollination of Flowers.* London: Collins.

Prósperi, C. H. and A. E. Cocucci. 1979. Importancia taxonómica de la calosa de los tubos polínicos en Tubiflorae (ensayo preliminar). *Kurtziana* 12–13: 75–81.

Pryer, K. M., D. M. Britton, and J. McNeill. 1983. A numerical analysis of chromatographic profiles in North American taxa of the fern genus *Gymnocarpium*. *Canad. J. Bot.* 61:2592–2602.

Prywer, C. 1965. Cytological evidence of natural intertribal hybridization of *Tripsacum* and *Manisuris*. *Amer. J. Bot.* 52:182–184.

Pullaiah, T. 1978. Embryology of *Tithonia*. *Phytomorphology* 28:437–444.

—— 1979a. Embryology of *Adenostemma*, *Elephantopus* and *Vernonia* (Compositae). *Bot. Not.* 132:51–56.

—— 1979b. Studies in the embryology of Compositae. IV. The tribe Inuleae. *Amer. J. Bot.* 66:1119–1127.

—— 1981. Studies in the embryology of Heliantheae (Compositae). *Pl. Syst. Evol.* 137:203–214.

—— 1982a. Embryology, seed coat and fruit wall of *Parthenium hysterophorus* L. (Compositae). *J. Jap. Bot.* 57:241–247.

—— 1982b. Studies in the embryology of Compositae II. The tribe—Eupatorieae. *Indian J. Bot.* 5:183–188.

—— 1983. Studies in the embryology of Senecioneae (Compositae). *Pl. Syst. Evol.* 142:61–70.

Punt, W., ed. 1976. *The Northwest European Pollen Flora*, vol. 1. Amsterdam: Elsevier.

Puri, V. 1951. The role of floral anatomy in the solution of morphological problems. *Bot Rev.* 17:471–553.

Purkinje, J. E. 1830. *De cellulis antherarum fibrosis nec non de granorum polinarum formis commentationis phytotomica.* Breslau.

Quasada, E., M. J. Valderrama, V. Bejar, A. Ventosa, F. Ruíz-Berraquero, and A. Ramos-Cormenzana. 1987. Numerical taxonomy of moderately halophytic gram-negative nonmotile eubacteria. *Syst. Appl. Microbiol.* 9:132–137.

Queiroz, K. de. 1985. The ontogenetic method for determining character polarity and its relevance to phylogenetic systematics. *Syst. Zool.* 34:280–299.

Quinn, J. A. and D. E. Fairbrothers. 1971. Habitat ecology and chromosome numbers of natural populations of the *Danthonia sericea* complex. *Amer. Midl. Naturalist* 85:531–536.

Radford, A. E. 1986. *Fundamentals of Plant Systematics.* New York: Harper & Row.

Radford, A. E., W. C. Dickison, J. R. Massey, and C. R. Bell. 1974. *Vascular Plant Systematics.* New York: Harper & Row.

Radlkofer, L. 1895. Sapindaceae. In A. Engler and K. Prantl, eds., *Die natürlichen Pflanzenfamilien*, Teil 3, Abt. 5, pp. 277–366. Leipzig: Engelmann.

Raffauf, R. F. 1970. *A Handbook of Alkaloids and Alkaloid-Containing Plants.* New York: Wiley.

Raghavan, V. 1976. *Experimental Embryogenesis in Vascular Plants.* London: Academic Press.

—— 1986. *Embryogenesis in Angiosperms: A Developmental and Experimental Study.* Cambridge: Cambridge Univ. Press.

Rahn, K. 1974. *Plantago section Virginica: A taxonomic revision of a group of American plantains, using experimental, taximetric and classical methods. Dansk Bot. Arkiv.* 30(2):1–180.

Rajendra, B. R., A. S. Tomb, K. A. Mujeeb, and L. S. Bates. 1978. Pollen morphology of selected Triticeae and two intergeneric hybrids. *Pollen et Spores* 20:145–156.

Raju, V. S. and P. N. Rao. 1977. Variation in the structure and development of foliar stomata in the Euphorbiaceae. *Bot. J. Linn. Soc.* 75:69–97.

Rak, Y. 1985. Australopithecine taxonomy and phylogeny in light of facial morphology. *Amer. J. Phys. Anthropol.* 66:281–287.

Raman, S. 1987. A code proposed for the classification of trichomes as applied to the Scrophulariaceae. *Beitr. Biol. Pflanzen* 62:349–367.

Ramsbottom, J. 1938. Linnaeus and the species concept. *Proc. Linn. Soc. London* 150:192–219.

Rao, T. A., J. Bhattacharya, and J. C. Das. 1978. Foliar sclereids in *Rhizophora* L. and their taxonomic implications. *Proc. Indian Acad. Sci.*, Sect. B, 87:191–195.

Rao, T. A. and S. Das 1979. Leaf sclereids—occurrence and distribution in the angiosperms. *Bot. Not.* 132:319–324.

Rao, V. S. 1968. Placentation in relation to anatomy. *Bot. Not.* 121:281–286.

—— 1971. The disk and its vasculature in the flowers of some dicotyledons. *Bot. Not.* 124:442–450.

Rasmussen, F. N. 1983. On "apomorphic tendencies" and phylogenetic inference. *Syst. Bot.* 8:334–337.

Raup, D. M. 1979. Biases in the fossil record of species and genera. *Bull. Carnegie Mus. Nat. Hist.* 13:85–91.

Raven, P. H. 1969. A revision of the genus *Camissonia* (Onagraceae). *Contr. U.S. Natl. Herb.* 37:161–396.

—— 1974. Plant systematics 1947–1972. *Ann. Missouri Bot. Gard.* 61:166–178.

—— 1975. The bases of angiosperm phylogeny: Cytology. *Ann. Missouri Bot. Gard.* 62:724–764.

—— 1976. The destruction of the tropics. *Frontiers* 40(4):22–23.

——, chm. 1980. *Research Priorities in Tropical Biology.* Washington, D.C.: National Academy of Sciences.

—— 1986. Modern aspects of the biological species in plants. In K. Iwatsuki, P. H. Raven, and W. J. Bock, eds., *Modern Aspects of Species*, pp. 11–29. Tokyo: Univ. Tokyo Press.

Raven, P. H. and D. I. Axelrod. 1974. Angiosperm biogeography and past continental movements. *Ann. Missouri Bot. Gard.* 61:593–673.

Raven, P. H., B. Berlin, and D. E. Breedlove. 1971. The origins of taxonomy. *Science* 174:1210–1213.

Raven, P. H. and D. W. Kyhos. 1961. Chromosome numbers in Compositae. II. Helenieae. *Amer. J. Bot.* 48:842–850.

Raven, P. H., D. W. Khyos, D. E. Breedlove, and W. W. Payne. 1968. Polyploidy in *Ambrosia dumosa* (Compositae: Ambrosieae). *Brittonia* 20:205–211.

Raven, P. H., S. G. Shetler, and R. L. Taylor. 1974. [Proposals for the simplification of infraspecific terminology]. *Taxon* 23:828–831.

Raven, P. H., O. T. Solbrig, D. W. Kyhos, and R. Snow. 1960. Chromosome numbers in Compositae. I. Astereae. *Amer. J. Bot.* 47:124–132.

Ray, J. 1686–1704. *Historia Plantarum*. London.

Real, L., ed., 1983. *Pollination Biology*. Orlando: Academic Press.

Reed, E. S. 1979. The role of symmetry in Ghiselin's "Radical solution to the species problem." *Syst. Zool.* 28:71–78.

Reed, H. S. 1942. *A Short History of the Plant Sciences*. Waltham, Chronica Botanica.

Reeder, J. R. 1957. The embryo in grass systematics. *Amer. J. Bot.* 44: 756–768.

—— 1962. The bambusoid embryo: A reappraisal. *Amer. J. Bot.* 49:639–641.

Rees, H. 1984. Nuclear DNA variation and the homology of chromosomes. In W. F. Grant, ed., *Plant Biosystematics*, pp. 87–96. Toronto: Academic Press.

Regan, C. T. 1926. Organic evolution. *Rep. Brit. Assoc. Advancem. Sci.* 1925:75–86.

Reichert, E. T. 1916. The specificity of proteins and carbohydrates in relation to genera, species and varieties. *Amer. J. Bot.* 3:91–98.

—— 1919. A Biochemic Basis for the Study of Problems of Taxonomy, Heredity, Evolution, etc., with Especial Reference to the Starches and Tissues of Parent-Stocks and Hybrid-Stocks and the Starches and Hemoglobins of Varieties, Species, and Genera Publ. Carnegie Inst. (Washington) 270:i–xii, 1–834. Washington, D.C.: Carnegie Inst.

Reitsma, T. 1969. Size modification of recent pollen grains under different treatments. *Rev. Palaeobot. Palynol.* 9:175–202.

—— 1970. Suggestions towards unification of descriptive terminology of angiosperm pollen grains. *Rev. Palaeobot. Palynol.* 10:39–60.

Renold, W. 1970. The chemistry and infraspecific variation of sesquiterpene lactones in *Ambrosia confertiflora* DC. (Compositae): A chemosystematic study at the population level. Ph.D. dissertation, Univ. Texas, Austin.

Rensch, B. 1954. *Neuere Probleme der Abstammungslehre: Die transspezifische Evolution.* 2d ed. Stuttgart: Ferdinand Enke.

—— 1959. *Evolution Above the Species Level.* New York: Columbia Univ. Press.

Reveal, J. L. 1969. The subgeneric concept in *Eriogonum* (Polygonaceae). In J. E. Gunckel, ed., *Current Topics in Plant Science*, pp. 229–249. New York: Academic Press.

Reynolds, J. F. and D. J. Crawford. 1980. A quantitative study of variation in the *Chenopodium atrovirens-desiccatum-pratericola* complex. *Amer. J. Bot.* 67:1380–1390.

Rhodes, A. M., S. G. Carmer, and J. W. Courter. 1969. Measurement and classification of genetic variability in horseradish. *J. Amer. Soc. Hort. Sci.* 94:98–102.

Ribéreau-Gayon, P. 1972. *Plant Phenolics.* New York: Hafner.

Rice, E. L. and A. R. O. Chapman. 1985. A numerical taxonomic study of *Fucus distichus* (Phaeophyta). *J. Mar. Biol. Assoc. U.K.* 65:433–459.

Richards, A. J., ed. 1978. *The Pollination of Flowers by Insects.* London: Academic Press.

—— 1986. *Plant Breeding Systems.* London: George Allen & Unwin.

Richardson, P. M. 1982. Anthocyanins of the Sterculiaceae: Flavonoid scores and Hennigian phylogenetic systematics. *Biochem. Syst. Ecol.* 10:197–199.

—— 1983a. Flavonoids and phylogenetic systematics. In N. I. Platnick and V. A. Funk, eds., *Advances in Cladistics*, 2:115–124. New York: Columbia Univ. Press.

—— 1983b. A bibliography of cladistics and plant secondary compounds. *Phytochem. Bull.* 15:32.

Richardson, P. M. and D. A. Young. 1982. The phylogenetic content of flavonoid point scores. *Biochem. Syst. Ecol.* 10:251–255.

Richter, R. 1938. Beobachtungen an einer gemischten Kolonie von Silbermowe (*Larus argentatus* Pont.) und Heringsmowe (*Larus fuscus graellsi* Brehm). *J. Ornithol.* 86:366–373.

Ricklefs, R. E. 1979. *Ecology.* 2d ed. New York: Chiron Press.

Ridley, M. 1983. *The Explanation of Organic Diversity: The Comparative Method and Adaptations for Mating.* Oxford: Clarendon Press.

—— 1986. *Evolution and Classification: The Reformation of Cladism.* London: Longman.

Riedl, R. 1984. *Biology of Knowledge: The Evolutionary Basis of Reason.* Translated by P. Foulkes. Chichester: Wiley.

Rieger, R. and S. Tyler. 1979. The homology theorem in ultrastructural research. *Amer. Zoologist* 19:655–664.

Rieppel, O. 1983a. *Kladismus oder die Lengende vom Stammbaum.* Basel: Birkhauser Verlag.

—— 1983b. The "tertium comparationis" of competing evolutionary theories. *Z. Zool. Syst. Evolut.-Forsch.* 21:1–6.

—— 1986. Species are individuals: A review and critique of the argument. *Evol. Biol.* 20:283–317.

Riley, D. and A. Young. 1966. *World Vegetation.* Cambridge: Cambridge Univ. Press.

Riley, R. and V. Chapman. 1958. Genetic control of the cytologically diploid behaviour of hexaploid wheat. *Nature* 182:713–715.

Risley, M. S., ed. 1986. *Chromosome Structure and Function.* New York: Van Nostrand Reinhold.

Ritland, K. and M. T. Clegg. 1987. Evolutionary analysis of plant DNA sequences. *Amer. Naturalist* 130 (suppl.):s74–s100.

Rivas, L. R. 1965. A proposed code system for storage and retrieval of information in systematic zoology. *Syst. Zool.* 14:131–132.

Robards, A. W., ed. 1974. *Dynamic Aspects of Plant Ultrastructure.* London: McGraw-Hill.

Robertson, C. 1928. *Flowers and Insects: Lists of Visitors of Four Hundred and Fifty-Three Flowers.* Carlinville, Ill.: published by the author.

Robichaux, R. H. and J. E. Canfield. 1985. Tissue elastic properties of eight Hawaiian *Dubautia* species that differ in habitat and diploid chromosome number. *Oecologia* 66:77–80.

Robinson, B. L. 1906. The generic concept in the classification of the flowering plants. *Science* n.s. 23:81–92.

Robinson, H. 1969. A monograph on foliar anatomy of the genera *Connellia*, *Cottendorfia* and *Navia* (Bromeliaceae). *Smithsonian Contr. Bot.* 2:1–41.

—— 1985. Comments on the cladistic approach to the phylogeny of the "bryophytes" by Mishler and Churchill. *Brittonia* 37:279–281.

Robinson, H. and R. M. King. 1977. Eupatorieae—systematic review. In V. H. Heywood, J. B. Harborne, and B. L. Turner, eds., *The Biology and Chemistry of the Compositae*, pp. 437–485. London: Academic Press.

—— 1985. Comments on the generic concepts in the Eupatorieae. *Taxon* 34:11–16.

Robinson, T. 1980. *The Organic Constituents of Higher Plants: Their Chemistry and Interrelationships.* 4th ed. Amherst, Mass: Cordus Press.

—— 1981. *The Biochemistry of Alkaloids.* 2d ed. New York: Springer-Verlag.

Robson, N. K. B., D. F. Cutler, and M. Gregory, eds. 1970. *New Research in Plant Anatomy.* London: Academic Press.

Rodman, J. E. 1976. Differentiation and migration of *Cakile* (Cruciferae): Seed glucosinolate evidence. *Syst. Bot.* 1:137–148.

Rodman, J. E., A. R. Kruckeberg, and I. A. Al-Shehbaz. 1981. Chemotaxonomic diversity and complexity in seed glucosinolates of *Caulanthus* and *Streptanthus* (Cruciferae). *Syst. Bot.* 6:197–222.

Rodman, J. E., M. L. Oliver, R. R. Nakamura, J. U. McClammer, Jr., and A. H. Bledsoe. 1984. A taxonomic analysis and revised classification of Centrospermae. *Syst. Bot.* 9:297–323.

Rodrigues, P. D. 1986. On the term character. *Syst. Zool.* 35:140–141.

Rodríguez, E., P. L. Healey, and I. Mehta, eds. 1984. *Biology and Chemistry of Plant Trichomes.* New York: Plenum Press.

Roe, K. E. 1971. Terminology of hairs in the genus *Solanum. Taxon* 20:501–508.

—— 1974. A simple technique for measuring phenetic similarity in *Solanum* using edge-punched cards. *Taxon* 23:707–713.

Rogers, C. M. and K. S. Xavier. 1972. Parallel evolution in pollen structure in *Linum. Grana* 12:41–46.

Rogers, D. J. 1963. Taximetrics—new name, old concept. *Brittonia* 15:285–290.

Rogers, J. S. 1972. Measures of genetic similarity and genetic distance. *Univ. Texas Publ.* 7213:145–153.

Rogers, S. O. and A. J. Bendich. 1985. Extraction of DNA from milligram amounts of fresh, herbarium and mummified plant tissues. *Pl. Mol. Biol.* 5:69–76.

Rohlf, F. J. 1968. Stereograms in numerical taxonomy. *Syst. Zool.* 17:246–255.

—— 1982. Consensus indices for comparing classifications. *Math. Biosci.* 59:131–144.

Rohlf, F. J. and D. R. Fisher. 1968. Tests for hierarchical structure in random data sets. *Syst. Zool.* 17:407–412.

Rohlf, F. J., A. J. Gilmartin, and G. Hart. 1983. The Kluge-Kerfoot phenomenon—a statistical artifact. *Evolution* 37:180–202.

Rohlf, F. J. and R. R. Sokal. 1967. Taxonomic structure from randomly and systematically scanned biological images. *Syst. Zool.* 16:246–260.

—— 1980. Comments on taxonomic congruence. *Syst. Zool.* 29:97–101.

—— 1981. Comparing numerical taxonomic studies. *Syst. Zool.* 30:459–490.

Rollins, R. C. 1953. Cytogenetical approaches to the study of genera. *Chron. Bot.* 14:133–139.

—— 1957. Taxonomy of the higher plants. *Amer. J. Bot.* 44:188–196.

—— 1958. The genetic evaluation of a taxonomic character in *Dithyrea* (Cruciferae). *Rhodora* 60:145–152.

—— 1965. On the bases of biological classification. *Taxon* 14:1–6.

—— 1981. Studies on *Arabis* (Cruciferae) of western North America. *Syst. Bot.* 6:55–64.

Rollins, R. C. and U. C. Banerjee. 1975. *Atlas of the Trichomes of Lesquerella (Cruciferae).* Cambridge, Mass.: Bussey Inst. of Harvard Univ.

Romberger, J. A. 1963. Meristems, growth, and development in woody plants: An analytical review of anatomical physiological, and morphogenetic aspects. *Tech. Bull. U.S.D.A.* 1293:i–vi, 1–214.

Romesburg, H. C. 1984. *Cluster Analysis for Researchers.* Belmont, Calif.: Lifetime Learning Publications.

Roose, M. L. and L. D. Gottlieb. 1976. Genetic and biochemical consequences of polyploidy in *Tragopogon. Evolution* 30:818–830.

—— 1978. Stability of structural gene number in diploid species with different amounts of nuclear DNA and different chromosome numbers. *Heredity* 40:159–163.

Rosanoff, S. 1866. Zur Kenntniss des Baues und der Entwickelungsgeschichte des Pollen der Mimoseae. *Jahrb. Wiss. Bot.* 4:441–450.

Rose, M. R. and W. F. Doolittle. 1983. Molecular biological mechanisms of speciation. *Science* 220:157–162.

Rosen, D. 1986. The role of taxonomy in effective biological control programs. *Agric. Ecosyst. Environm.* 15:121–129.

Rosen, D. E. 1975. A vicariance model of Caribbean biogeography. *Syst. Zool.* 24:431–464.

—— 1978. Vicariant patterns and historical explanation in biogeography. *Syst. Zool.* 27:159–188.

Rosendahl, C. O. 1949. The problem of subspecific categories. *Amer. J. Bot.* 36:24–27.

Rosenthal, G. A. and D. H. Janzen, eds. 1979. *Herbivores: Their Interaction with Secondary Plant Metabolites.* New York: Academic Press.

Ross, H. H. 1974. *Biological Systematics.* Reading, Mass.: Addison-Wesley.

Rossner, H. and M. Popp. 1986. Ionic patterns in some Crassulaceae from Austrian habitats. *Flora* 178:1–10.

Roth, I. 1984. *Stratification of Tropical Forests as Seen in Leaf Structure.* The Hague: Junk.

Roughgarden, J. 1979. *Theory of Population Genetics and Evolutionary Ecology: An Introduction.* New York: Macmillan.

Rowell, A. J. 1970. The contribution of numerical taxonomy to the genus concept. In T. W. Amsden, ed., *The Genus: A Basic Concept in Paleontology,* pp. 264–293. Lawrence, Kansas: Allen Press.

Rowley, J. R. 1981. Pollen wall characters with emphasis on applicability. *Nord. J. Bot.* 1:357–380.

Rowley, J. R. and J. J. Skvarla. 1986. Development of the pollen grain wall in *Canna. Nord. J. Bot.* 6:39–65.

—— 1987. Ontogeny of pollen in *Poinciana* (Leguminosae). II. Microspore and pollen grain periods. *Rev. Palaeobot. Palynol.* 50:313–331.

Roy, M. A., ed. 1980. *Species Identity and Attachment: A Phylogenetic Evaluation.* New York: Garland STPM Press.

Rudolph, E. D. 1982. The introduction of the natural system of classification of plants to nineteenth century American students. *Arch. Nat. Hist.* 10:461–468.

Runemark, H. 1961. The species and subspecies concepts in sexual flowering plants. *Bot. Not.* 114:22–32.

—— 1968. Critical comments on the use of statistical methods in chemotaxonomy. *Bot. Not.* 121:29–43.

Runemark, H. and W. K. Heneen. 1968. *Elymus* and *Agropyron,* a problem of generic delimitation. *Bot. Not.* 121:51–79.

Rury, P. M. and W. C. Dickison. 1984. Structural correlations among wood, leaves and plant habit. In R. A. White and W. C. Dickison, eds., *Contemporary Problems in Plant Anatomy,* pp. 495–540. Orlando: Academic Press.

Ruse, M. E. 1969. Definitions of species in biology. *Brit. J. Philos. Sci.* 20:97–119.

—— 1971. Gregg's paradox: A proposed revision to Buck and Hull's solution. *Syst. Zool.* 20:239–245.

—— 1973. On the supposed incoherence of numerical taxonomy. *J. Theor. Biol.* 40:603–605.

Russell, S. D. 1979. Fine structure of megagametophyte development in *Zea mays. Canad. J. Bot.* 57:1093–1110.

—— 1980. Participation of male cytoplasm during gamete fusion in an angiosperm, *Plumbago zeylanica. Science* 210:200–201.

—— 1982. Fertilization in *Plumbago zeylanica:* Entry and discharge of the pollen tube in the embryo sac. *Canad. J. Bot.* 60:2219–2230.

—— 1983. Fertilization in *Plumbago zeylanica:* Gametic fusion and fate of the male cytoplasm. *Amer. J. Bot.* 70:416–434.

Russell, S. D. and D. D. Cass. 1981. Ultrastructure of the sperms of *Plumbago zeylanica* 1. Cytology and association with the vegetative nucleus. *Protoplasma* 107:85–107.

—— 1983. Unequal distribution of plastids and mitochondria during sperm cell formation in *Plumbago zeylanica.* In D. L. Mulcahy and E. Ottaviano, eds., *Pollen: Biology and Implications for Plant Breeding,* pp. 135–140. New York: Elsevier.

Rutovitz, D. 1973. Pattern recognition by computer. *Proc. Roy. Soc.* (London), ser. B, 184:441–454.

Rydberg, P. A. 1922. Carduales: Ambrosiaceae, Carduaceae. *North Amer. Flora* 33:1–110.

—— 1924a. Some senecioid genera—I. *Bull. Torrey Bot. Club* 51:369–378.

—— 1924b. Some senecioid genera—II. *Bull. Torrey Bot. Club* 51:409–420.

Rzedowski, J. and R. McVaugh. 1966. La vegetación de Nueva Galicia. *Contr. Univ. Michigan Herb.* 9:i–iv, 1–123.

Sabrosky, C. W. 1955. The interrelations of biological control and taxonomy. *J. Econ. Entomol.* 48:710–714.

Sacarrão, G. F. 1980. Critical remarks on classification and species. *Arq. Mus. Bocage* n.s. 7:279–289.

Sachs, J. von. 1890. *History of Botany (1530–1860).* Translated by H. E. F. Garnsey; revised by I. B. Balfour. Oxford: Clarendon Press; reprinted, New York: Russell & Russell, 1967.

Sachs, T. 1978. Phyletic diversity in higher plants. *Pl. Syst. Evol.* 130:1–11.

Saether, O. A. 1979. Underlying synapomorphies and anagenetic analysis. *Zoologica Scripta* 8:305–312.

—— 1983. The canalized evolutionary potential: Inconsistencies in phylogenetic reasoning. *Syst. Zool.* 32:343–359.

—— 1986. The myth of objectivity—post-Hennigian deviations. *Cladistics* 2:1–13.

Sakamoto, S. 1974. Intergeneric hybridization among three species of *Heteranthelium, Eremopyrum* and *Hordeum*, and its significance for the genetic relationships within the tribe Triticeae. *New Phytol.* 73: 341–350.

Salthe, S. N. 1985. *Evolving Hierarchical Systems: Their Structure and Representation.* New York: Columbia Univ. Press.

Sampson, F. B. 1981. Synchronous versus asynchronous mitosis within permanent pollen tetrads of the Winteraceae. *Grana* 20:19–23.

Sanders, R. W. 1981. Cladistic analysis of *Agastache* (Lamiaceae). In V. A. Funk and D. R. Brooks, eds., *Advances in Cladistics*, pp. 95–114. New York: New York Botanical Garden.

—— 1987. Taxonomic significance of chromosome observations in Caribbean species of *Lantana* (Verbenaceae). *Amer. J. Bot.* 74:914–920.

Sanders, R. W., T. F. Stuessy, C. Marticorena, and M. Silva O. 1987. Phytogeography and evolution of *Dendroseris* and *Robinsonia*, tree-Compositae of the Juan Fernandez Islands. *Opera Bot.* 92:195–215.

Sanders, R. W., T. F. Stuessy, and R. Rodríguez. 1983. Chromosome numbers from the flora of the Juan Fernandez Islands. *Amer. J. Bot.* 70:799–810.

Sass, J. E. 1958. *Botanical Microtechnique.* 3d ed. Ames: Iowa State College Press.

Sattler, R. 1964. Methodological problems in taxonomy. *Syst. Zool.* 13:19–27.

—— 1966. Towards a more adequate approach to comparative morphology. *Phytomorphology* 16:417–429.

—— 1984. Homology—a continuing challenge. *Syst. Bot.* 9:382–394.

Savidan, Y. and J. Pernès. 1982. Diploid-tetraploid-dihaploid cycles and the evolution of *Panicum maximum* Jacq. *Evolution* 36:596–600.

Saville, D. B. O. 1954. Taxonomy, phylogeny, host relationship, and phytogeography of the microcyclic rusts of Saxifragaceae. *Canad J. Bot.* 32:400–425.

—— 1968. The rusts of Cheloneae (Scrophulariaceae): A study in the co-evolution of hosts and parasites. *Nova Hedwigia* 15:369–392.

—— 1971. Co-ordinated studies of parasitic fungi and flowering plants. *Naturaliste Canad.* 98:535–552.

—— 1975. Evolution and biogeography of Saxifragaceae with guidance from their rust parasites. *Ann. Missouri Bot. Gard.* 62:354–361.

Scadding, S. R. 1981. Do "vestigial organs" provide evidence for evolution? *Evol. Theory* 5:173–176.

—— 1982. Vestigial organs do not provide scientific evidence for evolution. *Evol. Theory* 6:171–173.

Schaeffer, B., M. K. Hecht, and N. Eldredge. 1972. Phylogeny and paleontology. *Evol. Biol.* 6:31–46.

[Schaffner, J. H.] 1937. Sub-species and varieties. *Amer. Botanist* 43:177.

Schieferstein, R. H. and W. E. Loomis. 1956. Wax deposits on leaf surfaces. *Pl. Physiol.* 31:240–247.

Schill, R., W. Barthlott, N. Ehler, and W. Rauh. 1973. Raster-elektronen-mikroskopische Untersuchungen an Cactaceen-Epidermen und ihre Bedeutung für die Systematik. *Trop. Subtrop. Pflanzenw.* 4:1–14.

Schill, R., A. Baumm, and M. Wolter. 1985. Vergleichende Mikromorphologie der Narbenoberflächen bei den Angiospermen; Zusammenhänge mit Pollenoberflächen bei heterostylen Sippen. *Pl. Syst. Evol.* 148:185–214.

Schlarbaum, S. E. and T. Tsuchiya. 1984. Cytotaxonomy and phylogeny in certain species of Taxodiaceae. Pl. Syst. Evol. 147:29–54.

Schleiden, M. J. 1837. Einige Blicke auf die Entwicklungsgeschichte des vegetabilischen Organismus bei den Phanerogamen. *Arch. Naturgesch.* 3(1):289–320.

—— 1838. Beiträge zur Phytogenesis. *Archiv. Anat. Physiol. Weiss. Med.* 2:137–176.

Schlichting, C. D. 1986. The evolution of phenotypic plasticity in plants. *Ann. Rev. Ecol. Syst.* 17:667–693.

Schlichting, C. D. and D. A. Levin. 1984. Phenotypic plasticity of annual *Phlox*: Tests of some hypotheses. *Amer. J. Bot.* 71:252–260.

—— 1986. Phenotypic plasticity: An evolving plant character. *Biol. J. Linn. Soc.* 29:37–47.

Schmid, R. 1972a. Floral bundle fusion and vascular conservatism. *Taxon* 21:429–446.

—— 1972b. A resolution of the *Eugenia-Syzygium* controversy (Myrtaceae). *Amer. J. Bot.* 59:423–436.

Schmid, R. and P. Baas. 1984. The occurrence of scalariform perforation plates and helical vessel wall thickenings in wood of Myrtaceae. *IAWA Bull.*, n.s., 5:197–215.

Schmidt, K. P. 1952. The "Methodus" of Linnaeus, 1736. *J. Soc. Bibliogr. Nat. Hist.* 2:369–374.

Schnarf, K. 1929. Embryologie der Angiospermen. In K. Linsbauer, ed., *Handbuch der Pflanzenanatomie*, vol. 10, pt. 2. Berlin: Borntraeger.

—— 1931. *Vergleichende Embryologie der Angiospermen*. Berlin: Borntraeger.

Schnepf, E. and A. Klasova. 1972. Zur Feinstruktur von Öl-und Flavon-Drüsen. *Ber. Deutsch. Bot. Ges.* 85:249–258.

Schoch, R. M. 1986. *Phylogeny Reconstruction in Paleontology*. New York: Van Nostrand Reinhold.

Schofield, E. K. 1968. Petiole anatomy of the Guttiferae and related families. *Mem. New York Bot. Gard.* 18(1):1–55.

Schopf, J. M. 1969. Systematics and nomenclature in palynology. In R. H. Tschudy and R. A. Scott, eds., *Aspects of Palynology*, pp. 49–77. New York: Wiley.

—— 1970. Relation of floras of the southern hemisphere to continental drift. *Taxon* 19:657–674.

Schröter, C. 1908. *Das Pflanzenleben der Alpen*. Zürich: Albert Raustein.

Schuh, R. T. and F. J. Farris. 1981. Methods for investigating taxonomic congruence and their application to the Leptopodomorpha. *Syst. Zool.* 30:331–351.

Schuh, R. T. and J. T. Polhemus. 1980. Analysis of taxonomic congruence among morphological, ecological, and biogeographic data sets for the Leptopodomorpha (Hemiptera). *Syst. Zool.* 29:1–26.

Schulz, G. E. 1977. Recognition of phylogenetic relationships from polypeptide chain fold similarities. *J. Mol. Evol.* 9:339–342.

Schulz, P. and W. A. Jensen. 1969. *Capsella* embryogenesis: The suspensor and the basal cell. *Protoplasma* 67:139–163.

—— 1971. *Capsella* embryogenesis: The chalazal proliferating tissue. *J. Cell Sci.* 8:201–227.

—— 1973. *Capsella* embryogenesis: The central cell. *J. Cell Sci.* 12:741–763.

—— 1974. *Capsella* embryogenesis: The development of the free nuclear endosperm. *Protoplasma* 80:183–205.

Schulz, R. and W. Jensen. 1968a. *Capsella* embryogenesis: The early embryo. *J. Ultrastruct. Res.* 22:376–392.

—— 1968b. *Capsella* embryogenesis: The synergids before and after fertilization. *Amer. J. Bot.* 55:541–552.

—— 1968c. *Capsella* embryogenesis: The egg, zygote, and young embryo. *Amer. J. Bot.* 55:807–819.

Schuyler, A. E. 1971. Scanning electron microscopy of achene epidermis in species of *Scirpus* (Cyperaceae) and related genera. *Proc. Acad. Nat. Sci.* (Philadelphia) 123:29–52.

Schweizer, D. 1983. Distamycin-DAPI bands: Properties and occurrence in species. In P. E. Brandham and M. D. Bennett, eds., *Kew Chromosome Conference II*, pp. 43–51. London: George Allen & Unwin.

Scogin, R. 1984. Anthocyanins of Begoniaceae: 2. Additional data and cladistic analysis. *Aliso* 11:115–120.

Scora, R. W. 1966. The evolution of the genus *Monarda* (Labiatae). *Evolution* 20:185–190.

—— 1967. Divergence in *Monarda* (Labiatae). *Taxon* 16:499–505.

Scott, P. J. 1973. A consideration of the category in classification. *Taxon.* 22:405–406.

Seaman, F. C. 1982. Sesquiterpene lactones as taxonomic characters in the Asteraceae. *Bot. Rev.* 48:121–595.

Seaman, F. C. and V. A. Funk. 1983. Cladistic analysis of complex natural products: Developing transformation series from sesquiterpene lactone data. *Taxon* 32:1–27.

Seavey, S. R., R. E. Magill, and P. H. Raven. 1977. Evolution of seed size, shape, and surface architecture in the tribe Epilobieae (Onagraceae). *Ann. Missouri Bot. Gard.* 64:18–47.

Sederoff, R. R. 1987. Molecular mechanisms of mitochondrial-genome evolution in higher plants. *Amer. Naturalist* 130 (suppl.):s30–s45.

Seigler, D. and P. W. Price. 1976. Secondary compounds in plants: Primary functions. *Amer. Naturalist* 110:101–105.

Seki, T. 1968. A revision of the family Sematophyllaceae of Japan with special reference to a statistical demarcation of the family. *J. Sci. Hiroshima Univ.* ser. B, Div. 2, Bot. 12:1–80.

Seong, L. F. 1972. Numerical taxonomic studies on North American lady ferns and their allies. *Taiwania* 17:190–221.

Serota, C. A. and B. W. Smith. 1967. The cyto-ecology of four species of *Trillium* from western North Carolina. *Amer. J. Bot.* 54:169–181.

Settle, T. W. 1979. Popper on "When is a science not a science?" *Syst. Zool.* 28:521–529.

Shaffer, H. B. 1986. Utility of quantitative genetic parameters in character weighting. *Syst. Zool.* 35:124–134.

Sharma, A. K. 1983. Additional genetic materials in chromosomes. In P. E. Brandham and M. D. Bennett, eds., *Kew Chromosome Conference II*, pp. 35–42. London: George Allen & Unwin.

Sharma, A. K. and A. Sharma. 1980. *Chromosome Techniques: Theory and Practice.* 3d ed. London: Butterworths.

Sharma, G. K. and D. B. Dunn. 1968. Effect of environment on the cuticular features in *Kalanchoe fedschenkoi. Bull. Torrey Bot. Club* 95:464–473.

—— 1969. Environmental modifications of leaf surface traits in *Datura stramonium. Canad. J. Bot.* 47:1211–1216.

Sharma, S. K., C. R. Babu, B. M. Johri, and A. Hepworth. 1977. SEM studies on seed coat patterns in *Phaseolus mungo-radiatus-sublobatus* complex. *Phytomorphology* 27:106–111.

Sharp, A. J. 1964. The compleat botanist. *Science* 146:745–748.

Sharp, L. W. 1943. *Fundamentals of Cytology*. New York: McGraw-Hill.

Shaw, G. 1971. The chemistry of sporopollenin. In J. Brooks, P. R. Grant, M. Muir, P. van Gijzel, and G. Shaw, eds., *Sporopollenin*, pp. 305–348. London: Academic Press.

Shaw, S. and R. M. Keddie. 1983. A numerical taxonomic study of the genus *Kurthia* with a revised description of *Kurthia zopfii* and a description of *Kurthia gibsonii* sp. nov. *Syst. Appl. Microbiol.* 4:253–276.

Shepard, R. N., A. K. Romney, and S. B. Nerlove, eds., 1972. *Multidimensional Scaling: Theory and Applications in the Behavioral Sciences.* 2 vols. New York: Seminar Press.

Sherff, E. E. 1940. The concept of the genus. IV. The delimitations of genera from the conservative point of view. *Bull. Torrey Bot. Club* 67:375–380.

Sherwin, P. A. and R. L. Wilbur. 1971. The contributions of floral anatomy to the generic placement of *Diamorpha smallii* and *Sedum pusillum. J. Elisha Mitchell Sci. Soc.* 87:103–114.

Shivanna, K. R. and B. M. Johri. 1985. *The Angiosperm Pollen: Structure and Function.* New Delhi: Wiley Eastern.

Shukla, P. and S. P. Misra. 1979. *An Introduction to Taxonomy of Angiosperms.* Sahibabad, India: Vikas.

Shull, G. H. 1923. The species concept from the point of view of a geneticist. *Amer. J. Bot.* 10:221–228.

Sibley, C. G. 1954. The contribution of avian taxonomy. *Syst. Zool.* 3:105–110, 125.

Sibley, C. G. and J. E. Ahlquist. 1983. Phylogeny and classification of birds based on the data of DNA-DNA hybridization. *Current Ornithology* 4:245–292.

—— 1984. The phylogeny of the hominoid primates, as indicated by DNA-DNA hybridization. *J. Mol. Evol.* 20:2–15.

Siegler, R. S. 1983. How knowledge influences learning. *Amer. Scientist* 71:631–638.

Simpson, B. B. 1973. Contrasting modes of evolution in two groups of *Perezia* (Mutisieae; Compositae) of southern South America. *Taxon* 22:525–536.

Simpson, B. B., J. L. Neff, and D. Seigler. 1977. *Krameria*, free fatty acids and oil-collecting bees. *Nature* 267:150–151.

Simpson, D. R. and D. Janos. 1974. *Punch Card Key to the Families of Dicotyledons of the Western Hemisphere South of the United States.* Chicago: Field Museum of Natural History.

Simpson, G. G. 1951. The species concept. *Evolution* 5:285–298.

—— 1953. *The Major Features of Evolution.* New York: Columbia Univ. Press.

—— 1961. *Principles of Animal Taxonomy.* New York: Columbia Univ. Press.

—— 1975. Recent advances in methods of phylogenetic inference. In W. P. Luckett and F. S. Szalay, eds., *Phylogeny of the Primates*, pp. 3–19. New York: Plenum Press.

—— 1980. *Why and How: Some Problems and Methods in Historical Biology.* New York: Pergamon Press.

Sims, L. E. and H. J. Price. 1985. Nuclear DNA content variation in *Helianthus* (Asteraceae). *Amer. J. Bot.* 72:1213–1219.

Sims, R. W., J. H. Price, and P. E. S. Whalley, eds. 1983. *Evolution, Time and Space: The Emergence of the Biosphere.* London: Academic Press.

Sinclair, W. A. 1951. *The Traditional Formal Logic: A Short Account for Students.* 5th ed. London: Methuen.

Singer, R. 1986. *The Agaricales in Modern Taxonomy.* 4th ed. Koenigstein: Koeltz.

Sinha, U. and S. Sinha. 1980. *Cytogenetics, Plant Breeding and Evolution.* 2d ed. Sahibabad, India: Vikas.

Sivarajan, V. V. 1980. Contributions of palynology to angiosperm systematics. In P. K. K. Nair, ed., *Advances in Pollen-Spore Research*, vols. 5–7, pp. 35–50. New Delhi: Today & Tomorrow's Printers and Publishers.

—— 1984. *Introduction to Principles of Plant Taxonomy.* New Delhi: Oxford & IBH.

Skalinska, M. and E. Pogan. 1973. A list of chromosome numbers of Polish angiosperms. *Acta Biol. Cracov., Bot.* 16:145–201.

Sklansky, J., ed. 1973. *Pattern Recognition: Introduction and Foundations.* Stroudsburg, Pa.. Dowden, Hutchinson & Ross.

Sklar, A. 1964. On category overlapping in taxonomy. In J. R. Gregg and F. T. C. Harris, eds., *Form and Strategy in Science: Studies Dedicated to Joseph Henry Wheeler on the Occasion of His Seventieth Birthday,* pp. 395–401. Dordrecht: D. Reidel.

Skvarla, J. J. and D. A. Larson. 1965. An electron microscopic study of pollen morphology in the Compositae with special reference to the Ambrosiinae. *Grana Palynol.* 6:210–269.

Skvarla, J. J., P. H. Raven, W. F. Chissoe, and M. Sharp. 1978. An ultrastructural study of viscin threads in Onagraceae pollen. *Pollen et Spores* 20:5–143.

Skvarla, J. J., P. H. Raven, and J. Praglowski. 1975. The evolution of pollen tetrads in Onagraceae. *Amer. J. Bot.* 62:6–35.

Skvarla, J. J. and J. R. Rowley. 1970. The pollen wall of *Canna* and its similarity to the germinal apertures of other pollen. *Amer. J. Bot.* 57:519–529.

—— 1987. Ontogeny of pollen in *Poinciana* (Leguminosae). I. Development of exine template. *Rev. Palaeobot. Palynol.* 50:293–311.

Skvarla, J. J., J. R. Rowley, and W. F. Chissoe. 1988. Adaptability of scanning electron microscopy to studies of pollen morphology. *Aliso* 12:119–175.

Skvarla, J. J. and B. L. Turner. 1966a. Pollen wall ultrastructure and its bearing on the systematic position of *Blennosperma* and *Crocidium* (Compositae). *Amer. J. Bot.* 53:555–563.

—— 1966b. Systematic implications from electron microscopic studies of Compositae pollen—a review. *Ann. Missouri Bot. Gard.* 53:220–256.

Skvarla, J. J., B. L. Turner, V. C. Patel, and A. S. Tomb. 1977. Pollen morphology in the Compositae and in morphologically related families. In V. H. Heywood, J. B. Harborne, and B. L. Turner, eds., *The Biology and Chemistry of the Compositae,* pp. 141–248. London: Academic Press.

Skvortsov, A. K. and I. I. Rusanovitch. 1974. Scanning electron microscopy of the seed-coat surface in *Epilobium* species. *Bot. Not.* 127:392–401.

Slatkin, M. 1985. Gene flow in natural populations. *Ann. Rev. Ecol. Syst.* 16:393–430.

Slingsby, P. and W. J. Bond. 1984. The influence of ants on the dispersal distance and seedling recruitment of *Leucospermum conocarpodendron* (L.) Buek (Proteaceae). *S. African J. Bot.* 51:30–34.

Slobodchikoff, C. N., ed. 1976. *Concepts of Species.* Stroudsburg, Pa.: Dowden, Hutchinson & Ross.

Sluiman, H. J. 1985. A cladistic evaluation of the lower and higher green plants (Viridiplantae). *Pl. Syst. Evol.* 149:217–232.

Small, E. 1978a. A numerical and nomenclatural analysis of morpho-geographic taxa of *Humulus. Syst. Bot.* 3:37–76.

—— 1978b. A numerical taxonomic analysis of the *Daucus carota* complex. *Canad. J. Bot.* 56:248–276.

—— 1981. A numerical analysis of major groupings in *Medicago* employing traditionally used characters. *Canad. J. Bot.* 59:1553–1577.

Small, E., I. J. Bassett, C. W. Crompton, and H. Lewis. 1971. Pollen phylogeny in *Clarkia. Taxon* 20:739–746.

Small, E. and B. S. Brookes. 1983. The systematic value of stigma morphology in the legume tribe Trifolieae with particular reference to *Medicago. Canad. J. Bot.* 61:2388–2404.

Small, E., P. Y. Jui and L. P. Lefkovitch. 1976. A numerical taxonomic analysis of *Cannabis* with special reference to species delimitation. *Syst. Bot.* 1:67–84.

Small, E., L. P. Lefkovitch, and D. Classen. 1982. Character set incongruence in *Medicago. Canad. J. Bot.* 60:2505–2510.

Smets, E. 1984. Dahlgren's systems of classification (1975 & 1980): Implications on taxonomical ordening and impact on character state analysis. *Bull. Jard. Bot. Natl. Belg.* 54:183–211.

Smith, A. C. 1969. Systematics and appreciation of reality. *Taxon* 18:5–13.

Smith, B. B. 1973. The use of a new clearing technique for the study of early ovule development, megasporogenesis, and megagametogenesis in five species of *Cornus* L. *Amer. J. Bot.* 60:322–338.

—— 1975. A quantitative analysis of the megagametophyte of five species of *Cornus* L. *Amer. J. Bot.* 62:387–394.

Smith, B. N. and B. L. Turner. 1975. Distribution of Kranz syndrome among Asteraceae. *Amer. J. Bot.* 62:541–545.

Smith, B. W. 1974. Cytological evidence. In A. E. Radford, W. C. Dickison, J. R. Massey, and C. R. Bell, *Vascular Plant Systematics*, pp. 237–258. New York: Harper & Row.

Smith, E. B. 1968. Pollen competition and relatedness in *Haplopappus* section *Isopappus*. *Bot. Gaz.* 129:371–373.

—— 1970. Pollen competition and relatedness in *Haplopappus* section *Isopappus* (Compositae). II. *Amer. J. Bot.* 57:874–880.

Smith, G. D., dir. 1960. *Soil Classification: A Comprehensive System: 7th Approximation.* U.S. Dept. Agric. Soil Conservation Service.

Smith, H. M. and F. N. White. 1956. A case for the trinomen. *Syst. Zool.* 5:183–190.

Smith, J. E. 1809. *An Introduction to Physiological and Systematical Botany.* 2d ed. London.

Smith, P. G. and J. B. Phipps. 1984. Consensus trees in phenetic analyses. *Taxon* 33:586–594.

Smith, P. M. 1976. *The Chemotaxonomy of Plants.* London: Arnold.

Smith-White, S. 1968. *Brachycome lineariloba:* A species for experimental cytogenetics. *Chromosoma* (Berlin) 23:359–364.

Smoot, E. L., R. K. Jansen, and T. N. Taylor. 1981. A phylogenetic analysis of the land plants: A botanical commentary. *Taxon* 30:65–67.

Smouse, P. E. and W.-H. Li. 1987. Likelihood analysis of mitochondrial restriction-cleavage patterns for the human-chimpanzee-gorilla trichotomy. *Evolution* 41:1162–1176.

Sneath, P. H. A. 1957. The application of computers to taxonomy. *J. Gen. Microbiol.* 17:201–226.

—— 1962. The construction of taxonomic groups. In G. C. Ainsworth and P. H. A. Sneath, eds., *Microbial Classification*, pp. 289–332. Cambridge: Cambridge Univ. Press.

—— 1968a. Goodness of intuitive arrangements into time trends based on complex pattern. *Syst. Zool.* 17:256–260.

—— 1968b. Numerical taxonomic study of the graft chimaera + *Laburnocytisus adamii (Cytisus purpureus* + *Laburnum anagyroides). Proc. Linn. Soc. London* 179:83–96.

—— 1971. Numerical taxonomy: Criticisms and critiques. *Biol. J. Linn. Soc.* 3:147–157.

—— 1976a. Phenetic taxonomy at the species level and above. *Taxon* 25:437–450.

—— 1976b. Some applications of numerical taxonomy to plant breeding. *Z. Pflanzenzücht.* 76:19–46.

—— 1979. Numerical taxonomy and automated identification: Some implications for geology. *Computers and Geosciences* 5:41–46.

Sneath, P. H. A. and A. O. Chater. 1978. Information content of keys for identification. In H. E. Street, ed., *Essays in Plant Taxonomy*, pp. 79–95. London: Academic Press.

Sneath, P. H. A. and R. R. Sokal. 1962. Numerical taxonomy. *Nature* 193:855–860.

—— 1973. *Numerical Taxonomy: The Principles and Practice of Numerical Classification.* San Francisco: Freeman.

Sober, E. 1983. Parsimony in systematics: Philosophical issues. *Ann. Rev. Ecol. Syst.* 14:335–357.

—— 1985. A likelihood justification of parsimony. *Cladistics* 1:209–233.

—— 1986. Parsimony and character weighting. *Cladistics* 2:28–42.

Sokal, R. R. 1965. Statistical methods in systematics. *Biol. Rev. Cambridge Philos. Soc.* 40:337–391.

—— 1966. Numerical taxonomy. *Scientific Amer.* 215(6):106–116, 155–156.

—— 1973. The species problem reconsidered. *Syst. Zool.* 22:360–374.

—— 1974. Classification: Purposes, principles, progress, prospects. *Science* 185:1115–1123.

—— 1983a. A phylogenetic analysis of the Caminalcules. I. The data base. *Syst. Zool.* 32:159–184.

—— 1983b. A phylogenetic analysis of the Caminalcules. II. Estimating the true cladogram. *Syst. Zool.* 32:185–201.

—— 1983c. A phylogenetic analysis of the Caminalcules. III. Fossils and classification. *Syst. Zool.* 32:248–258.

—— 1983d. A phylogenetic analysis of the Caminalcules. IV. Congruence and character stability. *Syst. Zool.* 32:259–275.

—— 1983e. Taxonomic congruence in the Caminalcules. In J. Felsenstein, ed., *Numerical Taxonomy*, pp. 76–81. Berlin: Springer-Verlag.

—— 1986. Phenetic taxonomy: Theory and methods. *Ann. Rev. Ecol. Syst.* 17:423–442.

Sokal, R. R., J. H. Camin, F. J. Rohlf, and P. H. A. Sneath. 1965. Numerical taxonomy: Some points of view. *Syst. Zool.* 14:237–243.

Sokal, R. R. and T. J. Crovello. 1970. The biological species concept: A critical evaluation. *Amer. Naturalist* 104:127–153.

Sokal, R. R., T. J. Crovello, and R. S. Unnasch. 1986. Geographic variation of vegetative characters of *Populus deltoides. Syst. Bot.* 11:419–432.

Sokal, R. R., K. L. Fiala, and G. Hart. 1984. OTU stability and factors determining taxonomic stability: Examples from the Caminalcules and the Leptopodomorpha. *Syst. Zool.* 33:387–407.

Sokal, R. R. and C. D. Michener. 1958. A statistical method for evaluating systematic relationships. *Univ. Kansas Sci. Bull.* 38:1409–1438.

Sokal, R. R. and N. Oden. 1978a. Spatial autocorrelation in biology. I. Methodology. *Biol. J. Linn. Soc.* 10:199–228.

—— 1978b. Spatial autocorrelation in biology. II. Some biological implication and four applications of evolutionary and ecological interest. *Biol. J. Linn. Soc.* 10:229–249.

Sokal, R. R. and F. J. Rohlf. 1962. The comparison of dendrograms by objective methods. *Taxon* 11:33–40.

—— 1966. Random scanning of taxonomic characters. *Nature* 210:461–462.

—— 1970. The intelligent ignoramus, an experiment in numerical taxonomy. *Taxon* 19:305–319.

—— 1981a. *Biometry: The Principles and Practice of Statistics in Biological Research.* 2d ed. San Francisco: Freeman.

—— 1981b. Taxonomic congruence in the Leptopodomorpha re-examined. *Syst. Zool.* 30:309–324.

Sokal, R. R. and P. H. A. Sneath. 1963. *Principles of Numerical Taxonomy.* San Francisco: Freeman.

Sokolovskaya, A. P. and N. S. Probatova. 1977. [On the least chromosome number ($2n = 4$) of *Colpodium versicolor* (Stev.) Woronow (Poaceae)]. *Bot. Zh.* 62:241–245. (In Russian)

Solbrig, O. T. 1968. Fertility, sterility and the species problem. In V. H. Heywood, ed., *Modern Methods in Plant Taxonomy*, pp. 77–96. London: Academic Press.

—— 1970a. The phylogeny of *Gutierrezia:* An eclectic approach. *Brittonia* 22:217–229.

—— 1970b. *Principles and Methods of Plant Biosystematics.* Toronto: Macmillan.

—— 1973. Chromosome cytology and arboreta: A marriage of convenience. *Arnoldia* 33:135–146.

——, ed. 1980. *Demography and Evolution in Plant Populations.* Oxford: Blackwell.

Solbrig, O. T., L. C. Anderson, D. W. Kyhos, and P. H. Raven. 1969. Chromosome numbers in Compositae VII: Astereae III. *Amer. J. Bot.* 56:348–353.

Solbrig, O. T., L. C. Anderson, D. W. Kyhos, P. H. Raven, and L. Rüdenberg. 1964. Chromosome numbers in Compositae V. Astereae II. *Amer. J. Bot.* 51:513–519.

Solbrig, O. T., D. W. Kyhos, M. Powell, and P. H. Raven. 1972. Chromosome numbers in Compositae VIII: Heliantheae. *Amer. J. Bot.* 59:869–878.

Solereder, H. 1908. *Systematic Anatomy of the Dicotyledons: A Handbook for Laboratories of Pure and Applied Botany.* Translated by L. A. Boodle and F. E. Fritsch; revised by D. H. Scott. 2 vols. Oxford: Clarendon Press.

Solomon, A. M. and H. D. Hayes. 1972. Desert pollen production I: Qualitative influence of moisture. *J. Arizona Acad. Sci.* 7:52–74.

Soltis, D. E. 1981. Variation in hybrid fertility among the disjunct populations and species of *Sullivantia* (Saxifragaceae). *Canad. J. Bot.* 59:1174–1180.

—— 1982. Allozymic variability in *Sullivantia* (Saxifragaceae). *Syst. Bot.* 7:26–34.

—— 1983. Supernumerary chromosomes in *Saxifraga virginiensis* (Saxifragaceae). *Amer. J. Bot.* 70:1007–1010.

—— 1984a. Autopolyploidy in *Tolmiea menziesii* (Saxifragaceae). *Amer. J. Bot.* 71:1171–1174.

—— 1984b. Karyotypes of *Leptarrhena* and *Tanakaea* (Saxifragaceae). *Canad. J. Bot.* 62:671–673.

—— 1988. Karyotypes of *Bensoniella, Conimitella, Lithophragma* and *Mitella*, and relationships in Saxifrageae (Saxifragaceae). *Syst. Bot.* 13:64–72.

Soltis, D. E. and B. A. Bohm. 1985. Chromosomal and flavonoid chemical confirmation of intergeneric hybridization between *Tolmiea* and *Tellima* (Saxifragaceae). *Canad. J. Bot.* 63:1309–1312.

Soltis, D. E. and L. H. Rieseberg. 1986. Autopolyploidy in *Tolmiea menziesii* (Saxifragaceae): Genetic insights for enzyme electrophoresis. *Amer. J. Bot.* 73:310–318.

Soria, J. and C. B. Heiser, Jr. 1961. A statistical study of relationships of certain species of the *Solanum nigrum* complex. *Econ. Bot.* 15:245–255.

Southworth, D. 1974. Solubility of pollen exines. *Amer. J. Bot.* 61:36–44.

Speta, F. 1977. Proteinkörper in Zellkernen: Neue Ergebnisse und deren Bedeutung für die Gefässpflanzensystematik nebst einer Literaturübersicht für die Jahre 1966–1976. *Candollea* 32:133–163.

—— 1979. Weitere Untersuchungen über Proteinkörper in Zellkernen und ihre taxonomische Bedeutung. *Pl. Syst. Evol.* 132:1–26.

Spooner, D. M. 1984. Reproductive features of *Dentaria laciniata* and *D. diphylla* (Cruciferae), and the implications in the taxonomy of the eastern North American *Dentaria* complex. *Amer. J. Bot.* 71:999–1005.

Spooner, D. M., T. F. Stuessy, D. J. Crawford, and M. Silva O. 1987. Chromosome numbers from the flora of the Juan Fernandez Islands, II. *Rhodora* 89:351–356.

Sporne, K. R. 1948. Correlation and classification in dicotyledons. *Proc. Linn. Soc. London* 160:40–47.

—— 1954. Statistics and the evolution of dicotyledons. *Evolution* 8:55–64.

—— 1956. The phylogenetic classification of the angiosperms. *Biol. Rev.* 31:1–29.

—— 1972. Some observations on the evolution of pollen types in dicotyledons. *New Phytol.* 71:181–186.

—— 1975. *The Morphology of Angiosperms: The Structure and Evolution of Flowering Plants.* New York: St. Martin's Press.

—— 1976. Character correlations among angiosperms and the importance of fossil evidence in assessing their significance. In C. B. Beck, ed., *Origin and Early Evolution of Angiosperms*, pp. 312–329. New York: Columbia Univ. Press.

Sprague, T. A. 1940. Taxonomic botany, with special reference to the angiosperms. In J. Huxley, ed., *The New Systematics*, pp. 435–454. Oxford: Oxford Univ. Press.

Sprengel, C. K. 1793. *Das entdeckte Geheimniss der Natur im Bau und in der Befruchtung.* Berlin.

Stace, C. A. 1965. Cuticular studies as an aid to plant taxonomy. *Bull. Brit. Mus. (Nat. Hist.), Bot.* 4:1–78.

—— 1966. The use of epidermal characters in phylogenetic considerations. *New Phytol.* 65:304–318.

—— 1969. The significance of the leaf epidermis in the taxonomy of the Combretaceae III. The genus *Combretum* in America. *Brittonia* 21:130–143.

—— 1980. *Plant Taxonomy and Biosystematics.* London: Arnold.

—— 1984. The taxonomic importance of the leaf surface. In V. H. Heywood and D. M. Moore, eds., *Current Concepts in Plant Taxonomy*, pp. 67–94. London: Academic Press.

—— 1986. The present and future infraspecific classification of wild plants. In B. T. Styles, ed., *Infraspecific Classification of Wild and Cultivated Plants*, pp. 9–20. Oxford: Clarendon Press.

Stafleu, F. A. 1971. Linnaeus and the Linnaeans: The spreading of their ideas in systematic botany, 1735–1789. *Regnum Veg.* 79:1–386.

Stanley, R. G. and H. F. Linskens. 1974. *Pollen: Biology, Biochemistry, Management.* Berlin: Springer-Verlag.

Stanley, S. M. 1979. *Macroevolution: Pattern and Process.* San Francisco: Freeman.

Starrett, A. 1958. What *is* the subspecies problem? *Syst. Zool.* 7:111–115.

Stearn, W. T. 1957. An introduction to the *Species Plantarum* and cognate botanical works of Carl Linnaeus. In C. Linnaeus, *Species Plantarum*, 1:v–xiv, 1–76. Facsimile ed. London: Ray Society.

—— 1961. Botanical gardens and botanical literature in the eighteenth century. In Allan Stevenson, comp., *Catalogue of Botanical Books in the Collection of Rachel McMasters Miller Hunt*, vol. 2, part 1: *Introduction to Printed Books 1701–1800*, pp. xli–cxl. Pittsburgh: Hunt Botanical Library.

—— 1964. Problems of character selection and weighting: Introduction. In V. H. Heywood and J. McNeill, eds., *Phenetic and Phylogenetic Classification*, pp. 83–86. London: Systematics Association.

—— 1971. Linnaean classification, nomenclature, and method. In W. Blunt. *The Compleat Naturalist: A Life of Linnaeus*, pp. 242–252. New York: Viking Press.

—— 1983. *Botanical Latin: History, Grammar, Terminology and Vocabulary.* 3d ed. Newton Abbot, England: David and Charles.

Stebbins, G. L. 1966. Chromosomal variation and evolution. *Science* 152:1463–1469.

—— 1967. Adaptive radiation and trends of evolution in higher plants. *Evol. Biol.* 1:101–142.

—— 1970a. Adaptive radiation of angiosperms, I: Pollination mechanisms. *Ann. Rev. Ecol. Syst.* 1:307–326.

—— 1970b. Biosystematics: An avenue towards understanding evolution. *Taxon* 19:205–214.

—— 1971a. Adaptive radiation of reproductive characters of angiosperms. II: Seeds and seedlings. *Ann. Rev. Ecol. Syst.* 2:237–260.

—— 1971b. *Chromosomal Evolution in Higher Plants.* London: Arnold.

—— 1971c. Relationships between adaptive radiation, speciation and major evolutionary trends. *Taxon* 20:3–16.

—— 1973. Morphogenesis, vascularization and phylogeny in angiosperms. *Breviora* 418:1–19.

—— 1974. *Flowering Plants: Evolution Above the Species Level.* Cambridge, Mass.: Harvard Univ. Press.

—— 1975. Deductions about transspecific evolution through extrapolation from processes at the population and species level. *Ann. Missouri Bot. Gard.* 62:825–834.

—— 1977. *Processes of Organic Evolution.* 3d ed. Englewood Cliffs, N.J.: Prentice-Hall.

—— 1984a. Chromosome pairing, hybrid sterility, and polyploidy: A reply to R. C. Jackson. *Syst. Bot.* 9:119–121.

—— 1984b. Polyploidy and the distribution of the arctic-alpine flora: New evidence and a new approach. *Bot. Helvet.* 94:1–13.

Stebbins, G. L. and F. J. Ayala. 1981. Is a new evolutionary synthesis emerging? *Science* 213:967–971.

Steere, W. C., ed. 1958. *Fifty Years of Botany: Golden Jubilee Volume of the Botanical Society of America.* New York: McGraw-Hill.

Steeves, T. A. and I. M. Sussex. 1972. *Patterns in Plant Development.* Englewood Cliffs, N.J.: Prentice-Hall.

Stein, W. E., Jr. 1987. Phylogenetic analysis and fossil plants. *Rev. Palaeobot. Palynol.* 50:31–61.

Stern, K. R. 1970. Pollen aperture variation and phylogeny in *Dicentra* (Fumariaceae). *Madroño* 20:354–359.

Stern, W. L. 1967. *Kleinodendron* and xylem anatomy of Cluytieae (Euphorbiaceae). *Amer. J. Bot.* 54:663–676.

Stern, W. L. and K. L. Chambers. 1960. The citation of wood specimens and herbarium vouchers in anatomical research. *Taxon* 9:7–13.

Sterner, R. W. and D. A. Young. 1980. Flavonoid chemistry and the phylogenetic relationships of the Idiospermaceae. *Syst. Bot.* 5:432–437.

Stevens, P. F. 1980a. Evolutionary polarity of character states. *Ann. Rev. Ecol. Syst.* 11:333–358.

—— 1980b. A revision of the Old World species of *Calophyllum* (Guttiferae). *J. Arnold Arbor.* 61:117–699.

—— 1981. On ends and means, or how polarity criteria can be assessed. *Syst. Bot.* 6:186–188.

—— 1984a. Homology and phylogeny: Morphology and systematics. *Syst. Bot.* 9:395–409.

—— 1984b. Metaphors and typology in the development of botanical systematics 1690–1960, or the art of putting new wine in old bottles. *Taxon* 33:169–211.

—— 1985. The genus in practice—but for what practice? *Kew Bull.* 40:457–465.

—— 1986. Evolutionary classification in botany, 1960–1985. *J. Arnold Arbor.* 67:313–339.

Steward, F. C. 1968. *Growth and Organization in Plants: Structure, Development, Metabolism, Physiology.* Reading, Mass.: Addison-Wesley.

Steyskal, G. C. 1965. Notes on uninominal nomenclature. *Syst. Zool.* 14:346–348.

—— 1968. The number and kind of characters needed for significant numerical taxonomy. *Syst. Zool.* 17:474–477.

Stinebrickner, R. 1984a. An extension of intersection methods from trees to dendrograms. *Syst. Zool.* 33:381–386.

—— 1984b. *s*-Consensus trees and indices. *Bull. Math. Biol.* 46:923–936.

Stix, E. 1960. Pollenmorphologische Untersuchungen an Compositen. *Grana* 2(2):41–104.

St. John, H. 1958. *Nomenclature of Plants: A Text for the Application by the Case Method of the International Code of Botanical Nomenclature.* New York: Ronald Press.

Stone, A. R. and D. L. Hawksworth, eds. 1986. *Coevolution and Systematics.* Oxford: Clarendon Press.

Stone, D. E. 1961. Ploidal level and stomatal size in the American hickories. *Brittonia* 13:293–302.

—— 1963. Pollen size in hickories *(Carya)*. *Brittonia* 15:208–214.

Stone, D. E., S. C. Sellers, and W. J. Kress. 1979. Ontogeny of exineless pollen in *Heliconia*, a banana relative. *Ann. Missouri Bot. Gard.* 66:701–730.

Stort, M. N. S. 1984. Phylogenetic relationship between species of the genus *Cattleya* as a function of crossing compatibility. *Revista Biol. Trop.* 32:223–226.

Stott, P. 1981. *Historical Plant Geography: An Introduction.* London: George Allen & Unwin.

Stoutamire, W. P. and J. H. Beaman. 1960. Chromosome studies of Mexican alpine plants. *Brittonia* 12:226–230.

Strasburger, E. A. 1877. Ueber Befruchtung und Zelltheilung. *Jenaische Z. Naturwiss.* 11:435–536.

—— 1879. *Die Angiospermen und die Gymnospermen.* Jena: Fischer.

—— 1884. *Neue Untersuchungen über den Befruchtungsvorgang bei den Phanerogamen.* Jena: Fischer.

Strathmann, R. R. and M. Slatkin. 1983. The improbability of animal phyla with few species. *Paleobiology* 9:97–106.

Strauch, J. G., Jr. 1978. The phylogeny of the Charadriiformes (Aves): A new estimate using the method of character compatibility analysis. *Trans. Zool. Soc. London* 34:263–345.

Strickland, R. G. 1974. The nature of the white colour of petals. *Ann. Bot.* 38:1033–1037.

Strother, J. L. 1969. Systematics of *Dyssodia* Cavanilles (Compositae: Tageteae). *Univ. California Publ. Bot.* 48:1–88.

Stuckey, R. L. 1968. Distributional history of *Butomus umbellatus* (flowering-rush) in the western Lake Erie and Lake St. Clair region. *Michigan Bot.* 7:134–142.

—— 1970. Distributional history of *Epilobium hirsutum* (great hairy willow-herb) in North America. *Rhodora* 72:164–181.

—— 1974. The introduction and distribution of *Nymphoides peltatum* (Menyanthaceae) in North America. *Bartonia* 42:14–23.

—— 1979. Distributional history of *Potamogeton crispus* (curly pondweed) in North America. *Bartonia* 46:22–42.

—— 1985. Distributional history of *Najas marina* (spiny naiad) in North America. *Bartonia* 51:2–16.

Stuckey, R. L. and J. L. Forsyth. 1971. Distribution of naturalized *Carduus nutans* (Compositae) mapped in relation to geology in northwestern Ohio. *Ohio J. Sci.* 71:1–15.

Stuckey, R. L. and W. L. Phillips. 1970. Distributional history of *Lycopus europaeus* (European water-horehound) in North America. *Rhodora* 72:351–369.

Stuckey, R. L. and E. D. Rudolph. 1974. History of botany 1947–1972 with a bibliographic appendix. *Ann. Missouri Bot. Gard.* 61:237–261.

Stuckey, R. L. and D. P. Salamon. 1987. *Typha angustifolia* in North America: A foreigner masquerading as a native [abstract]. *Ohio J. Sci.* 87(2):4.

Stucky, J. and R. C. Jackson. 1975. DNA content of seven species of Astereae and its significance to theories of chromosome evolution in the tribe. *Amer. J. Bot.* 62:509–518.

Stuessy, T. F. 1969. Re-establishment of the genus *Unxia* (Compositae-Heliantheae). *Brittonia* 21:314–321.

—— 1970. Chromosome studies in *Melampodium* (Compositae, Heliantheae). *Madroño* 20:365–372.

—— 1971a. Chromosome numbers and phylogeny in *Melampodium* (Compositae). *Amer. J. Bot.* 58:732–736.

—— 1971b. Systematic relationships in the white-rayed species of *Melampodium* (Compositae). *Brittonia* 23:177–190.

—— 1972a. Revision of the genus *Melampodium* (Compositae: Heliantheae). *Rhodora* 74:1–70, 161–219.

—— ed. 1972b. *References and Lists of Equipment for Systematic Studies with Flavonoids, Mustard Oils, Proteins (Electrophoresis and Serology), and Terpenoids.* Prepared for ASPT symposium "Biochemistry and the Taxonomist," AIBS Meetings, Minneapolis. iii, 24 pp. mimeographed.

—— 1973. Revision of the genus *Baltimora* (Compositae, Heliantheae). *Fieldiana Bot.* 36:31–50.

—— 1975. The importance of revisionary studies in plant systematics. *Sida* 6:104–113.

—— 1976. A systematic review of the subtribe Lagasceinae (Compositae, Heliantheae). *Amer. J. Bot.* 63:1289–1294.

—— 1977. Heliantheae—systematic review. In V. H. Heywood, J. B. Harborne, and B. L. Turner, eds., *The Biology and Chemistry of the Compositae*, pp. 621–671. London: Academic Press.

—— 1978. Revision of *Lagascea* (Compositae, Heliantheae). *Fieldiana Bot.* 38:75–133.

—— 1979a. Cladistics of *Melampodium* (Compositae). *Taxon* 28:179–192.

—— 1979b. Ultrastructural data for the practicing plant systematist. *Amer. Zoologist* 19:621–636.

—— 1980. Cladistics and plant systematics: Problems and prospects. Introduction. *Syst. Bot.* 5:109–111.

—— 1981. A new format for revisionary studies in systematic botany [abstract]. *Bot. Soc. Amer. Misc. Ser. Publ.* 160:79.

—— 1983. Phylogenetic trees in plant systematics. *Sida* 10:1–13.

—— 1985. Review of *Plant Biosystematics*, edited by W. F. Grant. *Syst. Zool.* 34:375–377.

—— 1987. Explicit approaches for evolutionary classification. *Syst. Bot.* 12:251–262.

—— In press. Comments on concepts of species in flowering plants. *Pl. Syst. Evol.*

Stuessy, T. F., and J. N. Brunken. 1979. Artificial interspecific hybridizations in *Melampodium* section *Zarabellia* (Compositae). *Madroño* 26:53–63.

Stuessy, T. F. and D. J. Crawford. 1983. Flavonoids and phylogenetic reconstruction. *Pl. Syst. Evol.* 143:83–107.

Stuessy, T. F. and J. V. Crisci. 1984a. Phenetics of *Melampodium* (Compositae, Heliantheae). *Madroño* 31:8–19.

—— 1984b. Problems in the determination of evolutionary directionality of character-state change for phylogenetic reconstruction. In T. Duncan and T. F. Stuessy, eds., *Cladistics: Perspectives on the Reconstruction of Evolutionary History*, pp. 71–87. New York: Columbia Univ. Press.

Stuessy, T. F. and G. F. Estabrook. 1978. Introduction [to a series of papers on cladistics and plant systematics]. *Syst. Bot.* 3:145.

Stuessy, T. F. and R. S. Irving. 1968. A morphological plant species: *Euonymus glanduliferus*. *Southw. Naturalist* 13:353–357.

Stuessy, T. F., R. S. Irving, and W. L. Ellison. 1973. Hybridization and evolution in *Picradeniopsis* (Compositae). *Brittonia* 25:40–56.

Stuessy, T. F. and H.-Y. Liu. 1983. Anatomy of the pericarp of *Clibadium, Desmanthodium* and *Ichthyothere* (Compositae, Heliantheae) and systematic implications. *Rhodora* 85:213–227.

Stuessy, T. F. and D. M. Spooner. 1988. The adaptive value and phylogenetic significance of receptacular bracts in the Compositae. *Taxon* 37:114–126.

Stuessy, T. F., D. M. Spooner, and K. A. Evans. 1986. Adaptive significance of ray corollas in *Helianthus grosseserratus* (Compositae). *Amer. Midl. Naturalist* 115:191–197.

Sturtevant, A. H. 1965. *A History of Genetics*. New York: Harper & Row.

Stutz, H. C. and L. K. Thomas. 1964. Hybridization and introgression in *Cowania* and *Purshia*. *Evolution* 18:183–195.

Sulinowski, S. 1967. Interspecific and intergeneric hybrids in grasses of the *Festuca* and *Lolium* genera. *Genet. Polon.* 8:17–30.

Sultan, S. E. 1987. Evolutionary implications of phenotypic plasticity in plants. *Evol. Biol.* 21:127–178.

Sumner, A. T. 1983. The role of protein sulphydryls and disulphides in chromosome structure and condensation. In P. E. Brandham and M. D. Bennett, eds., *Kew Chromosome Conference II*, pp. 1–9. London: George Allen & Unwin.

Sundberg, P. 1985. Numerisk kladistik. *Svensk Bot. Tidskr.* 79:205–217.

Sundberg, S. and T. F. Stuessy. In press. Isolating mechanisms and implications for modes of speciation in Heliantheae (Compositae). *Pl. Syst. Evol.*

Sutton, W. S. 1903. The chromosomes in heredity. *Biol. Bull.* 4:231–251.

Suzuki, D. T., A. J. F. Griffiths, J. H. Miller, and R. C. Lewontin. 1986. *An Introduction to Genetic Analysis*. 3d ed. New York: Freeman.

Svenson, H. K. 1945. On the descriptive method of Linnaeus. *Rhodora* 47:274–302, 363–388.

—— 1953. Linnaeus and the species problem. *Taxon* 2:55–58.

Swain, T., ed. 1963. *Chemical Plant Taxonomy*. London: Academic Press.

—— ed. 1966. *Comparative Phytochemistry*. London: Academic Press.

Swamy, B. G. L. and N. Parameswaran. 1963. The helobial endosperm. *Biol. Rev. Cambridge Philos. Soc.* 38:1–50.

Swanson, C. P. 1957. *Cytology and Cytogenetics*. Englewood Cliffs, N.J.: Prentice-Hall.

Swanson, C. P., T. Merz, and W. J. Young. 1981. *Cytogenetics: The Chromosome in Division, Inheritance and Evolution*. 2d ed. Englewood Cliffs, N.J.: Prentice-Hall.

Swift, L. H. 1974. *Botanical Classifications: A Comparison of Eight Systems of Angiosperm Classification*. Hamden, Conn.: Archon Books.

Swingle, D. B. 1946. *A Textbook of Systematic Botany*. 3d ed. New York: McGraw-Hill.

Sybenga, J. 1975. *Meiotic Configurations: A Source of Information for Estimating Genetic Parameters*. Berlin: Springer-Verlag.

Sylvester-Bradley, P. C. 1952. *The Classification and Coordination of Infra-Specific Categories*, 19 pp. mimeograph. London: Systematics Association.

—— 1956a. The new palaeontology. In P. C. Sylvester-Bradley, ed., *The Species Concept in Palaeontology*, pp. 1–8. London: Systematics Association.

——, ed. 1956b. *The Species Concept in Palaeontology*. London: Systematics Association.

—— 1968. The science of diversity. *Syst. Zool.* 17:176–181.

Systematics Association Committee for Descriptive Biological Terminology. 1960. Preliminary list of works relevant to descriptive biological terminology. *Taxon* 9:245–257.

—— 1962. Terminology of simple symmetrical plane shapes (Chart 1). *Taxon* 11:145–156.

Szalay, F. S. 1976. Ancestors, descendants, sister groups and testing of phylogenetic hypotheses. *Syst. Zool.* 26:12–18.

Takhtajan, A. 1969. *Flowering Plants: Origin and Dispersal*. Translated by C. Jeffrey. Edinburgh: Oliver & Boyd.

—— 1976. Neoteny and the origin of flowering plants. In C. B. Beck, ed., *Origin and Early Evolution of Angiosperms*, pp. 207–219. New York: Columbia Univ. Press.

—— 1980. Outline of the classification of flowering plants (Magnoliophyta). *Bot. Rev.* 46:225–359.

—— 1986. *Floristic Regions of the World*. Translated by T. J. Crovello; edited by A. Cronquist.

Berkeley: Univ. California Press. (See appendix, pp. 305–356, for outline of his system of classification.)

—— 1987. *Systema Magnoliophytorum.* Leningrad: Nauka.

Tanksley, S. D. 1987. Organization of the nuclear genome in tomato and related diploid species. *Amer. Naturalist* 130 (suppl.):s46–s61.

Taylor, P. 1955. The genus *Anagallis* in tropical and south Africa. *Kew Bull. 1955:* 321–350.

Taylor, R. L. and G. A. Mulligan. 1968. *Flora of the Queen Charlotte Islands,* Part 2: *Cytological Aspects of the Vascular Plants.* Canada Department of Agriculture, Research Branch Monograph No. 4, pt. 2.

Taylor, T. N. 1981. *Paleobotany: An Introduction to Fossil Plant Biology.* New York: McGraw-Hill.

Taylor, T. N. and D. A. Levin. 1975. Pollen morphology of Polemoniaceae in relation to systematics and pollination systems: Scanning electron microscopy. *Grana* 15:9–112.

Templeton, A. R. 1981. Mechanisms of speciation—a population genetic approach. *Ann. Rev. Ecol. Syst.* 12:23–48.

—— 1983a. Convergent evolution and non-parametric inference from restriction data and DNA sequences. In B. S. Weir, ed., *Statistical Analysis of DNA Sequence Data,* pp. 151–179. New York: Dekker.

—— 1983b. Phylogenetic inference from restriction endonuclease cleavage site maps with particular reference to the evolution of humans and the apes. *Evolution* 37:221–244.

Terzaghi, E. A., A. S. Wilson, and D. Penny, eds. 1984. *Molecular Evolution: An Annotated Reader.* Boston: Jones and Bartlett.

Tétényi, P. 1970. *Infraspecific Chemical Taxa of Medicinal Plants.* Budapest: Akademiai Kiado.

Thanikaimoni, G. 1972–1986. *Index bibliographique sur le morphologie des pollens d'angiospermes.* 5 vols. Institut Francais de Pondichéry. Pondichéry: All India Press.

Theobald, W. L. 1967. Anatomy and systematic position of *Uldinia* (Umbelliferae). *Brittonia* 19:165–173.

Theobald, W. L. and J. F. M. Cannon. 1973. A survey of *Phlyctidocarpa* (Umbelliferae) using the light and scanning electron microscope. *Notes Roy. Bot. Gard.* (Edinburgh) 32:203–210.

Theobald, W. L., J. L. Krahulik, and R. C. Rollins. 1979. Trichome description and classification. In C. R. Metcalfe and L. Chalk, eds., *Anatomy of the Dicotyledons.* Vol. 1: *Systematic Anatomy of Leaf and Stem, with a Brief History of the Subject,* 2d ed., pp. 40–53. Oxford: Clarendon Press.

Theophrastus. 1916. *Enquiry into Plants, and Minor Works on Odours and Weather Signs.* Translated by A. Hort. 2 vols. Cambridge, Mass.: Harvard Univ. Press.

Thompson, E. A. 1986. Likelihood and parsimony: Comparison of criteria and solutions. *Cladistics* 2:43–52.

Thompson, W. R. 1952. The philosophical foundations of systematics. *Canad. Entomol.* 84:1–16.

Thorne, R. F. 1968. Synopsis of a putatively phylogenetic classification of the flowering plants. *Aliso* 6:57–66.

—— 1972. Major disjunctions in the geographic ranges of seed plants. *Quart. Rev. Biol.* 47:365–411.

—— 1976. A phylogenetic classification of the Angiospermae. *Evol. Biol.* 9:35–106.

—— 1983. Proposed new realignments in the angiosperms. *Nord. J. Bot.* 3:85–117.

Thorpe, R. S. 1976. Biometric analysis of geographic variation and racial affinities. *Biol. Rev. Cambridge Philos. Soc.* 51:407–452.

Throckmorton, L. H. 1965. Similarity *versus* relationship in *Drosophila. Syst. Zool.* 14:221–236.

—— 1968. Biochemistry and taxonomy. *Ann. Rev. Entomol.* 13:99–114.

Tilden, J. W. 1961. Certain comments on the subspecies problem. *Syst. Zool.* 10:17–23.

Titz, W. 1982. Über die Anwendbarkeit biomathematischer und biostatistischer Methoden in der Systematik (mit besonderer Berücksichtigung multivariater Verfahren). *Ber. Deutsch. Bot. Ges.* 95:149–154.

Tobe, H. and P. H. Raven. 1983. An embryological analysis of Myrtales: Its definition and characteristics. *Ann. Missouri Bot. Gard.* 70:71–94.

—— 1984a. The embryology and relationships of *Rhynchocalyx* Oliv. (Rhynchocalycaceae). *Ann. Missouri Bot. Gard.* 71:836–843.

—— 1984b. The embryology and relationships of *Alzatea* Ruiz & Pav. (Alzateaceae, Myrtales). *Ann. Missouri Bot. Gard.* 71:844–852.

Tobe, H., W. L. Wagner, and H.-C. Chin. 1987. A systematic and evolutionary study of *Oenothera* (Onagraceae): Seed coat anatomy. *Bot. Gaz.* 148:235–257.

Togby, H. A. 1943. A cytological study of *Crepis fuliginosa, C. neglecta* and their F_1 hybrid, and its bearing on the mechanism of phylogenetic reduction in chromosome number. *J. Genet.* 45:67–111.

Tomb, A. S. 1975. Pollen morphology in tribe Lactuceae (Compositae). *Grana* 15:79–89.

Tomb, A. S., D. A. Larson, and J. J. Skvarla. 1974. Pollen morphology and detailed structure of family Compositae, tribe Cichorieae. I. Subtribe Stephanomeriinae. *Amer. J. Bot.* 61:486–498.

Tomlinson, P. B. 1961. *Anatomy of the Monocotyledons*, Vol. 2: *Palmae*. Oxford: Clarendon Press.

—— 1962. The leaf base in palms: Its morphology and mechanical biology. *J. Arnold Arbor.* 43:23–50.

—— 1969. *Anatomy of the Monocotyledons*, Vol. 3: *Commelinales-Zingiberales*. Oxford: Clarendon Press.

—— 1982. Chance and design in the construction of plants. *Acta Biotheor.* 31A:162–183.

—— 1984a. Homology: An empirical view. *Syst. Bot.* 9:374–381.

—— 1984b. Vegetative morphology—some enigmas in relation to plant systematics. In V. H. Heywood and D. M. Moore, eds., *Current Concepts in Plant Taxonomy*, pp. 49–66. London: Academic Press.

Tomlinson, P. B. and M. H. Zimmermann, eds. 1978. *Tropical Trees as Living Systems*. Cambridge: Cambridge Univ. Press.

Torrey, J. G. 1967. *Development in Flowering Plants*. New York: Macmillan.

Torrey, J. G. and D. T. Clarkson, eds. 1975. *The Development and Function of Roots*. New York: Academic Press.

Tothill, J. C. and R. M. Love. 1964. Supernumerary chromosomes and variation in *Ehrharta calycina* Smith. *Phyton* 21:21–28.

Tournefort, J. P. de. 1700. *Institutiones Rei Herbariae, editio altera*. Paris.

Trela-Sawicka, Z. 1974. Embryological studies in *Anemone ranunculoides* L. from Poland. *Acta Biol. Cracov.* 17:1–11.

Troll, W. 1937–1939. *Vergleichende Morphologie der höheren Pflanzen*, Vol. 1: *Vegetationsorgane*, parts 1–2. Berlin: Borntraeger.

Tschermak, E. 1900. Ueber künstliche Kreuzung bei *Pisum sativum*. *Ber. Deutsch. Bot. Ges.* 18:232–239. Expanded version under the title "Über Kreuzung von *Pisum sativum*." *Z. Landw. Versucheswesen Österreich* 3:465–555.

Tschudy, R. H. 1969. The plant kingdom and its palynological representation. In R. H. Tschudy and R. A. Scott, eds., *Aspects of Palynology*, pp. 5–34. New York: Wiley.

Tschudy, R. H. and R. A. Scott, eds. 1969. *Aspects of Palynology*. New York: Wiley.

Tsukada, M. 1967. Chenopod and amaranth pollen: Electron-microscopic identification. *Science* 157:80–82.

Tucker, S. C. 1977. Foliar sclereids in the Magnoliaceae. *Bot. J. Linn. Soc.* 75:325–356.

—— 1980. Inflorescence and flower development in the Piperaceae. I. *Peperomia*. *Amer. J. Bot.* 67:686–702.

—— 1982a. Inflorescence and flower development in the Piperaceae. II. Inflorescence development of *Piper*. *Amer. J. Bot.* 69:743–752.

—— 1982b. Inflorescence and flower development in the Piperaceae. III. Floral ontogeny in *Piper*. *Amer. J. Bot.* 69:1389–1401.

—— 1984. Origin of symmetry in flowers. In R. A. White and W. C. Dickison, eds., *Contemporary Problems in Plant Anatomy*, pp. 351–395. New York: Academic Press.

Tufte, E. R. 1983. *The Visual Display of Quantitative Information*. Cheshire, Conn.: Graphics Press.

Tukey, J. W. 1977. *Exploratory Data Analysis*. Reading, Mass.: Addison-Wesley.

Tuomikoski, R. 1967. Notes on some principles of phylogenetic systematics. *Ann. Ent. Fenn.* 33:137–147.

Turesson, G. 1922a. The species and variety as ecological units. *Hereditas* 3:100–113.

—— 1922b. The genotypical response of the plant species to the habitat. *Hereditas* 3:211–350.

—— 1923. The scope and import of genecology. *Hereditas* 4:171–176.

—— 1925. The plant species in relation to habitat and climate: Contributions to the knowledge of genecological units. *Hereditas* 6:147–236.

Turner, B. L. 1956. A cytotaxonomic study of the genus *Hymenopappus* (Compositae). *Rhodora* 58:163–186, 208–242, 250–269, 295–308.

—— 1957. The chromosomal and distributional relationships of *Lupinus texensis* and *L. subcarnosus* (Leguminosae). *Madroño* 14:13–16.

—— 1958. Chromosome numbers in the genus *Krameria*: Evidence for familial status. *Rhodora* 60:101–106.

—— 1966. Chromosome numbers in *Stackhousia* (Stackhousiaceae). *Austral. J. Bot.* 14:165–166.

—— 1967. Plant chemosystematics and phylogeny. *Pure Appl. Chem.* 14:189–213.

—— 1969. Chemosystematics: Recent developments. *Taxon* 18:134–151.

—— 1970a. Chromosome numbers in the Compositae. XII. Australian species. *Amer. J. Bot.* 57:382–389.

—— 1970b. Molecular approaches to populational problems at the infraspecific level. In J. B. Harborne, ed., *Phytochemical Phylogeny*, pp. 187–205. London: Academic Press.

—— 1971. Training of systematics for the seventies. *Taxon* 20:123–130.

—— 1972. Chemosystematic data: Their use in the study of disjunctions. *Ann. Missouri Bot. Gard.* 59:152–164.

—— 1974. The latest in chemosystematics. *Taxon* 23:402–404.

—— 1977a. Chemosystematics and its effect upon the traditionalist. *Ann. Missouri Bot. Gard.* 64:235–242.

—— 1977b. Summary of the biology of the Compositae. In V. H. Heywood, J. B. Harborne, and B. L. Turner, eds., *The Biology and Chemistry of the Compositae*, pp. 1105–1118. London: Academic Press.

—— 1985. A summing up [generic concepts in the Compositae]. *Taxon* 34:85–88.

Turner, B. L., J. Bacon, L. Urbatsch, and B. Simpson. 1979. Chromosome numbers in South American Compositae. *Amer. J. Bot.* 66:173–178.

Turner, B. L., J. H. Beaman, and H. F. L. Rock. 1961. Chromosome numbers in the Compositae. V. Mexican and Guatemalan species. *Rhodora* 63:121–129.

Turner, B. L. and W. L. Ellison. 1960. Chromosome numbers in the Compositae. I. Meiotic chromosome counts for 25 species of Texas Compositae including 6 new generic reports. *Texas J. Sci.* 12:146–151.

Turner, B. L., W. L. Ellison, and R. M. King. 1961. Chromosome numbers in the Compositae. IV. North American species, with phyletic interpretations. *Amer. J. Bot.* 48:216–223.

Turner, B. L. and O. S. Fearing. 1960. The basic chromosome number of the genus *Neptunia* (Leguminosae-Mimosoideae). *Madroño* 15:184–187.

Turner, B. L. and D. Flyr. 1966. Chromosome numbers in the Compositae. X. North American species. *Amer. J. Bot.* 53:24–33.

Turner, B. L. and H. S. Irwin. 1960. Chromosome numbers in the Compositae. II. Meiotic counts for fourteen species of Brazilian Compositae. *Rhodora* 62:122–126.

Turner, B. L. and M. C. Johnston. 1961. Chromosome numbers in the Compositae. III. Certain Mexican species. *Brittonia* 13:64–69.

Turner, B. L. and R. M. King. 1964. Chromosome numbers in the Compositae. VIII. Mexican and Central American species. *Southw. Naturalist* 9:27–39.

Turner, B. L. and W. H. Lewis. 1965. Chromosome numbers in the Compositae. IX. African species. *J. S. African Bot.* 31:207–217.

Turner, B. L., A. M. Powell, and J. Cuatrecasas. 1967. Chromosome numbers in Compositae. XI. Peruvian species. *Ann. Missouri Bot. Gard.* 54:172–177.

Turner, B. L., A. M. Powell, and R. M. King. 1962. Chromosome numbers in the Compositae. VI: Additional Mexican and Guatemalan species. *Rhodora* 64:251–271.

Turner, B. L., A. M. Powell, and T. J. Watson, Jr. 1973. Chromosome numbers in Mexican Asteraceae. *Amer. J. Bot.* 60:592–596.

Turner, V. 1982. Marsupials as pollinators in Australia. In J. A. Armstrong, J. M. Powell, and A. J. Richards, eds., *Pollination and Evolution*, pp. 55–66. Sydney: Royal Botanic Gardens.

Turrill, W. B. 1940. Experimental and synthetic plant taxonomy. In J. Huxley, ed., *The New Systematics*, pp. 47–71. Oxford: Oxford Univ. Press.

—— 1942. Taxonomy and phylogeny. *Bot. Rev.* 8:247–270, 473–532, 655–707.

Tyler, S. 1979. Contributions of electron microscopy to systematics and phylogeny: Introduction to the symposium. *Amer. Zoologist* 19:541–543.

Udvardy, M. D. F. 1969. *Dynamic Zoogeography with Special Reference to Land Animals.* New York: Van Nostrand Reinhold.

Uhl, C. H. 1972. Intraspecific variation in chromosomes of *Sedum* in the southwestern United States. *Rhodora* 74:301–320.

Urbanska-Worytkiewicz, K. 1980. Cytological variation within the family of Lemnaceae. *Veroff. Geobot. Inst. ETH Stiftung Rubel* (Zürich) 70:30–101.

Vaillant, S. 1718. *Sermo de Structura Florum.* Leiden.

Valentine, D. H. 1949. The units of experimental taxonomy. *Acta Biotheor.* 9:75–88.

—— ed. 1972. *Taxonomy, Phytogeography and Evolution.* London: Academic Press.

—— 1975. The taxonomic treatment of polymorphic variation. *Watsonia* 10:385–390.

—— 1978. Ecological criteria in plant taxonomy. In H. E. Street, ed., *Essays in Plant Taxonomy*, pp. 1–18. London: Academic Press.

Valentine, D. H. and A. Löve. 1958. Taxonomy and biosystematic categories. *Brittonia* 10:153–166.

Van Campo, M. 1978. La face interne de l'exine. *Rev. Palaeobot. Palynol.* 26:301–311.

Van Campo, M. and C. Millerand. 1985. *Bibliographie Palynologie, 1984–85. Pollen et Spores* (suppl.). Paris: Editions du Museum.

Van der Pijl, L. 1972. *Principles of Dispersal in Higher Plants.* 2d ed. New York: Springer-Verlag.

Van der Pijl, L. and C. H. Dodson. 1966. *Orchid Flowers: Their Production and Evolution.* Coral Gables, Fla.: Fairchild Tropical Garden and Univ. Miami Press.

Van der Steen, W. J. and W. Boontje. 1973. Phylogenetic versus phenetic taxonomy: A reappraisal. *Syst. Zool.* 22:55–63.

Van Emden, H. F., ed. 1973. *Insect/Plant Relationships.* New York: Halstead Press.

Van Son, G. 1955. A proposal for the restriction of the use of the term subspecies. *Lepidopterists' News* 9:1–3.

Van Valen, L. 1973. Are categories in different phyla comparable? *Taxon* 22:333–373.

—— 1976. Ecological species, multispecies, and oaks. *Taxon* 25:233–239.

—— 1978a. A price for progress in paleobiology. *Paleobiology* 4:210–217.

—— 1978b. Why not to be a cladist. *Evol. Theory* 3:285–299.

Varadarajan, G. S. and A. J. Gilmartin. 1983. Phenetic and cladistic analyses of North American *Chloris* (Poaceae). *Taxon* 32:380–386.

Vaughan, A. 1905. The palaeontological sequence in the Carboniferous limestone of the Bristol area. *Quart. J. Geol. Soc.* (London) 61:181–307.

Vaughan, J. G. 1970. Seed anatomy and seed microscopy. In N. K. B. Robson, D. F. Cutler, and M. Gregory, eds., *New Research in Plant Anatomy,* pp. 35–43. London: Academic Press.

Vaughan, J. G. and K. E. Denford. 1968. An acrylamide gel electrophoretic study of the seed proteins of *Brassica* and *Sinapis* species, with special reference to their taxonomic value. *J. Exp. Bot.* 19:724–732.

Verdoorn, F. 1953a. Editor's foreword. *Chron. Bot.* 14:93–101.

Verdoorn, F., ed. 1953b. Plant genera: Their nature and definition. *Chron. Bot.* 14:89–160.

Vickery, R. K., Jr. 1974. Crossing barriers in the yellow monkey flowers of the genus *Mimulus* (Scrophulariaceae). *Oregon State Univ. Genetics Lectures* 3:33–82.

Vignal, C. 1984. Etude phytodermologique de la sous-famille des Chloridoideae (Gramineae). *Bull. Mus. Nat. Hist.* (Paris) ser. 4, 6 (sect. B): 279–295.

Vijayaraghavan, M. R. and K. Prabhakar. 1984. The endosperm. In B. M. Johri, ed., *Embryology of Angiosperms,* pp. 319–376. Berlin: Springer-Verlag.

Vilgalys, R. 1986. Phenetic and cladistic relationships in *Collybia* sect. *Levipedes* (Fungi: Basidiomycetes). *Taxon* 35:225–233.

Vincent, P. L. D. and F. M. Getliffe. 1988. The endothecium in *Senecio* (Asteraceae). *Bot. J. Linn. Soc.* 97:63–71.

Vithanage, H. I. M. V. and R. B. Knox. 1977. Development and cytochemistry of stigma surface and response to self and foreign pollination in *Helianthus annuus. Phytomorphology* 27:168–179.

Vitt, D. H. 1971. The infrageneric evolution, phylogeny, and taxonomy of the genus *Orthotrichum* (Musci) in North America. *Nova Hedwigia* 21:683–711.

Vliet, G. J. C. M. van and P. Baas. 1984. Wood anatomy and classification of the Myrtales. *Ann. Missouri Bot. Gard.* 71:783–800.

Vogel, S. 1974. Ölblumen und ölsammelnde Bienen. *Trop. Subtrop. Pflanzenw.* 7:283–547.

Voigt, J. W. 1952. A technique for morphological analysis in population studies. *Rhodora* 54:217–220.

Von Rudloff, E. and M. S. Lapp. 1979. Populational variation in the leaf oil terpene composition of western red cedar, *Thuja plicata. Canad. J. Bot.* 57:476–479.

Voss, E. 1952. The history of keys and phylogenetic trees in systematic biology. *J. Sci. Lab. Denison Univ.* 43(1):1–25.

Wagenaar, E. B. 1968. Meiotic restitution and the origin of polyploidy. I. Influence of genotype on polyploid seedset in a *Triticum crassum* x *T. turgidum* hybrid. *Canad. J. Genet. Cytol.* 10:836–843.

Wagner, W. H., Jr. 1961. Problems in the classification of ferns. *Recent Advances in Botany* 1:841–844.

—— 1962. The synthesis and expression of phylogenetic data. In L. Benson, *Plant Taxonomy: Methods and Principles,* pp. 273–277. New York: Ronald Press.

—— 1968. Plant taxonomy and modern systematics. *BioScience* 18:96–100.

—— 1969. The construction of a classification [with discussion]. In C. G. Sibley, chm., *Systematic Biology,* pp. 67–103. Washington, D.C.: National Academy of Sciences.

—— 1980. Origin and philosophy of the groundplan-divergence method of cladistics. *Syst. Bot.* 5:173–193.

—— 1983. Reticulistics: The recognition of hybrids and their role in cladistics and classification. In N. I. Platnick and V. A. Funk, eds., *Advances in Cladistics*, 2:63–79. New York: Columbia Univ. Press.

—— 1984. Applications of the concepts of groundplan-divergence. In T. Duncan and T. F. Stuessy, eds., *Cladistics: Perspectives on the Reconstruction of Evolutionary History*, pp. 95–118. New York: Columbia Univ. Press.

Wain, R. P. 1982. Genetic differentiation in the Florida subspecies of *Helianthus debilis* (Asteraceae). *Amer. J. Bot.* 69:1573–1578.

—— 1983. Genetic differentiation during speciation in the *Helianthus debilis* complex. *Evolution* 37:1119–1127.

Waldeyer, W. 1888. Über Karyokinese und ihre Beziehungen zu den Befruchtungsvorgängen. *Arch. Mikr. Anat.* 32:1–222.

Waldrop, M. M. 1984. Natural language understanding. *Science* 224:372–374.

Walker, J. W. 1971a. Pollen morphology, phytogeography, and phylogeny of the Annonaceae. *Contr. Gray Herb.* 202:1–131.

—— 1971b. Unique type of angiosperm pollen from the family Annonaceae. *Science* 172:565–567.

—— 1974a. Evolution of exine structure in the pollen of primitive angiosperms. *Amer. J. Bot.* 61:891–902.

—— 1974b. Aperture evolution in the pollen of primitive angiosperms. *Amer. J. Bot.* 61:1112–1137.

—— 1976a. Comparative pollen morphology and phylogeny of the Ranalean complex. In C. B. Beck, ed., *Origin and Early Evolution of Angiosperms*, pp. 241–299. New York: Columbia Univ. Press.

—— 1976b. Evolutionary significance of the exine in the pollen of primitive angiosperms. In I. K. Ferguson and J. Muller, eds., *The Evolutionary Significance of the Exine*, pp. 251–308. New York: Academic Press.

—— 1979. Contributions of electron microscopy to angiosperm phylogeny and systematics: Significance of ultrastructural characters in delimiting higher taxa. *Amer. Zoologist* 19:609–619.

Walker, J. W. and J. A. Doyle. 1975. The bases of angiosperm phylogeny: Palynology. *Ann. Missouri Bot. Gard.* 62:664–723.

Walker, J. W. and A. G. Walker. 1979. Comparative pollen morphology of the American myristicaceous genera *Compsoneura* and *Virola. Ann. Missouri Bot. Gard.* 66:731–755.

—— 1984. Ultrastructure of Lower Cretaceous angiosperm pollen and the origin and early evolution of flowering plants. *Ann. Missouri Bot. Gard.* 71:464–521.

Walter, H. 1985. *Vegetation of the Earth and Ecological Systems of the Geo-Biosphere.* 3d ed. Translated by O. Muise. Berlin: Springer-Verlag.

Walker, K. S. 1975. A preliminary study of the achene epidermis of certain *Carex* (Cyperaceae) using scanning electron microscopy. *Michigan Bot.* 14:67–72.

Walters, S. M. 1961. The shaping of angiosperm taxonomy. *New Phytol.* 60:74–84.

Wang, R.-C., D. R. Dewey, and C. Hsiao. 1985. Intergeneric hybrids of *Agropyron* and *Pseudoroegneria. Bot. Gaz.* 146:268–274.

Warburton, F. E. 1967. The purposes of classifications. *Syst. Zool.* 16:241–245.

Warnock, M. J. 1987. Vicariant distribution of two *Delphinium* species in southeastern United States. *Bot. Gaz.* 148:90–95.

Wartenberg, D. 1985. Canonical trend surface analysis: A method for describing geographic patterns. *Syst. Zool.* 34:259–279.

Watanabe, K. and S. Smith-White. 1987. Phyletic and evolutionary relationships of *Brachyscome lineariloba* (Compositae). *Pl. Syst. Evol.* 157:121–141.

Waterman, P. G. and M. F. Grundon, eds. 1983. *Chemistry and Chemical Taxonomy of the Rutales.* London: Academic Press.

Watrous, L. E. and Q. D. Wheeler. 1981. The out-group comparison method of character analysis. *Syst. Zool.* 30:1–11.

Watson, J. D. and F. H. C. Crick. 1953. Molecular structure of nucleic acids: A structure for deoxyribose nucleic acid. *Nature* 171:737–738.

Watson, L. 1971. Basic taxonomic data: The need for organisation over presentation and accumulation. *Taxon* 20:131–136.

Watson, L. and P. Milne. 1972. A flexible system for automatic generation of special purpose dichotomous keys, and its application to Australian grass genera. *Austral. J. Bot.* 20:331–352.

Watson, L., W. T. Williams, and G. N. Lance. 1967. A mixed-data numerical approach to angiosperm taxonomy: The classification of Ericales. *Proc. Linn. Soc.* (London) 178:25–35.

Watt, J. C. 1968. Grades, clades, phenetics, and phylogeny. *Syst. Zool.* 17:350–353.

Watts, W. A. and R. C. Bright. 1968. Pollen, seed and mollusk analysis of a sediment core from Pickerel Lake, northeastern South Dakota. *Bull. Geol. Soc. Amer.* 79:855–876.

Weatherby, C. A. 1942. Subspecies. *Rhodora* 44:157–167.

Webb, A. and S. Carlquist. 1964. Leaf anatomy as an indicator of *Salvia apiana-mellifera* introgression. *Aliso* 5:437–449.

Webb, C. J. and A. P. Druce. 1984. A natural intergeneric hybrid, *Aciphylla squarrosa* x *Gingidia montana*, and the frequency of hybrids among other New Zealand apioid Umbelliferae. *New Zealand J. Bot.* 22:403–411.

Webster, G. L., W. V. Brown, and B. N. Smith. 1975. Systematics of photosynthetic carbon fixation pathways in *Euphorbia*. *Taxon* 24:27–33.

Weevers, T. 1943. The relation between taxonomy and chemistry of plants. *Blumea* 5:412–422.

Weimarck, G. 1972. On "numerical chemotaxonomy." *Taxon* 21:615–619.

Weinburg, W. 1908. Über den Nachweis der Vererbung beim Menschen. *Jahresh. Vereins Vaterl. Naturk. Württemberg* 64:368–382.

Weir, B. S., E. J. Eisen, M. M. Goodman, and G. Namkoong, eds. 1988. *Proceedings of the Second International Conference on Quantitative Genetics.* Sunderland, Mass.: Sinauer Associates.

Wells, H. 1980. A distance coefficient as a hybridization index: An example using *Mimulus longiflorus* and *M. flemingii* (Scrophulariaceae) from Santa Cruz Island, California. *Taxon* 29:53–65.

Wells, L. G. and R. A. Franich. 1977. Morphology of epicuticular wax on primary needles of *Pinus radiata* seedlings. *New Zealand J. Bot.* 15:525–529.

Went, J. L. van and M. T. M. Willemse. 1984. Fertilization. In B. M. Johri, ed., *Embryology of Angiosperms*, pp. 273–317. Berlin: Springer-Verlag.

Wernham, H. F. 1912. Floral evolution: With particular reference to the sympetalous dicotyledons: IX.—Summary and conclusion. Evolutionary genealogy; and some principles of classification. *New Phytol.* 11:373–397.

Wesley, J. P. 1974. *Ecophysics: The Application of Physics to Ecology.* Springfield, Ill.: Charles C. Thomas.

Westfall, R. H., H. F. Glen, and M. D. Panagos. 1986. A new identification and combining feature of a polyclave and an analytical key. *Bot. J. Linn. Soc.* 92:65–73.

Wetter, M. A. 1983. Micromorphological characters and generic delimitation of some New World Senecioneae (Asteraceae). *Brittonia* 35:1–22.

Whalen, M. D. 1978. Foliar flavonoids of *Solanum* section *Androceras*: A systematic survey. *Syst. Bot.* 3:257–276.

Wheeler, Q. D. 1981. The ins and outs of character analysis: A response to Crisci and Stuessy. *Syst. Bot.* 6:297–306.

—— 1986. Character weighting and cladistic analysis. *Syst. Zool.* 35:102–109.

Whiffin, T. and M. W. Bierner. 1972. A quick method for computing Wagner trees. *Taxon* 21:83–90.

Whiffin, T. and A. S. Tomb. 1972. The systematic significance of seed morphology in the neotropical capsular-fruited Melastomataceae. *Amer. J. Bot.* 59:411–422.

White, J. 1979. The plant as a metapopulation. *Ann. Rev. Ecol. Syst.* 10:109–145.

White, M. J. D. 1978. *Modes of Speciation.* San Francisco: Freeman.

White, R. A. and W. C. Dickison, eds. 1984. *Contemporary Problems in Plant Anatomy.* Orlando: Academic Press.

Whitehead, P. J. P. 1972. The contradiction between nomenclature and taxonomy. *Syst. Zool.* 21:215–224.

Whittaker, R. H. 1969. New concepts of kingdoms of organisms. *Science* 163:150–160.

Whittemore, A. T. 1987. Transition to a land flora: A critique. *Cladistics* 3:60–65.

Whittingham, A. D. and G. L. Stebbins. 1969. Chromosomal rearrangements in *Plantago insularis* Eastw. *Chromosoma* (Berlin) 26:449–468.

Wiens, D., M. Renfree, and R. O. Wooller. 1979. Pollen loads of honey possums *(Tarsipes spenserae)* and nonflying mammal pollination in southwestern Australia. *Ann. Missouri Bot. Gard.* 66:830–838.

Wiens, D., J. P. Rourke, B. B. Casper, E. A. Rickart, T. R. LaPine, C. J. Peterson, and A. Channing. 1983. Nonflying mammal pollination of southern African proteas: A non-coevolved system. *Ann. Missouri Bot. Gard.* 70:1–31.

Wilbur, R. L. 1970. Infraspecific classification in the Carolina flora. *Rhodora* 72:51–65.

Wild, H. and A. D. Bradshaw. 1977. The evolutionary effects of metalliferous and other anomalous soils in south central Africa. *Evolution* 31:282–293.

Wiley, E. O. 1975. Karl R. Popper, systematics, and classification: A reply to Walter Bock and other evolutionary taxonomists. *Syst. Zool.* 24: 233–243.

—— 1978. The evolutionary species concept reconsidered. *Syst. Zool.* 27:17–26.

—— 1980. Is the evolutionary species fiction?—A consideration of classes, individuals and historical entities. *Syst. Zool.* 29:76–80.

—— 1981a. *Phylogenetics: The Theory and Practice of Phylogenetic Systematics.* New York: Wiley.

—— 1981b. Remarks on Willis' species concept. *Syst. Zool.* 30:86–87.

—— 1987a. Approaches to outgroup comparison. In P. Hovenkamp, ed., *Systematics and Evolution: A Matter of Diversity*, pp. 173–191. Utrecht: Utrecht Univ.

—— 1987b. Methods in vicariance biogeography. In P. Hovenkamp, ed., *Systematics and Evolution: A Matter of Diversity*, pp. 283–306. Utrecht: Utrecht Univ.

Willemse, M. T. M. and J. L. van Went. 1984. The female gametophyte. In B. M. Johri, ed., *Embryology of Angiosperms*, pp. 159–196. Berlin: Springer-Verlag.

Williams, E. G., R. B. Knox, and J. L. Rouse. 1981. Pollen-pistil interactions and control of pollination. *Phytomorphology* 31:148–157.

Williams, H. P. 1987. Evolution, game theory and polyhedra. *J. Math. Biol.* 25:393–409.

Williams, N. H. and C. H. Dodson. 1972. Selective attraction of male euglossine bees to orchid floral fragrances and its importance in long distance pollen flow. *Evolution* 26:84–95.

Williams, R. F. 1974. *The Shoot Apex and Leaf Growth: A Study in Quantitative Biology.* Cambridge: Cambridge Univ. Press.

Williams, W. T. 1967. Numbers, taxonomy, and judgment. *Bot. Rev.* 33:379–386.

—— 1969. The problem of attribute-weighting in numerical classification. *Taxon* 18:369–374.

—— 1971. Principles of clustering. *Ann. Rev. Ecol. Syst.* 2:303–326.

Williams, W. T. and M. B. Dale. 1965. Fundamental problems in numerical taxonomy. In R. D. Preston, ed., *Advances in Botanical Research*, 2:35–68. London: Academic Press.

Willis, E. O. 1981. Is a species an interbreeding unit, or an internally similar part of a phylogenetic tree? *Syst. Zool.* 30:84–85.

Willis, J. C. 1922. *Age and Area: A Study in Geographical Distribution and Origin of Species.* Cambridge: Cambridge Univ. Press.

Willson, M. F. 1983. *Plant Reproductive Ecology.* New York: Wiley.

Wilson, A. C., S. S. Carlson, and T. J. White. 1977. Biochemical evolution. *Ann. Rev. Biochem.* 46:573–639.

Wilson, C. L. 1982. Vestigial structures and the flower. *Amer. J. Bot.* 69:1356–1365.

Wilson, E. B. 1896. *The Cell in Development and Heredity.* New York: Macmillan.

Wilson, E. O. 1965. A consistency test for phylogenies based on contemporaneous species. *Syst. Zool.* 14:214–220.

—— 1968. Recent advances in systematics. *BioScience* 18:1113–1117.

—— 1971. The plight of taxonomy. *Ecology* 52:741.

—— 1985. Time to revive systematics. *Science* 230:1227.

——, ed. 1988. *Biodiversity.* Washington, D.C.: National Academy Press.

Wilson, E. O. and W. L. Brown, Jr. 1953. The subspecies concept and its taxonomic application. *Syst. Zool.* 2:97–111.

Wilson, H. D. 1981. Genetic variation among South American populations of tetraploid *Chenopodium* sect. *Chenopodium* subsect. *Cellulata. Syst. Bot.* 6:380–398.

Winge, Ö. 1917. The chromosomes. Their numbers and general importance. *Compt.-Rend. Trav. Carlsberg Lab.* 13:131–275.

Wodehouse, R. P. 1935. *Pollen Grains: Their Structure, Identification and Significance in Science and Medicine.* New York: McGraw-Hill; reprinted, New York: Hafner, 1959.

Woese, C. R. and G. E. Fox. 1977. Phylogenetic structure of the prokaryotic domain: The primary kingdoms. *Proc. Natl. Acad. (U.S.A.)* 74:5088–5090.

Wright, H. E., Jr. 1970. Vegetational history of the Central Plains. In W. Dort, Jr. and J. K. Jones, Jr., eds., *Pleistocene and Recent Environments of the Central Great Plains*, pp. 157–172. Lawrence: Univ. Press of Kansas.

Wright, S. 1931. Evolution in Mendelian populations. *Genetics* 16:97–159.

Wülker, W., G. Lörincz, and G. Dévai. 1984. A new computerized method for deducing phylogenetic trees from chromosome inversion data. *Z. Zool. Syst. Evolut.-Forsch.* 22:86–91.

Wyatt, R. 1983. Pollinator-plant interactions and the evolution of breeding systems. In L. Real, ed., *Pollination Biology*, pp. 51–95. Orlando: Academic Press.

Yablokov, A. V. 1986. *Phenetics: Evolution, Population, Trait.* Translated by M. J. Hall. New York: Columbia Univ. Press.

Yakovlev, M. S. 1967. Polyembryony in higher plants and principles of its classification. *Phytomorphology* 17:278–282.

Yamazaki, T. 1982. [Recognized types in early development of the embryo and the phylogenetic significance in the dicotyledons]. *Acta Phytotax. Geobot.* 33:400–409. (in Japanese with English summary)

Yang, D., B. P. Kaine, and C. R. Woese. 1985. The phylogeny of Archaebacteria. *Syst. Appl. Microbiol.* 6:251–256.

Yeo, P. F. 1971. X *Solidaster*, an intergeneric hybrid (Compositae). *Baileya* 18:27–32.

Yochelson, E. L. 1966. Nomenclature in the machine age. *Syst. Zool.* 15:88–91.

Yoshioka, H., T. J. Mabry, and B. N. Timmermann. 1973. *Sesquiterpene Lactones: Chemistry, NMR and Plant Distribution.* Tokyo: Univ. Tokyo Press.

Young, D. A. 1979. Heartwood flavonoids and the infrageneric relationships of *Rhus* (Anacardiaceae). *Amer. J. Bot.* 66:502–510.

—— 1987. Concept of the genus: Introduction and historical perspective [abstract]. *Amer. J. Bot.* 74:718.

Young, D. A. and P. M. Richardson. 1982. A phylogenetic analysis of extant seed plants: The need to utilize homologous characters. *Taxon* 31:250–254.

Young, D. A. and D. S. Seigler, eds. 1981. *Phytochemistry and Angiosperm Phylogeny.* New York: Praeger.

Young, D. J. and L. Watson. 1970. The classification of dicotyledons: A study of the upper levels of the hierarchy. *Austral. J. Bot.* 18:387–433.

Yule, G. U. 1902. Mendel's laws and their probable relations to intra-racial heredity. *New Phytol.* 1:193–207, 222–238.

Zavada, M. S. 1984a. Angiosperm origins and evolution based on dispersed fossil pollen ultrastructure. *Ann. Missouri Bot. Gard.* 71:444–463.

—— 1984b. Pollen wall development of *Austrobaileya maculata. Bot. Gaz.* 145:11–21.

Zavada, M. S. and J. M. Benson. 1987. First fossil evidence for the primitive angiosperm family Lactoridaceae. *Amer. J. Bot.* 74:1590–1594.

Zavarin, E., W. B. Critchfield, and K. Snajberk. 1969. Turpentine composition of *Pinus contorta* x *Pinus banksiana* hybrids and hybrid derivatives. *Canad. J. Bot.* 47:1443–1453.

Zimmerman, J. R. and J. A. Ludwig. 1975. Multiple-discriminant analysis of geographical variation in the aquatic beetle, *Rhantus gutticollis* (Say) (Dytiscidae). *Syst. Zool.* 24:63–71.

Zimmermann, M. H. 1983. *Xylem Structure and the Ascent of Sap.* Berlin: Springer-Verlag.

Zimmermann, M. H. and C. L. Brown. 1971. *Trees: Structure and Function.* New York: Springer-Verlag.

Zinderen Bakker, E. M. van. 1953. *South African Pollen Grains and Spores,* vol. 1. Amsterdam: A. A. Balkema.

Zirkle, C. 1935. *The Beginnings of Plant Hybridization.* Philadelphia: Univ. Pennsylvania Press.

Zohary, D. and U. Nur. 1959. Natural triploids in the orchard grass, *Dactylis glomerata* L., polyploid complex and their significance for gene flow from diploid to tetraploid levels. *Evolution* 13:311–317.

Zuckerkandl, E. and L. Pauling. 1965. Molecules as documents of evolutionary history. *J. Theor. Biol.* 8:357–366.

AUTHOR INDEX

Page numbers in brackets refer to entries represented by "et al." in the text.

Aalders, L. E., 308
Abbott, H. C. de S., 329
Abbott, L. A., 65
Abbott, T. K., 29
Abrahamson, W. G., 351, 365, 373
Adams, E. N. III, 127
Adams, J., 15
Adams, R. P., 38, 60, 71, 78, 90, 92, [92], 127, 190, 332, 335, 348
Adanson, M., 44, 61, 62
Afonso, A., 338
Ahlquist, J. E., 15, 344
Ahlstrand, L., 254
Aiken, S. G., 279
Ainsworth, G. C., 202
Airy Shaw, H. K., 197
Al-Aish, M., 347
Al-Shehbaz, I. A., 340
Alberch, P., 111
Albrecht, D. G., 23
Alexopoulos, C. J., 219
Allen, R. T., 365
Allen, T. F. H., 22
Alston, R. E., 75, 330, 332, 397, 399
Amadon, D., 10, 188
Ambegaokar, K. B., 263, 264
Amici, J. -B., 252
Amsden, T. W., 194, 205
Andersen, S. T., 286
Anderson, E., 23, 162, 168, 169, 200, 202, 212, 397
Anderson, G. J., 358
Anderson, L. C., 200, 254, 258, 259, 266, 293, [293], 296, 324, 398
Anderson, R. S., 344
Anderson, S., 23
Anderson, W. R., 94, 97, 123, 124
Anfinsen, C. B., 331
Angell, B., 393
Anton, A. M., 55, 254, 256
Anuradha, S. M. J., 397
Arber, A., 44
Archie, J. W., 104, 140
Arends, J. C., 293
Armstrong, J. A., 352

Armstrong, J. E., 248
Arnheim, N., 41
Arnold, E. N., 106
Arroyo, M. T. K., 190
Ashlock, P. D., 95, 137, 142
Ashton, M., 279
Askew, R. R., 188
Asmussen, M. A., 32
Astolfi, P., 127
Atchley, W. R., 128
Atran, S., 195
Auger, P., 22
Averett, J. E., 330
Avise, J. C., 18, 32, 192, 193
Axelrod, D. I., 133
Ayala, F. J., 18, [40], 192, 212, 331, [385]
Ayensu, E. S., 234

Baagøe, J., 225, 228
Baas, P., 233, 240, 249, 250
Baba, M. L., 341
Babcock, E. B., 165, 166, 290, 314
Babu, C. R., [225]
Bachmann, K., 315, [344]
Bacon, J. D., 114, 225, [293]
Badr, A., 91
Bagavathi Subramanian, R., 222
Bailey, I. W., 198, 234, 235, 250
Bain, J. F., 332, 348
Baker, H. G., 173, 352, 357
Baker, I., 357
Baker, R. T., 329
Balch, W. E., [207]
Baldwin, J. T., Jr., 292
Balick, M. J., 332
Banarescu, P., 137
Banerjee, U. C., 222
Banks, R. C., 14
Baranova, M., 222, 235, 248
Barber, H. N., 40
Barbosa, P., 349
Barbour, M. G., 365
Barigozzi, C., 179
Barker, G., 284
Barkley, W. D., 225
Barlow, B. A., 308
Barrs, J., [79]

Bartcher, R. L., 114
Barth, F. G., 352
Barth, O. M., 270
Barthlott, W., 211, 222, [222], 225, 286
Bartlett, H. H., 194-97
Bassett, I. J., [287]
Bastmeijer, J. D., 293
Bate-Smith, E. C., 330, 332
Bates, D. M., 293
Bates, L. S., [200]
Bates, V. M., 396
Batten, J. W., 369
Bauhin, G., 44, 52, 53, 196
Baum, B. R., 60, 62, 69, 72, 91, 92, 117, 123, 126, 133, 178, 202
Baumm, A., 359, 360
Bayer, R. J., 192, 297, 324
Beachy, R. N., 345
Beaman, J. H., 292, 311
Beard, J. S., 368
Beattie, A. J., 360
Beatty, J., 99
Beaudry, J. R., 162
Beck, C. B., 205
Becker, K. M., 56
Beckner, M., 52
Becvar, J. E., 357
Behnke, H. -D., 211, 219, 236, 237, 240-42, 249
Beil, C. E., 231
Bejar, V., [91]
Bell, A. D., 223
Bell, C. R., [9], [31], [32], 149, 153, [219], [221], [235], [254], [261], 287, 292, [393]
Bell, E. A., 331
Bendich, A. J., 344
Bendz, G., 330
Bennett, M. D., 290, 299, 302
Benson, J. M., 150
Benson, L., 219
Bentham, G., 44, 53-55, 197, 208, 218, 234
Bentzer, B., 300
Benzing, D. H., 222, 231, 401
Bergeron, Y., 139
Berggren, W. A., 18
Bergqvis, F., 91

Berkowitz, N., 21
Berlin, B., 5, 21, 158, 161, 166, 195, 196
Berlyn, G. P., 234
Bermudes, D., 401
Bernier, R., 169
Bersier, J. D., 254
Bertalanffy, L. von, 22
Bessey, C. E., 44, 55, 161, 166
Beyer, W. A., 126
Bhandari, N. N., 254, 265
Bhatnagar, S. P., 252, 253, 260
Bhattacharya, J., 235
Bhojwani, S. S., 252, 253, 260
Bibel, D., 21
Bibelriether, H., 225
Bierhorst, D. W., 219
Bierner, M. W., 97, 114, 115
Biesboer, D. D., 236
Bigelow, R. S., 210, 212
Bird, R. M., 92
Birks, H. H., 375
Birks, H. J. B., 270, 286, 373, 375
Bisby, F. A., 65, 91, 330
Bisby, G. R., 202
Bissing, D. R., 250
Bittner, J. L., 286
Blackith, R. E., 60, 65
Blackmore, S., 211, 269-71, 273, 280, 285, 286
Blackwelder, R. E., 6, 9, 28, 29, 43, 137
Blair, W. F., 90
Blanchard, O. J., Jr., 293
Blasdell, R. F., 133
Bledsoe, A. H., [133], [212], [340]
Block, M., 190
Bloom, M., 332
Blumenberg, B., 14
Böcher, T. W., 11
Bock, W. J., 17, 24, 49, 95, 104, 105, 108, 137, 143, 159, 160, 175
Bocquet, G., 254
Boesewinkel, F. D., 264
Bogert, C. M., 188
Bogle, A. L., 240
Bohm, B. A., 84, 92, 199, 200, 283, 332
Bohm, D., 21, 22
Bohm, L. R., 231
Bohme, W., 189
Boivin, B., 182, 183, 206
Bold, H. C., 207, 219
Bolick, M. R., 202, 285, 287
Bolton, E. T., 344
Bond, W. J., 360
Boontje, W., 137
Booth, C., 202
Boquet, G., 10
Borgen, L., 311
Borrill, M., 296
Bossert, W., 74
Bostock, C. J., 290
Bothmer, R. von, [300], 302
Botosaneanu, L., 189
Boulter, D., 7, 124, 126, 331, 341, 343
Boulton, D., [92]

Bouman, F., 264
Boyden, A., 6
Bradfield, G. E., 81
Bradshaw, A. D., 369
Brady, R. H., 29, 99
Brandham, P. E., 222, [222], [235], 290
Braun-Blanquet, J., 365
Brazier, J. C., 240
Breedlove, D. E., 5, 21, 158, 166, 195, 196, [308]
Brehm, B. G., 225, 397
Bremekamp, C. E. B., 165
Bremer, K., 93, 94, 109, 114, 116, 128, 132-34, 177, 207, 212
Bresnan, J., 21
Brewbaker, J. L., 277
Briggs, B. G., 133, 144
Briggs, D., 190
Bright, R. C., 354, 374
Brim, S. W., [199]
Brittan, N. H., 235, 237
Britton, D. M., 92
Britton, N. L., 162
Brookes, B. S., 360
Brooks, D. R., 24, 42, 46, 93, 94, 98, 130, 153
Brooks, J., 270, 285
Broome, C. R., 293
Brothers, D. J., 187, 209
Brown, C. L., 234
Brown, D. F. M., 133
Brown, G. K., 360
Brown, H., 15
Brown, R., 268
Brown, R. C., [293]
Brown, W. L., Jr., 188, 189
Brown, W. M., 14
Brown, W. V., 235, 314, 346, 347
Brückner, E., 225
Brummitt, R. K., 183, 271
Brundin, L., 127
Brunfels, O., 49, 196
Brunken, J. N., 92, 322
Bryson, V., 331
Buck, R. C., 23, 157, 158, 159
Buck, W. R., 133
Bullini, L., 41
Burger, W. C., 46, 133, 171, 172
Burgman, M. A., 103
Burk, J. H., 365
Burma, B. H., 165, 166
Burnett, W. C., Jr., 349
Burnham, C. R., 314
Burns, J. M., 133, 152
Burns-Balogh, P., 133
Burt, W. H., 188
Burtt, B. L., 59, 61, 70, 138, 193
Bush, G. L., 179
Buss, P. A., Jr., 253
Buth, D. G., 192
Butterfield, B. G., 234
Byatt, J. I., 200
Bye, R. A., Jr., 292
Byerly, H. C., 125

Caccone, A., 344
Cain, A. J., 12-13, 28, 29, 33, 35,

37, 38, 59, 71, 76, 93, 141, 158, 166, 173, 178, 202
Cain, S. A., 132, 199, 365
Caira, J. N., [153]
Calderon, C. E., 198
Calvin, M., 331
Camerarius, R. J., 50, 252, 351
Cameron, D. R., 290
Cameron, H. D., 21
Camin, J. H., 94, 96, 97, 114, [137], 153
Camp, W. H., 11, 162, 178, 185, 187, 203
Campbell, A., 21
Campbell, C. S., 117
Campbell, I., 92
Candolle, A. L. P. P. de, 44, 53, 184, 218, 365
Candolle, A. P. de, 6, 34, 43, 44, 53, 184, 218, 385
Canfield, J. E., 347
Canne, J. M., 225
Cannon, J. F. M., 225
Canright, J. E., 271
Cantino, P. D., 109, 125, 127, 133, 137
Caputo, G., 236
Carlock, J. R., 22
Carlquist, S., 111, 112, 133, 198, 233, 234, 240, 243, 245, 248, 249, 250
Carlson, S. S., 341
Carmer, S. G., 77
Carpenter, J. M., 115, 127
Carr, D. J., 235
Carr, G. D., 311
Carr, P. H., 222
Carr, S. G. M., 235
Carroll, C. P., 296
Carter, S., [222], [235]
Cartmill, M., 99, 131, 138
Casey, J., 327
Casper, B. B., [357]
Caspersson, T., 290
Cass, D. D., 254, 258
Castle, W. E., 314
Caswell, H., 351
Cattell, R. B., 92
Cavalli-Sforza, L. L., 96, 97
Cave, M. S., 200, 251, 253, 258, 265, 290, 292, 293, 302
Celebioglu, T., 225
Chabot, B. F., 365
Chalk, L., 222, 234
Challice, J. S., 92
Chaloner, W. G., 14
Chamberlain, C. J., 252
Chambers, J. M., 393
Chambers, K. L., 248, 308, [315], 331, [344], 377
Chambers, T. C., 222, 223
Chance, G. D., 225
Chang, C. P., [240], 344
Channing, A., [357]
Chapman, D. J., 207
Chapman, R, O., 91
Chapman, R. W., 32
Chapman, V., 319
Chappill, J. A., 137
Charig, A. J., 99
Charlesworth, B., 17

Charlwood, B. V., 331
Chater, A. O., 10, 183
Chatfield, C., 65
Cheadle, V. I., 250
Chimal, A., 293, 302
Chimphamba, B. B., 200
Chin, H. -C., 246
Chinnappa, C. C., 305
Chiribog, D. A., 21
Chissoe, W. F., [284], 286
Chopra, R. N., 253
Chopra, S., 231, 246
Christensen, J. E., 285
Chuang, T. -I., 225, 228, 292, 308
Churchill, S. P., 97, 101, 133, [133], [134]
Chute, H. M., 246
Ciampolini, F., 285, [285]
Clark, C., 110
Clark, P. J., 77
Clark, R. B., 166
Clarke, A. E., 270, 285
Clarkson, D. T., 234
Classen, D., 67, 85, 92
Claugher, D., 286
Clausen, C. P., 16
Clausen, J., 11, 165, 177, 185, 313, 314, 376
Clausen, R. T., 183
Clayton, W. D., 91, 92, 190, 197, 199, 205, 209
Clegg, M. T., 345, 349
Cleland, R. E., 179, 290, 314
Clements, F. E., 184, 352, 365
Clench, W. J., 188
Cleveland, W. S., 393, [393], 397
Clifford, H. T., 65, 67, 75, 77-80, 90-92, 133, 212, 218, 277
Clowes, F. A. L., 234
Coates, D., 346, 347
Cocucci, A. E., 254, 256
Cody, V., 331
Cohen, D. M., 14
Cohn, N. S., 290
Cole, A. J., 65
Cole, G. T., 219
Colless, D. H., 60, 68, 91, 109, 125, 128, 137, 139
Collins, A. J., 65
Coluzzi, M., 41
Comer, C. W., 90
Constance, L., xvii, 3, 4, 7, 11, 292
Cook, O. F., 152, 153, 175
Cooper-Driver, G. A., 332
Core, E. L., 44, 49, 50, 53, 55, 219
Core, H. A., 234
Correns, C., 288, 313
Côté, W. A., Jr., 234
Cotthem, W. van, 222
Coulter, J. M., 252
Coulter, M. A., 92
Courter, J. W., 77
Covey, S. N., 290
Cowan, S. T., 166
Cowles, H. C., 365
Cracraft, J., 29, 94, 98, 111, 133
Crane, E., 252

Crane, P. R., 130, 133
Crang, R. E., 200
Craw, R. C., 132
Crawford, D. J., 82, 91, 97, 123, 133, 175, 192, 225, 292, [311], 315, 317, 318, 324, [327], 330, 332, 338, 341, 343, 350
Crawley, M. J., 365, 373
Crepet, W. L., 286, 362
Cressey, R. F., 14
Cresti, M., 285
Crété, P., 253, 260
Crick, F. H. C., 313
Crisci, J. V., 65, 69, 76-78, 81, 82, 91, 92, 106-10, 113, 126, 311, 328
Critchfield, W. B., 336
Croat, T. B., 15, 353
Croizat, L., 132, 199
Crompton, C. W., [287]
Cronquist, A., 39, 40, 42, 44, 55, 90, 91, 95, 126, 127, 137, 171, 184, 193, 196, 207, 212, 218, 349, 365
Crout, D. H. G., 331
Crovello, T. J., 42, 66, 67, 69, 89, 90, 92, 125, 126, 165, 174, 384, 393, 395
Crowson, R. A., 9, 28, 29, 37-39, 41, 126
Cruden, R. W., 285, 357, 360
Crum, H., 181
Cuatrecasas, J., 292
Cullen, J., 70
Cullimore, D. R., 91
Cullis, C. A., 346, 347
Curran, D. J., 110
Curtis, J. D., 222
Cutler, D. F., 222, 233-35
Cutter, E. G., 253
Cuvier, G., 61, 162-64
Czaja, A. T., 236
Czapik, R., 265, 266
Czelusniak, J., [341]

D'Amato, F., 292
D'Antoni, H. L., 270
D'Arcy, W. G., 235, 238, 331
Dabinett, P. E., 91
Dahl, E., 372
Dahlgren, R., 44, 55, 91, 133, 196, 208, 212, 218, 235, 240, 241, 277
Dakshini, K. M. M., 225
Dale, J. E., 234
Dale, M. B., 74, 78, 137, 221
Dallwitz, M. J., 10
Danser, B. H., 9, 106, 178
Dansereau, P., 365
Darga, L. L., [341]
Darlington, C. D., 290, 302, 304
Darlington, P. J., Jr., 9, 24, 25, 137
Darwin, C., 6, 35, 53, 54, 56-58, 135, 161, 163, 164, 351
Das, J. C., 235
Das, S., 235
Daumer, K., 354
Davey, J. C., 91
Davey, V. M., 11
Davidse, G., 352

Davidson, J. F., 166, 397
Davidson, R. A., 23
Davies, D. R., 290
Davis, G. L., 251, 252, 253, 263, 265
Davis, G. M., 133
Davis, M. B., 370
Davis, P. H., xiii, 9, 24, 25, 28, 30, 32, 34, 35, 39, 40, 41, 43, 50, 56, 57, 69, 95, 104, 159, 161, 163, 178, 183, 187, 188, 205, 206, 208, 210, 218, 219, 265, 266, 271, 290
Davis, S. A., 303
Day, A. C., 234
Day, W. H. E., 125, 127, 327
Dayhoff, M. O., 114
Dean, H. L., 200
DeBry, R. W., 115
Debuhr, L., 250
Deets, G. B., 105
De Jong, D. C. D., 311
DeLage, I., 51
Delevoryas, T., 108, 219
Denford, K. E., 332, 342, 348
Dennis, W. M., [115]
Denton, M. F., 91, 332
Deodikar, G. B., 303
De Robertis, E. D. P., 290
De Robertis, E. M. F., Jr., 290
Desch, H. E., 234
Deumling, B., 304
Dévai, G., 327
De Vries, H., 288, 313
De Wet, J. M. J., 173, 178, 296, 311, 319
Dewey, D. R., 49, 200, 201
Diaz-Piedrahita, S., [293]
Dickinson, H. G., 270, 284, 286
Dickinson, T. A., 221
Dickison, W. C. [9], [31], [32], [219], [221], 233, 235, [235], 239, 240, 250, [254], [261], [393]
Dickson, E. E., 343
Diels, L., 34
Dilcher, D. L., 221, 286
Dimas, C. T., 315, 319
Dirzo, R., 365
Dixit, S. N., 254
Dixon, W. J., 103
Dobzhansky, T., 40, 165, 173, 385
Dodson, C. H., 349, 352
Doebley, J. F., 91, 318
Dong, V., 15
Donoghue, M. J., 11, 104, 109, 110, 130, 133, 137, 177, 180, 202, 223
Doolittle, W. F., 179
Dormer, C., 234
Dormer, K. J., 254, 255
Douglas, M. E., 18
Dover, G. A., 290
Dowd, J. M., 341
Doyen, J. T., 180, 181
Doyle, J. A., 104, 130, 133, 205, 270, 286
Doyle, J. J., 343, 345, 348
Dreiding, A. S., 339
Dressler, R. L., 352

Drew, C. M., 270
Druce, A. P., 201
Du Rietz, G. E., 172, 191, 194
Ducker, S. C., 91, 352, 354
Duek, J. J., 91
Duke, J. A., 10
Dunbar, M. J., 126
Duncan, T., 24, 60, 62, 87, 91,
    94-99, 123, 128, 140, 144, 151,
    153, 384
Dunford, M. P., 308
Dunn, D. B., 222
Dunn, G., 65, 76
Dunn, R. A., 23
DuPraw, E. J., 23, 156
Dupuis, C., 96, 114
Durrant, S. D., 188
Dyer, A. F., 290, 304

Eames, A. J., 198, 233, 234, 249,
    250
East, E. M., 288
Echlin, P., 225
Eck, R. V., 114
Eckardt, M. J., 97
Eckenwalder, J. E., 68
Edmonds, J. M., 86, 90, 92
Edwards, A. W. F., 96, 97
Edwards, J. G., 188
Edwards, J. L., 108
Eggers Ware, D. M., 315, 316
Eggli, U., 222
Eglinton, G., 222
Ehler, N., 222, [222], 225, 286
Ehrendorfer, F., 114, 133, 181,
    290, 297, 298, 311
Ehrlich, P. R., 15, 165, 167, 172-
    74, 182, 365
Eichler, A. W., 44, 54, 55, 207
Eifert, I. J., 240, [240]
Eigen, M., 14
Eisen, E. J., [314]
El-Gadi, A., 92
El-Ghazaly, G., 275, 284
Eldredge, N., 17, 29, 93, 98,
    107, 111
Eleftheriou, E. P., 240
Elias, T. S., 15
Eliasson, U., 265
Elisens, W. J., 248
Elkington, T. T., 91, 92
Ellison, W. L., 75, 128, 292, 296,
    298, 308, 326, 332, 397, 399,
    400
Elsal, J. A., 92
Emboden, W. A., Jr., 190, 287
Emig, C. C., 101, 144
Endler, J. A., 132, 167
Endress, P. K., 246, 284
Engel, H., 163
Engler, A., 44, 55, 233
Engstrand, L., [300]
Erdtman, G., 268, 270, 271, 273,
    274, 275, 276, 278, 279
Erdtman, H. G. H., 330
Esau, K., 234
Eshbaugh, W. H., 90, [92], 190
Estabrook, G. F., [7], 24, 36, 37,
    65, 94, 97, 98, 114, 119, 123-
    26, [126], 133, 140, 143, 146,
    384

Estrada, A., 352
Ettlinger, M. G., 340
Etzler, M. F., 330
Evans, A. M., 133
Evans, K. A., 41, 225
Everitt, B. S., 65, 76
Evert, R. F., 240
Ewan, J., 217
Eyde, R. H., 106, 111, 221, 240,
    246, 250
Eyre, S. R., 369

Faden, R. B., 293
Faegri, K., 270, 271, 273, 274,
    276, 352
Fahn, A., 222, 234
Fair, F., 131
Fairbrothers, D. E., 211, 222,
    292, 330, 331, 343
Falconer, D. S., 314
Farber, P. L., 163, 164
Farkas, L., 331
Farris, F. J., 139
Farris, J. S., 24, 41, 42, 54, 60,
    68, 76, 84, 93, 94, 95, 97, 99,
    104, 110, 114-15, 125, 130,
    140, 159, 327
Farwell, O. A., 187
Favarger, C., 225, 298
Favre-Duchartre, M., 134, 266
Fearing, O. S., 298
Featherly, H. I., 219
Fedorov, A. A., 290
Fellows, C. E., 308
Felsenstein, J., 65, 97, 98, 113-
    15, 123-27
Fennah, R. G., 188
Fensholt, D. E., 22
Ferguson, A., 330
Ferguson, I. K., 200, 270, 271,
    285, 286
Fernald, M. L., 184, 200
Fernandes, A., 311
Feulgen, R., 313
Fiala, K. L., 97, 123, 128, 140
Findlay, G. W. D., 240
Fink, W. L., 11
Fioroni, P., 111
Firmage, D. H., 335
Fischer, C. A. H., 268
Fischer, H. D., 331
Fischer, N. H., 331
Fish, R. K., 256, 265
Fisher, D. B., 258
Fisher, D. R., 69, 158
Fisher, F. J. F., 231
Fisher, J. B., 223, 231, 233
Fisher, R. A., 314
Fitch, W. M., 92, 97, 126, 349
Flake, R. H., 90, 92, 190, 336
Flavell, R. B., 290, 291
Fleming, T. H., 352
Flemming, W., 288
Flyr, D., 292
Forsum, U., 91
Forsyth, J. L., 370
Fosberg, F. R., 9, 17, 184
Foster, A. S., 219, 234
Foster, C. E., 349
Fox, G. E., 207
Fox, L. R., [360]

Fox, R. M., 188
Franca, F., 311
Franich, R. A., 222
Franklin, T. B., 370
Fraser, M. A., [360]
Free, J. B., 352
Freeman, C. E., 354, 357
Freeman, P. W., 206
French, J. C., 254
Friedmann, H., 45
Friend, J. H., 27, 29
Friis, I., 198
Frisch, K. von, 352
Fritzsche, C. J., 268
Frohlich, M. W., 109, 110
Frohne, D., 331
Frost, F. H., 106, 113, 249
Frye, N., 46
Fryns-Claessens, E., 222
Fryxell, P. A., 97, 115
Fu, K. S., 23
Fuchs, H. P., 49, 183
Fujita, Y., 350
Fukuda, I., 302
Funk, V. A., 46, 93, 94, 97, 98,
    101, 103, 113, 114, 126, 128,
    130, 131, 133, 134, 139, 201,
    206, 292, [293], 327, 350
Funkhouser, J. W., 286
Furlow, J. J., 152

Gabor, M., 331
Gabriel, B. L., 234
Gabriel, K. R., 90
Gaffney, E. S., 131
Gajzago, C., [92]
Gangadhara, M., 222
Garay, L. A., 200
Gardner, E. J., 313, 314
Gardner, R. C., 6, 97, 123
Gardner, R. O., 279
Gasc, J. P., 137
Gasson, P., 240
Gastony, G. J., 126
Gates, D. M., 365
Gates, R. R., 165
Gatlin, L. L., 24
Gauld, I. D., 95
Geer, D., 270
Geesink, R., 103
Geissman, T. A., 331
Genermont, M., 144
Gensel, P. G., 92
Gentry, A. H., 287, 353
George, T. N., 175
Gerassimova-Navashina, H.,
    260
Gerbeth-Jones, S., 222
Germeraad, J. H., 271
Gershenzon, J., 211, 335
Geslot, A., 287
Getliffe, F. M., 254
Ghiselin, M. T., 29, 35, 169, 170,
    174, 209
Gibbs, R. D., 329, 331
Gibson, J. S., 369
Gifford, E. M., Jr., 219
Gijzel, P. van, [270]
Gilbert, L. E., 352

Gill, L. S., 305
Gillham, N. W., 188
Gillis, W. T., 203
Gilly, C. L., 11, 178, 185, 187
Gilmartin, A. J., 11, 41, 69, 88, 103, 137, 139, 190, 360
Gilmour, J. S. L., 25, 37, 57-59, 178, 190
Gingeras, T. R., 126
Gingerich, P. D., 92, 130
Giseke, P. D., 44, 391
Gisin, H., 141
Givnish, T. J., 401
Glassman, S. F., 235
Gleason, H. A., 365
Gleaves, J. T., [7], [126], [343]
Gleeson, P. A., 270, 285
Glen, H. F., 10
Goddijn, W. A., 165
Godley, E. J., 352
Goldberg, R. B., 344
Goldblatt, P., 128, 190, 297, 327
Goldman, N., 104
Goldstein, I. J., 330
Gómez-Pompa, A., 15, 293, 302
Good, R., 113, 365
Goodall, D. W., 77
Goodman, M., 126, 331, [341]
Goodman, M. M., 92, [314], [318]
Goodspeed, T. H., 314
Goodwin, T. W., 331
Gordon, A. D., 65, 286, 373
Goronzy, F., 21
Gosline, W. A., 188
Gottlieb, L. D., 40, 173, 175, 192, 296, 298, 302, 315, 316, 317, 326
Gottlieb, O. R., 350
Gottsberger, G., 356
Gould, F. W., 287, 308
Gould, S. J., 17, 111
Gower, J. C., 24, 73, 76, 140
Graham, A., 270, 279, 280, 284, 286
Graham, B. F., Jr., [270]
Graham, S. A., 270, 279
Grant, K. A., 352, 354
Grant, P. R., [270]
Grant, V., 18, 107, 126, 128, 164, 165, 167, 173, 178, 179, 190, 204, 211, 250, 297, 314, 324, 327, 352, 354, 385
Grant, W. F., 162, 178, 290, 293, 298, 302, 305, 307
Grashoff, J. L., 33, 203, 204
Gray, A., 53, 184, 187
Gray, J., 286
Gray, J. R., 222
Grayum, M. H., 277, 279
Greene, E. L., 184, 196, 217
Greenman, J. M., 198, 204, 211
Greger, H., [114], [133], 349
Gregg, J. R., 23, 158, 159, 165
Gregor, J. W., 11, 190, 313
Gregory, M., 233
Greig-Smith, P., 77
Greilhuber, J., 290, 297, 302-4
Gresson, R. A. R., 290
Greuter, W., 10, 15, 157, 182, 194, 287, 393

Greven, B., 343
Grew, N., 233, 252, 267
Grierson, D., 290
Griffith, J. G., 92
Griffiths, A. J. F., [314]
Griffiths, G. C. D., 9, 29, 37, 156, 179
Groves, R. H., [221]
Grundon, M. F., 331
Guédès, M., 93, 219
Guevara, S., 15
Guignard, L., 252
Guise, A., [7], [126]
Gunn, C. R., 225
Gunning, B. E. S., 234
Guralnik, D. B., 27, 29
Gustafsson, M., [300], 325
Guttman, S. I., [92]

Haapala, O., 291
Habeler, E., [298]
Haeckel, E., 34, 364, 391, 392
Hagen, J. B., 11, 365
Hagerman, A., 335
Hagmeier, E. M., 189
Haldane, J. B. S., 314
Hale, W. G., 188
Hall, A. V., 146
Hall, H. M., 11, 184, 315
Hall, I. V., 308
Hall, J. B., 81
Hallam, N. D., 222, 223
Halle, F., 233
Halle, M., 21
Halmos, P. R., 22
Hamby, R. K., 345
Hamerton, J. L., 299
Hamilton, R. J., 222
Hansell, R. I. C., 23
Hansen, B., 10
Hanson, E. D., 143
Harborne, J. B., 330-34, 336, 338-40, 342, 344, 353, 377
Hardin, J. W., 97, 114, 133, 222, 226
Hardy, G. H., 314
Harlan, J. R., 173, 178, 312, 319
Harling, G.,. 254, 265
Harms, L. J., 293
Harper, R. A., 161
Harrington, H. D., 219
Harris, S. J., [222], [235]
Harrison, G. A., 28, 29, 37, 59, 71, 76, 93, 141
Hart, G., 41, 140
Hartigan, J. A., 65
Hartl, D., 314
Hartman, H., 14
Hartman, R. L., 204, 292, 366, 398
Hasegawa, M., 345
Hashimoto, N., 91
Haskell, G., 290
Haslett, B. G., [343]
Hassall, D. C., 200, 202
Hatch, M. H., 179
Hattersley, P. W., 347
Hauber, D. P., 314, 327
Hauke, R., 133
Hauser, E. J. P., 286, 319
Hauser, L. A., 97

Haverbeke, D. F. van, 90
Hawkes, J. G., 15, 330, 331
Hawksworth, D. L., 362
Hayat, M. A., 234
Hayes, H. D., 286
Hayley, D. E., 311
Hayward, H. E., 234
Healey, P. L., 222
Hebda, R. J., [199]
Hecht, M. K., 107, 108
Heckard, L. R., 225, 228, 279, 308
Hedberg, I., 16, 311
Hedberg, O., 15, 16, 28, 311
Hedglin, F. L., 200
Hegnauer, R., 330, 331
Heimburger, M., 299
Heinrich, B., 55, 352
Heintzelman, C. E., Jr., 198, 235
Heiser, C. B., Jr., 77, 89, 161, 162, 173, 200, 201, 324
Helfend, L., 21
Hemleben, V., 290
Henderson, K., [222]
Heneen, W. K., 200, 201, 302
Hengeveld, P., 173
Hennig, W., 34, 36, 44, 66, 94-98, 100, 101, 113, 114, 119, 128, 130, 137, 155, 156
Henrickson, J. A., Jr., 137
Hepworth, A., [225]
Herr, J. M., Jr., 251, 253, 254, 256, 258, 264, 265
Herrnstadt, I., 198, 225
Hershey, B. J., 351
Heslop-Harrison, J., 24, 111, 178, 190, 270, 284, 285, 350
Heslop-Harrison, Y., 222, 358
Heusden, A. W. van, [315]
Heusi, *see* Hiesey
Heusser, C. J., 270
Heyn, C. C., 198, 225
Heywood, V. H., xiii, xvii, 4, 9, 16, 17, 20, 24, 25, 28, 30, 32, 34, 35, 39-41, 43, 50, 56, 57, 60, 65, 69, 95, 104, 142, 159, 162, 163, 172, 178, 183, 187, 188, 201, 205, 206, 218, 219, 225, 265, 266, 271, 290, 331, 365, 379
Hickey, L. J., 205, 221
Hickman, J. C., 190
Hideux, M., 271, 286
Hiesey, W. M., 11, 165, 177, 185, 313, 314, 376
Higgins, C. A., 21
Hill, C. R., 130, 133
Hill, J., 233
Hill, M. O., 73
Hill, R. J., 225, 228
Hill, R. S., 92, 221
Hillis, D., 386
Hillson, C. J., 249
Hilu, K. W., 91, 192, 315
Hitchcock, A. S., 9, 179
Hochachka, P. W., 331
Hoefert, L. L., 236
Hoenigswald, H. M., 21, 98
Hofler, K., 234
Hofmeister, W., 252
Hogge, L., 190

Holdridge, L. R., 366, 368
Hollingshead, L., 290
Holm, R. W., 67
Holmgren, A. H., 200
Holmgren, N. H., 393
Honda, H., 223, 233
Honegger, R., 284
Hooke, R., 267, 288
Hooker, J. D., 44, 53, 55, 197, 208, 218, 234
Hooper, S. S., 81
Hopfinger, J. A., 335
Hopwood, A. T., 20
Hopwood, R. A., 290
Hori, H., 345
Horn, H. S., 223
Horner, H. T., Jr., 285
Hoshizaki, B. J., 133
Howard, R. A., 198, 235, 240
Howe, T. D., 254, 265
Hsiao, C., 200
Hsiao, J. -Y., 92
Hsu, C. -C., 311
Hu, C. C., 67
Hubbell, T. H., 188
Hubbs, C. L., 12
Hudson, G. E., 73
Hudson, H. J., 184
Hughes, N. F., 18, 205
Hull, D. L., 10, 21, 23, 24, 29, 35, 95, 124, 137, 157, 158, 159, 160, 167, 170-71, 172, 174, 209
Hull, V. J., [221]
Humphrey, H. B., 217
Humphries, C. J., 93, 114, 132, 133, [133], [134], 137, 350
Hunt, D. R., 307
Hunt, E., 23
Hunter, I. J., 104
Huntley, B., 270
Hunziker, J. H., [76], [77], [82], 240, 317
Hutchinson, A. H., 397
Hutchinson, J., 291
Hutchinson, John, 44, 46, 55, 253
Hutchinson, S., 222
Huxley, J., 12, 93, 314
Huynh, K. -L., 225, 279
Hyde, H. A., 267
Hyland, F., 270
Hymowitz, T., 225

Iltis, H. H., 16
Inamdar, J. A., 222
Inglis, W. G., 68, 147
Ingrouille, M. J., 344
Irving, R. S., 92, 128, 175, 296, 308, 326, 335, 400
Isely, D., 20, 225
Iversen, J., 270
Iversen, T. -H., 236, 237
Izco, J., 364

Jaccard, P., 75
Jackson, R. C., 125, 292, 296-99, 314, 315, 319, 327, 328
Jacobs, M., 61, 210, 332
Jacobsen, N., 293

James, F. C., 365
James, M. T., 202
James, S. H., 292
Jancey, R. C., 81, 156
Jane, F. W., 234
Jansen, R. K., 69, 113, 133, 134, 198, 293, 297, 326, 328, 345
Janssen, C. R., 286
Janvier, P., 128
Janzen, D. H., 374, 377
Jardine, C. J., 158
Jardine, N., 10, 23, 28, 65, 68, 90, 140, 158
Jeanmonod, D., 385
Jeffrey, C., 10, 345
Jeffrey, D. W., 369
Jeffrey, E. C., 234
Jenik, J., 225
Jenkins, I. A., 290
Jennings, A. C., 341
Jensen, R. G., 285
Jensen, R. J., 81, 90, 92, 97, 115, 190
Jensen, U., 211, 331, 343
Jensen, W. A., 234, 253, 258, 260, 279
Johansen, D. A., 234, 251-53, 260
John, B., 314
Johnson, A. M., 219, 256
Johnson, C., 235, 346
Johnson, C. S., Jr., 36, 97, 119
Johnson, G. B., 41
Johnson, H. B., 222
Johnson, L. A. S., 11, 89, 90, 133, 137, 144
Johnson, M. P., 67, 190
Johnson, R., 126
Johnson, R. W., 70, 71
Johnson, W. S., 222
Johnston, M. C., 292
Johri, B. M., [225], 253, 254, 264, 270
Jones, A., 200
Jones, C. E., 352
Jones, K., 288, 290, 293, 307, 327
Jones, R. N., 290, 297
Jones, S. B., Jr., 9, 28, 33, 34, 43, 219, 280, [349]
Jong, R. de, 106
Jonsell, B., 158
Jopling, C., 288, 293
Jordan, D. E., 21, 164
Jordon, K., 164
Joswiak, G. R., [92]
Joysey, K. A., 98
Judd, W. S., 114, 133, 206
Jui, P. Y., 92
Jukes, T. H., 41, 331
Juniper, B. E., 222
Jussieu, A. L. de, 43, 44, 51-53, 56, 57, 61, 196, 208, 218, 385
Just, T., 32, 33, 34, 194

Kaine, B. P., 207
Kallay, F., 331
Kallunki, J. A., 310
Kam, Y. K., 231

Kanal, L. N., 23
Kaneko, K. -I., 91
Kapadia, Z. J., 185, 287
Kapil, R. N., 265
Kaplan, D. R., 35, 104, 111, 112, 212, 231, 250
Kapoor, M., 292, 293
Kapp, R. O., 270, 271, 373
Kasha, K. J., 296
Kaul, R. B., 200, 201
Kavanaugh, D. H., 43, 96
Keating, R. C., 235, 238, 240, 267, 270
Keck, D. D., 11, 165, 177, 185, 313, 314, 376
Keddie, R. M., 91
Kee, D. W., 21
Keefe, J. M., 240
Keener, C. S., 133
Keil, D. J., 184, 293
Kemp, T. S., 99
Kendrew, W. G., 369, 371
Kendrick, W. B., 27, 28, 32, 71, 92, 194
Kenton, A., 293, 307
Kerfoot, W. C., 41
Kershaw, A. P., 283
Kershaw, K. A., 365
Kerster, H. W., 173
Kerwin, J. L., 332
Kessel, B., [222]
Kethley, J. B., 114
Kevan, P. G., 353
Kidd, K. K., 127
Kimura, M., 41
King, D. S., 91
King, G. J., 344
King, J. L., 41
King, P. J. H., 60
King, R. M., 32, 203, 204, 292, 293, 298, 353
Kirkbride, M. C. G., 109
Kirkpatrick, J. B., 92
Kishino, H., 345
Kitts, D. B., 131, 160
Kjaer, A., 340
Kjellqvist, E., 302
Klasova, A., 222
Kleiner, B., [393]
Kliphuis, E., 311
Kluge, A. G., 41, 78, 93, 95, 97, 109, 111, 114, 126
Knight, D., 20
Knobloch, I. W., 200, 222, 324
Knox, R. B., 270, 285, 352, 354
Knuth, P., 351
Koch, M. F., 112
Kolreuter, J. G., 252
Konig, A., [315]
Koponen, T., 133
Kostler, J. N., 225
Kowal, R. R., 310
Krahulik, J. L., 222
Krantz, V. E., 353
Kraus, F., 111
Kraus, O., 352
Krebs, C. J., 365
Krell, D., 225
Kremp, G. O. W., 271
Krendl, F., [298]

Kress, W. J., 275, 277
Krischik, V. A., 349
Kruckeberg, A. R., 173, 200, [293], 306, 340, 369
Kruskal, J. B., Jr., 96, 97, 114, 125
Krystal, S., 21
Kubitzki, K., 331, 352
Kubler, G., 32, 45
Kugler, H., 352
Kujat, R., 225
Kukkonen, I., 248
Kumamoto, J., 335
Kupchan, S. M., 338
Kuprianova, L. A., 287
Kutschera, L., 225, 227
Kyhos, D. W., 200, 293, [293], [298], [308], 327, 378

Lackey, J. A., 235
La Cour, L., 290, 302, 304
Ladiges, P. Y., 222
Ladizinsky, G., 296
La Duke, J. C., 11, 81, 91, 97, 123, 133
Lakshmanan, K. K., 263
Lakshminarayana, K., *see* K. L. Narayana
Lam, H. J., 141
Lambert, J. M., 79
Lambinon, J., 187
Lammers, T. G., 149, 212, 354, 357
Lance, G. N., 78, 79, 91, 92
Lande, R., 17
Landrum, L. R., 119, 123, 126, 133
Lane, M. A., 194, 198
Lang, J. M. S., 11
Langenheim, J. H., 335, 349
Lanham, U., 10
Lanzillotti, P. J., [73]
LaPine, T. R., [357]
Lapp, M. S., 190
Larson, D. A., 268, 280
Larson, J. L., 158, 163, 197
Laszlo, E., 19, 22
Lavania, U. C., 327
Lawless, J. G., 14
Lawrence, G. H. M., 9, 10, 43, 55, 199, 219
Lawrence, M. E., 302
Le Quesne, W. J., 117, 121
Leaver, C. J., 290
Lefkovitch, L. P., 67, 92
Legendre, P., 179, 180, 194, 195
Lehman, H., 168
Leins, P., 284
Leitao, M. T., 311
Lemen, C. A., 206
Lennox, J. G., 162, 196
Leone, C. A., 330, 331
Leppik, E. E., 113, 228, 362
Leroi-Gourhan, A., 286
Lersten, N. R., 222, 253
Les, D. H., 376
Levin, D. A., 161, 165-67, 173, 211, 222, 280, 292, 323, 324, 326, 341, 349, 376
Levins, P. A., 225

Levinton, J., 18
Levy, J. F., 240
Lewin, B., 291
Lewin, R. A., 207
Lewis, H., 161, 179, 190, [287], 288, 290, 292, 296, 312, 327
Lewis, W. H., 270, 290, 292, 293, 294, 295, 297, 300, 303, 345
Lewonton, R. C., [314]
Li, H. -L., 195
Li, N., 297
Li, W. -H., 15
Li, Y. -H. C., 92
Lin, B. -L., 345
Lincoln, D. E., 335
Linde-Laurson, I., 302
Lindley, J., 5, 56, 208, 221, 268
Link, H. F., 183, 184
Linnaeus, C., 5, 29, 33, 44, 50-53, 61, 135, 156-58, 162-64, 183, 196, 197, 203, 208, 218, 221, 364, 391
Linskens, H. F., 270
Lionni, L., 3
Liston, J., 65
Little, F. J., Jr., 10
Little, R. J., 352
Littlejohn, M. J., 173
Liu, H. -Y., 246
Lloyd, A. T., 14
Lockhardt, W. R., 65
Lohmann, G. P., 18
Lomax, A., 21
Long, F. L., 352
Long, R. W., 201
Loomis, W. E., 222
Looney, J. H. H., 365
López, A. M. F., 65
Lorence, D. H., 246
Lörincz, G., 327
Lotsy, J. P., 178
Lott, J. N. A., 248
Louie, A. H., 23
Löve, A., 173, 178, 179, 200, 290, 292, 293, 302, 308, 311
Löve, D., 290, 302, 308, 311, 397
Love, R. M., 297
Lovtrup, S., 111, 177
Lowrey, T. K., 193
Lubke, R. A., 92, 236
Luchsinger, A. E., 7, 28, 33, 34, 43, 219
Ludwig, J. A., 190
Lundberg, J. G., 125
Luteyn, J. L., 109, 115
Lyon, D. L., 360

Mabry, H., 331
Mabry, T. J., 211, 240, [240], 330, [330], 331, [336], 339, 344, [349]
MacBryde, B., 294, 295
McClammer, J. U., Jr., [133], [212], [340]
McComb, J. A., 200
McCully, M. E., 234
MacDaniels, L. H., 233, 234
MacDonald, N., 125
McDowall, R. M., 132

McGill, L. A., [293], 397
McGinley, R. B., 349
McGrath, S., [360]
McGuire, R. F., 91
Machol, R. E., 92
McIntosh, R. P., 364, 365
Macior, L. W., 362, 364
McKelvey, B., 147
McKusick, V. A., 56
McLeod, M. G., 293, [293]
McLeod, M. J., [92]
McMorris, F. R., 36, 97, 119, 125, 127
McMullen, C. K., 357
McNair, J. B., 329
McNeill, J., 54, 60, 65, 67, 71, 92, 142, 152, 153
McVaugh, R., 7, 33, 203, 204
McWilliams, E. L., 347
Maddison, D. R., 110
Maddison, W. P., 110
Magill, R. E., 225
Magrum, L. J., [207]
Mahe, J., 271
Maheshwari, P., 252-54, 260
Maheswari Devi, H., 253, 265
Mahlberg, P. G., 236
Maksymowych, R., 234
Malecka, J., 263
Malmgren, B. A., 18
Malpighi, M., 233, 267
Manischewitz, J. R., 140
Manitz, H., 183
Mann, W. F., Jr., 225
Manton, I., 178
Marceau, L., 286
Marchi, E., 23
Marcus, L. F., 65, 80
Margoliash, E., 97, 126
Markgraf, V., 270
Markham, K. R., 331, 332
Marks, G. E., 296
Marticorena, C., 125, [125], 270, 283
Martin, A. C., 225
Martin, J. T., 222
Martin, P. G., 341
Martin, P. S., 270
Marx, H., 109
Mascherpa, J. -M., 10
Maslin, T. P., 105, 106
Mason, H. L., 5, 9, 159, 199
Massey, J. R., [9], [31], [32], [219], [221], [235], [254], [261], [393]
Mathew, L., 235
Mathew, P. M., 279
Matthews, J. F., 225
Mauseth, J. D., 234
Maynard, Smith, J., 23
Mayr, E., 6, 7, 9, 12-13, 16, 18, 25, 29, 34, 35, 37, 39-41, 43, 45, 58, 60, 70, 93-95, 130, 137, 138, 161-66, 169, 172, 173, 178, 188, 392
Maze, J., 81, 184, 196, 231
Meacham, C. A., 33, 95, 97, 110, 117, 119, 122, 123, 125, 128, 133, 144
Meacock, S. E., [79]

Medus, J., 287
Meeuse, A. D. J., 70, 137, 212
Meeuse, B., 352
Mehta, I., 222
Meikle, R. D., 189
Meisel, W. S., 23
Melhert, T. E., 190
Melchior, H., 55
Melkó, E., 91
Melville, R., 221, 285
Mendel, G., 313
Mendelson, D., 297
Menzel, M. Y., 201
Mercer, E. I., 331
Meronk, D. B., 127
Merxmuller, H., 4, 11
Merz, T., 314
Metcalfe, C. R., 222, 233, 234, 249, 388
Meyer, D. E., 290, 394
Meyer-Abich, A., 25
Mez, C., 329
Michener, C. D., 10, 25, 28, 29, 46, 61, 63, 67, 75, 77, 84, 94, 137, 140, 143
Mickel, J. T., 97, 133
Mickevich, M. F., 68, 95, 131, 133, 139
Middleton, E., Jr., 331
Miescher, F., 313
Miksche, J. P., 234
Milkovits, L., 235
Miller, G. A., 21
Miller, H. E., 336
Miller, J. H., [314]
Miller, J. M., 308
Miller-Ward, S., 357
Millerand, C., 270
Milne, P., 10
Milne, R. G., 161
Milthorpe, F. L., 234
Mirov, N. T., 330
Mishler, B. D., 133, [133], [134], 177, 180
Misra, S. P., 219
Miyamoto, M. M., 127
Mohan, J. S. S., 222
Mohan Ram, H. Y., 258
Mohl, H., 233, 268
Molisch, H., 234
Mooney, H. A., 190, 365
Moore, D. M., 290, 299, 365
Moore, G. W., 126
Moore, I. M., 188
Moore, J. A., 20
Moore, P. D., 270, 271, 273, 274, 275, 277, 278
Moore, R. J., 290
Moore, W. S., [92]
Mooring, J. S., 292
Moral, R. del., 91
Morgan, T. H., 313
Mori, S. A., 310
Morishima, H., 77
Morris, S., 352
Morrison, J. H., 287, 319
Moorison, P., 14
Morse, L. E., 10
Morton, A. G., 217
Morton, A. J., 81

Moseley, M. F., Jr., 240
Mosquin, T., 15, [293], 308, 311, 321, 322
Moss, W. W., 60, 67, 80, 82, 92, 137, 138, 143, 153
Mound, L. A., 95
Mueller, L. E., 222
Muhammad, A. F., 244
Muir, J. W., 160
Muir, M., [270]
Mujeeb, K. A., [200]
Mulcahy, D. L., 270, 324
Müller, H. L. H., 351
Muller, J., 270, 271, 279, 285, 286
Mulligan, G. A., 222, 311
Muncer, S., 21
Muniyamma, M., 297
Müntzing, A., 11
Murata, T., 347
Murray, B. G., 200
Murray, D. R., 352
Musselman, L. J., 225
Muxica, L., 91
Myers, G. S., 9, 10
Myers, N., 15
Myint, T., 133

Nadeau, L., 397
Nagl, W., 290
Nair, P. K. K., 267, 270
Nairn, A. E. M., 370
Nakamura, R. R., [133], [212], [340]
Namkoong, G., [314]
Nanzetta, P., 22
Naranjo, C. A., [76], [77], [82]
Narayana, K. L., 253, 265
Narayana, L. L., 397
Nastansky, L., 97
Natesh, S., 262
Nawaschin, S., 252
Naylor, B. G., 112
Neff, J. L., 357
Neff, N. A., 65, 75, 80, 90, 103
Nei, M., 175, 192, 314, 317
Nelson, C. H., 76, 97, 114, 115, 118, 125
Nelson, G. J., 13, 25, 61, 94, 96, 98, 108, 111, 124, 132, 136, 137, 199, 210, 391
Nerlove, S. B., 65
Nesom, G. L., 296, 396
Neticks, F. G., 135
Neumann, D. A., 127
Newcomb, W., 258, 260
Newell, C. A., 225
Newell, I. M., 10
Newman, A. A., 331
Newton, L. E., 222, 225
Niehaus, T. F., 225
Nielson, B. J., 218
Niklas, K. J., 92, 126, 332, 350
Nilsson, S., 270, 271, 280
Nitecki, M. H., 17
Nogler, G. A., 263
Nordenstam, B., 203
Northington, D. K., 184
Norton, S. A., 354

Nowicke, J. W., 279, 280, 283, 284, 286
Nozzolillo, C., 85, 92
Nur, U., 296
Nuttall, T., 5

O'Brien, T. P., 234
O'Callaghan, J. F., [221]
O'Dell, M., 291
O'Grady, R. T., 105
Oden, N., 90
Ohsugi, R., 347
Okamura, J. K., 344
Oldeman, R. A. A., 233
Oliver, M. L., [133], [212], [340]
Oliver, R. L., 294
Olivier, E. J., 331
Olsen, J. S., 114
Olson, E. C., 38
Orchard, A. E., 198
Ornduff, R., 283, 290, 293
Osawa, S., 345
Ottaviano, E., 270
Owens, S. J., 360
Ownbey, M., 128, 296, 326, 397
Oxford, G. S., 331

Pacheco, P., 327
Pacini, E., [285]
Pacqué, M., 286
Padolina, W. G., [349]
Page, R. D. M., 132
Pahlson, C., 91
Palacios, R. A., [76], [77], [82]
Palma-Otal, M., 92
Palmer, J. D., 40, 175, 211, 345
Palmer, P. G., 222
Palser, B. F., 253, 261
Panagos, M. D., 10
Panchen, A. L., 126
Pandey, A. K., 231, 246
Pankhurst, R. J., 10
Panshin, A. J., 234
Parameswaran, N., 258
Parenti, L. R., 132, 133, 212
Park, C. -W., 348
Parker, R. A., [73]
Parker, W. H., 81, 84, 92, 221, 332
Parkes, K. C., 188
Parkhurst, D. F., 41, 222
Parra, O., 270, 283
Pasteur, G., 6, 60
Patel, V. C., [270], 284
Patil, V. P., 303
Paton, D. C., 354, 357
Pattee, H. H., 22
Patterson, C., 98, 99, 108, 130, 212
Patterson, R., 198
Pauling, L., 329
Payne, W. W., 201, 222, 279, 285, 293, [308], 336, [336]
Peacock, D., [7], [126], [343]
Pearson, C., 285
Pearson, K., 314
Pechere, J., 126
Pennell, F. W., 186, 197

Penny, D., 126, 331, 341
Percival, M. S., 352, 357
Pernès, J., 296
Perry, D., [92]
Persoon, C. H., 183
Peters, D. S., 140
Petersen, F. P., 211, 343
Petersen, R. H., 133
Peterson, C. J., 25, [357]
Pettigrew, C. J., 91
Philbrick, C. T., 358
Philip, O., 279
Philipson, M. N., 235
Philipson, W. R., 205, 234
Phillips, R. B., 151, 153
Phillips, W. L., 376, 377
Phipps, J. B., 92, 127, 208, 236, 297
Pianka, E. R., 365
Piazza, A., 127
Pimentel, R. A., 65, 188
Pinkava, D. J., 293
Pitts, W. D., 365
Pizzolongo, P., 236
Platnick, N. I., 13, 21, 29, 94, 95, 98, 99, 115, 131, 132, 136, 140, 169, 201, 205, 210, 391
Platt, T. R., [153]
Playfair, W., 391
Pleszczynska, J., 236, [236]
Plumier, C., 197
Pogan, E., 311
Pohl, R. W., 200
Pojar, J., 311
Polhemus, J. T., 139
Polhill, R. M., 91
Ponzi, R., 236
Poole, M. M., 271
Pop, L., 240, 336
Popp, M., 369
Popper, K. R., 93, 131, 132
Porter, C. L., 9, 43, 219
Porter, D. M., 361
Porter, L. J., 332
Posluszny, U., 231
Potvin, C., 139
Powell, A. M., 200, 292, 293, [293], 311
Powell, J. M., 352
Powell, J. R., 344
Powell, R. A., 335
Power, D. M., 67
Prabhakar, K., 258
Prager, E. M., [14], 126
Praglowski, J., 271
Prakash, N., 265
Prance, G. T., 15, 92, 202, 357
Prantl, K., 44, 55, 233
Pratt, V., 66, 137, 164
Prendergast, H. D. V., 347
Presch, W., 125
Price, H. J., 315, [315], 344
Price, J. H., 365
Price, P. W., 349
Pridgeon, A. M., 222
Pridham, J. B., 331
Prim, R. C., 96, 97, 114, 125
Prior, M., 92
Pritchard, M. R., [153]
Probatova, N. S., 292

Proctor, J. R., 73
Proctor, M., 352, 354, 355
Prósperi, C. H., 254
Pryer, K. M., 92
Prywer, C., 200
Pullaiah, T., 253, 265
Punt, W., 270
Puri, V., 240
Purkinje, J. E., 268

Quasada, E., 91
Queirós, M., 311
Queiroz, K. de, 111
Quinn, J. A., 222, 292

Rabb, G. B., 109
Radford, A. E., xiii, 9, 31, 32, 219, 221, 235, 254, 261, 393
Radhakrishnaiah, M., 397
Radlkofer, L., 233
Raffauf, R. F., 331
Rafinski, J. H., 225
Raghavan, V., 253
Rahn, K., 10, 91
Rajendra, B. R., 200
Raju, V. S., 222
Rak, Y., 15
Raman, S., 222
Ramos-Cormenzana, A., [91]
Ramsbottom, J., 163, 183
Rao, P. N., 222
Rao, P. S., 253
Rao, T. A., 235
Rao, V. S., 240
Rasmussen, F. N., 127, 218
Rasmussen, H. P., 222
Rau, M. A., 262
Rauh, W., [222], [236]
Raup, D. M., 107
Raven, P. H., 5, 15, 21, 133, 158, 165-67, 182, 185, 186, 195, 196, 217, 225, 265, 266, 271, 284, [284], 292, 293, [293], 298, 308, 310, 352
Ray, J., 44, 53, 162, 163, 218, 364, 391
Raymond, M., 302, 308
Real, L., 352
Reed, E. S., 169
Reed, H. S., 217
Reeder, J. R., 260, 263, 265
Rees, H., 290, 297, 302
Regan, C. T., 171
Reichert, E. T., 329
Reid, W. H., 357
Reitsma, T., 271, 274, 287
Renfree, M., 357
Renfrow, A., [222]
Renold, W., 337
Rensch, B., 34, 93, 212
Reveal, J. L., 200, 205, 324
Reyment, R. A., 60, 65
Reynolds, J. F., 82, 91
Rhodes, A. M., 77
Ribéreau-Gayon, P., 331
Rice, E. L., 91
Richards, A. J., 352, 358, 363
Richardson, P. M., 69, 133, 134, 350
Richter, R., 6

Rickart, E. A., [357]
Ricklefs, R. E., 365
Ridley, M., 98, 110
Riedl, R., 158
Rieger, R., 68
Rieppel, O., 18, 94, 98, 169
Rieseberg, L. H., 296
Riggs, J., [344]
Riley, D., 366, 367
Riley, R., 319
Risley, M. S., 291
Ritland, K., 345, 349
Robards, A. W., 234
Roberts, M. R., 327
Roberts, R. J., 126
Robertson, C., 352
Robichaux, R. H., 347
Robinson, B. L., 194, 204
Robinson, H., 32, 134, 198, 203, 204, 235 [293]
Robinson, T., 331
Robson, N. K. B., 233
Rock, H. F. L., 292
Rodman, J. E., 133, 212, 340
Rodrigues, P. D., 29
Rodríguez, E., 222
Rodríguez, R., 311
Roe, K. E., 78, 222
Rogers, C. M., 287
Rogers, D. J., 60, 65, 92, 202
Rogers, J. S., 317
Rogers, S. O., 344
Rohlf, F. J., 41, 67, 69, 80, 84, 127, [137], 138-40, 153, 158, 384, 390
Rollins, R. C., xvii, 7, 11, 25, 40, 62, 72, 184, 192, 200, 201, 222, 315
Rollinson, D., 331
Romberger, J. A., 234
Romesburg, H. C., 65
Romney, A. K., 65
Roose, M. L., 296, 302
Rosanoff, S., 268
Rose, M. R., 179
Rosen, D., 16
Rosen, D. E., 132, 169, 199
Rosendahl, C. O., 187
Rosendal-Jensen, S., 218
Rosenfeld, A., 23
Rosenthal, G. A., 377
Ross, H. H., 28, 30, 105, 108, 113
Rossenbeck, H., 313
Rossner, H., 369
Roth, I., 234
Roughgarden, J., 314, 365
Rourke, J. P., [357]
Rouse, J. L., 285
Rowell, A. J., 202
Rowley, J. R., 277, 284-85, 286
Roy, M. A., 167
Rubery, P. H., 332
Rudenberg, L., [293]
Rudolph, E. D., 53, 217, 252, 351
Ruíz-Berraquero, F., [91]
Runemark, H., 78, 172, 192, 200, 201, 302, 332
Rury, P. M., 250

Rusanovitch, I. I., 225
Ruse, M. E., 159, 162
Russell, S. D., 252, 254, 258
Rutovitz, D., 23
Rydberg, P. A., 203, 210
Rzedowski, J., 7

Sabrosky, C. W., 16
Sacarrão, G. F., 172
Sachar, R. C., 253
Sachs, J. von, 217
Sachs, T., 213
Saether, O. A., 99, 108, 127
Safayeni, F. R., 21
St. John, H., 10
Sakamoto, S., 200
Salamon, D. P., 376
Saleh, N. A. M., 283
Salthe, S. N., 22
Sampson, F. B., 285
Sanders, R. W., 125, 133, 293, 311
Sankoff, D., 327
Santesson, J., 330
Santos, M. de F., 311
Sarfatti, G., 285, [285]
Sarukhan, J., 365
Sass, J. E., 234
Sattler, R., 68, 104, 212, 244
Sauer, W., [298]
Savidan, Y., 296
Saville, D. B. O., 113
Scadding, S. R., 112
Schaal, B. A., 341
Schaeffer, B., 107
Schaffner, J. H., 187
Schieferstein, R. H., 222
Schill, R., 222, 286, 359, 360
Schlarbaum, S. E., 292
Schleiden, M. J., 252, 288
Schlichting, C. D., 376
Schmid, R., 198, 235, 240, 249, 250
Schmidt, K. P., 163
Schnarf, K., 252, 253
Schnepf, E., [236]
Schoch, R. M., 98, 143
Schofield, E. K., 235, 236
Schopf, J. M., 18, 287
Schröter, C., 365
Schuh, R. T., 139
Schulz, G. E., 126
Schulz, P., 260
Schulz, R., 260
Schuyler, A. E., 225
Schweizer, D., [114], [133], 291
Scogin, R., 133, 357
Scora, R. W., 97, 133, 335
Scott, P. J., 158
Scott, R. A., 270
Seaman, F. C., 133, 336, 337, 350
Seemann, J., [222]
Seavey, S. R., 225
Sederoff, R. R., 345
Seigler, D., 331, 349, 357
Seki, T., 91
Selkow, S. M., 97
Sellars, S. C., 277
Semple, J. C., 292

Seong, L. F., 133
Serota, C. A., 302
Settle, T. W., 131
Shaffer, H. B., 103, 104
Shah, G. L., 235
Sharma, A., 290, 304
Sharma, A. K., 290, 291, 304
Sharma, G. K., 222
Sharma, S. K., 225
Sharp, A. J., 388
Sharp, L. W., 290
Sharp, M., [284]
Shaw, G., 270, [270], 285
Shaw, S., 91
Sheerin, A., 14
Shelton, N., 15
Shepard, R. N., 65
Sherff, E. E., 205
Sherwin, P. A., 198, 249
Shetler, S. G., 186
Shivanna, K. R., 270
Short, L. L., 188
Shukla, P., 219
Shull, G. H., 161, 165, 166, 171
Sibley, C. G., 15, 188, 344
Sibson, R., 65, 140, 158
Sidhu, B. S., 298
Siegler, R. S., 23
Silva, O. M., [125], [311], [327]
Simon, J. -P., 139
Simpson, B. B., 93, [293], 357
Simpson, D. R., 10
Simpson, G. G., 6, 7, 45, 47, 58, 93, 94, 96, 137, 164, 176, 177, 210
Sims, L. E., 344
Sims, R. W., 365
Sinclair, W. A., 22, 29, 196
Singer, R., 92, 197
Singh, R. P., 231, 246
Sinha, S., 314
Sinha, S. P., 91
Sinha, U., 314
Sivarajan, V. V., xiii, 240, 270
Skalinska, M., 311
Sklansky, J., 23
Sklar, A., 159
Skvarla, J. J., 268, 270, 271, 277, 279, 280, 283-87
Skvortsov, A. K., 225
Slade, N. A., 115
Slatkin, M., 17, 173, 209
Slingsby, P., 360
Sloan, S. A., 311
Slobodchikoff, C. N., 160-62, 180, 181
Sluiman, H. J., 133, 134
Small, E., 67, 85, 92, 287, 360
Smartt, P. F. M., [79]
Smets, E., 218
Smith, A. C., 9, 14, 25
Smith, A. J. E., 73
Smith, B. B., 264, 265, 266
Smith, B. N., 347
Smith, B. W., 290, 302
Smith, D. M., 353
Smith, E. B., 192, 318, 358
Smith, G. D., 369
Smith, G. R., 75, 80, 90
Smith, H. G., 329

Smith, H. M., 188
Smith, J. E., 5
Smith, P. G., 127
Smith, P. M., 330, 344
Smith, T. F., [126]
Smith, W., 286
Smith-White, S., 292
Smoot, E. L., 69, 134
Smouse, P. E., 15
Snajberk, K., 336
Sneath, P. H. A., 10, 24, 25, 28, 34, 37, 42, 44, 52, 57, 59-62, 65-78, 80, 84-86, 89-95, 137, [137], 141, 155, 172
Snogerup, S., [300]
Snow, R., [293], [298]
Snustad, D. F., 313, 314
Sober, E., 103, 104, 126
Soderstrom, T. R., 198
Sokal, R. R., 24, 25, 28, 29, 34, 37, 42, 44, 46, 52, 57, 59-63, 65, 66-80, 82, 84-86, 89-97, 114, 127, 128, 137, 138, 139-41, 150, 153, 155, 165, 171, 174, 190, 384, 390
Sokolovskaya, A. P., 292
Solbrig, O. T., 60, 97, 125, 173, 293, [293], [298], 310, 365
Solereder, H., 233
Solomon, A. M., 286
Soltis, D. E., 192, 200, 296, 297, 308, 345
Soltis, P. S., 345
Somero, G. N., 331
Soria, J., 77
Southworth, D., 285
Speta, F., 200, 235
Spooner, D. M., 41, 263, 292, 311
Sporne, K. R., 33, 106, 110, 113, 219, 286
Sprague, T. A., 34
Sprengel, C. K., 351
Stace, C. A., xiii, 7, 11, 32, 35, 37, 222, 235, 299, 301, 405
Stafleu, F. A., 163
Stanley, R. G., 17
Stanley, S. M., 270
Starr, T. B., 22
Starrett, A., 188
Stearn, W. T., 45, 50, 61, 158, 219, 221, 225
Stebbins, G. L., 11, 17, 18, [40], 55, 109, 110, 112, 126, 161, 212, 213, 231, 250, 290, 292, 297, 311, 314, 319, 321, 327, 385, [385], 386, 388
Steer, M. W., 234
Steere, W. C., 217
Steeves, T. A., 253
Stein, M. L., [126]
Stein, W. E., Jr., 130
Steinmetz, F. H., [270]
Stephenson, W., 65, 75, 78-80
Stern, K. R., 286
Stern, W. L., 240, 248
Sterner, R. W., 348
Stevens, P. F., 54, 106-10, 137, [199], 206, 212, 223
Steward, F. C., 234

Stewart, N. F., 97
Steyskal, G. C., 10, 69
Stinebrickner, R., 127
Stix, E., 268
Stone, A. R., 362
Stone, D. E., 275, 277, 287
Stone, S. J. L., 341
Stort, M. N. S., 324
Stott, P., 365
Stoutamire, W. P., 311
Strasburger, E. A., 252
Strathmann, R. R., 209
Strauch, J. G., Jr., 97, 113, 123
Strauss, R. E., 111, 221
Strecker, G. E., 22
Strickland, R. G., 225
Strother, J. L., 186, 193
Stuckey, R. L., 217, 370, 373, 376, 377, 402
Stucky, J., 298
Stuessy, T. F., 7, 8, 17, 33, 41, 48, 69, 70, 74, 77, 81, 91, 92, 93, 94, 96-101, 103, 104, 106-10, 112-15, 119, 125, [125], 126, 128, 130, 131, 133, 134, 139, 141, 144-49, 172, 174, 175, 184, 192, 199, 204, 205, 211, 212, 219, 246, 265, 279, 292, 293, [293], 296, 298, 308, 311, [311], 317, 322, 326-28, [327], 348, 350, 366, 376, 395, 398, 400, 403, 404
Sturtevant, A. H., 313
Stutz, H. C., 200
Suda, Y., 293-95, 300
Sulak, J., [222]
Sulinowski, S., 200
Sultan, S. E., 376
Sumner, A. T., 290, 291
Sundberg, P., 152
Sundberg, S., 376
Sussex, I. M., 253
Sutton, W. S., 313
Suzuki, D. T., 314
Svenson, H. K., 163, 168, 179, 197
Swain, T., 330
Swamy, B. G. L., 235, 253, 258
Swanson, C. P., 288, 290, 314
Sweet, H. R., 200
Swift, L. H., 56
Swingle, D. B., 9, 33, 41
Sybenga, J., 290, 314
Sylvester-Bradley, P. C., 9, 12, 161, 176, 187
Szalay, F. S., 142

Takhtajan, A., 44, 55, 90, 111, 196, 207, 212, 366
Tanksley, S. D., 345
Taylor, C. E., 41
Taylor, P., 205, 394
Taylor, R. L., 186, 311
Taylor, T. N., 69, 134, 233, 280
Templeton, A. R., 126, 179, 385
Terrell, E. E., 10
Terzaghi, E. A., 331
Tétényi, P., 331
Thanikaimoni, G., 270

Theobald, W. L., 222, 225, 249
Theophrastus, 6, 44, 49, 196, 208, 218, 351, 364
Thomas, L. K., 200
Thomas, M. B., 331
Thompson, E. A., 126
Thompson, W. F., 291
Thompson, W. R., 163, 164
Thorne, R. F., 18, 44, 55, 196, 207
Thorpe, R. S., 190
Throckmorton, L. H., 96, 128, 330
Tilden, J. W., 188
Timmermann, B. N., 331
Titz, W., 80
Tobe, H., 246, 265, 266
Togby, H. A., 327
Tomb, A. S., [200], 225, [270], 280, 286, 287
Tomlinson, P. B., 212, 219, 222, 223, 231-34
Torrey, J. G., 234
Tothill, J. C., 297
Tournefort, J. P. de, 44, 50, 51, 196, 197, 208, 218
Trela-Sawicka, Z., 258
Trench, R. K., 207
Troll, W., 219
Tschermak, E., 288, 313
Tschudy, R. H., 270, 272, 276
Tseng, C. C., 250
Tsuchiya, T., 292
Tsukada, M., 270
Tucker, S. C., 213, 231, 235
Tufte, E. R., 391, 393, 398, 401-3
Tukey, J. W., 390, 393
Tukey, P. A., [393]
Tuomikoski, R., 127
Turesson, G., 11, 177
Turner, B. L., 15, 33, 75, 90, 92, 184, 190, 194, 203, 204, 240, 268, [270], 283, 292-93, 298, 308, 310, 330, [330], 331, 332, 333, 334, 336, [336], 338, 339, 340, 342, 344, 347, 381, 397, 399
Turner, V., 357
Turrill, W. B., 55, 314
Tyler, S., 68, 219

Udvardy, M. D. F., 132
Uhl, C. H., 302
Ulam, S. M., [126]
Unnasch, R. S., 90
Urbanska-Worytkiewicz, K., 293
Urbatsch, L., [293]

Vaillancourt, P., 180, 194
Vaillant, S., 50
Valderrama, M. J., [91]
Valentine, D. H., 177, 178, 187, 365, 378
Valentine, J. W., [40], [385]
Van Campo, M., 270, 275
Vanden Berge, J., [73]
Van der Pijl, L., 352, 360
Van der Steen, W. J., 137
Van Emden, H. F., 352

Van Horn, G. S., 76, 97, 114, 115, 118, 125
Vani, R. S., 265
Van Son, G., 188
Van Valen, L., 99, 137, 160, 176, 178, 209
Varadarajan, G. S., 103
Vaughan, A., 176
Vaughan, J. G., 247, 330
Vázquez-Yanes, C., 15
Ventosa, A., [91]
Verdoorn, F., 194
Vickers, M. A., [270]
Vickery, R. K., Jr., 319
Vignal, C., 225
Vijayaraghavan, M. R., 258
Vilgalys, R., 110, 133
Villalobos-Pietrini, R., 293, 302
Vinay, P., 270
Vincent, P. L. D., 254
Vithanage, H. I. M. V., 285
Vitt, D. H., 133
Vliet, G. J. C. M. van, 240
Vogel, H. J., 131
Vogel, S., 357
Voigt, J. W., 397
Von Rudloff, E., 90, 92, 190, 336
Voss, E., 54, 64

Wadhi, M., 258
Wagenaar, E. B., 296
Wagner, W. H., Jr., 36, 94, 96-97, 101, 108, 114, 127, 133, 141, 144, 151, 153, 388
Wagner, W. L., 246
Wain, R. P., 192
Waldeyer, W., 288
Waldrop, M. M., 23
Walker, A. G., 280, 283, 286
Walker, J. W., 211, 240, 242, 270, 271, 280, 283, 286
Walker, P., 352
Walter, H., 366
Walter, K. S., 225
Walters, S. M., 208
Wang, A., [14]
Wang, R. -C., 200
Wanntorp, H. -E., 93, 94, 114, 117, 207, 212
Warburton, F. E., 20, 24-26
Ward, J. M., 234, 357
Warnock, M. J., 369
Wartenberg, D., 90
Watanabe, K., 292
Waterman, P. G., 331
Watrous, L. E., 95, 106, 108-10
Watson, J. D., 313
Watson, L., 10, 17, 66, 74, 90, 91, 92, 212
Watson, T. J., Jr., 293
Watt, J. C., 137
Watts, W. A., 374
Weatherby, C. A., 183-85, 187, 213
Webb, A., 249
Webb, C. J., 201
Webb, J. A., 270, 271, 273, 274, 275, 277, 278
Webster, G. L., 187, 347
Weevers, T., 330

Weimarck, G., 78, 332
Weinburg, W., 314
Weir, B. S., 314
Wellman, A. M., 91
Wells, H., 397
Wells, L. G., 222
Went, J. L. van, 256, 258, 260
Weresub, L. K., 92
Wernham, H. F., 34
Wesley, J. P., 365
Westfall, R. H., 10
Westwood, M. N., 92
Wetter, M. A., 204
Whalen, M. D., 332
Whalley, P. E. S., 365
Wheeler, Q. D., 95, 103, 106-12
Whiffin, T., 114, 115, 225
White, F., 92, 202
White, F. N., 188
White, J., 219
White, M. J. D., 107, 128, 179, 385
White, R. A., 233
White, R. R., 167
White, T. J., 341
Whitehead, P. J. P., 10
Whittaker, R. H., 207
Whittemore, A. T., 134
Whittingham, A. D., 319, 321
Wieffering, J. H., 311
Wiener, L. F., 98
Wiens, D., 357
Wilbur, R. L., 193, 198, 249
Wild, H., 369
Wiley, E. O., 29, 34-36, 58, 93-98, 101, 110, 128, 131, 132, 164, 177, 209
Wilkins, A. S., 331

Willemse, M. T. M., 256, 258, 260
Williams, D. A., 267
Williams, E. G., 285
Williams, H. P., 23
Williams, N. H., 349
Williams, R. F., 234
Williams, W. T., 71, 78, 79, 91, 137
Willis, E. O., 177
Willis, J. C., 199
Wills, A. B., 290
Willson, M. F., 351, 352
Wilmott, A. J., 175
Wilson, A. C., [14], 126, 292, 341
Wilson, C. L., 112
Wilson, E. B., 313
Wilson, E. O., 6, 7, 18, 96, 117, 188, 189
Wilson, H. D., 192, 317
Winge, Ö., 314
Winkler-Oswatitsch, R., 14
Wodehouse, R. P., 267, 268, 270, 271
Woese, C. R., 207, [207]
Wofford, B. E., [115]
Wolfe, J., 221
Wolfe, R. S., [207]
Wolter, M., 359, 360
Woodruff, D. S., 128
Wooller, R. O., 357
Wright, C. A., 330
Wright, H. E., Jr., 374
Wright, K., 91
Wright, S., 314
Wülker, W., 327
Wyatt, R., 362

Wylie, A. P., 290
Wynne, M. J., 207

Xavier, K. S., 287

Yablokov, A. V., 65
Yakovlev, M. S., 263
Yamazaki, T., 260
Yang, D., 207
Yano, T., 345
Yeo, P. F., 133, 200, 212, 218, 352, 354, 355
Yochelson, E. L., 10
Yoshioka, H., 331
Young, A., 366, 367
Young, D. A., 69, 133, 134, 194, 206, 331, 348
Young, D. J., 66, 90, 92, 212
Young, W. J., 314
Yule, G. U., 313

Zamir, D., 40, 175
Zanoni, T. A., 190
Zavada, M. S., 150, 285, 286
Zavarin, E., 336
Zech, L., 290
Zeeuw, C. de, 234
Zenger, V. E., 270
Zimmer, E. A., 345
Zimmerman, J. H., 338
Zimmerman, J. R., 190
Zimmermann, M. H., 233, 234
Zimmermann, W., 207
Zinderen Bakker, E. M. van, 270
Zirkle, C., 314
Zohary, D., 296, 297
Zuckerkandl, E., 329

TAXON INDEX

Abronia angustifolia, 284
Acacia, 35, 111, 212, 374
Acanthaceae, 201, 258, 271, 279
Acanthospermum, 298
Acer, 275, 361
Aceraceae, 361
Achillea, 343
Achyropappus anthemoides, 399
Acmella, 297, 326, 328; *oppositi-folia*, 326; *papposa*, 326
Aconitum, 277
Adoxa, 257
Agaricales, 92
Agastache, 133
Agavaceae, 258, 293, 302
Agave, 302
Aizoaceae, 265, 340
Alismatidae, 212
Allium, 92, 257
Alnus, 275
Aloe keayi, 225; *macrocarpa* var. *major*, 225; *schweinfurthii*, 225
Alzatea, 265
Alzateaceae, 265
Amaranthaceae, 284, 298, 340
Ambrosia, 201, 279, 336, 337, 374; *confertiflora*, 336, 337; *psilostachya*, 336
Ambrosiinae, 395
Amentiferae, 55, 343
Amoracia, 77
Amphibolis antarctica, 354
Anacyclus, 133
Anagallis, 205, 394; *serpens* subsp. *meyer-johannis*, 394
Andromedeae, 206
Andropogon virginicus, 117
Anemia, 133
Anemone nemorosa, 355; *pulsa-tilla*, 355; *quinque folia*, 299; *richardsonii*, 299; *rivularis*, 299
Anisochaeta, 283
Annonaceae, 271
Antennaria, 297, 324
Anthemideae, 343
Anthemis, 265
Anthurium, 353
Antirrhineae, 248
Apocynaceae, 265
Arabis, 265, 266; *hirsuta*, 266; *planisiliqua*, 266

Araceae, 254, 277, 293, 353
Araliaceae, 250
Archaeopteris, 205
Arctium, 343
Armeria maritima var. *sibirica*, 280
Arrabidaea, 353
Artemisia, 374
Asarum, 242
Asclepiadaceae, 69
Asplenium plenum, 126
Astereae, 200, 298, 343, 395
Athyrium, 133
Atriplex, 308, 325; *canescens*, 308; *longipes* subsp. *praecox*, 325
Austrobaileyaceae, 284, 285
Avena, 117, 359; *sativa*, 359

Baccharis, 204
Bahia, 397, 399; *bigelovii*, 399; *dissecta*, 399
Bahiinae, 395
Balsaminaceae, 305, 360
Banisteriopsis latifolia, 356
Baptisia, 330
Basellaceae, 265, 340
Bellis, 343
Berberidaceae, 133, 170
Betula, 275, 374
Betulaceae, 354
Bignoniaceae, 249, 287, 353
Blastemanthus gemmiflorus, 236
Blennosperma, 280, 283
Boottia cordata, 201
Boraginaceae, 279
Bouteloua, 287
Boykinia, 308
Brachyscome lineariloba, 292
Brackenridgea australiana, 236; *nitida*, 236
Brassica, 247, 342; *carinata*, 247; *juncea*, 247; *nigra*, 247
Bromeliaceae, 69, 222, 235, 347, 401
Bryophytes, 55, 91, 133, 207, 332
Bulbostylis, 81
Bulnesia, 76, 77, 82
Bursera, 190
Burseraceae, 190
Butomus, 273

Byrsonima coccolobifolia, 356; *intermedia*, 356; *vaccinifolia*, 356; *verbascifolia*, 356

Cachrys, 198
Cactaceae, 211, 284, 340
Caesalpinioideae, 284
Cakile, 340
Calendula, 228, 343
Calenduleae, 343
Callisia fragrans, 288
Callixylon, 205
Calophyllum, 223, 238; *angulare*, 238; *brasiliense*, 238; *inophyl-lum*, 238; *longifolium*, 238; *nubicola*, 238; *rekoi*, 238; *sou-lattri*, 238
"Caminalcules," 150
Campanula, 287
Campanula sect. *Heterophylla*, 287
Campanulaceae, 287, 345
Cannabaceae, 92
Cannabis, 92
Cannaceae, 277
Capparaceae, 236
Capparales, 236, 237, 340
Caprifoliaceae, 133, 202
Cardionema ramosissima, 284
Carduus, 369, 370; *nutans*, 369, 370
Carex, 302
Carya, 287
Caryophyllaceae, 200, 208, 256, 284, 306, 340, 355
Caryophyllales, 207, 211, 240, 249, 339, 340
Caryophyllidae, 242
Caryophyllineae, 340
Castilleja, 279
Castillejinae, 279, 308
Casuarinaceae, 283
Cattleya, 324
Caulanthus, 340
Cavendishia, 115
Celosia, 298
Centauria, 343
Centaurium, 293
Centrolepidaceae, 265
Centrospermae, 212, 284, 339, 344
Cerastium, 277

Ceratostigma plumbaginoides, 359
Cercidiphyllaceae, 265
Chaenactis, 327, 378; *douglasii,* 292; *fremontii,* 378; *glabriuscula,* 378; *stevioides,* 378
Chenopodiaceae, 82, 192, 256, 275, 308, 317, 325, 340
Chenopodiineae, 340, 374
Chenopodium, 82, 192, 256, 317; *atrovirens,* 82; *dessicatum,* 82; *pratericola,* 82
Chionopappus, 283
Chlorophyta, 207
Chrysanthemum cinerariaefolium, 257
Chrysobalanaceae, 92, 202
Chrysothamnus, 258, 259, 266; *linifolius,* 259; *nauseosus* ssp. *leiospermus,* 259; *nauseosus* ssp. *nauseosus,* 259; *parryi* affin ssp. *nevadensis,* 259; *pulchellus,* 259; *viscidiflorus* ssp. *humilis,* 259; *viscidiflorus* ssp. *lanceolatus,* 259; *viscidiflorus* ssp. *viscidiflorus,* 259
Cichorieae, 343
Circaea lutetiana, 245
Cirsium, 343
Cladothamneae, 199
Clarkia, 287, 321, 322, 327; *rhomboidea,* 321, 322
Claytonia, 292, 294, 295, 300; *cordifolia,* 300; *perfoliata,* 300; *sarmentosa,* 300; *virginica,* 292, 294, 295, 300
Clematis, 133
Clermontia, 357
Coelorhachis, 92
Colchicum, 275
Colpodium versicolor, 292
Commelina sp., 288
Commelinaceae, 288, 293, 307, 327
Commelinales, 234
Compositae, 32, 52, 69, 77, 81, 92, 97, 112, 113, 116, 118, 119, 128, 133, 145, 149, 168, 173, 179, 190, 192-94, 198-205, 208, 210, 228, 248, 254, 255, 258, 259, 265, 266, 268-70, 279, 280, 283-85, 287, 290, 292, 296-98, 308, 314, 315, 319, 322, 324, 326-28, 331, 336, 337, 341, 345, 347, 349, 358, 362, 366, 369, 370, 372, 376, 378, 395, 397-400, 403, 404
Comptonia peregrina, 244
Convolvulaceae, 133
Conyza, 204
"Cookophytes," 152
Cordylanthus, 225, 228, 279
Coreopsidinae, 395
Coreopsis, 192, 318; *grandiflora* var. *grandiflora,* 318; *grandiflora* var. *harveyana,* 318; *grandiflora* var. *longipes,* 318; *grandiflora* var. *saxicola,* 318
Cornaceae, 256, 265, 266

Cornus, 256, 265, 266
Corydalis bulbosa, 359
Crambe hispanica, 247
Crassulaceae, 249, 292, 302
Crataegus, 59, 243, 289, 297
Crepis, 290, 314, 327
Crocidium, 283
Crossandra stenostachya, 271
Cruciferae, 40, 52, 53, 77, 192, 236, 247, 265, 266, 315, 340, 342
Crusea, 124
Crypteroniaceae, 265
Cryptocoryne, 293
Cucurbitaceae, 232
Cuphea, 279
Cupressaceae, 71, 190, 335
Cymbopetalum odoratissimum, 271
Cymodoceaceae, 354
Cynareae, 343
Cyperaceae, 81, 200, 229, 234, 248, 277, 293, 302, 374
Cystopteris, 133
Cytisus, 73

Dahlia, 255
Danthonia, 292; *sericea,* 292
Daucus, 92; *carota,* 92
Delphinium alabamicum, 369; *treleasei,* 369
"Dendrogrammaceae," 151
Diapensiaceae, 296
Dicentra, 286
Didiereaceae, 340
Digitaria, 308; *adscendens,* 308; *sanguinalis,* 308
Dillenia indica, 239
Dilleniaceae, 239
Dilleniidae, 91, 212
Dioscoreales, 234
Dithyrea, 40, 192, 315; *griffithsii,* 315; *wislizenii,* 315
Doronicum, 228
Drusa, 257
Dubautia, 347
Dyssodia, 193, 399; *setifolia,* 399

Ecliptinae, 395
Eleocharis, 293
Elymus, 201, 302; *striatulus,* 302
Endodontaceae, 133
Endymion, 257
Engelmanniinae, 395
Equisetum, 133
Erica carnea, 359
Ericaceae, 115, 133, 199, 206, 235, 273, 359, 373
Ericales, 92
Erigeron, 193, 204
Eriocaulon, 273
Eriogonum, 205
Eriophorum comosum, 229; *crinigerum,* 229; *japonicum,* 229; *latifolium,* 229; *microstachyum,* 229
Erythrina, 280, 286
Eucalyptus, 92, 223, 235, 329; *citriodora,* 223; *planchoniana,* 223; *terminalis,* 223

Eugenia, 249
Eupatorieae, 32, 203, 204
Euphorbia, 204, 347, 355; *paralias,* 355
Euphorbiaceae, 200, 202, 236, 347, 355
Euphorbieae, 200

Fagaceae, 81, 172, 226
Fagopyrum, 277
Feddea, 283
Festucoideae, 260, 347
Fitchiinae, 395
Floerkea, 84
Frankeniaceae, 265
Franseria, 201
Fraxinus pennsylvanica, 374
Fritillaria, 257
Fumariaceae, 286, 359

Gaillardiinae, 395
Galax, 296
Galinsoginae, 395
Gentianaceae, 280, 293
Geraniaceae, 355
Geranium pratense, 355; *pyrenaicum,* 355
Gilia, 173, 324
Ginkgo biloba, 159, 170
Gnaphalium norvegicum, 372; *supinum,* 372
Gnetaceae, 244
Gnetum gnemon, 244; *montanum,* 244, 298
Gossypium, 115, 258
Gramineae, 53, 77, 92, 113, 117, 168, 197, 199-201, 210, 231, 234, 235, 258, 260, 263, 265, 273, 285, 287, 292, 302, 308, 318, 324, 345, 347, 359, 374
Grimmiaceae, 133
Grindelia, 228, 265
Guizotia, 343
Gutierrezia, 97
Guttiferae, 223, 230, 235, 236, 238

Haloragaceae, 279
Hamamelidaceae, 240, 265
Hamamelidae, 354, 362
Hamamelidales, 265
Haplopappus, 292, 315, 319, 358; *gracilis,* 292; *phyllocephalus,* 315; sect. *Blepharodon,* 315; sect. *Isocoma,* 315
Helenieae, 283, 395
Heleniinae, 395
Heliantheae, 265, 343, 395
Helianthinae, 395
Helianthus, 192, 201, 258, 285, 343
Heliconia, 275
Heliconiaceae, 275
Helleborus foetidus, 355
Herniaria glabra, 284
Heuchera, 308
Hibbertia cuneiformis, 239; *dentata,* 239; *tontoutensis,* 239
Hieracium, 343

Hippocastanaceae, 133
Hippuris, 275
Hydrocharitaceae, 200, 201, 354
Hydrophyllaceae, 292
Hymenaea, 349
Hyoseridinae, 280
Hypericum chapmanii, 230; *fasciculatum*, 230; *lissophloeus*, 230; *lloydii*, 230; *reductum*, 230
Hyphomycetes, 194
Hypochaeris radicata, 255

Icacinaceae, 235
Idiospermaceae, 348
Illiciaceae, 243
Illicium cubense, 243; *floridanum*, 243
Impatiens, 305, 360
Inula, 343
Inuleae, 283, 284, 343
Iridaceae, 200
Iva, 202
Ixodia, 198

Jaumeinae, 395
Juglandaceae, 287
Juncales, 234
Juniperus, 60, 71, 190, 334, 335

Kalanchoe, 292
Krameriaceae, 310, 357

Labiatae, 53, 133, 208, 275, 287, 334, 350, 377
Lactoridaceae, 148-50, 212
Lactuca, 342, 343
Lactuceae, 280, 342
Lagascea, 112, 128, 145, 149, 265, 403; *decipiens*, 403; sect. *Nocca*, 128
Lagasceinae, 395
Lantana, 293
Larix, 374
Lasthenia, 228
Laurales, 284
Lecocarpus, 298
Lecythidaceae, 310
Leguminosae, 53, 67, 69, 73, 85, 92, 111, 200, 235, 280, 284, 286, 293, 298, 308, 330, 331, 349, 374
Leiphaimos, 280
Lemnaceae, 293
Leptarrhena, 308
Liabeae, 283
Liliaceae, 92, 169, 222, 225, 237, 258, 265, 303-5, 339, 359
Limnanthaceae, 84
Limnanthes, 84
Linaceae, 256, 287, 346, 347
Linum, 256, 287, 346, 347; *alpinum*, 346; *bienne*, 346; *boreale*, 346; *grandiflorum* var. *rubrum*, 346; *lewisii*, 346; *narbonense*, 346; *perenne*, 346; *usitatissimum*, 346
Loasaceae, 228
Lobelioideae, 357
Lopezia suffrutescens, 245

Loranthaceae, 254
Lotus, 293, 298
Lupinus, 308; *subcarnosis*, 308; *texensis*, 308
Lychnis, 256
Lycopersicon, 285
Lycopus europaeus, 377
Lysimachia vulgaris, 355
Lythraceae, 265, 279

Madiinae, 248, 314, 395
Magnolia tripetala, 373
Magnoliaceae, 168, 235, 248, 265, 373
Magnolialean complex, 287
Magnoliales, 212, 265, 284
Magnoliidae, 343
Magnoliophyta, 207
Malpighiaceae, 356
Malva, 277
Malvaceae, 115, 258, 293
Medicago, 67, 85, 92; subg. *Spirocarpus*, 85
Melampodiinae, 395
Melampodium, 69, 77, 81, 119, 191, 192, 205, 292, 296, 298, 308, 322, 376, 404; *cinereum*, 404; *gracile*, 322; *leucanthum* var. *argophyllum*, 404; *leucanthum* var. *leucanthum*, 404; *microcephalum*, 322; *montanum*, 244; *paniculatum*, 322
Melastomataceae, 208, 211, 232
Mentzelia, 228
Menyanthes, 277
Microseris, 315
Milleriinae, 395
Mimosa, 273
Mimosoideae, 268
Mimulus, 319
Mniaceae, 133
Molluginaceae, 340
Monarda, 133
Monimiaceae, 246
Montanoa, 133, 292; *guatemalensis*, 292
Mosla, 350
Mutisieae, 283, 345
Myrceugenia, 133
Myricaceae, 244
Myriophyllum, 279
Myristicaceae, 280, 283
Myrtaceae, 92, 133, 223, 249
Myrtales, 235, 265, 266, 284

Nassauviinae, 92
Neptunia, 298
Nerium indicum, 265
Neurolaeninae, 395
Nicotiana, 290, 314
Nitella, 332
Notoptera, 204
Nyctaginaceae, 279, 284, 340
Nyctanthes, 265
Nymphaea, 277
Nymphoides, 273, 277

Oenothera, 179, 257, 290, 314
Oleaceae, 265

Onagraceae, 179, 245, 271, 284, 287, 290, 314, 321, 322, 327
Onopordon tauricum, 255
Ophiocephalus, 279
Ophioglossum, 292; *reticulatum*, 292
Opuntia lindheimeri, 284
Orchidaceae, 168, 200, 210, 222, 268, 273, 324, 347, 352
Orthocarpus, 228, 279
Orthotrichum, 133
Oryza, 77; *perennis*, 77
Oryzopsis, 231
Otopappus, 204, 366, 398; *acuminatus*, 366; *australis*, 398; *epaleaceus*, 366; *imbricatus*, 366; *koelzii*, 366; *microcephalus*, 366; *robustus*, 366; *tequilanus*, 366
Ottelia alismoides, 201

Palmae, 234, 235
Panicoideae, 347
Papaveraceae, 355
Papaver rhoeas, 355
Paris quadrifolia, 359
Parnassia, 275
Parthenice mollis, 287
Passifloraceae, 286
Pectidinae, 395
Pedicularis, 273
Penaea, 257
Pentachaeta, 118
Peperomia, 257
Perezia, 228
Petasites, 343
Petrobiinae, 395
Phaeophyta, 207
Phlox aspera, 323; *floridana*, 323; *stolonifera*, 373; *villosissima*, 323
Phormium, 258
Phyodina navicularis, 288
Physostegia, 133
Phytolaccaceae, 292, 340
Picea, 374
Picradeniopsis, 128, 296, 308, 326, 399, 400; *oppositifolia*, 308, 399; *woodhousei*, 308
Pieris, 133
Pinaceae, 81, 242
Pinus, 273, 330
Piperales, 231
Plantaginaceae, 321
Plantago, 275, 277, 321; *insularis*, 321
Platycerium, 133
Plumbagella, 257
Plumbaginaceae, 258, 280, 359
Plumbago, 257, 258; *zeylanica*, 258
Poaceae, 256
Podophyllum peltatum, 170
Podostemaceae, 265
Polemoniaceae, 173, 280, 323, 324, 352, 373
Polemoniales, 254
Polemonium, 277
Polygala, 303
Polygalaceae, 303

Polygonaceae, 205, 242, 256, 283, 293, 396
Polygonella fimbriata, 396; *robusta*, 396
Polygonum, 256, 257, 275, 280, 283; *amphibium*, 275, 283; *cilinode*, 283; *convolvulus*, 283; *forrestii*, 283; *glaciale*, 283; *orientale*, 283; *raii*, 275
Polypodium, 133
Populus, 68, 277, 361
Portulaca oleracea, 252
Portulacaceae, 92, 252, 292, 294, 295, 300, 340
Potamogeton, 273, 402; *crispus*, 402
Potamogetonaceae, 402
Potentilla erecta, 354; *reptans*, 354; *verna*, 354
Prangos, 198
Primulaceae, 205, 355, 394
Prochlorophyta, 207
Proteaceae, 133, 268
Prunus, 256
Psilotrichum amplum, 284
Pteridophytes, 91, 133

Quercus, 81, 172, 226, 374; *arkansana*, 226; *chapmanii*, 226; *lyrata*, 226; *margaretta*, 226; *rolfsii*, 226

Ranales, 55, 231
Ranunculaceae, 87, 123, 133, 140, 208, 256, 258, 265, 275, 299, 355, 369, 372
Ranunculoides, 258
Ranunculus, 87, 123, 140, 256, 355, 372; *acris*, 355; *glacialis*, 372; *hispidus*, 87, 123, 140
Relhania, 116
Rhamnaceae, 256
Rhamnus, 256
Rheum, 256
Rhiziridium, 92
Rhizophoraceae, 235
Rhododendron maximum, 373
Rhodophyta, 207
Rhynchocalycaceae, 265
Rhynchocalyx, 265
Rhytachne, 92
Riddelliinae, 395
Robinsonia, 327
Rosaceae, 162, 200, 208, 256, 297, 354, 355
Rosidae, 91, 212
Rubiaceae, 124, 275, 279, 345
Rubus fruticosus, 355; *idaeus*, 355
Rudbeckia, 228, 343
Ruellia, 201
Rumex, 275, 293
Rutaceae, 139, 211, 331, 334, 335

Salicaceae, 68, 69, 256, 361
Salix, 69, 256, 277
Salvadoraceae, 266

Sanguisorba, 275
Sapindaceae, 233
Satureja, 335
Saxifragaceae, 192, 200, 271, 296, 308
Saxifraga oppositifolia, 277
Saxifrageae, 308
Scheuchzeria, 273
Scheuchzeriaceae, 231
Schkuhria multiflora, 399; *pinnata* var. *virgata*, 399; *schkuhrioides*, 399
Scilla bisotunensis, 303; *hohenackeri*, 303; *mischtschenkoana*, 304; *siberica*, 303, 304; *siberica* "Spring Beauty," 304
Scorzonera hispanica, 269; *laciniata*, 269; *lanata*, 269
Scorzonerinae, 269
Scrophulariaceae, 197, 200, 222, 228, 248, 249, 279, 308, 319
Senecio, 194, 204, 343; *jacobaea*, 343; *vulgaris*, 343
Senecioneae, 204, 283, 343, 395
Silene bridgesii, 306; *campanulata*, 306; *douglasii*, 306; *menziesii*, 306; *nuda*, 306; *oraria*, 306; *oregana*, 306; *parishii*, 306; *parryi*, 306; *scouleri*, 306; *souleri* subsp. *grandis*, 306
Simsia, 292
Sinapis, 342
Siphonychia americana, 284
Solanaceae, 77, 86, 92, 201, 285, 290, 314, 331
Solanum, 77, 86, 92; *nigrum*, 77; sect. *Solanum*, 86, 92
Solidago, 343
Sonchus, 342, 343
Sonneratia, 279
Sonneratiaceae, 279
Sorghum, 285
Spergula, 275
Sphagnum, 273
Spilanthes, 133
Spiraea, 256
Stachyuraceae, 266
Stackhousiaceae, 310
Stellaria holostea, 355
Stephanomeria, 173
Stephanomeriinae, 280
Stipa, 231
Streptanthus, 340
Stylidiaceae, 292
Stylidium, 292
Stylisma, 133
Sullivantia, 192
Syzygium, 249

Tagetininae, 395
Tambourissa, 246
Tanacetum, 255, 343; *vulgare*, 255
Tanakaea, 308
Taraxacum, 59, 179, 342, 343
Taxodiaceae, 292
Tetramolopium, 193
Tetraplasandra, 250
Tetrapterys ramiflora, 356

Thelesperma, 190
Thelycrania, 277
Thysanotus arenarius, 237; *formosus*, 237; *multiflorus*, 237; *tuberosus*, 237
Tilia, 374
Tithonia, 81, 133
Tofieldia, 275
Tolmiea, 296
Tournefortia, 279
Tradescantia commelinoides, 307
Tragopogon, 128, 296, 326, 343, 397; *dubius*, 397; *mirus*, 397; *miscellus*, 397; *porrifolius*, 397; *pratensis*, 397
Trifolium, 277
Trillium grandiflorum, 305
Triticeae, 92, 396
Tropaeolaceae, 266
Tussilago, 343
Typha, 273, 376; *angustifolia*, 376; *latifolia*, 376
Typhaceae, 376

Uldinia, 249
Ulmaceae, 348, 350, 354
Ulmus, 275, 374
Umbelliferae, 52, 92, 198, 201, 208, 249, 292, 331, 334
Uncinia, 248
Unxia, 199, 308
Ursinia, 343
Urtica, 275
Urticales, 242
Utricularia, 275
Uvularia grandiflora, 169; *perfoliata*, 169

Vaccinium, 308; *angustifolium*, 308; *boreale*, 308
Valerianaceae, 227, 315, 316
Valerianella, 227, 315, 316; *bushii*, 316; *bushii* × *ozarkana*, 316; *carinata*, 227; *ozarkana*, 315, 316; *ozarkana* forma *bushii*, 316; *ozarkana* forma *ozarkana*, 316; *rimosa*, 227
Vallisneria, 354
Veratrum, 338
Verbenaceae, 235, 293
Verbesininae, 395
Vernonia, 280, 349
Vernonieae, 395
Viburnum, 133, 202
Viguiera, 201; *porteri*, 201
Viola, 275, 359; *lutea*, 359; *odorata*, 359
Violaceae, 359
Virola, 280, 283; *calophylloidea*, 280, 283; *carinata*, 283; *glaziovii*, 283; *malmei*, 280, 283; *minutiflora*, 280; *multinervia*, 280, 283; *pavonis*, 283; *surinamensis*, 280; *weberbaueri*, 280, 283
Voyria, 280

Winteraceae, 285
Woodsia, 133

Zea, 258, 318; *diploperennis,* 318; *luxurians,* 318; *mays* subsp. *mays,* 318; *mays* subsp. *mexicana* Central Plateau, 318; *mays* subsp. *mexicana* Chalco, 318; *mays* subsp. *mex-*

icana Nobogame, 318; *mays* var. *huehuetenangensis,* 318; *mays* var. *parviglumis* Balsas, 318; *mays* var. *parviglumis* Jalisco, 318; *perennis,* 318

Zelkova, 350
Zexmenia columbiana, 398; *mi-kanioides,* 398
Zingiberales, 212, 234, 277
Zinniinae, 395
Zygophyllaceae, 76, 77, 82

SUBJECT INDEX

Numbers in **boldface** indicate definitions of terms when more than one page is given; those in *italics* refer to major discussions.

Abiotic factors, 364, 369, 378
Accident, 29, 30
Accidental quality, 30
Adaptation, 34, 97, 109, 110, 115, 213, 222, 250, 285, 313, 346, 347, 351, 354
Agameon, 178
Agamic reproduction, 297
Algae, 55, 91, 207, 332; blue-green, 207; green, 207
Algorithms, **74**, 78, 97, 98-100, 113-16, 120, 123, 124, 125, 126, 138, 143, 144, 190, 193, 202, 327, 332, 349, 384; Camin and Sokal Parsimony, 114; character compatibility, 116; Dollo parsimony, 115; Farris parsimony, 114, 115, 125; Manhattan (*City-block*) distance, 76, 97, 114, 115, 125; parsimony, 114, 115, 120, 123, 125; polymorphism parsimony, 115
Alkaloids, 211, 329-31, *338*, 339, 349, 377
Allopatry, 132, 181, 190, 352
Alloploidion, 178
Allopolyploidy, 127, 168, 296, 297, 298, 326, 327; segmental, 297
Ament, 55, 354
Amino acid sequences, 7, 97, 211, 341, 349
Anagenesis, **93**, 142
Analogue, 35
Anatomy, 3, 9, 11, 13, 14, 17, 26, 53, 63, 67, 92, 103, 133, 172, 198, 199, 204, 211, 217, 219, 222, 225, 231, *233-50*, 251, 252, 261, 263, 264, 345, 346, 362, 382; characters, 48, 204, 235, 249; comparative, 13, 233, 234; floral, 235, 249, 250; nodal, 240; wood, 234, 240, 249, 250
Aneuploidy, *293-96*, 298, 308, 327
Apogameon, 178

Apomixis, 162, 178, 179, 208, 253, 262, 263, 357
Apomorphy, 36, **94**, 95, 101, 108, 127, 143, 145, 147, 177, 201
Association coefficients, 63-65, 75, 76, 78, 82, 86
Attributes, 29, 30; serially dependent, 71
Autapomorphy, 36, **94**, 100, 128, 143, 145
Autoallopolyploidy, 297
Autonym, 191
Autopolyploidy, 126, **296**, 297, 327
Autoploids, interracial, 297

Banding patterns, 302
Bar graphs, 397
Basic (base) chromosome number, *see* Chromosome numbers
Basic data matrix, 30, 31, 39, *63*, 74, 89, 99-101, *102*, 113, 121, 122, 143, 378
Bats, 352, 354, 361
B-chromosomes, 290, 297
Bees, 61, 85, 349, 352, 354
Betalains, 211, 240, 284, 331, *339*, *340*, 344, 348
Biochemistry, taxonomic, 330
Biodiversity, 9
Biogeography, 13, 18, 92, 98, 132, 133, 364, 374, 378, 379; cladistic, 132; tracks, 132, 187
Biology, comparative, 13, 46
Biosystematics, xiii, **11**, 61, 177, 178, 187, 287
Birds, 46, 92, 166, 352, 354, 357, 360-62, 373
Botanical gardens, 51, 52, 53, 55, 61, 290, 310
Breeding systems, 40, 262, 270, 324, 351, 352, 357, 358, 362, 363
Bryophytes, 91, 133, 332
Buds, 292, 303, 311, 321

Camin and Sokal parsimony algorithm, 114
Canonical correlation analysis, 80
Canonical trend surface analysis, 90
Canonical variate analysis, 80
Carposphere, 225
Categorical system theory, 23
Categories, xiii, 23, 25, 48, 57, 62, 65, 84, 85, 86, 87, 88, 90, 128, 130, 155, 156-60, 161, 169, 170, 174, 177-80, 182-88, 191, 193-96, 206-13, 405
Catkin, *see* Ament
Centromere, 299, 300
Character-bearing semaphoront, 66, **94**
Character states, **28**-30, 33, 36, 38, 42, 47-49, 54-56, 59, 60, 62, 63, 65-67, 71-74, 76, 77, 89, 92, 94-97, 99-102, 105-15, 120, 121, 122, 124, 125, 127, 130, 132, 143, 147, 155, 176, 206, 212, 213, 233, 240, 249, 250, 335, 349, 362, 363, 378, 382, 383; advanced, 36; apotypic, 36; basimorphic, 94; clear-cut, 33; co-occurrence of primitive states, 110; derived, 36, 55, 94, 96, 97, 100-2, 114, 115, 124, 132, 143, 349, 378; generalized, 36; intergrading, 33; major, 33; minor, 33; networks, 99, 100, **105**, 106, 113; nonrandom, 33; plesiomorphic, 36; plesiotypic, 36; primitive, 36, 94, 97, 106, 109, 111; specialized, 36; trees, 99, 100, **106**, 112, 113, 122; unique, 36
Characteristic, 29, 30
Characters, **27**; a posteriori, 38; a priori, 38; adaptive, 34-36, 37, 67, 89; admissible, 37; advanced, 36; analogous, 34; analytic, 33; anatomical, 53, 204, 235, 240, 249, 250; bad,

509

Characters (*continued*)
34, 204; binary, 72, 75; biological, 34; character essentialis, 197; chemical, 67, 329; cladistic, 31, 36; compatibility algorithm, 116; compatible, 36, 97, 122-24; constitutive, 34; continuous quantitative, 32; correlation of, 33, 34, 37, 89; cryptic, 32; cytological, 309; descriptive, 34; developmental, 231; diagnostic, 34, 52; discontinuous quantitative, 32; embryological, 251; empirically correlated, 34; endomorphic, 250; epharmonic, 34; essential, 33; evolutionary, 34, 42; factitious, 197; familial, 33; fixed, 33; fortuitous, 34; fruit, 67, 315; functional, 34; fundamental, 33; generic, 33, 197, 198, 200; good, 34; homologous, 34-37, 90, 104; inadmissible, 37; individual, 37, 39; invariant, 33, 37; isocratic, 60; key, 34; logically correlated, 37; macrocharacters, 32, 33; mean difference, 76; meaningful, 34, 89; meaningless, 34, 37; measurement of, 59, 62, 71; meristic, 32; microcharacters, 32, 33, 198, 204; morphological, 171, 203, 219, 222, 225, 233, 250, 376; multistate, 32, 72, 73; natural, 197; nonadaptive, 36; nonconstitutive, 34; noncorrelated, 37; ontogenetic, 35; partially logically correlated, 37; phaneritic, 32; phyletic, 34, 35; phylogenetic, 34, 35; plastic, 33; pollen, 271; primitive, 36, 94, 97, 106, 109, 111; qualitative, 31, 32, 42, 73; quantitative, 31, 32, 77, 89, 104, 297; quantitative multistate, 32, 73; reliable, 34, 41; selection of, 38, 39, 41, 42, 56, 62, 66, 68, 103, 104, 119, 131; single, 37, 61, 67, 203, 378; specialized, 36; specific, 33, 49; structural, 363; synthetic, 33; systematic, 28; "tautological," 37; taxonomic, 27, 28, 37-41, 47, 89, 219, 225, 234, 240, 261, 293, 315; true cladistic, 36; two-state, 72, 73; unit, 37, 39, 42, 46, 62, 63, 66-68, 70; unreliable, 34; unweighted, 37, 96; use of, 31, 95; variability of, 31, 33, 71, 382; variant, 33; variation of, 152, 188, 382, 398; vegetative, 69, 219; vegetative morphological, 219, 222; weighting of, 37, 38, 59, 60, 67, 70-72, 80, 143
Chemical constituents, 26, 68
Chemical race, 190, 331

Chemo-deme, 190
Chemosystematics, 330, 386
Chemotaxonomy, 330
Chemotype, 190
Chromatin, 291, 303
Chromosome numbers, 63, 67, 199, 200, 211, 288, 290-96, 295, 297-299, 302, 308, 310; ancestral, 298; basic (base), 199, 200, **297**, 298, 309; immediate, 298
Chromosomes, 63, 67, 199, 200, 211, 235, 266, 288, 290-99, 302-5, 308-10, 313, 317, 319, 321, 322, 327, 328; banding patterns, 302; behavior, 303, 322, 328; morphology, 299; supernumerary (B-chromosomes), 297; variation, 293, 308
Chronospecies, 175
Cisternae, dilated, 236, 237
Cladistics, xiii, 31, 34, 36, 37, 41, 43, 46, 48, 56, 58, 68, 76, 88, **93**-*134*, 137, 138, 139-47, 149, 152, 176, 177, 195, 201-3, 205, 206, 207, 210-12, 231, 286, 311, 327, 340, 345, 349, 378, 384, 405; basic methodology, 99; biogeography, 132; character compatibility, 116-24; character state networks, 105; characters, 31, 36; efficacy of algorithms, 126-28; evolutionary assumptions, 102-3; example, 99-102; formal classification, 128-30; history, 96-99; homology, 104; impact, 130-34; maximum likelihood, 124; neocladistics, 99; networks, 125-26; parsimony, 114-16; polarity, 106-13; selection of characters, 103-4; species, 176, 177; tree construction, 113-28
Cladogenesis, **93**, 127, 142
Cladograms, 94, 95, 96, 99, 100, 102, 104, 109, 125, 126, 128, 130-33, 138-40, 142, 144, 145, 147, 177, 345, 349, 395
Classification, **6**; artificial, 49, 135; cladistic, 94-134; congruent, 67; evaluation, 136-42; evolutionary, 18, 35, 210, 405; generic, 200, 202; hierarchical, 22-25, 46, 54, 156; horizontal, 210; monothetic, 52, 135; natural, 52, 53, 56-*58*, 136, 200; numerical, 62; phenetic, 36, 37, 57, *59-82*, 139, 140, 142; phyletic, xiii, 43, *54-56*, 59, 60, 64, 69, 71, 95, 97, 132, 135-38, 140, *142-50*, 152, 207, 378, 384, 405; phylogenetic, 9, 143; polythetic, **52**, 135; predictive, 42, 148, 388; systems of, 29, 47, 49, 53, 61, 157, 196, 197, 369, 396; vertical, 210
Cleistogameon, 178

Climatype, 187
Clinal variation, 189, 336
Cline, 188
Cluster analysis, 81, 87, 89, 200, 271, 384
Clusters, 64, 66, 78, 80-82, 84, 87, 89, 143, 144, 157, 172, 200, 271, 384
Code of nomenclature, 193
Coding: additive, 72; binary, 72; nonadditive, 72
Coefficients, 63-65, 70, 71, *75-78*, 82, 86, 87, 139, 327, 384; association, 63-65, 75, 76, 78, 82, 86; canonical correlation analysis, 80; canonical trend surface analysis, 90; canonical variate analysis, 80; cophenetic correlation, 82, 84; correlation, 71, 75, 77, 78, 82, 84, 139; distance, 75, 76; divergence, 77; Gower's general similarity, 76; Jaccard (Sneath), 75; paired affinity index, 75; Pearson product-moment correlation, 77; probabilistic similarity, 75, 77; simple matching, 63, 75
Coenogamodeme, 178
Coenospecies, 178
Coevolution, 362
Commiscuum, 178
Comparative: anatomy, 13, 233, 234; biology, 13, 46; embryology, 252; morphology, 211; phytochemistry, 330
Comparium, 178
Compilospecies, 178
Consensus trees, 127, 145
Conspecific populations, 192
Contour diagrams, 400
Contoured factor analysis, 90
Contoured surface trend analysis, 90
Contour mapping, 90, 190
Convergence, 38, 68, 94, 95, 112, 131, 312, 357
Cophenetic correlation coefficient, 82, 84
Correlation coefficients, 71, 75, 77, 78, 82, 84, 139
Crossability, 201, 327
Crossing diagrams, 319, 397, 398
Cryptotaxonomy, 13
Cultivated plants, 92, 290
Cuticle, 222
Cytochrome *c*, 97, 126, 341, 349
Cytoecorace, 308
Cytoform, 308
Cytogenetics, 3, 11, 16, 29, 165, 184, 199, 200-2, 211, 217, 288, 293, 298, 303, *313-28*, 341, 378, 398
Cytological variation, 308, 336
Cytology, 3, 9, 12, 14, 165, 172, 178, 184, 189, 190, 199, 200, 211, 217, 235, 251, 265, 267, *288-312*, 314, 317, 321, 324, 327, 329, 336, 378, 382

Cytorace, 308
Cytotaxonomy, 293, 304
Cytotype, 173, 178, 190, 297, 308

Data-built data measures, 398
Data matrix, *see* Basic data matrix
Deme, 190, 376; chemodeme, 190; coenogamodeme, 178; syngamodeme, 178; topodeme, 188
Deviation, 73
Diagrams: contour, 400; crossing, 319, 397, 398; pie, 398; scatter, 395; Wells' distance diagram, 397
Dichotomy, 95
Difference, 29, 30
Differentiae, 162, 196
Differentiation, 21, 91, 92, 179, 190, 192, 195, 222, 234, 253, 256, 338; populational, 92
Discontinuity, 56, 139, 152, 168, 172, 197, 199, 202, 206, 213, 364, 381, 385
Disjunctions, 328, 331
Dispersal, 18, 132, 133, 172, 199, 217, 279, 351, 352, 360-62, 373, 378
Dissimilarity, 74, 76, 309
Distance coefficient, 75, 76
DNA, 343-47
Dollo parsimony algorithm, 115
Dot graphs, 398
Drift, 18, 38, 40, 132, 199, 378
Dysploidion, 178

Eco-element, 188
Ecology, 7, 11, 12, 16, 18, 22, 39, 109, 132, 161, 173, 176, 177, 180, 181, 187, 189, 217, 233, 250, 292, 297, 298, 308, 311, 314, 329-31, 347, 349, 351, 362, *364-79*, 397
Ecophysics, 365
Ecospecies, 176-78
Ecotype, 177, 178, 187, 188, 376; geo-ecotype, 188; subecotype, 188
Ecovar, 188
Eidos, 22, 162
Ektexine, 274, 275
Electrophoresis, 32, 40, 41, 175, 192, 298, 317, 324, 341, 348
Embryo, 53, 251-53, 256, 258, 260, 263, 265, 266; sac, 251-53, 256, 258, 263
Embryology, 13, 200, *251-66*, 267, 285, 288; comparative, 252
Endemism, 132, 148
Endexine, 274, 275
Epidermal cells, 222, 225, 235, 236
Epidermis, 222, 225, 235, 253, 353
Essence, 22, 29, 30, 56, 162, 163, 334
Essentialism, 16, 21, 57, 93

Essential oils, *see* Terpenoids
Ethological isolation, 173
EU, *see* Evolutionary unit
Euchromatin, 291
Euclidean distance, 76
Euploidion, 178
Euploidy, 293, 296, 298, 308
Eurypalynous taxa, 279
Evolution: coevolution, 362; macroevolution, 17, 18; microevolution, 65; parallel, 369; pathways, 27, 65, 141; patterns, 25; phyletic, 93; rates, 128; sequence, 105
Evolutionary transformation series, 96
Evolutionary unit (EU), 94, 100, 102, 115, 122, 125, 128, 165, 167, 184, 206, 298
Ex-group analysis, 108
Exine, 268, 271, 273, 275, 277, 285, 286; adaptive significance, 285
Exomorphic features, 219
Extinction, 17, 107, 168, 199, 203, 209, 212, 385

Factor analysis, 80, 90, 332; contoured, 90
Family, *207-13*
Farris parsimony algorithm, 114, 115, 125
Farris tree, 97
Feature, 29, 30
Fertility, 189, 201, 296, 317, 319
Fertilization, 252, 253, 258, 357, 362; double, 252
Flavonoids, 67, 75, 92, 152, 190, 199, 225, 283, 331, *332, 333*, 339, 348, 350, 353, 377
Flora (floristic), 15, 16, 17, 53, 55, 133, 172, 184, 194, 205, 210, 270, 279, 310, 311, 350, 357, 361
Form (forma), *182-**187***, *188-193*
Form genus, 205
Fossil record, 97, 107, 175, 205
Fundamentum divisionis, 22, 158, 162

Game theory, 17, 23
Gene flow, 167-170, 173-176, 180
Genecology, 11
Genus (genera), 5, 10, 25, 30, 33, 50, 52, 53, 55, 57, 61, 62, 65, 66, 84, 85, 86, 90, 99, 100, 111, 113, 157-59, 162, 163, 168, 169, 170, 173, 180, 183, *194-206*, 208, 209, 211, 212, 234, 240, 249, 250, 258, 265, 267, 268, 277, 279, 280, 283, 287, 292, 293, 297, 298, 302, 308, 315, 324, 326, 336, 341, 347, 369, 376, 382; characters, 33, 197, 198, 200; classification, 200, 202; delimitation, 194, 199, 200, 202, 206; form, 205; grade, 50; monotypic,

195, 205, 308; names, 196, 197; organ, 205
Genetic: analysis, 40, 315, 322; barrier, 324; drift, 38, 40; variation, 34, 170, 315, 324, 341
Genotype, 40, 316, 388
Gens, 176
Geo-ecotype, 188
Geographic: isolation, 168; plots, 398; variation, 184, 188, 193
Geography, 187, 190, 199, 250, 365
Geology, 14, 18, 92, 130, 132, 144, 209, 285, 369, 373, 374
Geometric transformation, 73
Gestalten, 91
Glucosinolates, 211, 237, 331, 339, *340*, 348
Goodall's similarity index, 77
Gower's general similarity coefficient, 76
Gradualism, 18
Graphs: bar, 397; dot, 398; polygonal, 332, 397
Groundplan/divergence (Wagner groundplan/divergence), 97, 101, *102*, 114, *115*, 119, 124, 144, 286
Gymnosperms, 53, 240, 251, 268, 334

Habitats, 15, 112, 113, 285, 312, 347, 365, 366, 376, 378
Herbaria, 6, 12, 55, 198, 210, 211, 218, 248, 266, 305, 332, 343, 358
Heredity, 313, 314
Heterochromatin, 291, 297; constitutive, 291; facultative, 291
Heterogameon, 178
Hierarchical inflation, 210
Hierarchy, 18, 20-25, 29, 32, 33, 41, 45, 47, 48, 54, 62, 66, 68, 84, 90, 92, 100, 128, 133, 137, 139, 148, 155, 156-61, 164, 174, 175, 177, 182, 185, 187, 188, 190, 193-95, 198, 199, 207, 208, 210, 212, 213, 222, 228, 231, 232, 248, 251, 258, 265-67, 279, 287, 307, 308, 315, 324, 332, 336, 340, 343, 345, 348, 362, 375, 382, 385, 388, 405, 406; theory, 22
Histograms, 103, 397
Holomorphy, 94
Holophyly, **95**, 128, 131, 140, 142, 155, 195, 203, 209, 213
Homologues, 35, 47, 96
Homology, *35, 68*, 69, 71, 94, 99, 100, *104*, 105, 111, 133, 134, 138, 212, 231, 233, 250, 266, 275, 297, 317, 324, 327, 363; structural correspondence, 68
Homoplasy, 94
Horticulture, 184, 193, 265
Hybridization, 90, 126, 127, 162, 164, 165, 172, 173, 176-78, 185, 190, 206, 208, 211,

Hybridization (*continued*) 249, 286, 297, 302, 314, 317, 324, 326, 330, 332, 336, 341, 344, 345, 376, 382, 384, 385, 397; artificial, 206, 317; in situ, 302

Hybrids, 173, 189, 200, 201, 286, 296, 303, 313, 317, 319, 322, 326, 353, 376; intergeneric, 200, 201

Identification, 8, **10**, 16, 20, 24, 25, 27, 31, 33, 34, 52, 138, 266, 287, 290, 332, 340, 361
Idiograms, 299
Inbreeding, 59, 357, 362
Inflorescence, 61, 225, 248, 280, 354
Infraspecific, 90, 92, 133, 152, 175, *182-93*, 194, 199, 222, 280, 292, 302, 303, 308, 324, 331, 332, 336, 340, 348, 362, 366, 370, 375, 378, 379, 383, 398, 406
In-group analysis, 108
Insects, 16, 66, 69, 94, 98, 188, 349, 351, 352, 354, 361, 362, 373, 374
Intergradations, 385
International Code of Botanical Nomenclature, 10, 157, 182, 184, 194, 287, 393
Intine, 268, 274, 275
Introgression, 190, 286, 287, 385, 397
Isolating mechanisms, 11, 173, 376, 378
Isolation, 17, 163, 167, 168, 172-75, 181, 189, 190, 293, 308, 324, 332, 341, 351, 366, 376; ethological, 173; geographic, 168
Isozyme analysis, 302, 315, 316

Jaccard (Sneath) coefficient, 75

Karyotype, 299, 302, 308; natural, 302
Kew Record of Taxonomic Literature, 193
Keys: diagnostic, 49; dichotomous, 78; tabular, 10; taxonomic, 22, 31

Labeling, 24, 163, 361
Latin, 391, 393
Leaf: anatomy, 235; arrangement, 25, 26, 28; blade, 111, 221, 271; shape, 31, 32, 42, 47, 201; surfaces, 273; venation, 52
Life forms, 167
Linear transformation, 73
Linkage: average, 78, 80; complete, 78; single, 78
Logarithmic transformation, 73
Logical division, 21, *22*, 52, *158*, *162*, 163, 191, 196
"Lumpers," 56, 59
Lux-obscuritas analysis, 275

Macrocharacters, 32, 33
Macroevolution, 17, 18
Macromolecules, xvii, 92, 126, 139, 211, 331, 332, 340, 345, 348, 349
Macromorphology, 219
Manhattan (or *City-block*) distance, **76**, 97, 114, 115, 125
Mapping, 90, 155, 190, 335; contour, 90, 190
Maps, 345, 349, 364, 366, 398, 400
Matches asymptote hypothesis, 69, 70
Mathematical transformation, 73
Maximum likelihood, 98, *124*, 126, 345, 349
Mean character difference, 76
Megaspore, 256
Megasubspecies, 188
Meiosis, 40, 254, 256, 263, 290, 291, 296, 297, 303, 311, 313, 314, 319, 322, 328
Metasystematics, 7
Microcharacters, 32, 33, 198, 204
Microevolution, 65
Micromolecules, 211, 331, 332, 339-41, 345, 348, 349
Micromorphology, 219
Microspecies, 178
Microspores, 253
Micton, 178
Minimum spanning networks, 125
Monographs, 24, 25, 234, 366
Monophyly, **95**, 99, 103, 131, 155, 176, 213
Monotypic: genus, 195, 205, 308; taxa, 159, 169, 170
Morphocline, 105
Morphology, 218-32; comparative, 211; differentiation, 179, 190; external, 67; floral, 232; reproductive, 225; variation, 7, 11, 56, 92, 182, 186, 187, 198, 308; vegetative, 219, 222, 225
Morphometrics, multivariate, 60
Morphospecies, *see* Species, morphological
Multispecies concept, 178
Multivariate morphometrics, 60
Mutation, 168, 171, 175, 187, 324, 385

Natio, 188
Natural: groups, 57, 58, 166, 308; populations, 9, 164, 172, 184, 187, 324; selection, 6, 23, 37, 41, 54, 135, 156, 168, 170, 171, 313; system of classification, 23, 34, 43, *52-54*, 55, 57, 61
Neocladistics, 99
Neoteny, 111, 251
Networks, 96, 97, 99, 100, 105, 106, 113-15, *125*, 126, 143,

144, 327, 378; character state, 99, 100, **105**, 106, 113; minimum spanning, 125; taxon, 125
New Phyletics, xiii, 135, *142-50*, 405; efficacy, 149-50; methodology, 143-48
New Systematics, **12**, 314
New Taxonomy, **12**, 13
Nexus hypothesis, 67
Nixus, 208
Nodes, 59, 144
Nomenclature, 9, 10, 61, 86, 157, 163, 182, *184-88*, 191-95, 198, 203-5, 286, 287, 299, 393, 405; code, 193; type, 191
Nonmetric multidimensional scaling, 80
Nonspecificity hypothesis, 67, 68
Nucleic acids, 291, 331, 341, 343
Nucleosome, 291
Numerical phenetics, 60
Numerical taxonomy, 13, 36, **60**, 86, 88, 89, 91, 98, 222

Ontogenetic state, earliest, 111
Ontogeny, 29, 35, 69, 70, 94, 111, 231, 266, 297, 335, 348, 350
Operational taxonomic unit (OTU), 62-69, 71-78, 80-82, 84-86, 89, 94, 139, 140
Order, 207-13
Organ genus, 205
Organogenesis, 111
Origin of angiosperms, 133
OTU, *see* Operational Taxonomic Unit
Out-group analysis, 107-111
Ovules, 236, 251-54, 256, 263, 264, 357, 358

Paedomorphosis, 111
Paganae, 188
Paired affinity index, 75
Paleospecies, 175, 176
Palynology, 133, *267-87*
Parageneon, 178
Parallel evolution, 369
Parallelism, 38, 41, 68, 94, 107, 109, 113, 115, 126, 131, 133, 287, 341, 347
Paraphyly, **95**, 128, 148, 205-7
Parenchyma, 236
Parology, 104
Parsimony, 78, 97, 98, 100, 105, 109, 110, 114, 115, 119, 120, 123-27, 131, 141, 349; algorithm, 114, 115, 120, 123, 125
Pattern recognition, 23
PCA, *See* principal components analysis
Pearson product-moment correlation coefficient, 77
Phenetics, xiii, 31, 36, 37, 43, 46, 48, **59-92**, 93, 96, 98, 103, 107, 125, 126, 128, 130-32, 133, 135-47, 149, 152, 155,

167, 168, 171, 172, 174-76, 180, 181, 190, 197, 200, 201-3, 206, 210, 211, 212, 297, 327, 340, 349, 357, 378, 384, 405; basic methodology, 62; calculation of affinity, 75-78; comparison of character states, 74; description and measurement of characters and states, 71-74; determination of phenetic structure, 74, 78-88; example, 63-65; history, 61-62; homology, 68-69; impact, 88; number of characters, 69-70; numerical, 60; phenetic relationship, 60, 62, 63, 66-69, 71, 72, 74, 80, 82, 84, 91, 130, 144, 146, 147, 202, 210; selection of characters, 66-68; selection of OTUs, 65-66; stratophenetics, 92, 130; weighting of characters, 70-71

Phenocline, 105

Phenograms, **64**-66, 69, 80, 82, 84, 86, 125, 139, 145, 147, 395

Phenolics, 330-32

Phenon, 85-88, 147, 178; lines, 85, 88, 147

Phenotype, 37, 59, 60, 160, 312, 315, 317, 385

Phloem, 234, 236, 340

Phyletic evolution, 93

Phyletics, xiii, 31, 34, 35, 37, 41, 43, 45, 51, 54-56, 58-62, 64, 67, 68, 69, 71, 74, 88, 91, 93-95, 97, 103, 106, 107, 111, 131, 132, 135, 136, 137, 138, 140-48, 150, 152, 175, 176, 206, 207, 250, 343, 378, 384, 405; characters, 34, 35; classification, xiii, 43, *54-56*, 59, 60, 64, 69, 71, 95, 97, 132, 135-38, 140, *142-50*, 152, 207, 378, 384, 405; grouping of taxa, 147; quantitative, 97; *see also* New Phyletics

Phylogenetics, 34, **94**, 98; characters, 34, 35; classification, 9, 143; relationships, 127, 251; species concept, 177; trees, 127, 142, 177, 341, 391

Phylogenetic systematics, 94, 96, 98

Phylogeny, 6-9, 18, 25, 31, 34, 35, 43, 54, 55, 57, 58, 69, 88, 91-98, 103, 104, 107-9, 111, 113-15, 123-28, 130, 131, 134, 139, 141-43, 148, 152, 166, 167, 177, 203, 205, 211, 233, 240, 250, 251, 270, 285, 287, 329-31, 341, 349, 350, 360, 391, 400

Physiological race, 376

Physiology, 9, 13, 74, 94, 161, 217, 251, 347

Phytochemistry, 11, 330, 331, 336, 362, 377; comparative, 330

Phytography, 219

Pie diagrams, 398

Pillar complexes, *see* Polyploid Pillar Complexes

Placentation, 240

Plasticity, 92, 184, 190, 376

Plastids, 211, 237, 240, 249, 288, 340; sieve-tube plastids, 249, 288, 340

Plesiomorphy, 36, **94**, 96, 108, 143, 147, 201, 327

Polarity, 106-13; association, 112; co-occurrence of primitive states, 110; correlation, 112; earliest ontogenetic state, 111; ex-group analysis, 108; fossils, 107; group trends, 113; in-group analysis, 108; minor abnormalities of organogenesis, 111; outgroup analysis, 107-11; vestigial organs, 112

Pollen, 67, 150, 199, 205, 211, 252, 254, 258, 264, *267-87*, 305, 319, 324, 325, 328, 353, 354, 357, 358, 362, 373, 375; fertility, 319; fossil, 270, 286; morphology, 268

Pollination, 89, 168, 201, 217, 252, 258, 267, 279, 285, 314, 349, 351, 352, 353, *354-57*, 358, 360, 362, 364; bat, 354; bee, 349, 352, 354; beetle, 354, 362; mechanisms, 168; pollinators, 285, 349, 351, 352, 354, 357, 361, 362, 364, 373, 375; self-pollination, 89; wind, 354, 362

Polychotomy, 95, 128

Polyclaves, 10

Polygonal graphs, 332, 397

Polymorphism, 40, 114, 115, 116, 317; parsimony algorithm, 115

Polynomials, 53

Polyploid: complex, 326; pillar complexes, 297

Polyploidy, 12, 113, 126, 128, 287, 290, 293, 295-98, 302, 311, 314, 319, 326, 328

Polytomy, 95

Population: conspecific, 192; differentiation, 92; local, 18, 182; natural, 9, 164, 172, 184, 187, 324; variation, 17, 40, 153, 175, 315, 335

Predictive value, 25, 26, 31, 52, 141, 155

Principal components analysis, 80, 81

Principal coordinate analysis, 80

Priority, 25, 185, 193

Probabilistic similarity coefficients, 75, 77

Progenesis, 111

Prole, 187

Property, 29, 30

Protein, 41, 126, 192, 211, 236, 237, 240, 248, 291, 317, 324, 331, 341, 342, 348

Pteridophytes, 91, 133

Pure systematist, 9

Q-technique, 89

Qualitative morphological differences, 308

Quantitative phyletics, 97

Quality, 29-30

Race, 16, 161, 178, 184, 187-90, 296, 297, 308, 331, 376; chemical, 190, 331; cytocorace, 308; cytorace, 308; physiological, 376

Ranging, 73

Rank (ranking), 10, 20, 23, 25, 33, 47, 48, 49, 52, 56, 59, 61, 62, 65, 69, 84-88, 99, 100, 108, 128, 130, 140, 147-48, 155, 156, 157, 160, 163, 185, 187, 191, 192, 193, 194, 209, 210, 265, 363

Rassengrupp, 188

Recombination, 168, 171, 187

Revisions, 7, 12, 25, 96, 200, 398

Rheogameon, 178

R-technique, 89

Scaling, nonmetric multidimensional, 80

Scanning electron microscope (SEM), 32, 198, 219, 222, 225, 231, 232, 234, 240, 248, 264, 270, 273, 275, 279, 286, 362

Scatter diagrams, 395

Secondary metabolites, 329, 331, 349

Section (sectio), 57, 65, 100, 128, 157, 180, 194, 204, 265, 280, 298, 324

Seeds, 17, 71, 89, 183, 184, 196, 219, 225, 234, 246, 248, 253, 258, 261, 265, 266, 273, 311, 317, 319, 341, 360, 361, 362, 373; coat, 225, 261, 273

SEM, *see* Scanning Electron Microscope

Semaphoront, 66, 94

Semispecies concept, 178

Serially dependent attributes, 71

Serology, 211, 229-331, 341, 342-44, 348; systematic, 342

Sesquiterpene lactones, 336-38

Set theory, 22, 23, 271

Sexual systems, 43, 50-53, 61

Shading by overprinting, 90

Sieve-tube plastids, 249, 288, 340

Similarity index, 76

Simple matching coefficient, 63, 75

Simultaneous test procedures, 90

Sister group, **94**, 108, 110, 111, 112, 142, 205

Soil, 183, 364, 366, 369, 374

Spatial autocorrelation analysis, 90

Specialization, 17, 109, 213
Speciation, 17, 108, 109, 120,
127, 128, 168, 173, 179, 199,
205, 209, 310, 327, 369, 385
Species, 5-7, 10, 11, 12, 14-17,
20, 21, 23, 25, 30, 40, 50, 53,
54, 57, 60, 62, 65, 66, 69, 71,
84, 85, 99, 100, 103, 109, 111,
127, 128, 132, 153, 156-59,
161-81, 187-206, 208, 209,
212, 231, 235, 240, 249, 258,
265-68, 279, 280, 283, 285,
286, 288, 292, 293, 294, 296-
98, 302, 308, 311, 313, 324,
326, 330, 340, 344, 347, 351,
352, 354, 357, 358, 362, 364,
366, 369, 370, 373, 374, 376,
378, 382, 383, 393; aggregate,
184; biological, **164**, 165, 167,
168, *172-74*, 180, 324, 352,
364; chronospecies, 175; cladistic, 176, 177; classical concept, 171; classical phenetic,
171; coenospecies, 178; compilospecies, 178; delimitation,
324; ecological concept, 176;
ecospecies, 176-78; evolutionary, **164**, 176, 177; genetic
concept, 174, 176; hypermodern, 169; Linnaean, 171; microspecies, 178; morphological (morphospecies), 171, 172,
176; multispecies, 178;
names, 160; nominalistic,
162, 165; nondimensional,
164; numerical, 180; numerical phenetic concept, 174,
175, 180; paleospecies, 175,
176; phylogenetic concept,
177; semispecies, 178; species
aggregate, 178; species complex, 184; species concepts,
162-77, 179, 180, 187, 195,
324, 352, 364; superspecies,
178; sympatric, 178, 308; typological, 162-64, 173, 196;
variation, 184
"Splitters," **56**, 59, 93, 132, 184,
200, 203, 209
Spores, 126, 205, 267, 270
Staminodes, 112
Standardization, 60, 73, 85,
271, 299
Stasigenesis, 93
Statistical systematics, 60
Stenopalynous taxa, 279
Sterility, 164, 173, 189
Stigma, 50, 252, 264, 267, 285,
354, 357, 358
Stratophenetics, 92, 130
Structural correspondence, 68
Styles, 50, 145, 252, 357, 358
Subecotype, 188

Subform (subforma), 182
Subgenus, 65, 85, 100, 157, 194,
204, 206, 265, 280, 324
Subnatio, 188
Subspecies, 12, 65, 157, 169,
178, 180, *182-93*, 297, 308,
324, 325, 376, 383, 405, 406;
megasubspecies, 188
Subvariety (subvarietas), 182,
187
Superspecies, 178
Surface trend analysis, 90
Symplesiomorphy, 36, **94**, 109,
143
Synapomorphy, 36, **94**, 96, 100,
109, 131, 145, 156, 177, 203,
327, 345, 349
Syngameon, 178
Syngamodeme, 178
Synonymy, 6, 9, 12, 29, 36, 182,
189
Systematic botany, 5, 9, 53, 114
Systematics: xiii, 4-6, **7**, 8-18,
20, 25, 29, 30, 46, 60-62, 64,
88, 90, 93-96, 98, 104, 108,
126, 130, 133, 142, 161, 166,
193, 219, 221, 233, 270, 290,
312, 314, 326, 330, 331, 341,
344, 405; areas, 8; biosystematics, xiii, **11**, 61, 177, 178,
187, 287; chemosystematics,
330, 386; metasystematics, 7;
methodology, 7; New Systematics, **12**, 314; phylogenetic,
94, 96, 98; statistical, 60
Systems theory, 22, 23; categorical, 23

Taximetrics (taxometrics), 60
Taxon (taxa), 25
Taxon networks, 125
Taxonomic hierarchy, 32, 33,
47, 48, 62, 156-58, 161, 174,
175, 177, 182, 188, 194, 195,
198, 232, 248, 251, 265, 279,
307, 324, 336, 348, 362, 375,
385, 405, 406
Taxonomy, **6**; chemotaxonomy,
330; classical, 379; cryptic,
13; cryptotaxonomy, 13; evolutionary, 174; experimental,
11, 185; "hiatus," 197; mathematical, 60; New Taxonomy,
12, 13; numerical, 13, 36, **60**,
86, 88, 89, 91, 98, 222; quantitative, 60; statistical, 60
TEM, *see* Transmission electron
microscope
Tendril, 232
Terpenoids, 71, 329, 331, *332*,
334-38, 348, 349, 377, 400
Test procedures, simultaneous,
90

Tetrachotomy, 95
Topodeme, 188
Topotype, 187
Transformation: sequence, 69;
series, 96, 105, 240
Transformations, 73, 169; cytological, 327
Transitio, 188
Transmision electron microscope (TEM), 32, 222, 234,
235, 248, 254, 258, 264, 268,
270, 273, 275, 279, 286
Trees (diagrams), 23, 54, 64, 85,
96, 97, 99, 100, 106, 112-15,
119, 122-28, 131, 141, 142,
145, 177, 341, 349, 391; character state, 99, 100, **106**, 112,
113, 122; consensus, 127, 145;
Farris, 97; phylogenetic, 54,
127, 142, 177, 341, 391
Trichomes, 222, 249
Trichotomy, 95
Type: nomenclatural, 191; specimens, 198

Variability, intraspecific, 185
Variable features, 376
Variation: character, 152, 188,
382, 398; chromosome, 293,
308; clinal, 189, 336; cytological, 308, 336; genetic, 34, 170,
315, 324, 341; geographic,
184, 188, 193; individual, 17;
morphological, 7, 11, 56, 92,
182, 186, 187, 198, 308; populational, 17, 40, 153, 175, 315,
335; range of, 73, 266, 292;
species, 184
Variety (varietas), 35, 65, 82,
157, 163, 164, 169, 171, 178,
182-93, 195, 232, 297, 308,
315, 330, 253, 376, 405
Vascularization, 240
Vegetative morphology, 219,
222, 225
Venation, 52, 221, 235
Vestigial organs, 112
Vicariance, 18, 98, 132, 133,
378; biogeography, 132,
378
Vouchers, 248, 266, 287, 398

Waagenon, 188
Wagner, groundplan/divergence
(see groundplan/divergence)
Wagner tree, *see* Farris tree
Weighting: a posteriori, 71; a
priori, 70; character, 37, 38,
59, 60, 67, 70-72, 80, 143; extrinsic, 70; residual, 37
Wells' distance diagram, 397
Wood anatomy, 234, 240, 249,
250